ORGANIC CHEMISTRY SERIES

Series Editors: J E Baldwin, FRS & P D Magnus, FRS

VOLUME 7

Synthesis of Optically Active α-Amino Acids

Related Pergamon Titles of Interest

BOOKS

Organic Chemistry Series

CARRUTHERS: Cycloaddition Reactions in Organic Synthesis
DAVIES: Organotransition Metal Chemistry: Applications to Organic Synthesis
DEROME: Modern NMR Techniques for Chemistry Research
DESLONGCHAMPS: Stereoelectronic Effects in Organic Chemistry
GIESE: Radicals in Organic Synthesis: Formation of Carbon–Carbon Bonds
GIESE: Radicals in Organic Synthesis—2: Formation of Carbon–Heteroatom Bonds
HANESSIAN: Total Synthesis of Natural Products: The 'Chiron' Approach
PAULMIER: Selenium Reagents and Intermediates in Organic Synthesis
PERLMUTTER: Conjugate Addition Reactions in Organic Synthesis
WONG & WHITESIDES: Enzymes in Synthetic Organic Chemistry

Major Works of Reference

ALLEN & BEVINGTON: Comprehensive Polymer Science
BARTON & OLLIS: Comprehensive Organic Chemistry
HANSCH *et al*: Comprehensive Medicinal Chemistry
KATRITZKY & REES: Comprehensive Heterocyclic Chemistry
TROST & FLEMING: Comprehensive Organic Synthesis
WILKINSON *et al*: Comprehensive Organometallic Chemistry

Also of Interest

KATRITZKY: Handbook of Heterocyclic Chemistry
NICKON & SILVERSMITH: Organic Chemistry—The Name Game
PERRIN & ARMAREGO: Purification of Laboratory Chemicals, 3rd edition
RIGAUDY & KLESNEY: Nomenclature of Organic Chemistry ('The Blue Book')
SUSCHITZKY & SCRIVEN: Progress in Heterocyclic Chemistry

JOURNALS

TETRAHEDRON (rapid publication primary research journal for organic chemistry)
TETRAHEDRON LETTERS (rapid publication preliminary communications journal for
    organic chemistry)
TETRAHEDRON COMPUTER METHODOLOGY (new international electronic journal
    for rapid publication of original research in computer chemistry)

*Full details of all Pergamon publications, and a free specimen copy of any Pergamon
journal, are available on request from any Pergamon office*

Synthesis of Optically Active α-Amino Acids

ROBERT M. WILLIAMS
Colorado State University
Fort Collins, Colorado, USA

PERGAMON PRESS

OXFORD · NEW YORK · BEIJING · FRANKFURT
SÃO PAULO · SYDNEY · TOKYO · TORONTO

U.K.   Pergamon Press, Headington Hill Hall, Oxford OX3 0BW,
                        England

U.S.A.   Pergamon Press, Inc., Maxwell House, Fairview Park, Elmsford, New
                        York 10523, U.S.A.

PEOPLE'S REPUBLIC   Pergamon Press, Room 4037, Qianmen Hotel, Beijing,
OF CHINA   People's Republic of China

FEDERAL REPUBLIC   Pergamon Press GmbH, Hammerweg 6, D-6242 Kronberg,
OF GERMANY   Federal Republic of Germany

BRAZIL   Pergamon Editora Ltda, Rua Eça de Queiros, 346, CEP 04011,
                        Paraiso, São Paulo, Brazil

AUSTRALIA   Pergamon Press Australia Pty Ltd., P.O. Box 544, Potts Point, N.S.W.
                        2011, Australia

JAPAN   Pergamon Press, 5th Floor, Matsuoka Central Building, 1-7-1
                        Nishishinjuku, Shinjuku-ku, Tokyo 160, Japan

CANADA   Pergamon Press Canada Ltd., Suite No. 271, 253 College Street,
                        Toronto, Ontario, Canada M5T 1R5

First edition 1989

Library of Congress Cataloging in Publication Data

Williams, Robert Michael.
Synthesis of optically active alpha-amino acids.
(Organic chemistry series; v. 7)
Includes index.
1. Amino acids—Synthesis.   2. Optical isomers.
I. Title.   II. Series: Organic chemistry series
(Pergamon Press); v. 7.
QD431.W53   1989   547.7'5   89-15944

British Library Cataloguing in Publication Data

Williams, Robert M.
Synthesis of optically active alpha-amino acids
1. Organic compounds. Synthesis. Use of amino acids
I. Title   II. Series   547'.2

ISBN 0-08-035940-X Hardcover
ISBN 0-08-035939-6 Flexicover

Printed in Great Britain by BPCC Wheatons Ltd, Exeter

Dedicated to Cathy, Edith, William and the Memory of my Father

CONTENTS

PREFACE

α-Amino Acids are enjoying an unprecedented renaissance in virtually all disciplines of biology, medicine, biochemistry and chemistry. Amino acids are finding increasing utility in the synthesis of pharmaceuticals, agricultural products, the food industry and materials science. The recent revolution in molecular biology and protein engineering technologies have opened vast new vistas for the intelligently designed use and incorporation of amino acids in numerous proteinaceous and non-proteinaceous materials. Major advances in the understanding of enzyme mechanisms, protein conformations and properties related to molecular recognition, protein-nucleic acid interactions amongst countless other important regulatory interactions involving the basic twenty or so proteinogenic amino acids have placed the study of amino acid chemistry at the forefront of modern chemical science. The number of naturally-occurring , non-proteinogenic α-amino acids is rapidly approaching 1,000; the number of non-natural or totally synthetic amino acids could in principle be almost infinite. With the now routine automated technology to synthesize polypeptides on solid supports, the possibilities for synthesizing new enzymes, hormones, synthetic immunostimulants, drugs and countless other valuable materials *with non-proteinogenic amino acids* having *predetermined* properties have started to become recognized and represent emerging, exciting and creative fields. The benefits of these endeavors will undoubtedly reach far into the future.

This purpose of this monograph is to review and critically evaluate the *best* new methods to synthesize α-amino acids in *optically active* form. There is so much new literature on amino acid synthesis that the experimentalist will undoubtedly be confused and bewildered when selecting the *most appropriate* methodology for constructing the amino acid of immediate interest. Suffice it to say that , of all the methodologies extant, there is no panacea or one single best method that a laboratory may acquire for solving *every* amino problem that may be encountered. The almost limitless types of functionality that can be incorporated onto the glycine skeleton, with their attendant stabilities, solubilities and other chemical properties, neccessarily limits the applicability of *every* given method to a specific niche or niches. It is thus a job of matching the specific needs at hand with the most appropriate solution. The requirements of an industrial process chemist for example, can differ substantially from those of the medicinal research chemist; even when the amino acid is the same. This book is meant to be a guide in steering the amino acid scientist through the maze of existing reports on the preparation of amino acids. The reader will detect a certain bias in this monograph toward reports that provide full experimental details for preparing , characterizing and isolating the amino acid in a reliable experimental protocol. In areas that are relatively new conceptually and less studied experimentally, an effort has been made to review the most salient works with an eye toward future development.

Since a large body of these works have been conducted by synthetic organic chemists, it should not be lost upon the reader that many of the new methods that have been developed are cast in classically 'organic' protocol; that is, asymmetric derivatization of an organic-soluble template that is subsequently detached from the (usually) water-soluble amino acid. It is very often this very simple change in solubility properties that make the synthesis of amino acids so challenging at the laboratory bench. The scientist must become familiar with a broader range of experimental techniques; often ones that are not immediately familiar, whether they be on the 'organic' side or the 'aqueous' side. The synthesis of optically active amino acids therefore becomes a special and brutally challenging testing ground for methods in asymmetric synthesis. The emphasis of this book is placed in the context of asymmetric synthesis and not on more classical racemic syntheses requiring a subsequent resolution. Some of the newer enzymatic resolution techniques are however, presented in Chapter 7.

Throughout this monograph, reference is constantly made to the terms percent *ENANTIOMERIC EXCESS* (%ee); percent *DIASTEREOMERIC EXCESS* (%de); percent *DIASTEREOSELECTIVITY* (%ds); and percent *OPTICAL PURITY* (op). In some instances, authors of specific works have chosen to express the optical purity of a series of compounds by the older optical purity (op) data that is seldom employed in the newest contributions; %ee being the now universally accepted manner. This is particularly important for amino acids since, the specific optical rotation of amino acids (used to calculate the op) can be drastically affected by impurities, salts, metal ions, etc. rendering this criterion largely unreliable. These terms are all defined below for reference; %de and %ds (which is used by several authors) are equivalent.

ENANTIOMERIC EXCESS = [R]-[S]/[R]+[S] x 100 (expressed as %ee), where R and S are the amounts of individual enantiomers produced.

DIASTEREOMERIC EXCESS = [A]-[B]/[A]+[B] x 100 (expressed as %de), where A and B are the amounts of individual diastereoisomers produced.

DIASTEREOSELECTIVITY = [A]-[B]/[A]+[B] x 100 (expressed as %ds), where A and B are the amounts of individual diastereoisomers produced.

OPTICAL PURITY = [α]observed/ [α]maximum x 100 (expressed as %op), where [α]observed is the specific optical rotation of the sample under consideration and [α]maximum is the specific optical rotation of the optically pure substance recorded under identical conditions.

Since the basic twenty proteinogenic α-amino acids are all readily available, it is often wise to consider the semi-synthetic manipulation of these substances. Indeed, an entire book much larger than the present monograph, could easily cover the vast literature on amino acid modification. This topic will only be brushed upon in the context of the main chapter headings. As a reference point, the structures, names and IUPAC-IUB abbreviations of the twenty L-amino acids derived from common proteins is presented. The rough relative retail costs of these substances are plotted alongside their considerably more costly D-isomers. Since numerous new applications of D-configured amino acids have started to become appreciated, an effort has been made to highlight methods that give equally facile access to both L- and D-absolute configurations as well as an unambiguous protocol for establishing the absolute configuration of the final amino acid derivative. This last is particularly significant for new, unknown amino acids since rapid, unambiguous determination of the absolute stereochemistry is not neccessarily a straightforward task.

Numerous excellent reviews on the chemistry and biochemistry of the amino acids and their utilization have appeared ; some of these sources are tabulated at the end of this section. Very recently a Symposium-in-Print appeared (*Tetrahedron* Symposium-in-Print No. 33 (1988)) and many of these papers have been included in the present monograph. These reviews have been utilized in the preparation of this book as a background; the large majority of the works covered herein came directly from careful reading of the primary literature. In a work of this magnitude, mistakes, misinterpretations and omissions are inevitable; every effort to minimize these have been made through cross-referencing and proof reading. Extra time has been spent inserting the literature citations directly in the Schemes throughout in addition to the citations that are cross-referenced with the text. This format is intended to save the reader valuable time in quickly retrieving the primary

RELATIVE COSTS PER GRAM OF L- AND D- PROTEINOGENIC α–AMINO ACIDS

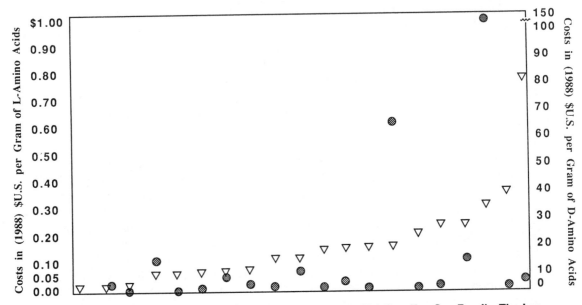

Abbreviations of α-Amino Acids (▽ =L-isomers; ◉ =D-isomers)

literature at a glance. It is the authors' sincere hope that this monograph will be a useful handbook and desk reference to the field as well as serving as a vehicle for stimulating better science and new breakthroughs through the exploitation of the rich chemistry of α-amino acids.

THE PROTEINOGENIC α-AMINO ACIDS

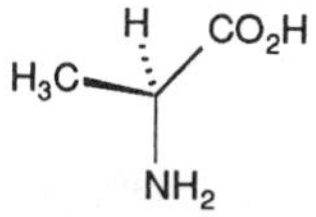

L-ALANINE
(L-Ala / A)

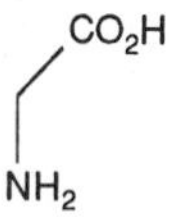

GLYCINE
(Gly / G)

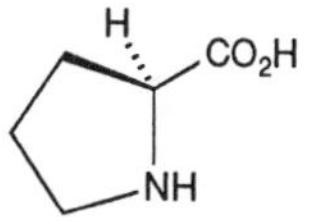

L-PROLINE
(L-Pro / P)

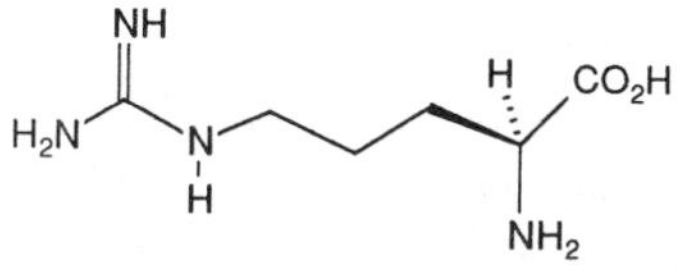

L-ARGININE
(L-Arg / R)

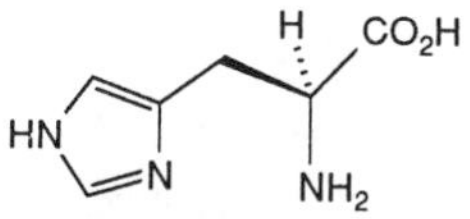

L-HYSTIDINE
(L-His / H)

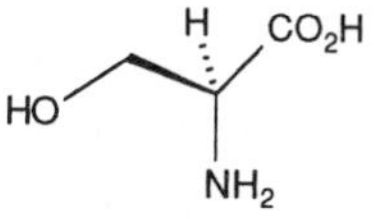

L-SERINE
(L-Ser / S)

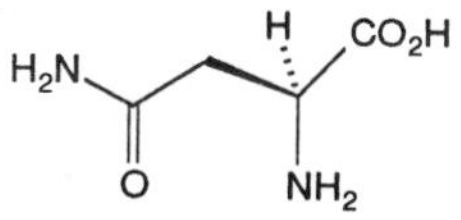

L-ASPARAGINE
(L-Asn / N)

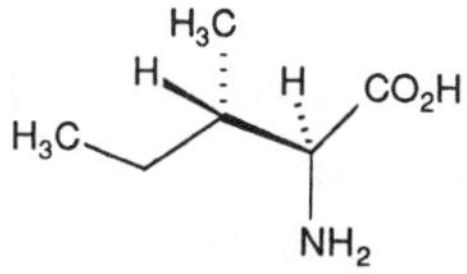

L-ISOLEUCINE
(L-Ile / I)

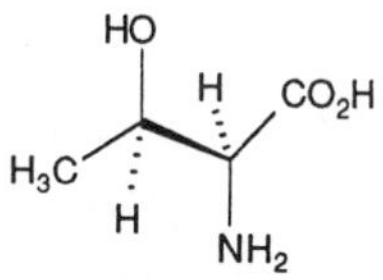

L-THREONINE
(L-Thr / T)

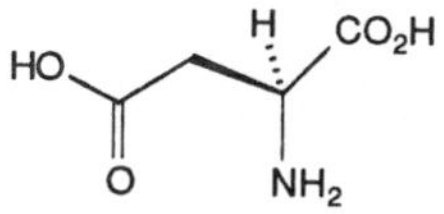

L-ASPARTIC ACID
(L-Asp / D)

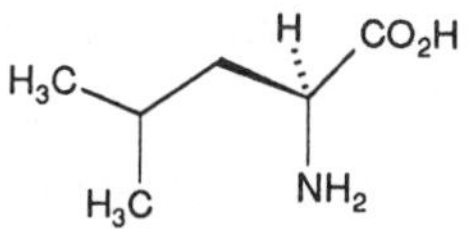

L-LEUCINE
(L-Leu / L)

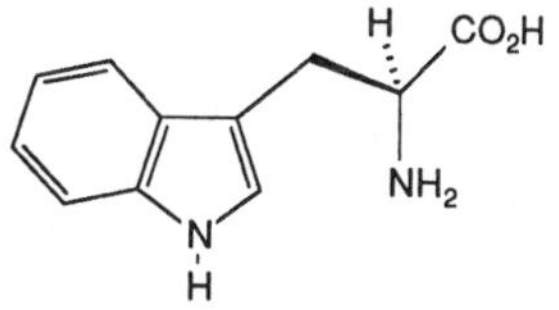

L-TRYPTOPHAN
(L-Trp / W)

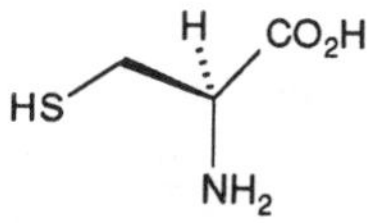

L-CYSTEINE
(L-Cys / C)

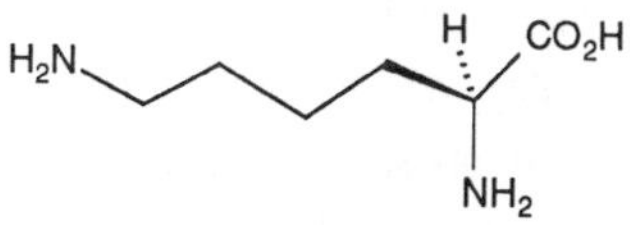

L-LYSINE
(L-Lys / K)

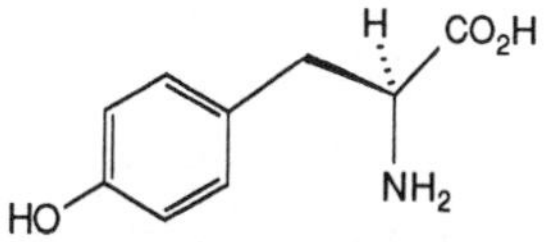

L-TYROSINE
(L-Tyr / Y)

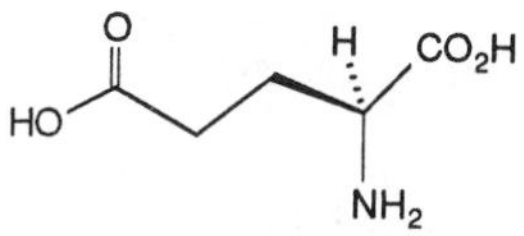

L-GLUTAMIC ACID
(L-Glu / E)

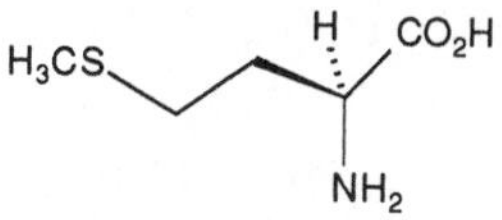

L-METHIONINE
(L-Met / M)

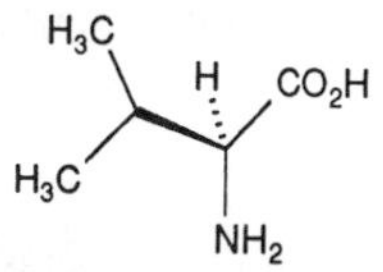

L-VALINE
(L-Val / V)

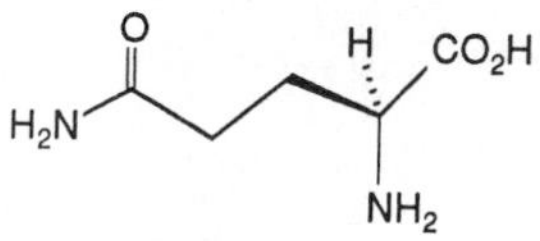

L-GLUTAMINE
(L-Gln / Q)

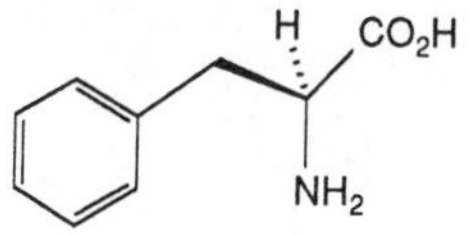

L-PHENYLALANINE
(L-Phe / F)

Reviews on α-Amino Acids

1.   Barrett, G.C., *Chemistry and Biochemistry of the Amino Acids*, (1985) Chapman
      and Hall, London.

2.   Greenstein, J.P.; Winitz, M., *Chemistry of the Amino Acids*, Vols.1,2 and
      3(1984) Robert E. Krieger, Malabar, Florida.

3.   Coppola, G.M.; Schuster, H.F., *Asymmetric Synthesis: Construction of Chiral
      Molecules Using Amino Acids* (1987) John Wiley & Sons, New York.

4.   Hermann, K.M.; Somerville, R.L., *Amino Acids: Biosynthesis and Genetic
      Regulation* (1983) Addison-Wesley, London.

5.   Morrison, J.D.,Ed., *Asymmetric Synthesis, Vol 5., Chiral Catalysts* (1985)
      Academic Press, Orlando.

6.   Morrison, J,D.; Scott, J.W., Eds., *Asymmetric Synthesis, Vol. 4, The Chiral Pool*
      (1984) Academic Press, Orlando.

7.   Morrison, J.D.,Ed., *Asymmetric Synthesis, Vol 2 and 3, Stereodifferentiating
      Addition Reactions* (1983 and1984) Academic Press, Orlando.

8.   Morrison, J.D.; Mosher, H.S., *Asymmetric Organic Reactions* (1971) American
      Chemical Society, Washington, D.C.

9.   Kochetkov, K.A.; Belikov, V.M., *Modern Asymmetric Synthesis of α–Amino Acids*,
      in *Russian Chemical Reviews* (1987) **56**, 1045.

10.   Martens, J., *Asymmetric Synthesis with Amino Acids*, in *Topics Curr.Chem.*
      (1984) **125**, 167.

11.   Wagner, I.; Musso, H., *New Naturally Occurring Amino Acids*, in
      Angew.Chem.Int.Ed.Engl. (1983) **22**, 816.

12.   Drauz, K.; Kleeman, A.; Martens, J., *Induction of Asymmetry by Amino Acids* in
      Angew. Chem.Int.Ed.Engl. (1982) **21**, 584.

13.   Valentine, D.; Scott, J.W., *Asymmetric Synthesis* in, *Synthesis* (1978) 329.

14.   Izumi, Y.; Chibata, I.; Itoh, T., *Production and Utilization of Amino Acids* , in
      Angew.Chem.Int.Ed.Engl. (1978) **17**, 176.

15.   O'Donnell, M.J. Ed. , *α-Amino Acid Synthesis* (*Tetrahedron* Symposium-in-
      Print) *Tetrahedron* (1988) **44**, 5253-5614.

16.   Bosnich, B., Ed., *Asymmetric Catalysis* (1986) Martinus Nijhoff Publishers,
      Dordrecht.

ABBREVIATIONS

4A	4 angstrom molecular sieves
AcO	acetoxy
Ac$_2$O	acetic anhydride
AIBN	azo-(*bis*)-isobutyronitrile
Bn	benzyl
BSA	bis(trimethylsilyl)acetamide
n-Bu	*normal*-butyl
t-Bu	*tert*-butyl
Bz	benzoyl
(BOC)$_2$O	di-*tert*-butyl dicarbonate
t-BOC	*tert*-butyloxycarbonyl
BOC-ON	2-(*tert*-butoxycarbonyloxyimino)-2-phenylacetonitrile
n-BuLi	n-butyl lithium
CAN	cerric ammonium nitrate
CBz	benzyloxycarbonyl
CBzCl	benzyl chloroformate
CSA	camphorsulphonic acid
DBU	1,8-diazabicyclo[5.4.0]undec-7-ene
DCC	N,N'-dicyclohexyl carbodiimide
de	diastereomeric excess
DEAD	N,N'-diethyl azodicarboxylate
DEPC	diethyl phosphorocyanidate
DET	diethyl tartrate
DHP	dihydropyran
DIBAH	diisobutyl aluminum hydride
diox.	dioxane
DIPA	N,N-diisopropyl ethylamine
DMAP	N,N-dimethylamino pyridine
DME	dimethoxyethane
DMF	N,N-dimethyl formamide
DMPA	2,2-*bis*(hydroxymethyl)propionic acid
DMSO	dimethyl sulfoxide
DPPA	diphenylphosphoryl azide
EDC	1-(3-dimethylaminopropyl)-3-ethylcarbodiimide (also WSC)
ee	enantiomeric excess
Et	ethyl
Et$_3$N	triethyl amine
EtOH	ethanol
EtOAc	ethyl acetate
Δ	heat

c-hex.	cyclohexyl
HMPA	hexamethylphosphoramide
HOAc	acetic acid
HOBT	N-hydroxybenzotriazole
im$_2$CO	carbonyl diimidazole
KOt-Bu	potasium *tert*-butoxide
LDA	lithium diisopropylamide
LDEA	lithium diethylamide
LICA	lithium isopropyl (cyclohexyl) amide
2,6-lut.	2,6-lutidine
Mbh	*para*-methoxybenzhydryl
m-CPBA	*meta*-chloroperoxybenzoic acid
Me	methyl
MeCN	acetonitrile
MeOH	methanol
MOM	methoxymethyl
Ms	methane sulfonyl (mesyl)
MTPACl	α-methoxy-α-(trifluoromethyl)phenylacetyl chloride (Mosher's acid chloride)
NB	*ortho*-nitrobenzyl
NBS	N-bromosuccinimide
NCS	N-chlorosuccinimide
NMM	N-methyl morpholine
PCC	pyridinium chlorochromate
PDC	pyridinium dichromate
Ph	phenyl
phth	phthalimido
pMB	*para*-methoxybenzyl
pMP	*para*-methoxyphenyl
Pr	propyl
py	pyridine
pyH-OTs	pyridinium tosylate
Ra-Ni	Raney-Nickel
Red-Al	sodium *bis*(2-methoxy-ethoxy)aluminum hydride in toluene
SWERN	oxidation using DMSO, oxalyl chloride and Et3N
TBDMS	*tert*-butyldimethylsilyl
TFA	trifluoro acetic acid
TFAA	trifluoro acetic anhydride
TfO	triflate (also OTf)
THF	tetrahydrofuran
THP	tetrahydropyranyl
TMEDA	N,N,N',N'-tetramethylethylenediamine
TMS	trimethylsilyl

TMSCl	trimethylsilyl chloride
tol.	toluene
trisyl	2,4,6-triisopropylbenzenesulfonyl
trityl	triphenylmethyl
Troc	trichloroethylcarboxy
p-TsOH	*para*-toluene sulfonic acid
Ts	tosyl
TsCl	*para*-toluene sulfonyl chloride
TTN	thallium trinitrate
WSC	water-soluble carbodiimide (1-(3-dimethylamino)propyl-3-ethylcarbodiimide hydrochloride (also EDC)

CHAPTER 1

ASYMMETRIC DERIVATIZATION OF GLYCINE

This section will outline the most recent and preparatively useful methods to homologate various glycine derivatives or glycine equivalents into a variety of optically active amino acids. Being the simplest amino acid, glycine derivatization potentially offers the greatest versatility in a conceptual sense and will only be limited by the imagination of the chemist and the efficiency of the reactions employed. Older methods that produce racemic amino acids from glyine derivatives will not be covered here since most of these early concepts have now been adapted to asymmetric versions. Since a significant emphasis has recently been placed on accessing α,α-disubstituted amino acids, the formal 'double derivatization' of a glycine template can be tackled in numerous ways. To unify the presentation, this section embraces various modes of functionalizing the α-carbon and is therefore, not restricted to glycine exclusively . All modes of chemically replacing the α-hydrogen with a new functional group in an *asymmetric* sense is therefore, reviewed.

A. GLYCINE AND RELATED ENOLATES
1. Bis-Lactim Ethers

Schollkopf and collaborators[1] have devised a versatile and useful method to prepare a large variety of amino acids based on the metallation and subsequent alkylation of bis-lactim ethers. The general protocol involves peptide coupling of two amino acids, piperazinedione formation and bis-lactim ether formation with trimethyloxonium tetrafluoroborate. The first system reported[2] on was the symmetrical bis-lactim ether of cyclo-L-Ala-L-Ala (**2**). Metallation with n-BuLi in THF at low temperature followed by electrophilic quench furnished the homologated bis-lactim ethers (**3**). The electrophile was shown to add *anti-* to the methyl group in a highly stereoselective fashion with typical diastereomeric excesses of

SCHEME 1

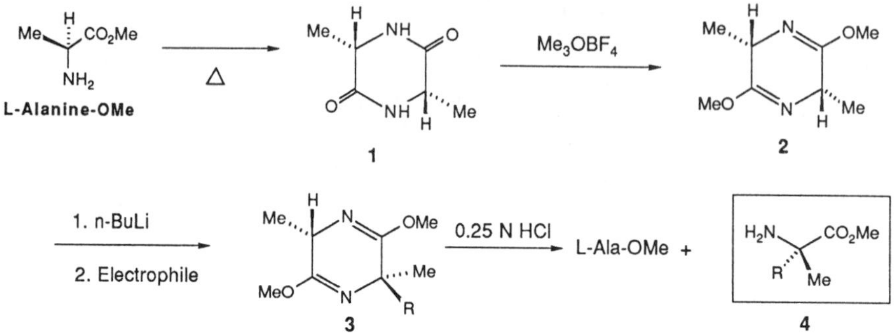

Schollkopf,U.; Hartwig, W.; Groth, U., *Angew.Chem.Int.Ed.Engl.* (1979) **18**, 863.

Schollkopf, U., Hartwig, W., Groth, U., Westphalen, K-O., *Liebigs Ann. Chem.*(1981)696.

R	ALKYLATION YIELD	%de	ABSOLUTE CONFIG.	AMINO ACID %ee
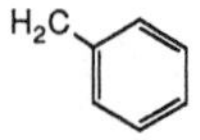 H₂C–phenyl	80-90	93	R	93
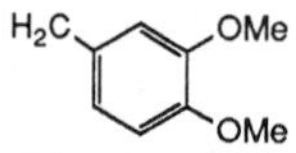 H₂C–dimethoxyphenyl	85-90	95	R	90
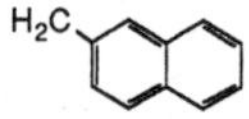 H₂C–naphthyl	85	93	R	90
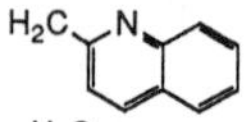 H₂C–quinolinyl	78	94	R	93
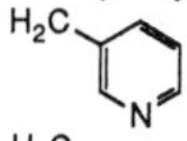 H₂C–pyridyl	90	>95	R	90
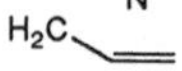 H₂C–CH=CH₂	80-85	92	R	90
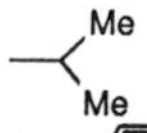 isopropyl (Me₂CH)	75-80	92	R	
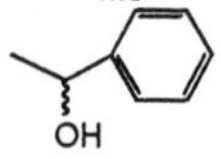 CH(Me)(phenyl)(OH)	85	95	R	
CH₂CH₃	81	95	R	
n-C₈H₁₇	83	92	R	
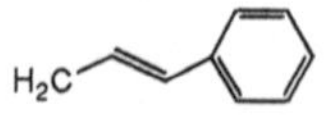 H₂C–CH=CH–phenyl	85	91	R	

greater than 90%. The homologated bis-lactim ether prduct can be hydrolyzed with 0. 5 N HCl to the new amino acid methyl ester(**4**) and L-alanine methyl ester. Bulb-to-bulb distillation is then employed to separate the L-alanine methyl ester (usually the most volatile component) from the desired product. As shown in Scheme 1, a variety of benzylic, allylic, primary and secondary electrophiles give very good yields and high asymmetric induction. Two subsequent reports[3,4] utilizing the same bis-ala template demonstrated that aldol condensations with both aldehydes and ketones gives high asymmetric induction at the alanine α-carbon but modest diastereoselection at the β-carbon (Schemes 2 and 3). Fortunately, the authors have been conscientious in all of these studies to follow the initial communications with full experimental details included in full papers. The amino acids may be obtained as either the free zwitterionic amino acids or as the corresponding methyl esters. Hydrolysis of the α-methylated derivatives (**3,5**) to the free zwitterions requires relatively vigorous conditions (refluxing 6 N HCl, 1h); the methyl esters are obtained under much milder conditions (0.25 ~ 0.5 N HCl 1h, room temp.). The L-alanine methyl ester can be recovered and reused but no mention is made of the optical purity of the recovered material.

SCHEME 2

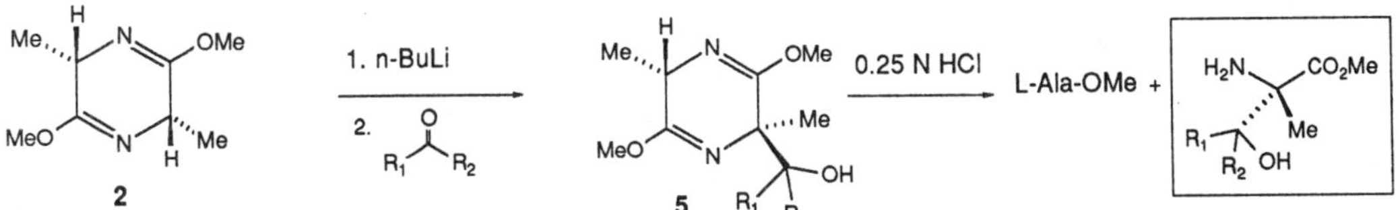

Schollkopf, U.; Hartwig, W.; Groth, U., *Angew Chem.Int Ed. Engl.* (1980) **19**, 212.

R₁	R₂	ALKYLATION YIELD	%de	ABSOLUTE CONFIG.	%de (c-3)
H	H	70	81	R	
Me	Me	80	85	R	
Ph	Ph	89	>95	R	
Ph	Me	70	>95	R	41 (R)
Ph	H	72	88	R	52 (R)
4MeO-Ph	H	69	89	R	74 (R)

SCHEME 3

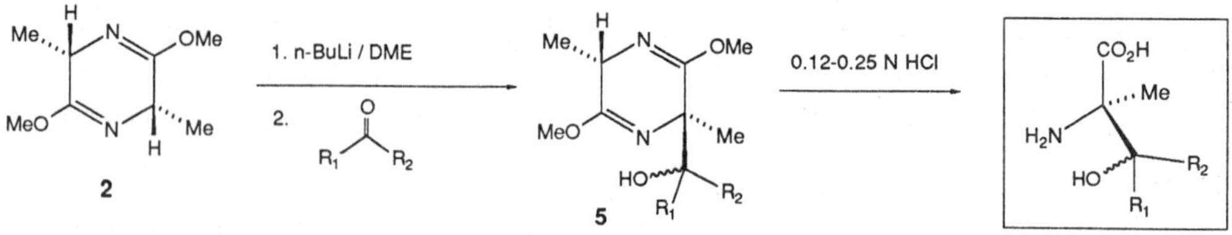

Schollkopf, U.; Groth, U.; Hartwig, W., *Liebigs AnnChem.* (1981) 2407.

R_1	R_2	ALDOL YIELD	%de (c-3)	%de (c-7)	AMINO ACID YIELD	METHYL ESTER
H	H	70	81 (R)		61	
Me	Me	83	85 (R)		82	73
C_6H_5	C_6H_5	89	>95 (R)			
C_6H_5	H	72	88 (R)	52 (R)		36 (2R,3R) 64 (2R,3S)
C_6H_4p-OMe	H	69	89 (R)	74 (R)		
C_6H_5	Me	70	>95 (R)	41 (R)		

By far the most popular and extensively studied bis-lactim ether is that derived from L-valine and glycine[5]. As shown in Scheme 4, L-valine is converted into the N-carboxyanhydride (**7**) and condensed with glycine ethyl ester. Heating the dipeptide (**8**) furnishes cyclo-L-Val-Gly (**9**) which is in turn, converted into the bis-lactim ether **10**. This reagent has become the most widely used glycine template and is now commercially available in both enantiomeric forms. As above, metallation of the bis-lactim ether with butyllithium in THF at low temperature followed by alkylation with a variety of electrophiles proceeds with a high degree of stereoselectivity furnishing the *anti*- adducts **11**. Hydrolytic cleavage of the heterocycle with dilute HCl at ambient temperature furnishes the new amino acid methyl ester (**12**) and L-valine methyl ester which must be separated. In most cases, it is possible to conveniently remove the volatile L-Val-OMe by bulb-to-bulb distillation from the less volatile product (**12**) ; if the boiling points of the two products are similar, this may become problematic and require more tedious chromatographic separations. Again, the L-valine methyl ester may be recovered and recycled, but this is rarely done in practice due to the commercial availability of **10**. As can be seen from the table accompanying Scheme 4,

SCHEME 4

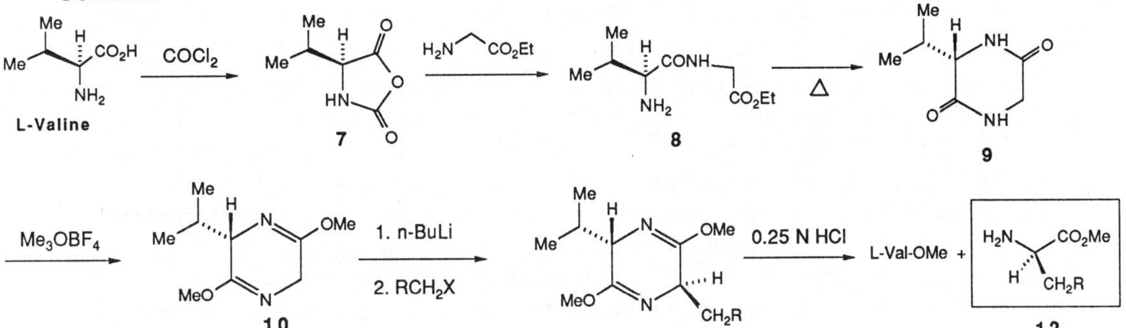

Schollkopf,U.; Groth,U.; Deng, C.,*Angew.Chem.Int.Ed.Engl.* (1981) **20**, 798.

CH₂R	ALKYLATION YIELD	AMINO ACID YIELD	AMINO ACID %ee
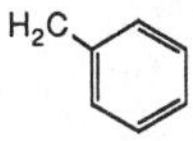 H₂C—⟨benzyl⟩	81	73	91-95
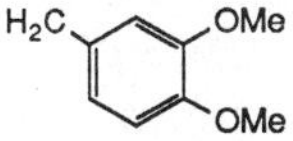 H₂C—⟨3,4-dimethoxyphenyl⟩ OMe OMe	91	78	85
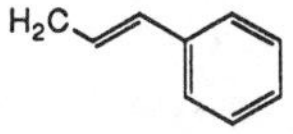 H₂C—CH=CH—⟨phenyl⟩	90	89	>95
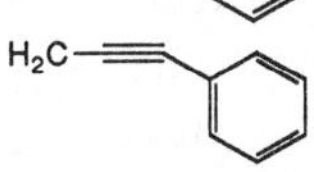 H₂C—C≡C—⟨phenyl⟩	92	86	>95
H₂C—C≡C—H	88	52	60-65
CH₂(CH₂)₅CH₃	62	75	75-80
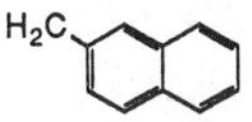 H₂C—⟨naphthyl⟩	89	78	92

the yields for the alkylations are excellent and the yields for the hydrolyses to **12** are quite good.

Several other bis-lactim ethers have been prepared following the standard protocol described above. Thus the systems   derived from (S)-

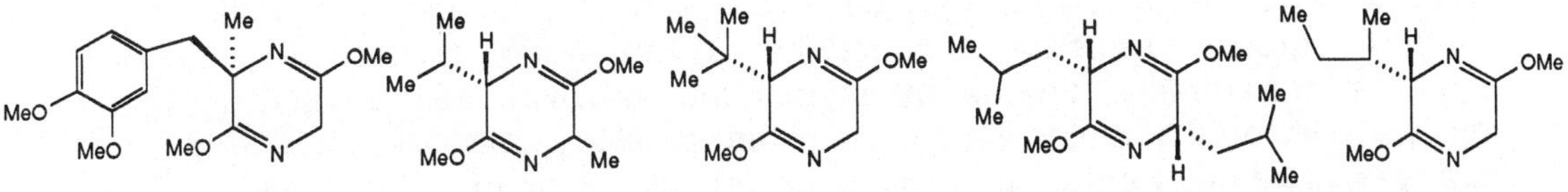

O,O-dimethyl-α-methyldopa and glycine [6](13), L-valine and alanine [7](14), L-terleucine and glycine [8](15), L-leucine -L-leucine [9](16), and L-isoleucine and glycine [10](17) have all been reported as useful templates for synthesizing optically active amino acids. Both the Val-Gly and the Val-Ala systems are commercially available in both enantiomeric forms. The syntheses of the bis-lactim ethers (13-17) are described in Schemes 5-9 with summaries of the specific amino acids prepared in each paper.

In search of chiral auxilliaries that were both inexpensive and that would result in high asymmetric induction in the alkylations, the authors essentially found that few advantages to the valine-derived templates are observed with the other systems (13-17). However, in some cases better results with particularly troublesome alkylations can be realized. For example, the diastereomeric excess obtained by alkylating the L-Val-Gly template (10) with 3-bromopropyne was only 60~65 %de but was at least 85% de with the dopa system 13[6]. It is quite noteworthy that several hindered electrophiles give good to excellent alkylation yields from the bis-lactim ether 13 (Scheme 5). The free zwitterionic amino acids can be obtained by hydrolyzing the methyl esters (24) in 6N HCl at reflux temperature for 1 h followed by treatment with propylene oxide in hot ethanol for a short time. The overall limitation of this methodolgy is in the preparation of amino acids containing acid labile side chains.

The Val-Ala bis-lactim ether 14 [7] is one of the most useful reagents available for preparing a wide array of α-methylated amino acids that are of considerable current biomedical interest. The reagent is prepared in the usual way from optically active L- or D- valine and D,L-alanine ethyl ester. The bis-lactim ether then, is a 1:1 diastereomeric mixture; this is of no consequence since, the metallation step converts this d,l- center to a planar carbanion. It is quite interesting that the metallation is highly *regioselective*   resulting in exclusive deprotonation at the alanine stereogenic center. Since the carbanion formation is done under kinetic control, this result is undoubtedly attributable to the relative steric accessibility of the Ala methine over the Val methine.

The terleucine-derived bis-lactim ether 15 [8] was observed to give exceptionally high diastereomeric excess in alkylations (typically >95% de) of the lithiated derivative. Since the bis-lactim ethers are all relatively flat molecules [1]lacking significant 1,3 diaxial interactions, the requisite axial orientation of the chiral directing group can only be attributed to A-1,3 strain between this substituent and the adjacent methoxy group. Since *anti*-selectivity is observed throughout these systems, the relative degree of asymmetric induction will depend solely on the bulk of the amino acid side chain auxilliary and the population of the pseudo-axial conformer. As a compelling example, the difficult 3-

SCHEME 5

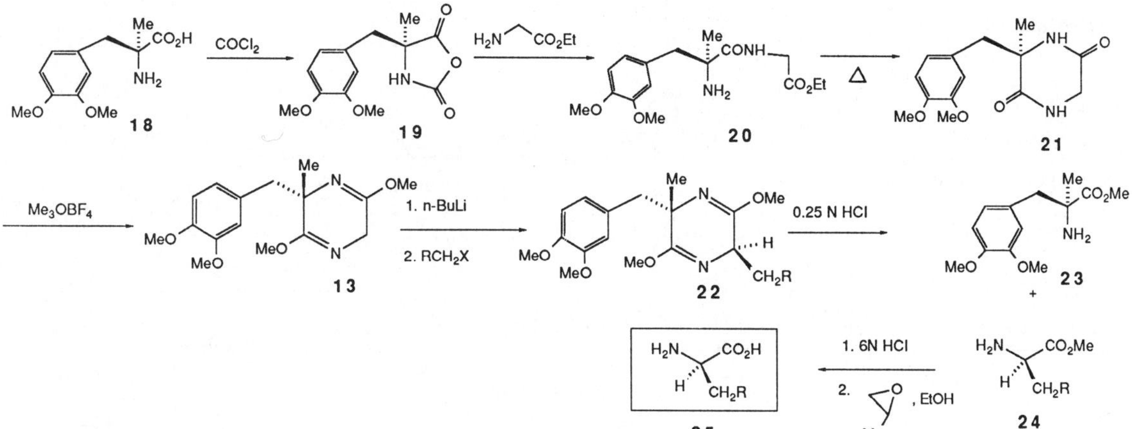

Schollkopf,U.; Hartwig, W.; Pospischil, K-H.; Kehne, H., *Synthesis* (1981) 966.

R	ALKYLATION YIELD	%de	AMINO ACID (OMe) YIELD	AMINO ACID %ee
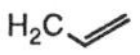 H$_2$C=/ (allyl)	90	~80	70	80
 H$_2$C /	89		65	80
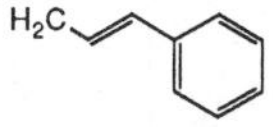 H$_2$C /phenyl	90		70	70
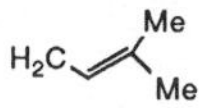 H$_2$C / Me Me	85		60	80
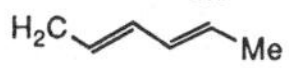 H$_2$C / Me	78		54	80
H$_2$C≡H	95		64	85
CH$_2$(CH$_2$)$_5$CH$_3$	88		45	85
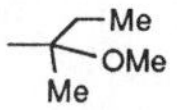 Me OMe Me	88	95(C-6) 22(C-7)	62	95 (21)
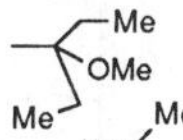 Me OMe Me	95	89	73	84
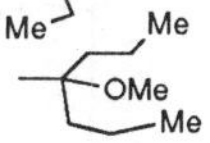 Me Me OMe Me	95	95	65	95
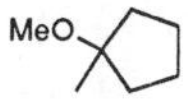 MeO (cyclopentyl)	68	88	75	85
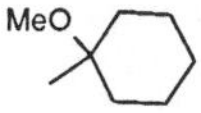 MeO (cyclohexyl)	21	90	68	87

bromopropyne alkylation in the terleucine case [8] is >95% de and only 60-65% for **10** (Scheme 4)[5] and 85% for **13** (Scheme 5)[6]. The relative cost and availability of terleucine will undoudtedly limit the utilization of this excellent bis-lactim ether.

SCHEME 6

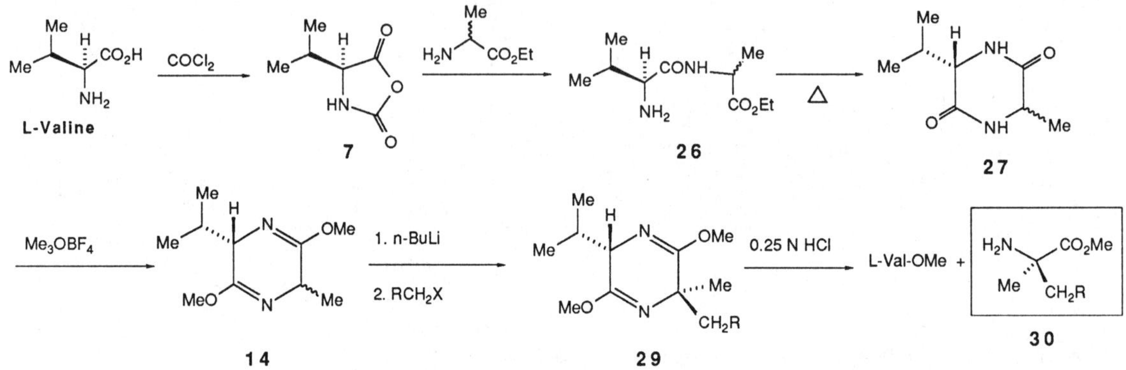

Schollkopf, U.; Groth, U.; Westphalen, K-O.; Deng, C., Synthesis (1981) 969.

CH₂R	ALKYLATION YIELD	%de
H₂C—⟨benzyl⟩	68	>95
H₂C—⟨3,4-dimethoxybenzyl⟩	76	>95
H₂C—⟨cinnamyl⟩	80	~85
H₂C—CH=CH₂	43	>95
CH₂(CH₂)₅CH₃	90	>95
H₂C—≡—H	81	>95
H₂C—C(Me)=CMe₂	94	>95

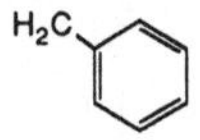

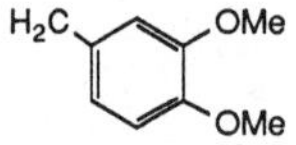

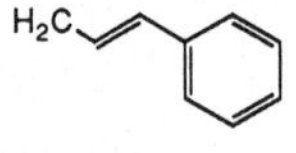

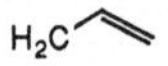

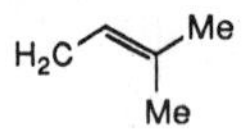

SCHEME 7

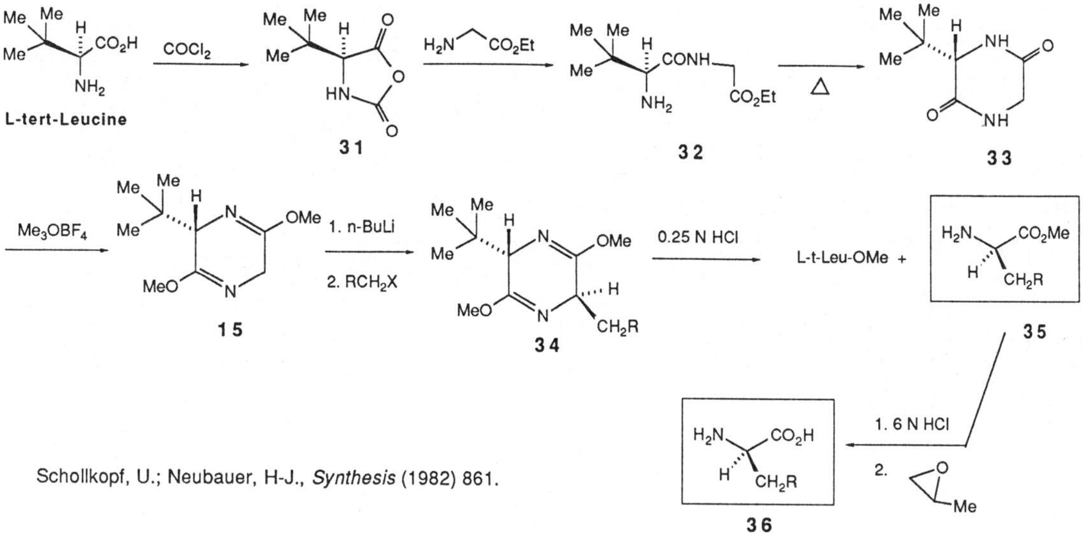

Schollkopf, U.; Neubauer, H-J., *Synthesis* (1982) 861.

CH$_2$R	ALKYLATION YIELD (34	AMINO ACID (OMe) YIELD (35)	%ee	AMINO ACID YIELD (36)
H$_2$C—⌬ (benzyl)	87	76	>95	73
CH$_2$(CH$_2$)$_5$CH$_3$	58	68	>95	81
H$_2$C—CH=CH$_2$	84	87	94	90
H$_2$C—≡—H	86	59	>95	
H$_2$C—C(=O)—Ot-Bu	89	82	>95	

The number of possible combinations of amino acids to prepare bis-lactim ethers that could conceiveably be utilized in making α- and α,α-disubstituted amino acids is formidable. As another example[9] of a symmetrical system, L-leucine was cyclo-dimerized and converted into the corresponding bis-lactim ether **16.** Metallation and alkylation in the usual way afforded **37** which was hydrolyzed to the corresponding α-alkylated leucine methyl ester derivatives in good yields and high enantiomeric excess. It should be noted that with these very hindered

SCHEME 8

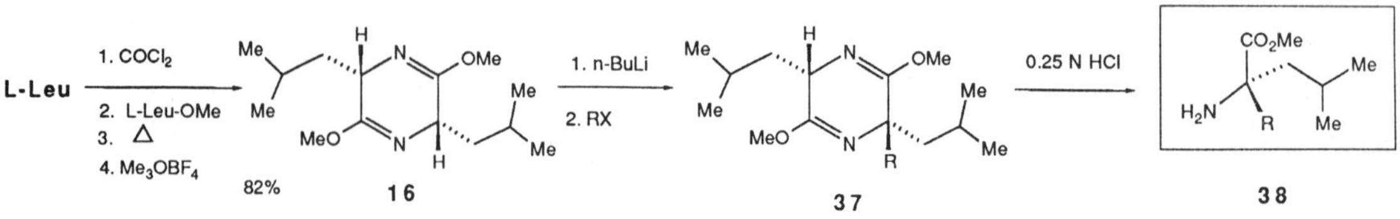

L-Leu
1. COCl₂
2. L-Leu-OMe
3. △
4. Me₃OBF₄

82% → **16**

1. n-BuLi
2. RX

→ **37**

0.25 N HCl → **38**

Schollkopf, U.; Busse, U.; Kilger, R.; Lehr, P. *Synthesis* (1984) 271.

R	ALKYLATION % (37)	METHYL ESTER % (38)	%ee	ABS. CONFIG.
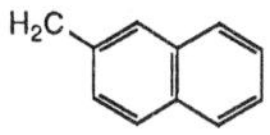 H₂C‑naphthyl	78	5		
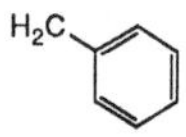 H₂C‑phenyl	82	50	>95	R
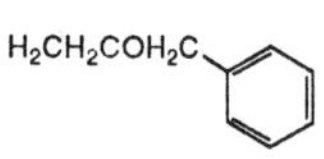 H₂CH₂COH₂C‑phenyl	85	52	>95	R
H₂C━≡━H	91	80	>95	R
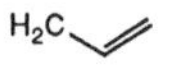 H₂C‑vinyl	88	74	>95	R
n-C₄H₉	87	74	85	S
CH₃	90	82	~85	S
H₂CSCH₃	89	54	88	R

lactim ether adducts, hydrolysis to the amino acid methyl esters proceeds rather slowly relative to the monosubstituted systems and was noted to even fail in some cases. The authors note [9] that in these difficult cases, dipeptide methyl esters form as side products under the standard conditions (dilute HCl, RT) and that piperazinedione formation is observed at higher temperatures which are very stable to further hydrolysis.

The *iso*-leucine-glycine-derived system[10] was prepared according to the standard protocol but gave relatively disappointing results when compared to the simpler valine-derived system (**10**). It would be interesting to examine if the additional stereogenic center present in **17** would provide remote asymmetric induction in aldol condensations (for example) at the β-carbon; a problem that is only poorly addressed by the other templates (see above Schemes 2 and 3).

A vast number of publications demonstrating the utility of the bis-lactim ether methodology have appeared and are summarized in detail below. Full experimental details are available for most and this should prove quite valuable to workers preparing the same or related amino acids.

SCHEME 9

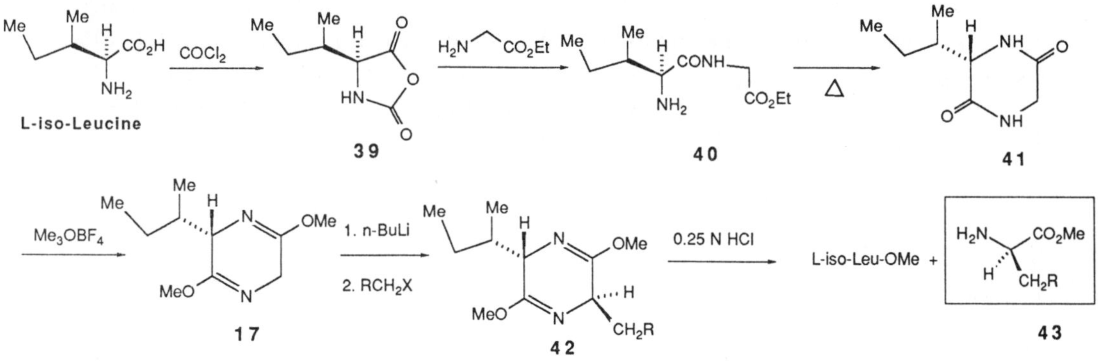

Jiang, Y.; Schollkopf, U.; Groth, U., *Scientia Sinica B* (1984) **27**, 566.

CH$_2$R	ALKYLATION YIELD (42)	AMINO ACID (OMe) YIELD (43)	%ee
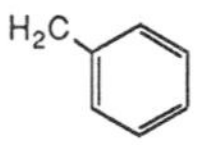 H$_2$C—⟨phenyl⟩	85	76	90
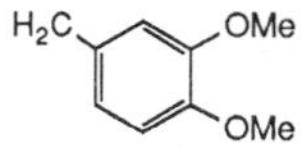 H$_2$C—⟨2,4-dimethoxyphenyl⟩ (OMe, OMe)	70	62	82
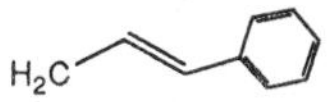 H$_2$C—CH=CH—⟨phenyl⟩	92	90	>95
CH$_2$(CH$_2$)$_5$CH$_3$	74	71	72

The direct alkylation of the bis-lactim ethers with allylic and propargylic electrophiles provides a very useful entry to the biologically important γ,δ-unsaturated amino acids. A substantially more challenging substitution pattern is that of the β,γ-unsaturated amino acids (vinyl glycine derivatives). This is due to the ease of racemization of these substances and the additional propensity for isomerization of the β,γ-unsaturation to the corresponding α,β-dehydro amino acids. Several examples of β,γ-unsaturated amino acids have been prepared via the aldol adducts of the bis-lactim ethers. The first example reported[11] was the preparation of (R)-β-methylenephenylalanine methyl ester (**47**) as shown in Scheme 10. Aldol condensation of **10** and acetophenone furnishes the carbinol **44** in 91% yield. Dehydration with thionyl chloride in the presence of 2,6-lutidine gave an 80:20 mixture of the desired   exo-methylene derivative **45** and the α,β-dehydro isomer **46** (88% yield). Hydrolysis of the mixture with 0.25 N HCl gave (R)-β-methylenephenylalanine methyl ester (**47**, in 64 % yield, >95% ee) and the α-keto acid **48** resulting from cleavage of **46**. It is quite significant that the standard hydrolysis conditions did not lead to any significant racemization or double bond isomerization.

SCHEME 10

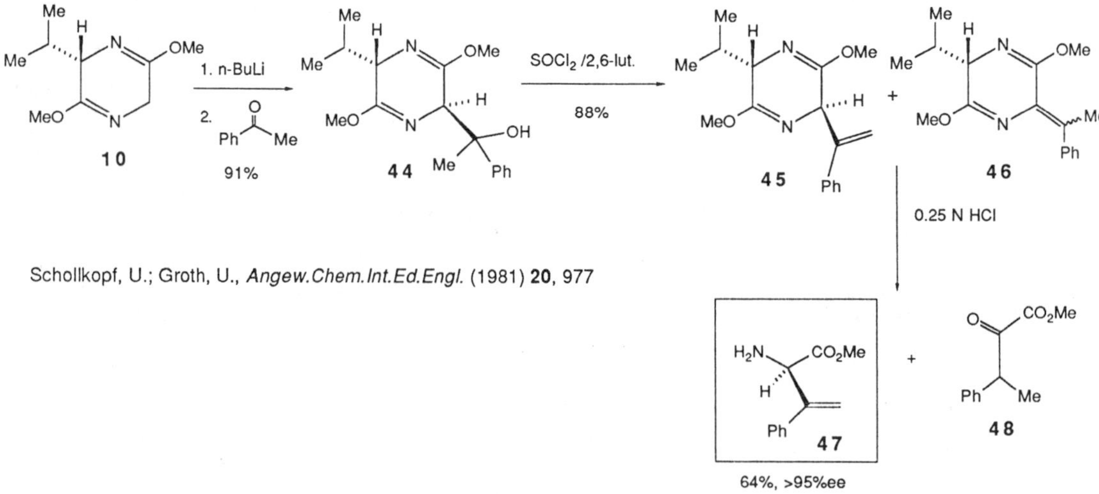

Schollkopf, U.; Groth, U., *Angew.Chem.Int.Ed.Engl.* (1981) **20**, 977

64%, >95%ee

A subsequent, and more detailed report [12] on the preparation of vinyl glycine derivatives employed a Pedersen olefination strategy and a Raney -Nickel dehydrosulfurization to install the unsaturation. As shown in Scheme 11, aldol condensation of **10** with α-trimethylsilyl acetaldehyde and propionaldehyde furnished the carbinols **49** in excellent diastereomeric excess. Hydrolysis of the bis-lactim ether to corresponding methyl esters (**50**) followed by heating in 5 N HCl effected hydrolysis and Pedersen elimination to the zwitterionic amino acids; vinyl glycine itself (**51**) was obtained in 62% yield of unspecified optical purity (assumed to be relatively high due to the observed optical rotation). The α-propenyl glycine (**52**) was obtained in 70% yield as a 1:1 E:Z mixture; again the exact %ee was not determined but was presumably high due to the optical rotation of the synthetic material. In an alternative procedure, **10** underwent aldol condensation with several thionoketones to furnish, after methyl iodide treatment, the sulfides **53**. Raney-Nickel desulfurization proceeded in a non-stereoselective fashion (E:Z mixtures) furnishing the olefins **54** in modest yields. The %de was >95% in all cases and provided access to the β,γ-disubstituted vinyl glycine methyl esters

(55) which could be further hydrolyzed in 2N HCl at 90°C to the corresponding free acids. The primary limitation of this technology is control of the olefin geometry where E/Z isomerism is possible.

SCHEME 11

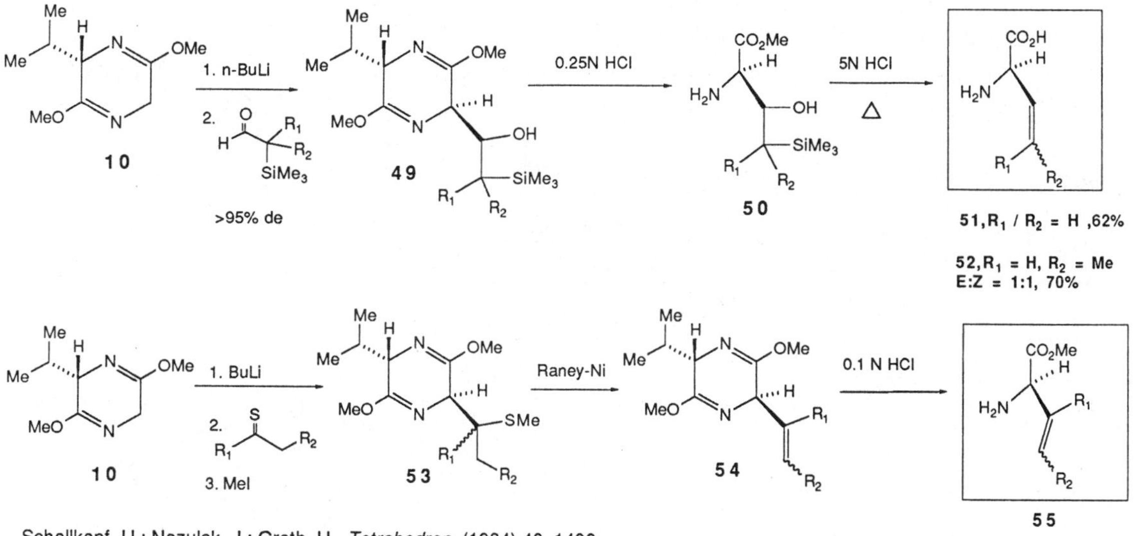

Schollkopf, U.; Nozulak, J.; Groth, U., *Tetrahedron* (1984) 40, 1409.

R₁	R₂	YIELD (53)	CONFIG.	%de	E:Z	Ra-Ni (54)
Et	Me	76	S	>95	2.5 : 1	88
-(CH₂)₄		40	S	~97		93
n-Pr	Et	36	S	>95	2:1	84
Me	Me	52	S	>95	1.5 : 1	86

The α-alkylated β,γ-unsaturated amino acids are also accessible by this technology [13] as illustrated in Scheme 12. The presence of the α-alkyl moiety renders these substances considerably more stable than the above-mentioned vinyl glycine derivatives since both racemization and double bond isomerization is precluded. It has not been shown that these α-alkylated β,γ-unsaturated amino acids can function as pyridoxal-dependent suicide enzyme inhibitors. Since the well-accepted mechanisms [14] for the inactivation of such enzymes involves abstraction of the amino acid α-methine proton bound as the Schiffs base to the PLP, it is quite unlikely that these amino acids will display the same types of biological activity exhibited by the mono-substituted congeners. As with some previously mentioned α-alkylated amino acids, hydrolysis of the hindered

SCHEME 12

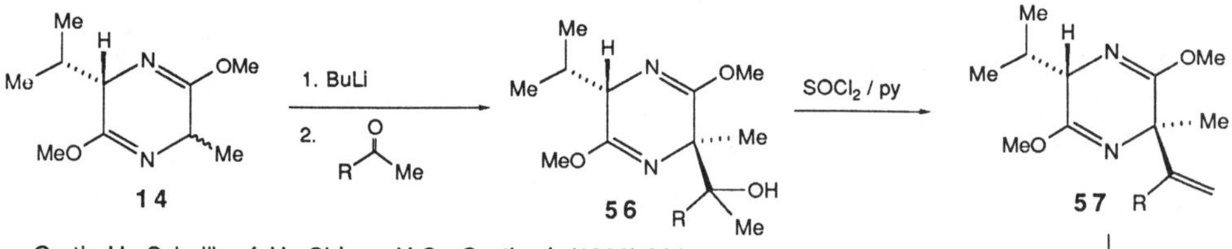

Groth, U.; Schollkopf, U.; Chiang, Y.-C., *Synthesis* (1982) 864.

R	ALDOL YIELD	%de	AMINO ACID(OMe) YIELD	%ee
Me	94	>95		>95
Ph	93	>95	68	>95
H	85	>95		

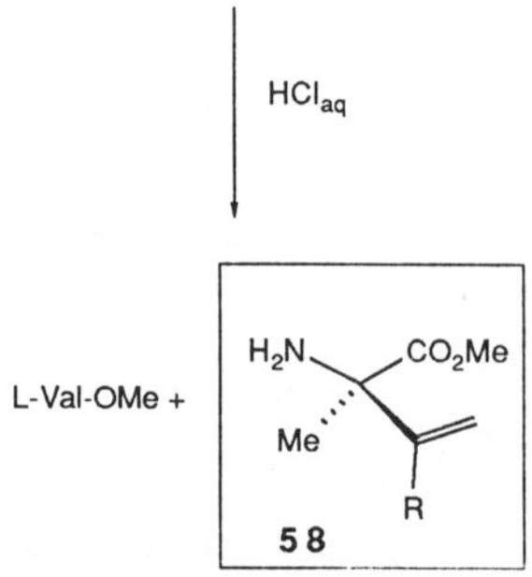

L-Val-OMe +

adducts **56** to the amino acid methyl esters is sluggish; the diastereomeric excess in the aldol condensation however, is excellent.

Serine was prepared [15] in ~ 95% ee by condensing chloromethyl benzyl ether with the bis-lactim ether **13**. As expected, the alkylation is highly diastereoselective giving selectivity for attack of the electrophile *anti*-to the dimethoxybenzyl group. Stepwise hydrolysis of **59** furnishes O-benzyl serine methyl ester; O-benzyl serine; and finally serine. It is interesting that the benzyl ether is deprotected under the conditions of refluxing 6 N HCl.

SCHEME 13

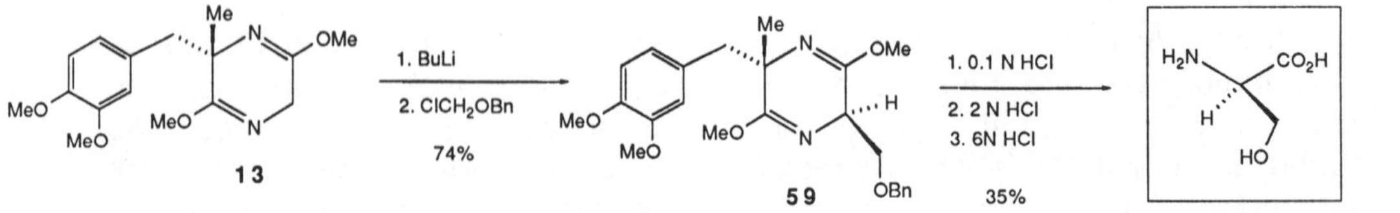

Nozulak, J.; Schollkopf, U., *Synthesis* (1982) 866.

60, SERINE ~95%ee

SCHEME 14

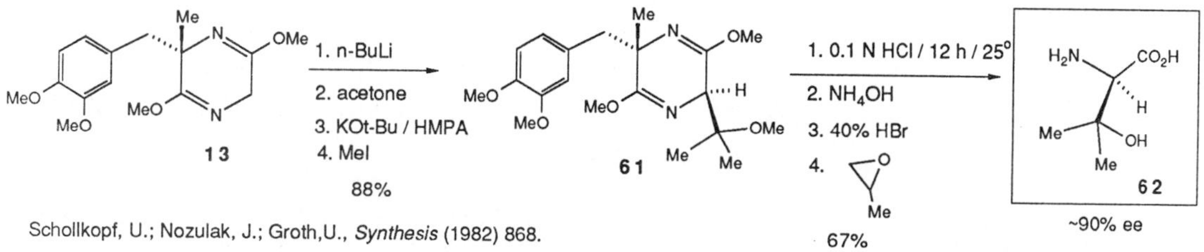

Schollkopf, U.; Nozulak, J.; Groth,U., *Synthesis* (1982) 868.

(R)-(-)-β-Hydroxyvaline (**62**) has been isolated as a constituent of a biologically active peptide and is also reported as being biologically active [16]. Utilizing either bis-lactim ether **10** or **13**, aldol condensation with acetone furnished the carbinol **61** in >93% de. Methylation and careful hydrolysis furnishes **62** in 90% ee or greater.

In a strictly similar fashion to that used for serine, the corresponding α-methyl serine **64** was prepared [17] (Scheme 15) from chloromethyl benzyl ether alkylation of the Val-Ala system **14**. The synthetic **64** is reported as being essentially enantiomerically pure , but an exact % ee was not determined.

SCHEME 15

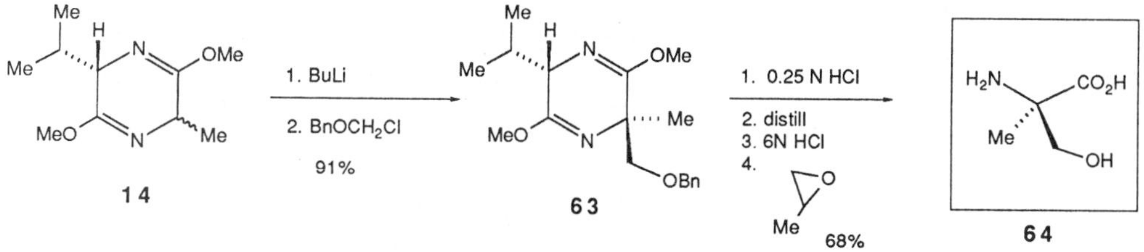

Groth, U.; Chiang, Y-C.; Schollkopf, U., *Liebigs Ann.Chem.* (1982) 1756.

Alkylation[18] of **14** with dibromomethane furnished the α-methyl bromomethyl derivative **65** in 79% yield. This compound can serve as a versatile intermediate for preparing a number of differentially S-protected α-methyl cysteine derivatives. Specifically, displacement of the bromide with benzyl mercaptan and tert-butyl mercaptan furnished the corresponding sulfides (**66**) in 88% and 93% yields, respectively. Standard hydrolysis processing provided the S-protected α-methyl cysteine derivatives (**67**) in good chemical yields and >95% ee.

SCHEME 16

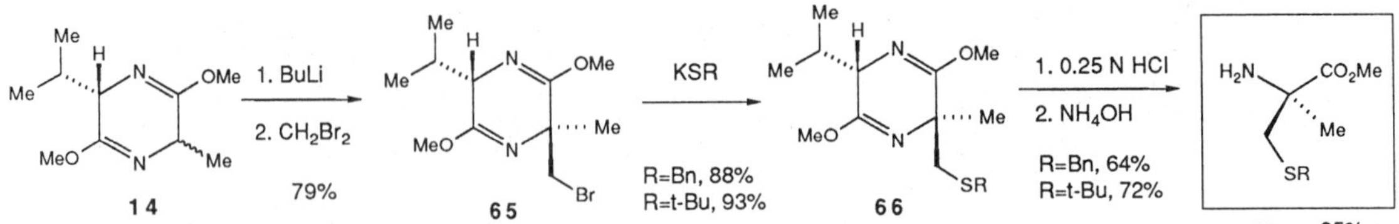

Groth,U; Schollkopf, U.; *Synthesis* (1983) 37.

Aldol condensation of acetaldehyde with the titanium enolate derived from transmetallation of lithio-**10** afforded [19] the secondary carbinol **68** (Scheme 17). The use of titanium increases the *threo/erythro* ratio to ~ 92:1. Hydrolysis all the way to the free amino acid and separation from L-valine by HPLC afforded the optically active D-threonine (**69**) with no detectable amount of *allo*-threonine.

SCHEME 17

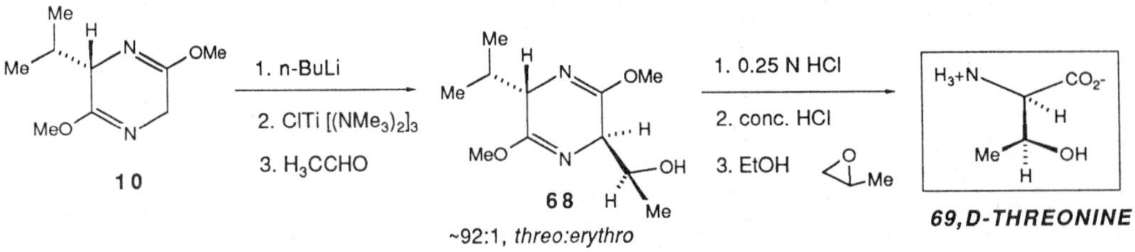

Schollkopf, U.; Nozulak, J.; Grauert, M., *Synthesis*, (1985), 55.

Further evidence [20] for the selectivity of the titanium enolate derived from **10** is illustrated in Scheme 18. α-Alkoxy aldehydes were observed to give significantly improved diastereoselection with the titanium salt relative to the same condensations with the lithio derivative. *Threo*-selectivity is displayed and is as high as 128:1 with (S)-glyceraldehyde and the L-Val system **10**. Some degree of double diastereodifferentiation is displayed by these systems since, (R)-glyceraldehyde and L-**10** gave a ~19:1 *threo:erythro* ratio. Acetylation of the secondary hydroxyl group of the adducts **70-72**, followed by the standard dilute acidic hydrolysis gave the N-acetyl amino acid methyl esters **73-75** in good overall yield (exact %ee not determined).

SCHEME 18

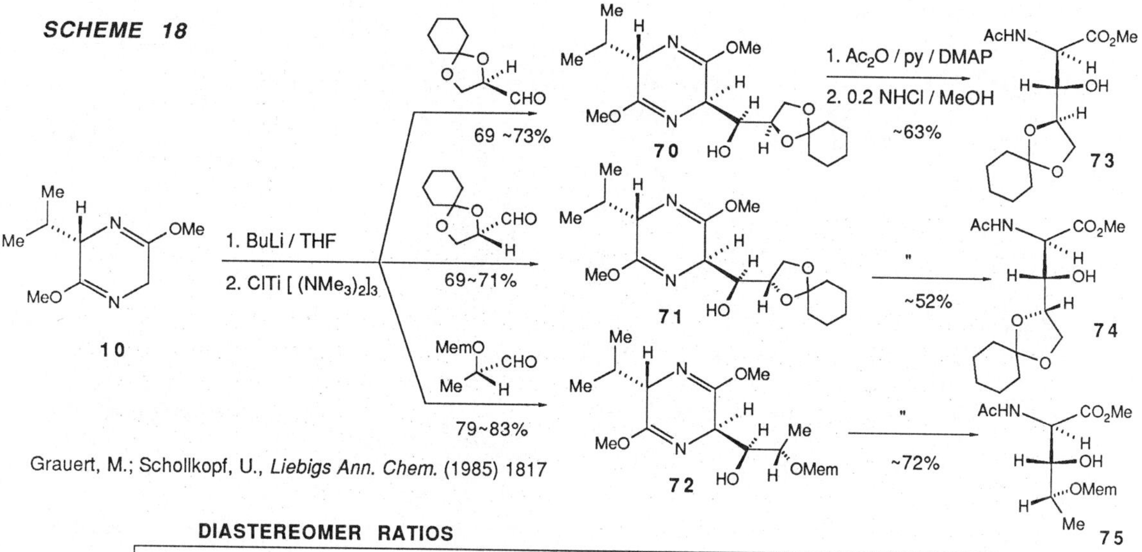

Grauert, M.; Schollkopf, U., *Liebigs Ann. Chem.* (1985) 1817

DIASTEREOMER RATIOS

METAL COUNTERION	ALDEHYDE	3R,1'R	3R,1'S	3S,1'R	3S,1'S
Li$^+$	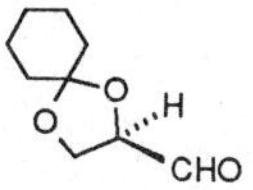	6.1	6.4	1.0	1.5
Ti$^+$(NMe$_2$)$_3$		78.5	4.2	0.5	1.1
Li$^+$	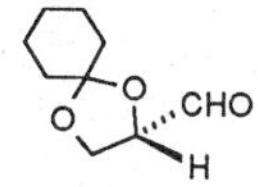	15.3	2.0	0.9	1.2
Ti$^+$(NMe$_2$)$_3$		64.0	0.5	0.3	0.5
Li$^+$	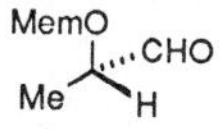	33.0	2.4	0.4	0.7
Ti$^+$(NMe$_2$)$_3$		32.5	0.4	0.2	0.2

Further utility of the aldol methodology is illustrated [21] in the preparation of the β-fluoro-substituted amino acids. These unnatural amino acids have been shown to be suicide substrates of various pyridoxal-dependent enzymes [14] and have been the subject of vigorous research activity. As shown in Scheme 19, aldol condensation of **10** with acetone and benzaldehyde furnished the corresponding carbinols **76** and **77**, respectively. Treatment of these compounds with diethylaminosulfur trifluoride (DAST) furnished a mixture of the desired fluoro derivatives **78** and **80**, respectively and products resulting from dehydration (**79** and **81**, respectively). Hydrolysis of these mixtures and separation of the products by chromatography afforded methyl(R)-β-fluorovalinate (**82**) and methyl (2R)-β-fluorophenylalaninate (**83**) in > 95% ee. The phenylalanine derivative (**83**) is obtained as an 8.2:8:1 ratio of diastereomers.

SCHEME 19

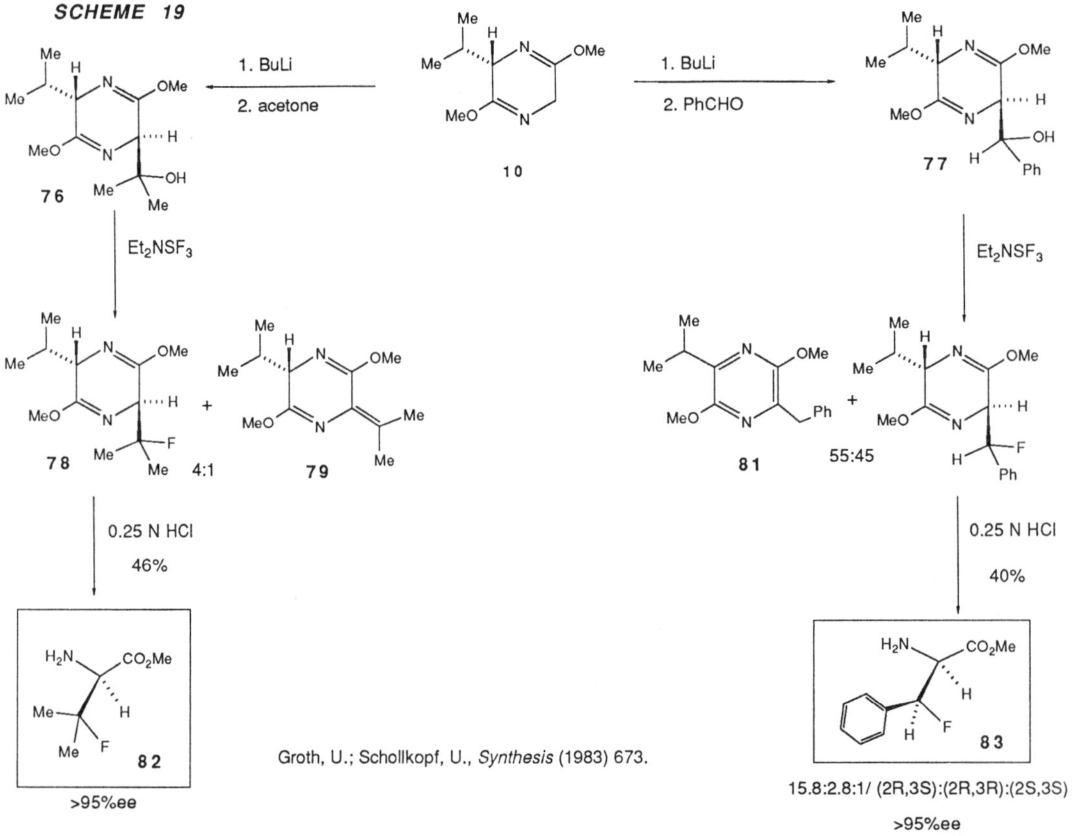

Groth, U.; Schollkopf, U., *Synthesis* (1983) 673.

15.8:2.8:1/ (2R,3S):(2R,3R):(2S,3S)

>95%ee

An additional use of aldol methodology is the approach[22] to 2-amino-3-hydroxy-4-alkenoates, a functional array found in MeBmt, the unusual C$_9$- amino acid constituent of the immunosuppressive peptide cyclosporine (discussed in more detail later in this chapter). As illustrated in Scheme 20, aldolization of various α,β-unsaturated aldehydes with the titanium enolate of either **10** or **14** furnishes the 1,2-addition products which are directly protected as either the corresponding acetates or MEM ethers **84**. In the case of the acetate derivatives, cleavage of the bis-lactim ether with dilute acid results in O-to N-acyl migration, furnishing the N-acetyl methyl esters **85**. The corresponding MEM derivatives give the amino acid methyl esters **86** in good yield. In one case, (R$_1$ = R$_2$ = Me) epoxidation of the double bond of **85** with *tert*-butylhydroperoxide and titanium isopropoxide furnished a mixture of epoxides **87** that spontaneously rearranged into a mixture of the N-acetyl carbomethoxy pyrrollidines **88** and **89** (50%).

SCHEME 20

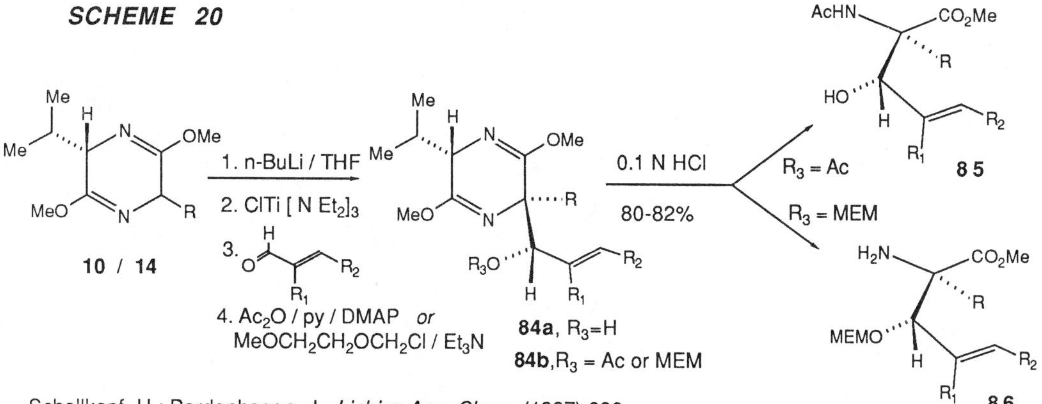

Schollkopf, U.; Bardenhagen, J., *Liebigs Ann. Chem.* (1987) 393.

R	R_1	R_2	(84a) ALDOL YIELD %	(84b) PROTECTION %
H	Me	H	85	
H	Me	Me	82	87 (R_3 = Ac)
H	H	Me	75	
H	H	⟨phenyl⟩	81	
H	H	H	78	89 (R_3 = Ac), 85 (R_3 = MEM)
H	H	CO_2Et	80	
Me	Me	H	70	
Me	Me	Me	75	
Me	H	H	78	

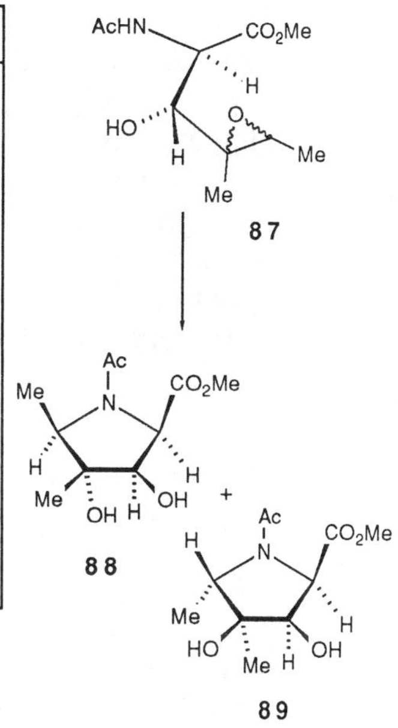

A route to β,γ-epoxy amino acids was reported[23] that employed the condensation of α-chloro ketones with lithio-**10** and subsequent treatment of the initial aldol adduct with base (Scheme 21). The epoxides (**90**) derived from chloroacetone and chloroacetophenone are obtained favoring the 3R, 1'S diastereomers. Hydrolysis of the bis-lactim ether in the usual way provided epoxides **91** and **92** . In the chloroacetone case, treatment of the epoxide with sodium azide followed by hydrolysis gave the azido carbinol **93** as a 1.2 : 1 mixture of diastereomers.

SCHEME 21

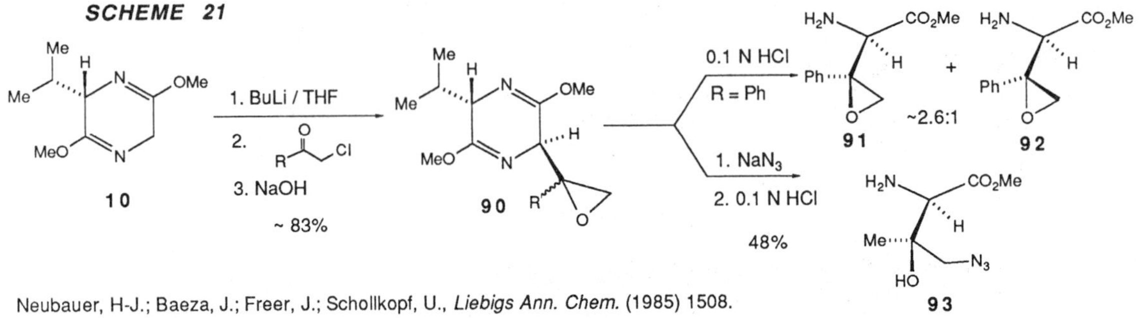

Neubauer, H.-J.; Baeza, J.; Freer, J.; Schollkopf, U., *Liebigs Ann. Chem.* (1985) 1508.

Homoserine derivatives can be prepared[24] by the addition of substituted epoxides to the lithium enolates of **10** and **14** in the presence of BF$_3$-etherate (Scheme 22). The carbinol adducts are directly protected as the corresponding Mem ethers in 55-81% yields. The %de is excellent (>95%) in most cases with the exception of ethylene oxide and propylene oxide with **10** (64-76 % de). Hydrolysis of the bis-lactim ethers in the usual way gives the O-Mem-protected homoserine derivatives (**95**) in good yields. Removal of the Mem protection with 8N HCl at reflux gives the corresponding amino acids which can be cyclized to the α-amino-γ-butyrolactones **96** (four examples).

SCHEME 22

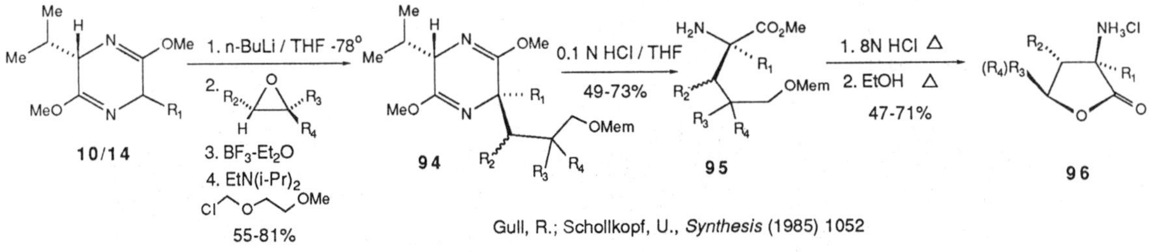

Gull, R.; Schollkopf, U., *Synthesis* (1985) 1052

MEM-PROTECTED HOMOSERINE DERIVATIVES

R$_1$	R$_2$	R$_3$	%de (C-3)	EPIMER RATIOS
H	H	H	64	(3R) : (3S) = 4.6 : 1
H	H	Me	76	(3R, 2'RS) : (3S, 2'RS) = 7.6 : 1
H	Me	Me	>95	(3R,1'R,2'R) : (3R, 1'S,2'S) = 6.6 :1
H	-(CH$_2$)$_4$-		>95	(3R,1'R,2'S) : (3R,1'S,2'R) = 7.8 :1
Me	H	H	>95	(3R) : (3S) = >97.5 : 2.5
Me	Me	Me	>95	(3R,1'R,1'S) : (3R,1'S,2'R) = >97.5 : 2.5
Me	Me	Me	>95	(3R,1'R,2'S) : (3R,1'S,2'R) = >97.5 : 2.5
Me	-(CH$_2$)$_4$-		>95	(3R,1'R,2'S) : (3R,1'R,2'S) = > 97.5 : 2.5

SCHEME 23

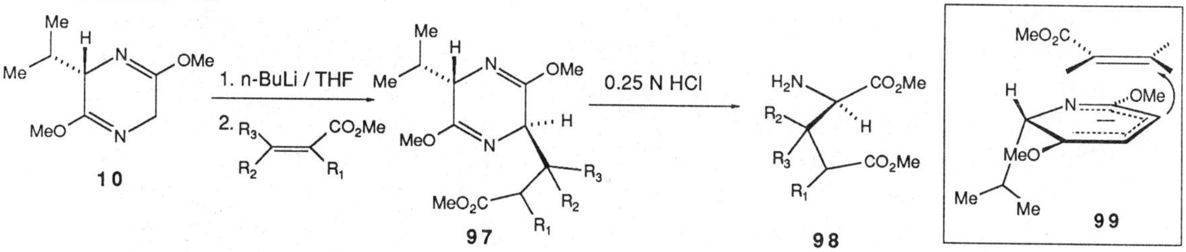

Schollkopf, U.; Pettig, D.; Busse, U., *Synthesis* (1986) 737.

R₁	R₂	R₃	% (97)	%de (C-3)	EPIMER RATIOS
H	H	H	42	>98	(3R) : (3S) = 200 : 1
H	Me	H	65	>98	(3R, 1'S) : (3R, 1'R) : (2S,1'S)= 170:59: 1
H	Me	Me	56	>98	(3R) : (2S) = 140 : 1
H	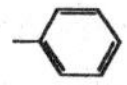 ⬡	H	92	>98	(3R,1'S) : (3R,1'R) : (3S, 1'S) = 200: 20 :1
H	H	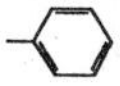 ⬡	88	>98	(3R,1'S) : (3R, 1'R) : (3S,1'S) =46 : 148: 1
Me	H	H	68	98	(3R,1'S) : (3R,1'R) : (3S, 1'S) = 98 : 98 : 1
H	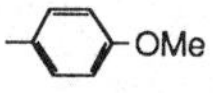 ⬡-OMe	H	85	98	(3R,1'S) : (3R,1'R) : (3S,1'S) = 88 : 8 : 1
H	 ⬡N	H	90	>98	(3R,1'S) : (3R,1'R) : (3S,1'S)= 95 : 7 : ~

The conjugate (1,4) addition of metallated bis-lactim ethers to unsaturated esters, ketones and nitro compounds also provides a valuable protocol for preparing non-protienogenic amino acids. In the case of conjugate addition to substituted acrylates, an entry [25] to the substituted glutamic acid manifold can be realized. As shown in Scheme 23, addition of lithio-**10** to various unsaturated esters furnishes the 1,4-addition products **97**. In most cases there was a reasonably high diastereomeric induction at the β-center; this stereoselectivity was rationalized by considering that the ester should fold over the bis-lactim ether in an orientation that places the carbomethoxy group proximal to the nitrogen that is presumed to carry the lithium cation (structure **99**). In this way, as negative charge begins to develop on the carbomethoxy group in the transition state, the proximal lithium can facilitate the Michael addition in a least motion process relative to the alternative conformer. Selectivities as high as 100 : 1 were observed with E-methylcinnamate. This is an extremely important type of reactivity that needs further exploration since, in general there are very few good ways to control the relative stereochemistry of β-substituted amino acids. Hydrolysis of the Michael-adducts (**97**) with dilute acid provides the substituted dimethylglutamates (**98**).

In a more recent permutation[26] on the Michael reaction methodology, it was found that reaction of the lithiated Val-Ala bis-lactim ether **14** with methyl acrylates bearing a leaving group in the β-position, gives 1,4-addition/elimination to furnish the β,γ-unsaturated esters **101** or **102**. The degree of diastereoselectivity is uniformly high giving, after hydrolysis, the α-methylated dehydroglutamates **103**; complete hydrolysis of the E-methyl esters **103** to the free dehydroglutamates is accomplished with refluxing 4 N HCl. The corresponding Z-isomers **104** are labile to lactam formation **106** upon mild acid hydrolysis of the bis-lactim ethers. A significant drawback of this approach is the failure to obtain the corresponding dehydroglutamates derived from the Val-Gly template **10**. Under the Michael reaction conditions, the initially formed adducts **107** undergo base-catalyzed olefin isomerization to the α,β-dehydro systems **108**. As mentioned previously, the presence of the α-alkyl residue (such as in **101 / 102** where R_1 =Me) renders β,γ-unsaturated amino acids quite stable relative to the biologically more interesting mono-substituted systems; the problems associated with the facile and unwelcome conversion of **107** to **108** further highlights the synthetic challenges inherent in β,γ-unsaturated amino acids.

Michael addition [27] of the titanium enolate of **10** to nitroolefins (**109**) gives the γ-nitro adducts **110**. Hydrolysis with dilute acid furnishes the corresponding γ-nitro amino acid methyl esters **111**. It was noted that the titanium enolate gives substantially higher asymmetric induction at the β-carbon than the lithium derivative. The relative configurations of the adducts were determined by NMR and secured through comparison to an X-ray structure of **110** (R_1 =3,4-methylenedioxy phenyl).

Reaction of methyl 2,4-pentadienoates with **10** occurs regioselectively in a 1,6-mode to furnish the adducts **112** (Scheme 26) in excellent diastereoselectivity [28]. There is also excellent asymmetric induction at the β-carbon (>95%) but the relative configuration remains to be elucidated. Hydrolysis of **112** in the usual manner affords the β-substituted dehydro homoglutamates **113**.

In a related study [29], the conjugate addition of various ketones with the cuprate derived from either **10** or **14** gives primarily the 1,4-adducts **114** with minor to trace amounts of the 1,2-adducts **115** (Scheme 27). The relative amount of asymmetric induction at the β-carbon varied with the structure of the ketone; cyclopentenone and cyclohexenone giving the best results. In these two cases the adducts **114** were hydrolyzed to the corresponding methyl esters and isolated as the corresponding t-BOC derivatives **116** and **117**. It is noteworthy that substitution at the β-

SCHEME 24

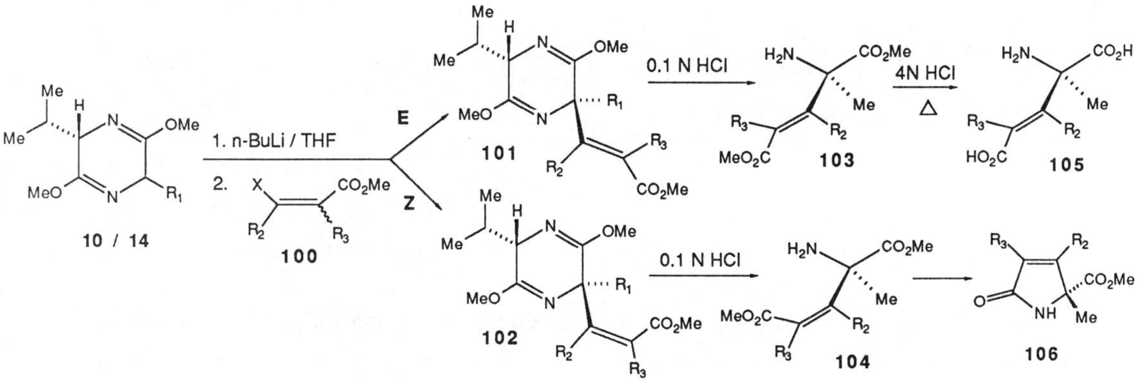

Schollkopf, U.; Schroder, J., *Liebigs Ann. Chem.* (1988) 87.

X	R₁	R₂	R₃	% (101 / 102)	ANTI : SYN	(103) %	(104) %	(105) %	(106) %
Cl	Me	H	H	(Z) 75	97.5 : 2.5				38
Cl	Me	H	H	(E) 34	99 : 1	42			
Cl	Me	Me	H	(Z) 73	99.5 : 0.5				47
Cl	Me	Me	H	(E) 63	99.5 : 0.5				
Br	Me	Me	H	(E) 71	99.5 : 0.5				
OP(O)(OEt)₂	Me	Me	H	(Z) 62	99.5 : 0.5				
Cl	Me	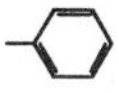 ⬡	H	(Z) 66	>99.5 : 0.5				43
Cl	Me	H	 ⬡	(E) 61	99 : 1	55		55	
Cl	Me	-(CH₂)₃-		(Z) 69	99.5 : 0.5		48		

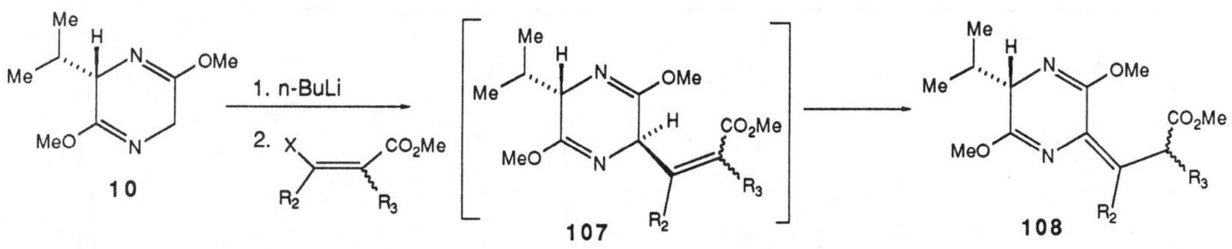

SCHEME 25

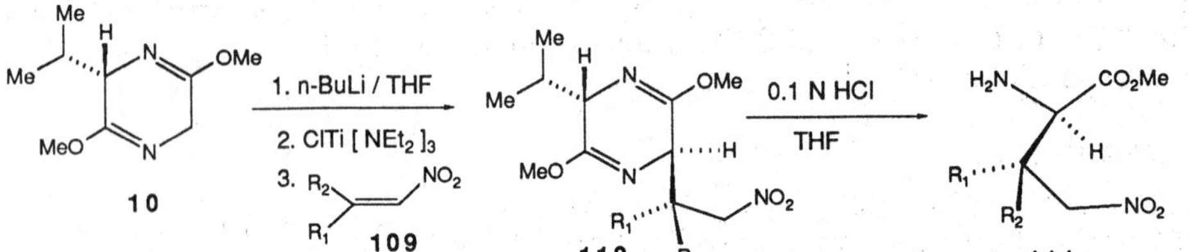

Schollkopf, U.; Kuhnle, W.; Egert, E.; Dyrbusch, M.,*Angew.Chem.Int.Ed. Engl.* (1987) **26**, 480.

R_1	R_2	Michael adduct (%) Ti$^+$ (Li$^+$)	RATIO (3R,1'S): (3R,1'R): (3S,1'S): (3S,1'R)
Me	H	51 (81)	97.4 : 0.8 : 1 : 0.8 (86 : ~ : 10 : 6)
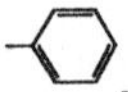 (phenyl)	H	57 (78)	94 : 2.7 : 3.3 : ~ (45 : 38 : 11 : 6)
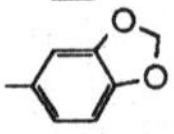 (methylenedioxyphenyl)	H	60 (52)	99 : ~ : 1 : ~ (49 : 41 : 6 : 4)
Me	Me	(80)	(98 : 2; 2R:2S)

SCHEME 26

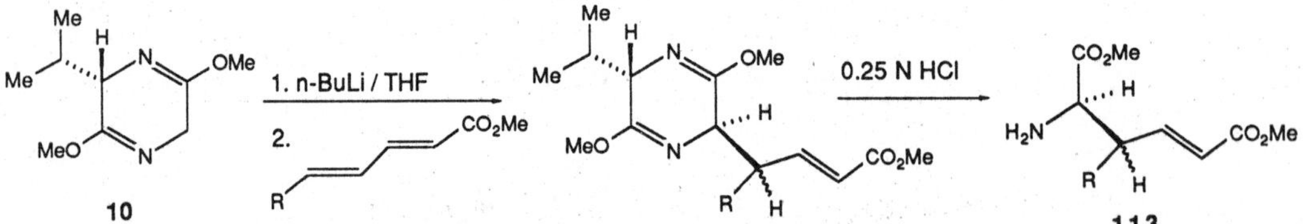

Pettig, D.; Schollkopf, U., *Synthesis* (1988) 173.

R	% (112)	%de (C-3)	Hydrolysis % (113)	%de (C-3, 113)
H	52	>98	60	
Me	78	>98	54	>95
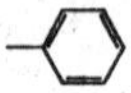 (phenyl)	70	>98	58	>95
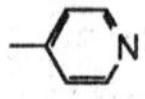 (pyridyl)	68	>98		

position of the enone or utilizing E-enones drastically affects the diastereoselectivity at the β-position in the adduct **114**. For example, cyclopentenone gives 100:2 selectivity whereas 2-methylcyclopentenone gives essentially 1:1.

SCHEME 27

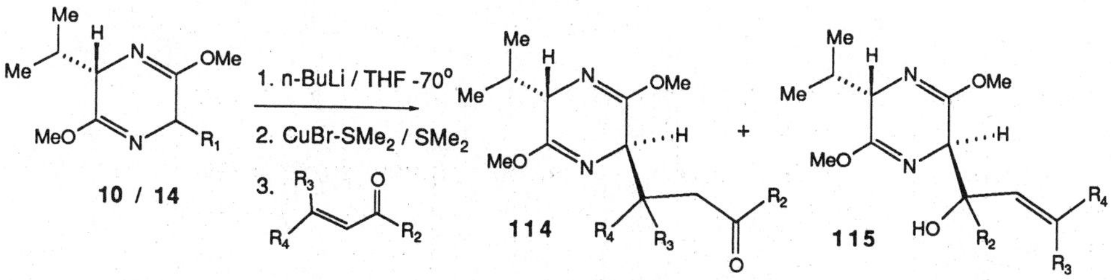

Schollkopf, U.; Pettig, D.; Schulze,E.; Klinge, M.; Eggert, E.; Benecke, B; Noltemeyer, M.,*Angew.Chem.Int.Ed.Engl.* (1988)**27**, 1194.

R$_1$	R$_2$	R$_3$	R$_4$	YIELD (%114/115)	3R,1'R : 3R, 1'S	114 : 115
H	-(CH$_2$)$_2$-		H	71	100 : 2	100 <1
H	-(CH$_2$)$_3$-		H	71	100 : 6	100 : 3
H	-(CH$_2$)$_4$-		H	66	100 : 12	100 <1
H	-(CH$_2$)$_2$-		Me	52	100 : 100	100 : 19
H	Me	H	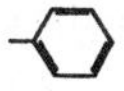 ⬡	62	83 : 100	100 : 6
H	Me	H	 ⬠O	60	100 : 64	100 : 22
H	Me	H	H	39	-	100 : 5

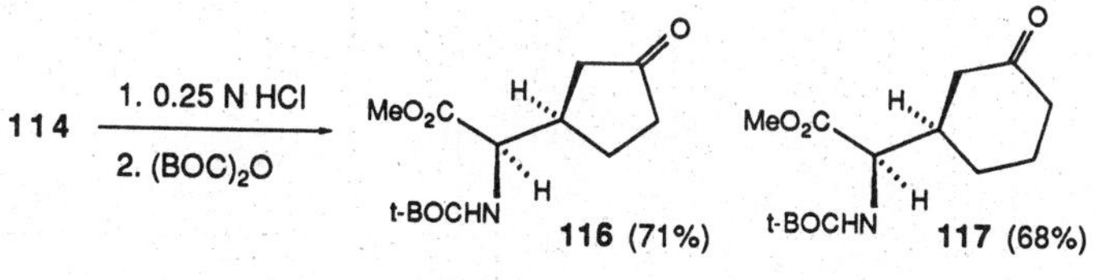

Aldol condensation of α,β-unsaturated aldehydes with the titanium derivative of **10** furnished the adducts **118** (Scheme 28)[30] which were obtained primarily as the 1'S isomers. Epoxidation of these allylic alcohols with *tert*-butyl hydroperoxide / titanium isopropoxide and either the (+)- or (-)-diethyl tartrates according to Sharpless gave the *syn*- and *anti*- epoxides **119** and **120**, respectively. The use of the (-)-tartrate was not consistently required for high *anti*-selectivity but the (+)-isomer had

SCHEME 28

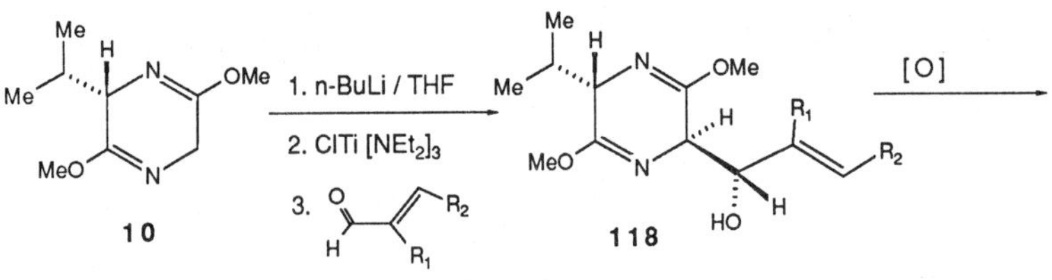

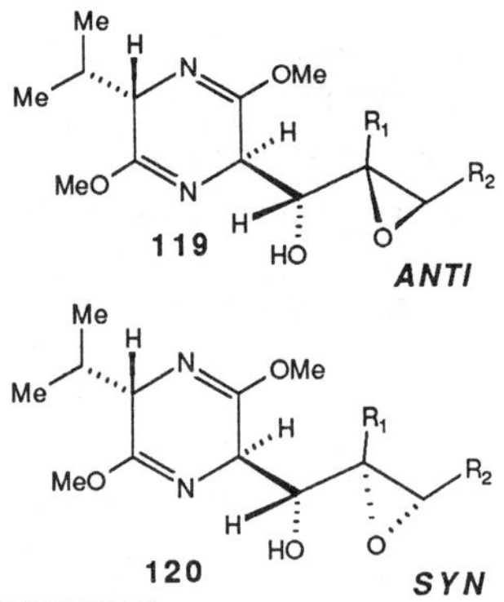

1. n-BuLi / THF
2. ClTi [NEt₂]₃
3. (acrylaldehyde)

[O]

10   **118**   **119**   *ANTI*

120   *SYN*

Schollkopf, U.; Tiller, T.; Bardenhagen, J. *Tetrahedron* (1988) **44**, 5293.

R₁	R₂	EPOXIDATION METHOD	% (119 / 120)	ANTI : SYN
H	Me	Ti(Oi-Pr)₄ / TBHP	74	1 : 2
		Ti(Oi-Pr)₄ / TBHP / (-)-DET	61	4.7 : 1
		Ti(Oi-Pr)₄ / TBHP / (+)-DET	71	1 : >99
H	⬡	Ti(Oi-Pr)₄ / TBHP	65	32 : 1
		Ti(Oi-Pr)₄ / TBHP / (+)-DET	55	1 : 39
Me	H	Ti(Oi-Pr)₄ / TBHP	74	66 : 1
		Ti(Oi-Pr)₄ / TBHP / (-)-DET	68	66 : 1
		Ti(Oi-Pr)₄ / TBHP / (+)-DET	70	32 : 1
Me	Me	Ti(Oi-Pr)₄ / TBHP	83	199 : 1
		Ti(Oi-Pr)₄ / TBHP / (-)-DET	76	>199 : 1
		Ti(Oi-Pr)₄ / TBHP / (+)-DET	79	39 : 1
H	H	Ti(Oi-Pr)₄ / TBHP	64	1.6 : 1

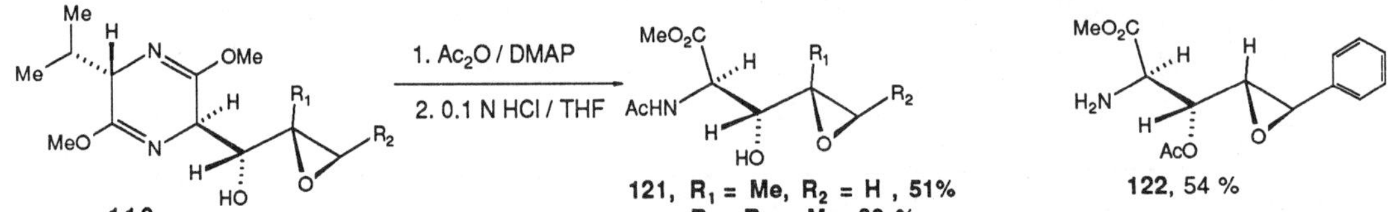

1. Ac₂O / DMAP
2. 0.1 N HCl / THF

119

121, R₁ = Me, R₂ = H , 51%
R₁ =R₂ = Me, 60 %

122, 54 %

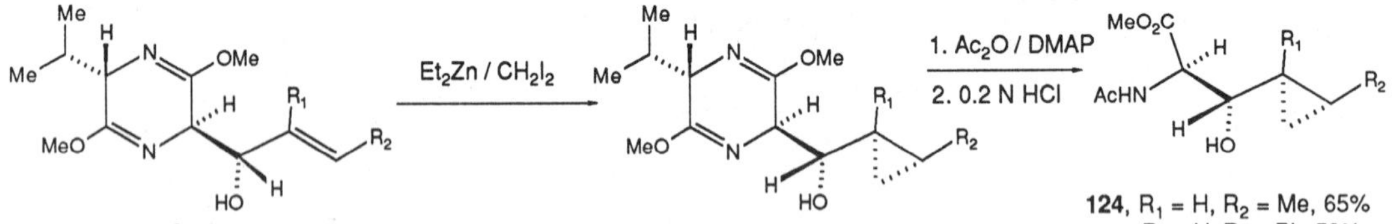

Et₂Zn / CH₂I₂

1. Ac₂O / DMAP
2. 0.2 N HCl

118

123, R₁ = H, R₂ = Me, 68%
R₁ = H, R₂ = Ph, 61%
R₁ = Me, R₂ = H, 76%

124, R₁ = H, R₂ = Me, 65%
R₁ = H, R₂ = Ph, 53%

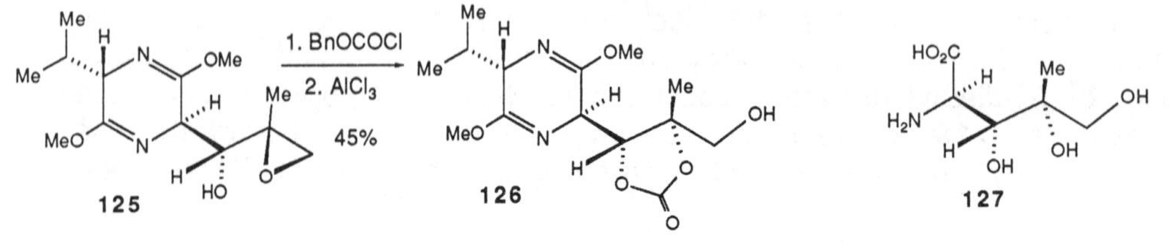

1. BnOCOCl
2. AlCl₃

45%

125   **126**   **127**

a marked effect on the highly selective production of the corresponding *syn*-epoxides in the case of E-substituted olefins. In several cases , the epoxy alcohol adduct (**119**) was hydrolyzed to the corresponding N-acetyl amino acid methyl esters (**121**) or methyl ester **122**. Various ring-opening reactions well-precedented for epoxy alcohols could also be achieved, such as azide or thiolate opening. Epoxy alcohol **125** was converted into the mixed carbonate and ring-opened with aluminum chloride to the cyclic carbonate **126**; this substitution pattern is the same as that found in the oxidized leucine derivative **127** found in the antibiotic bicyclomycin. Cyclopropanation of allylic alcohols **118** with diethyl zinc and methylene iodide proceeded stereospecifically giving the single diastereoisomers **123** in good yield; N-acetylation and hydrolysis of the bis-lactim ether afforded the interesting cyclopropanes **124**.

Direct acylation of lithio-**14** with acid chlorides furnishes the labile ketones **128** (Scheme 29) [31]; these substances can alternatively be prepared via Swern oxidation of the aldol-derived carbinols (**129**) discussed above. Wittig olefination with methylene triphenylphosphorane gave good yields of the *exo*-methylene derivatives **130**; subsequent hydrolysis of the bis-lactim ethers gave the α-methyl-β-substituted vinyl glycine methyl esters **58**. An extremely interesting permutation of this protocol involved the first preparation of the ethynylated amino acid methyl ester **135**. Aldol condensation of formaldehyde with lithio-**14** gave **131** which was oxidized to the labile aldehyde **132**. Olefination with carbon tetrabromide / triphenylphosphine gave in 81% yield the dibromoolefin **133**. Treatment of this material with n-butyllithium effected elimination to the acetylene **134** in good yield. Hydrolysis of the bis-lactim ether gave in 49% yield and >95% ee the α-methyl ethynyl glycine **135**. Ethynyl glycine itself is a naturally occurring, biologically active and very unstable amino acid that has not yet yielded to synthesis. Again, the presence of the α-methyl group in **135** renders this compound stable to epimerization, rearrangement and other decomposition modes that undoubtedly plague ethynylglycine as well as the previously mentioned vinyl glycines. It seems quite unlikely that the methodology used to prepare **135** would be directly applicable to the ethynyl glycine problem.

Employment of bis-alkylating reagents has led to the synthesis[32] of cyclic amino acids as illustrated in Scheme 30. Condensation of lithio-**14** with α,ω-dibromides provides the adducts **136**. Treatment with sodium iodide in hot DMF effects selective deblocking of the requisite lactim ether with concomitant cyclization to afford the bicyclic systems **137** in good yield. Cleavage of these systems to the α-methylated proline and

SCHEME 29

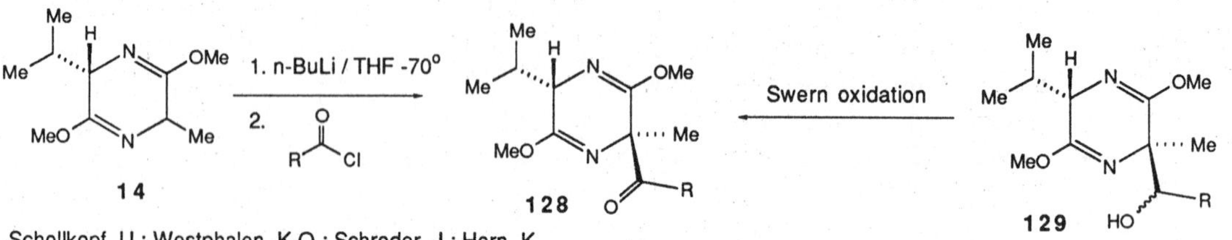

Schollkopf, U.; Westphalen, K-O.; Schroder, J.; Horn, K., *Liebigs Ann.Chem.* (1988) 781.

R (128)	% 128 (via 14)	%128 (via 129)
Me	86	
(cyclohexyl)	88	
(cinnamyl)	15	80
(propenyl)—Me		89
H		95

130, R = Me, 67%
R = Ph, 73%

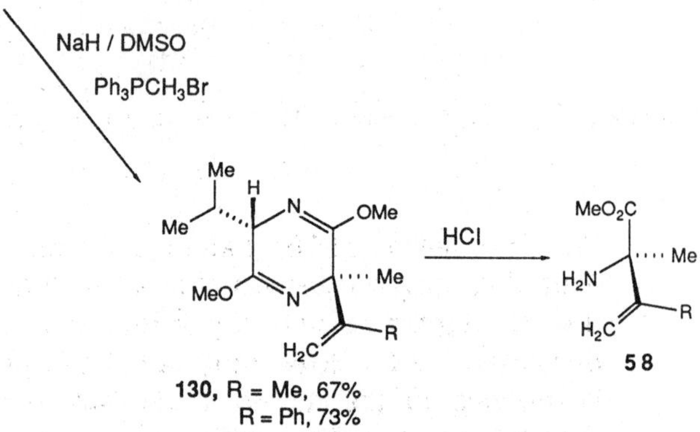

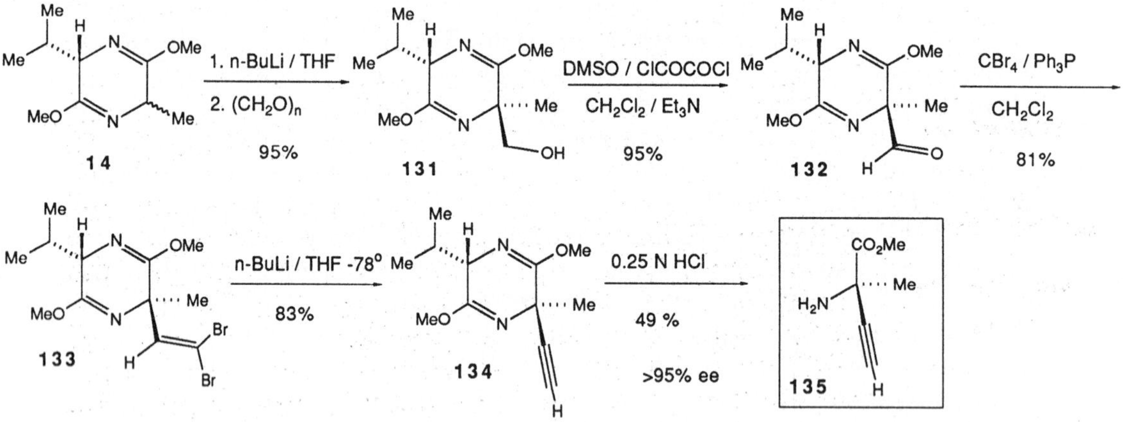

pipicolinic acids requires relatively vigorous conditions due to the presence of the newly formed tertiary amide. The stereoselectivity of the key alkylations is excellent (>97%) and the corresponding amino acids can be obtained in essentially optically pure form. Likewise, *ortho*-dibromomethyl benzene alkylation followed by cyclization and hydrolysis furnishes the tetrahydroisoquinoline derivative **141**.

SCHEME 30

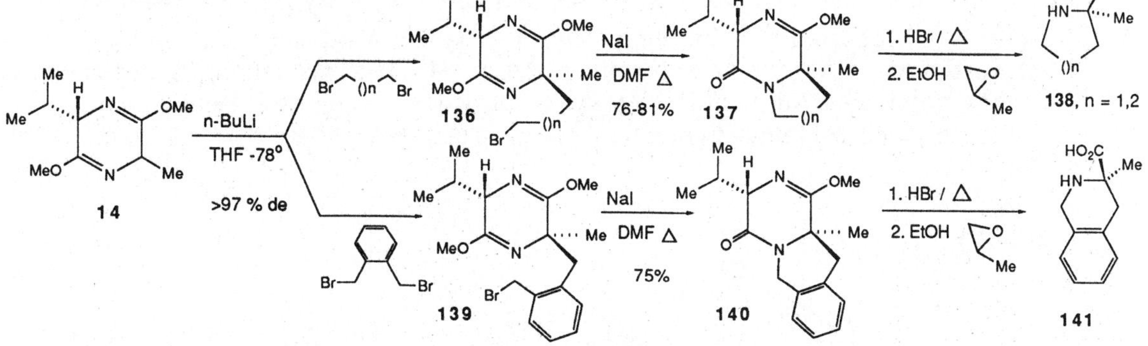

Schollkopf, U.; Hinrichs, R.; Lonsky, R., *Angew. Chem. Int. Ed. Engl.* (1987) **26**, 143.

R-Isovaline (**145**) has been prepared [33] by the alkylation of lithio-**14** with 1-benzylthio-2-iodoethane to furnish the adduct **142**. Hydrolysis of the bis-lactim ether furnishes the S-benzyl α-methyl homocysteine derivative **143** that may also be useful in other peptide applications. Hydrolysis to the amino acid (**144**) and Raney-Nickel reduction provides R-isovaline in good overall yield and >95% ee. Certain amphiphilic peptide antibiotics such as trichotoxine are mentioned to contain **145** .

SCHEME 31

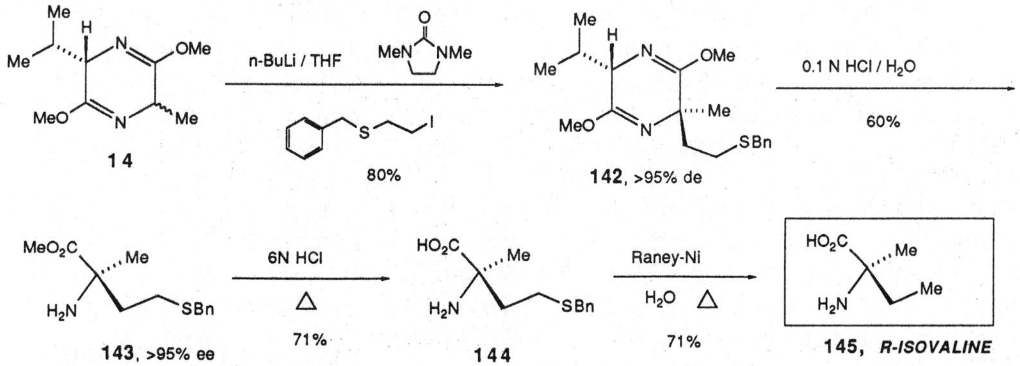

Schollkopf,U; Lonsky, R., *Synthesis* (1983) 675.

A useful synthesis of tryptophan and α-methyl tryptophan methyl esters was reported [34] by employing the labile alkylating reagent N-t-BOC-3-bromomethylindole with either **10** or **14**. The alkylations proceed in very good yield and after hydrolysis, provides tryptophan methyl ester (~90% ee) and α-methyl tryptophan methyl ester (>95% ee) **147** (Scheme 32).

SCHEME 32

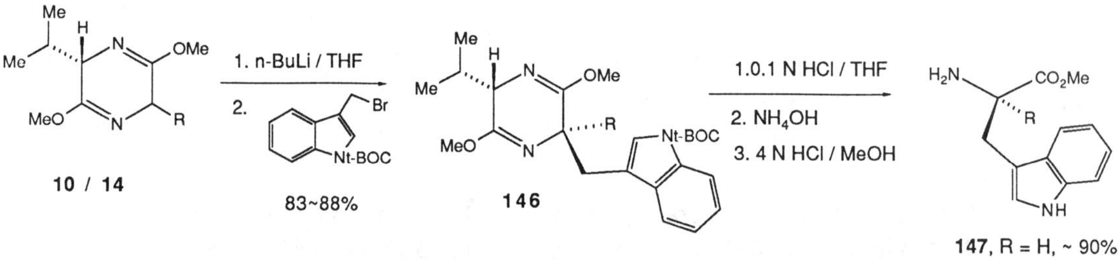

10 / 14   83~88%   146

147, R = H, ~ 90% ee
R = Me, >95% ee

Schollkopf, U.; Lonsky, R.; Lehr, P., *Liebigs Ann. Chem.* (1985) 413.

Woodard and Subramanian[35] have adapted the Schollkopf technology to prepare some stereospecifically labelled cyclopropyl amino acids as shown in Scheme 33. The new bis-lactim ether **148** was prepared from phenylacetone using a modified Strecker synthesis developed by Weineges (discussed separately in this monograph). The authors do not comment on why they chose this synthetically derived auxilliary instead of the previously reported α-methyldopa template **13** which also lacks the additional α-methine. Alkylation of lithio-**148** with regiochemically distinct *gem*-dideuterio 2-bromo ethyltriflates gives selective reaction via triflate displacement on the more hindered face of the bis-lactim ether to furnish the adducts **149** and **150** in 56-60% de. The authors note that only triflates seem to give this curious "*anti*-Schollkopf" diastereoselectivity. Treatment with a second equivalent of butyllithium results in spiro-alkylation furnishing **151** and **152** in modest diastereoselective excess. Mild hydrolysis of the bis-lactim ethers furnishes the methyl esters (**153 / 154**) which are then completely hydrolyzed to the amino acids in 44-46% ee. The stereospecifically labelled 1-aminocyclopropane-1-carboxylic acids (ACCs) have been important mechanistic probes in ethylene biosynthetic studies; alternate syntheses of these simple, yet challenging amino acids will be detailed elsewhere in this monograph.

SCHEME 33

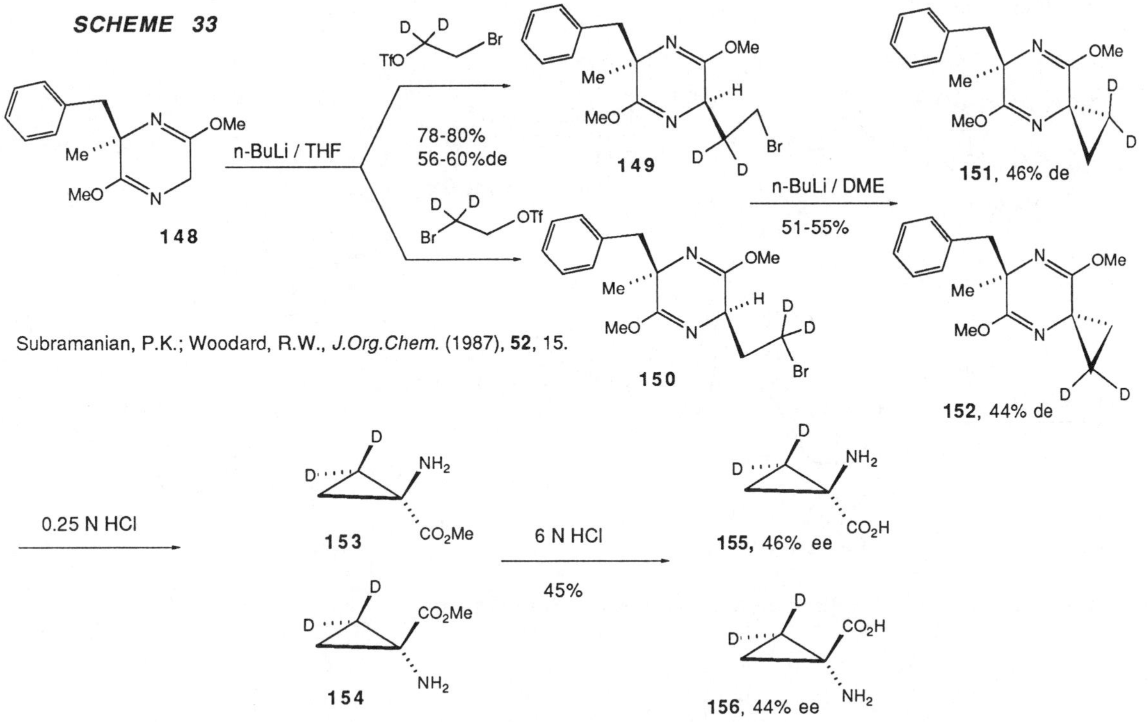

Subramanian, P.K.; Woodard, R.W., *J.Org.Chem.* (1987), **52**, 15.

Employing the Schollkopf template **10**, Hopkins and collaborators [36] prepared the unusual histidine derivatives ovothiol A and C (**158**) which are found in the eggs of certain marine invertabrates. Alkylation of lithio-**10** with N-methyl-4-(*p*-methoxybenzylthio)-5-chloromethyl imidazole furnishes **157** as a 5:1, *anti:syn* mixture of diastereomers. Chromatographic separation followed by hydrolysis and deprotection furnished ovothiol A . Reductive methylation of this compound with formaldehyde and sodium cyanoborohydride provided ovothiol C.

SCHEME 34

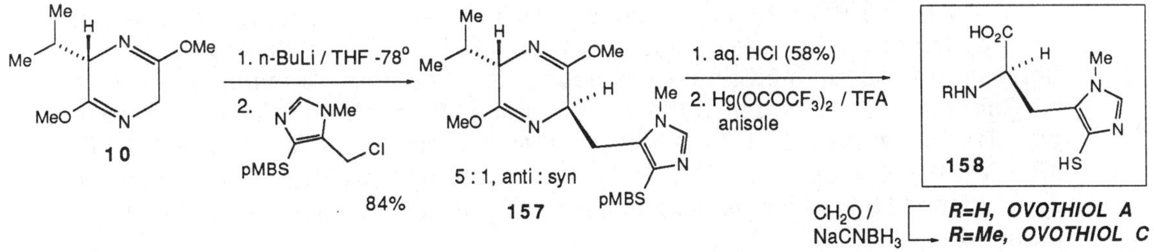

Holler, T.P.; Spaltenstein,A.; Turner, E.; Klevit, R.E.; Shapiro, B.M.; Hopkins, P.B., *J.Org.Chem.* (1987), **52**, 4421.

SCHEME 35

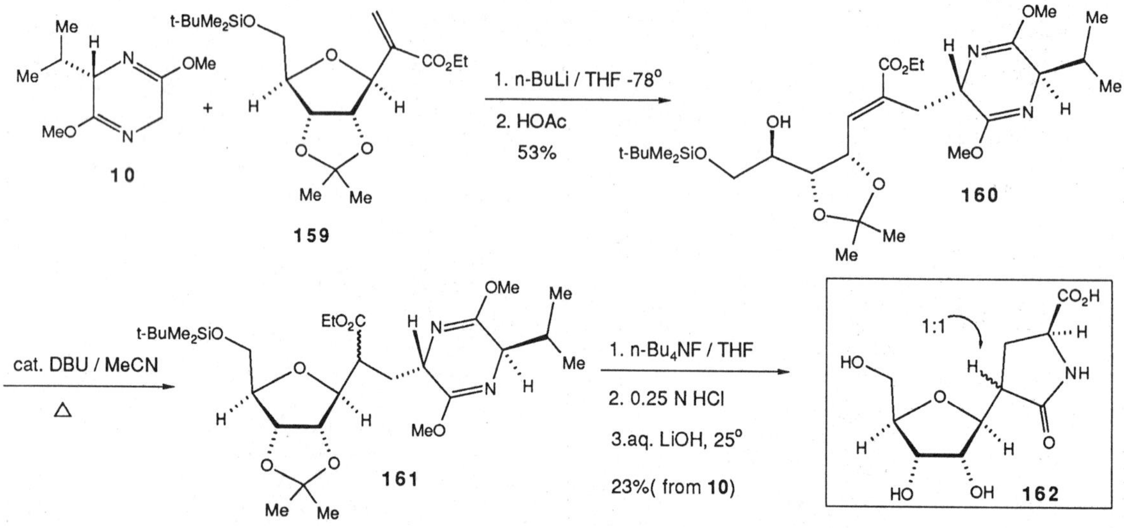

Baldwin, J.E.; Adlington, R.M.; Robinson, N.G., *Tetrahedron Lett.* (1988) **29**, 375.

   In an elegant extension of the Michael addition strategy presented in Scheme 23, Baldwin and associates [37] coupled the ribosyl acrylate **159** with lithio-**10** to furnish the adduct **160** in 53% yield as a single geometric and stereoisomer (Scheme 35). Treatment of **160** with a catalytic amount of DBU in hot acetonitrile effected reclosure of the ribosyl moiety to furnish **161** as a 1:1 mixture of diastereomers. Desilylation, hydrolysis of the bis-lactim ether and base-catalyzed ring closure furnished a 1:1 diastereomeric mixture of β-ribofuranosyl pyroglutamic acids (**162**). This complex amino acid has been proposed to be a possible biosynthetic precursor of the natural C-nucleosides such as showdomycin.

   Hartwig and Born [38] employed the Michael addition strategy to prepare the hepatoprotective agent clausenamide (**167**) which is isolated from the Chinese traditional medicine extracted from *Clausena lansium*. As shown in Scheme 36, Michael addition of lithio-**10** to *cis*-methylcinnamate provides the adduct **163** as a major product (epimeric ratio= 3S,1'S:3S,1'R: 3R,1'S: 3R,1'R = 384 : 70 : 2.7 : 1). Hydrolysis and cyclization furnishes the pyrrollidone **164** which is subsequently converted to (+)-clausenamide as illustrated.

SCHEME 36

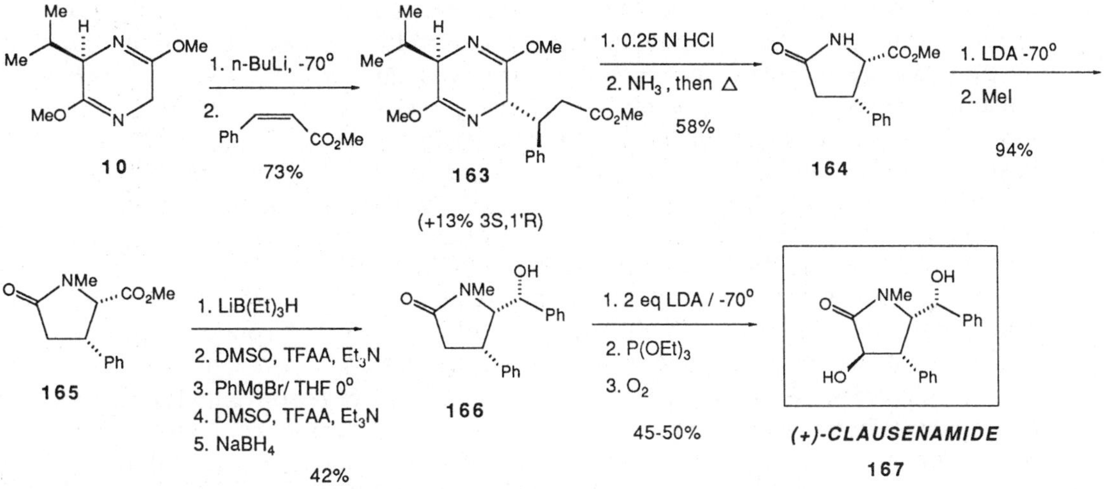

(+13% 3S,1'R)

Hartwig, W.; Born, L., *J.Org.Chem.* (1987) **52**, 4352.

The patent literature also contains a variety of additional examples of the utility of the Schollkopf method for the synthesis of some important amino acids; these will not be covered in this review. It is sufficient to note that the bis-lactim ether method provides a powerful and versatile tool for preparing a large array of α-amino acids and α,α-disubstituted amino acids in optically active form. The experimental protocol for the alkylations is thoroughly and carefully delineated in the volume of full papers published on this subject and provides the research chemist with a confident and predictable tool for preparing amino acids. The major weaknesses of the bis-lactim ether method involve some of the difficulties associated with the hydrolysis of the derivatized bis-lactim ethers to the amino acid methyl esters and finally (under harsher conditions) to the free amino acids. For the case of the α,α-disubstituted systems where racemization is precluded, the bis-lactim ether method is probably the method of choice in spite of the somewhat harsher hydrolysis conditions. The separation of the chiral auxilliary from the new amino acid can also be problematic especially in instances where a time-intensive chromatographic separation is required. The commercial availability of the bis-lactim ethers renders the reuse of the recovered auxilliary inconsequential.

2. Asymmetric Enolate Alkylation of Schiff Bases

Numerous groups have investigated the enolate alkylation of Schiff bases derived from glycine and other amino acids to access optically active amino acids and α-substituted amino acids. Strategies pertaining to the use of a chiral, non-racemic aldehyde to form the Schiff base; optically active esters of simple Schiff bases; combinations of the above as well as employing optically active alkylating reagents on achiral Schiff bases have all been examined. Both acyclic as well as cyclic versions have appeared and will be reviewed in turn in this section. The major conceptual advantage of this general approach is the intrinsic ability to recover and reuse the chiral auxilliary as well as the inherent versatility of having access to a wide range of amino acids from a single template. The effectiveness with which each of the various specific methods address these criteria are critically evaluated below.

Schollkopf[39] has recently devised the enolate alkylation of galactodialdehyde aldimines derived from valine, leucine and isoleucine. As shown in Scheme 37, condensation of Val, Leu, or Ile with **168** in the presence of molecular sieves furnishes the aldimines **169**. Enolate formation with LDA in THF followed by alkylation provides the homologated derivatives **170**; subsequent hydrolysis with methanolic HCl provides the amino acid methyl esters **171**. The diastereoselectivities of the alkylations are generally good to excellent with a broad range of 23->95%. Since the products are racemization-free α,α-disubstituted amino acids, the %de =%ee in every case reported.

One of the first reports [40] on the stereoselective alkylation of glycine imines was that by Yamada and associates [40d,e] in 1976. As shown in Scheme 38, condensation of (1S,2S,5S)-2-hydroxypinan-3-one (**172**) with glycine *t*-butyl ester in the presence of BF$_3$-Et$_2$O furnishes the ketimine **173**. Treatment with 2 equivalents LDA in THF gives the dilithio species that the authors formulate as the internally chelated species **174**; subsequent alkylation proceeds stereoselectively to furnish **175**. Hydrolysis of the ketimine with aqueous citric acid in THF provides the *t*-butyl esters (**176**) that can be completely hydrolyzed to the free amino acid hydrochlorides or isolated as the corresponding N-benzoates. The authors note that the ketol **172** can be recovered and is available in both enantiomeric forms via the precursors (+)- and (-)-α-pinene.

Numerous analogous reports on this same strategy have subsequently appeared. McIntosh and coworkers reported[41] on the utility of camphor-derived imines of glycine in enolate alkylations. Camphor itself was found to be too sterically hindered to condense directly with glycine esters to form the Schiff base; conversion of camphor to the reactive thione **177** and subsequent imine formation proceeds in good yields to furnish the key

SCHEME 37

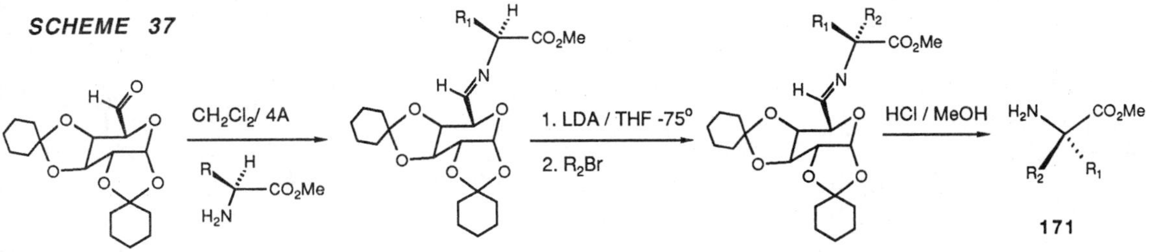

Schollkopf, U.; Tolle, R.; Egert, E.; Nieger, M., *Liebigs Ann.Chem.* (1987) 399.

R₁	R₂	% (170)	%de	Abs.Config.	% (171)	%ee
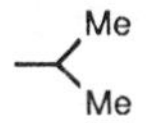 isopropyl (Me, Me)	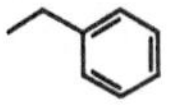 benzyl	91	80	S	57	80
	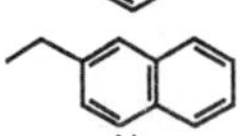 naphthylmethyl	91	>95	S	62	>95
	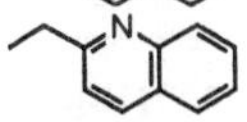 quinolinylmethyl	92	74	S	75	74
	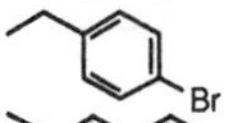 p-bromobenzyl	91	75	S	63	75
	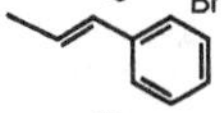 o-bromocinnamyl	92	76	S	62	76
	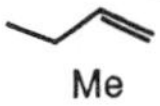 2-butenyl (Me)	80	76	S	75	76
		79	23	S	61	23
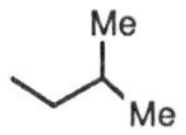 sec-butyl (Me, Me)	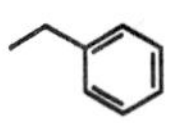 benzyl	93	75	S	74	75
	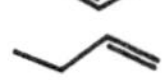 pentenyl	90	57	S	78	57
	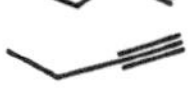 pentynyl	83	48	S	68	48
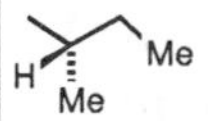 chiral (H, Me, ethyl)	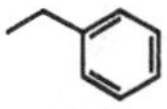 benzyl	94	88	S	70	88
	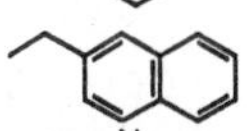 naphthylmethyl	96	96	S	68	96
	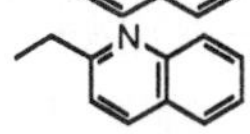 quinolinylmethyl	92	90	S	74	90
	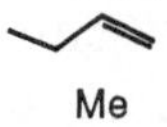 benzyl (isomer)	87	75	S	76	75
	2-butenyl (Me)	82	29	S	86	29

derivative **178.** Enolate formation with LDA and alkylation in the presence of HMPA provides the homologated imines **179.** Based on the wide range of stereoselectivities observed (0~>98%), the authors postulate that only electrophiles that can enter into pi-pi interactions with the enolate give good stereoselectivity (such as benzyl bromide). Hydrolysis of the imine by exchange with hydroxyl amine furnishes the *t*-butylesters **180.** Again, steric hindrance of the camphor imine makes the cleavage somewhat problematic and only one example (*t*-butyl phenylalaninate) is reported.

SCHEME 38

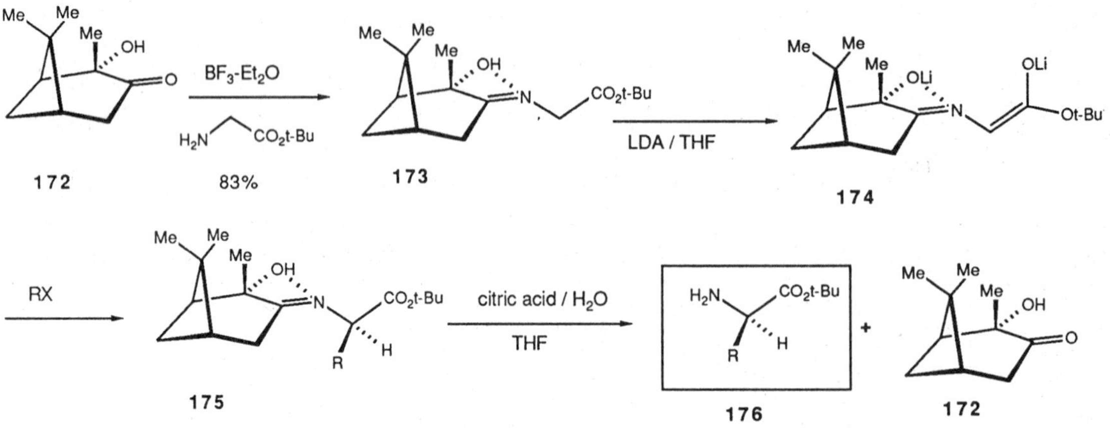

Yamada, S-I.; Oguri, T.; Shioiri, T., *J.C.S.Chem.Comm.* (1976) 136.

RX	PRODUCT	OVERALL %	OPTICAL PURITY
MeI	H-D-Ala-OH-HCl	52	83
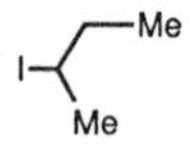 I—CH(Me)—Me	Bz-D-Leu-Ot-Bu	50	83
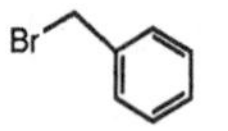 Br—CH2—Ph	H-D-Phe-Ot-Bu	79	72
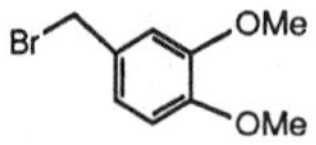 Br—CH2—C6H3(OMe)2	H-D-3,4-(OMe)$_2$-Phe-Ot-Bu	62	66

The resulting camphor oxime is separated from the amino acid ester by a simple extraction procedure. In its present state of development, this particular system is unlikely to enjoy widespread utility due to the need to prepare the thione, the inconsistent stereoselectivities which are dependent on the structure of the electrophile and the final hydrolysis of the hindered imine. A virtually identical study of the camphor-derived imine system has been independently developed by a Chinese group[42] in preliminary form.

Nature utilizes the Schiff base concept in the stereospecific tautomerization of pyridoxamine-derived imines in the second half-reaction of transaminations. This concept, as applied to the production of optically active amino acids will be discussed in detail in the chapter dealing with enzymatic syntheses. A number of groups have been involved in designing synthetic mimics of the pyridoxamine catalytic group with an eye toward developing a practical and general asymmetric synthesis of amino acids from α-keto acids.

SCHEME 39

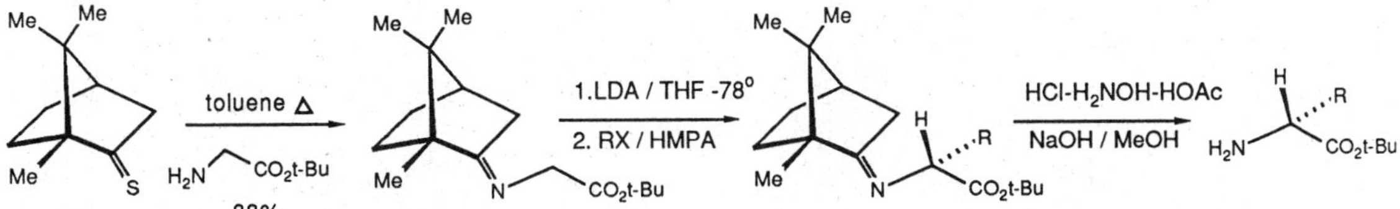

McIntosh, J.M.; Leavitt, R.K.; Mishra, P.; Cassidy, K.C.; Drake, J.E.; Chadha, R. *J.Org.Chem.* (1988)**53**, 1947.
McIntosh, J.M.; Mishra, P., *Can.J.Chem.* (1986) **64**, 726. ; McIntosh, J.M.; Leavitt, R.K., *Tetrahedron Lett.* (1986) **27**, 3839.

RX	YIELD (%179)	%de	RX	YIELD (%179)	%de
MeI	82	0	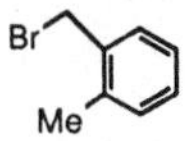 Br-CH2-(o-Me)C6H4	74	>98
MeTs	37	0	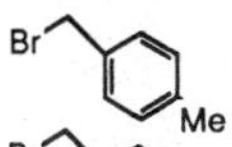 Br-CH2-(m-Me)C6H4	71	>98
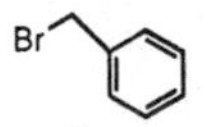 Br-CH2-C6H5	89	>98	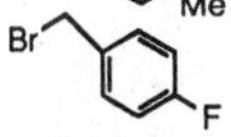 Br-CH2-(p-F)C6H4	89	>98
Br~CO2Me	78	46	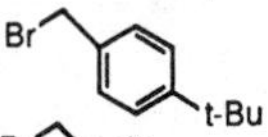 Br-CH2-(p-t-Bu)C6H4	71	85
Br~CO2Et	86	32	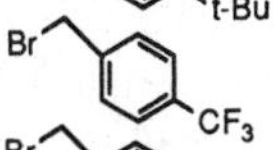 Br-CH2-(p-CF3)C6H4	71	79
Cl~CO2t-Bu	32	10	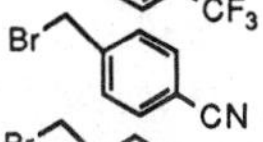 Br-CH2-(p-CN)C6H4	75	76
n-BuI	45	50	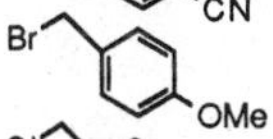 Br-CH2-(p-OMe)C6H4	83	88
i-BuI	48	52	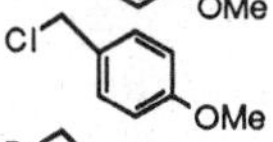 Cl-CH2-(p-OMe)C6H4	49	73
 Br~ (allyl)	85	76	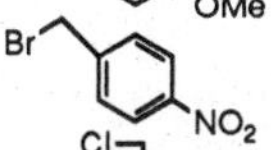 Br-CH2-(p-NO2)C6H4	69	51
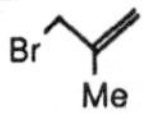 Br~Me (methallyl)	79	76	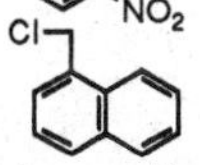 Cl-CH2-naphthyl	31	93
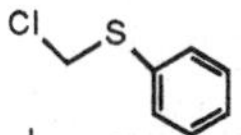 Cl-CH2-S-C6H5	38	-	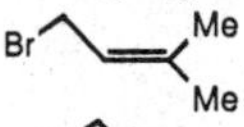 Br-CH2-CH=C(Me)Me	80	33
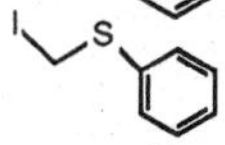 I-CH2-S-C6H5	54	64	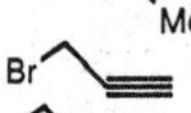 Br-CH2-C≡CH	78	67
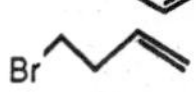 Br~~	68	34	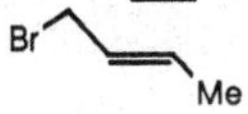 Br-CH2-CH=CH-Me	76	60
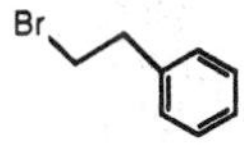 Br-CH2CH2-C6H5	58	34			
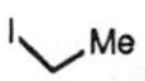 I~Me	38	33			
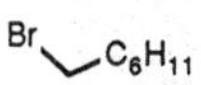 Br~C6H11	0	-			

    Kuzuhara and associates [43] have developed an interesting glycine enolate equivalent that is obtained from the condensation of glycine with the novel, optically active pyridoxal macrocycle **181** (Scheme 40). In a preliminary account, they found that reaction of this complex with a half molar equivalent of a Zinc(II) salt furnished a 3:1 mixture of Zinc complexes which were separated by Sephadex column chromatography; the

SCHEME 40

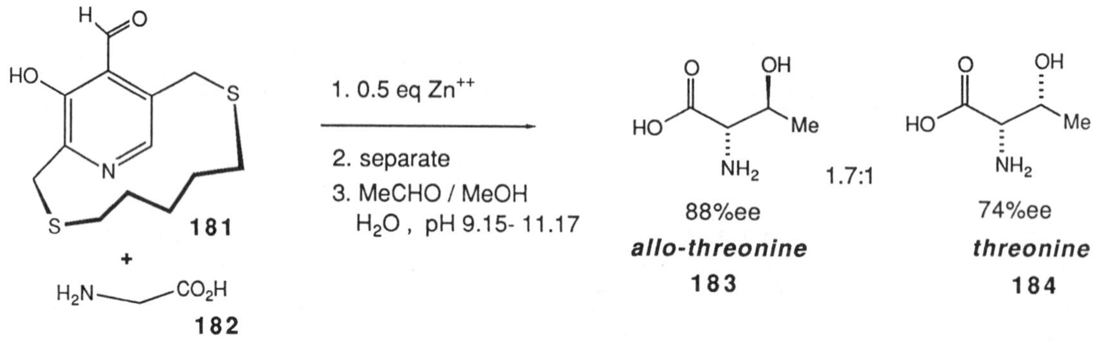

Kuzuhara, H.; Watanabe, N.; Ando, M., *J.C.S.Chem.Comm.* (1987) 95.

major product being used in the subsequent aldol reaction. Treatment of the major Zinc complex with a large excess of acetaldehyde in methanol/water at ambient temperature for 1 day followed by decomposition of the resulting product complex with 1N HCl, extraction and repeated ion-exchange chromatography afforded a 1.7:1 mixture of allo-threonine : threonine in 88%ee and 74%ee, respectively (73% yield).

The above, and related work derives from a series of reports by Breslow and co-workers [44] on the use of modified pyridoxamine mimics for stereospecific transaminations. Two salient examples will be noted here. In one approach, pyridoxamine was attached to a modified β-cyclodextrin (185) containing both a binding domain and an attached basic group to effect stereospecific tautomerization (Scheme 41). It was found that aromatic ketoacids would bind effectively inside the cavity and resulted in highly stereoselective transaminations to the L-amino acids 187 (>90% ee) although in modest chemical yields. Simple aliphatic keto acids gave poor stereoselectivities. It is quite significant however, that the racemization-prone phenylglycine could be obtained in 96% ee which is a powerful testament to the mildness of the reaction conditions and should provide motivation to develop this concept more fully for practical applications. In another study (Scheme 42), the bicyclic system 188a was shown to lead to stereoselective transaminations with a wider substrate specificity than the cyclodextrin model; D-alanine, D-norvaline, and D-tryptophan being obtained in 86-92% ee and 68-89% yields. The success of this system is predicated on a stereoselective tautomerization of the ketimine to the aldimine via the dimethylamino moiety on one face of a planar intermediate. Neither of these systems (185/188) is catalytic since, the pyridoxamine moiety is converted into the corresponding

SCHEME 41

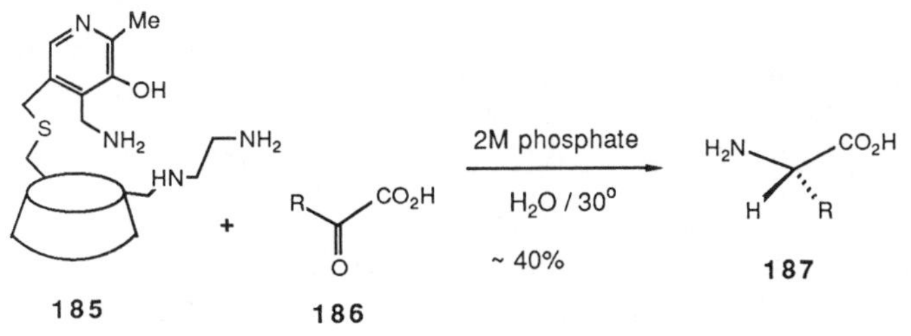

R	%ee (187)	Config.
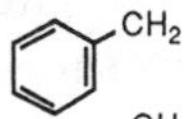 benzyl (CH₂)	96	L
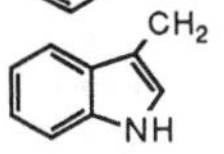 indolyl-CH₂	90	L
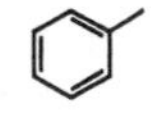 phenyl	96	L

185 **186** → **187**

2M phosphate / H₂O / 30° / ~ 40%

Tabushi, I.; Kuroda, Y.; Yamada, M.; Higashimura, H.; Breslow, R., *J.Am.Chem.Soc.* (1985) **107**, 5545.

SCHEME 42

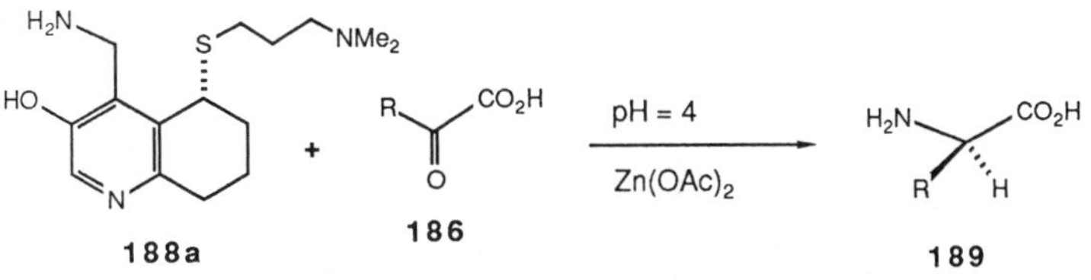

R	%ee (189)	Config.
Me	86	D
n-Pr	92	D
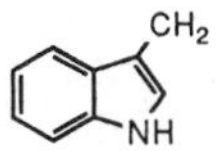 indolyl-CH₂	92	D

188a + **186** → pH = 4 / Zn(OAc)₂ → **189**

188b **188c** **188d**

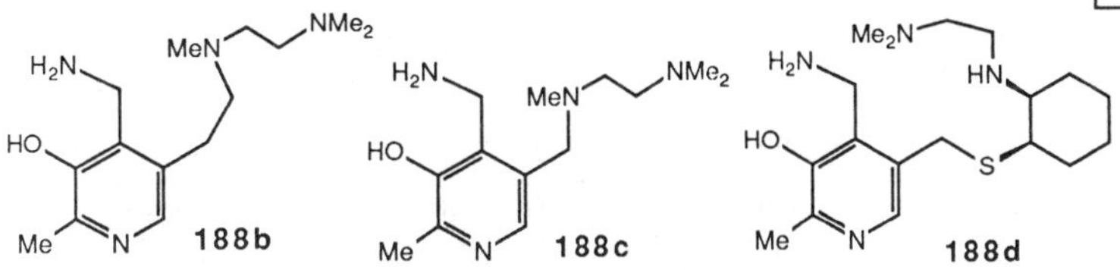

Breslow, R.; Chmielewski, J.; Foley, D.; Johnson, B.; Kumabe, N.; Varney, M.; Mehra, R., *Tetrahedron* (1988) **44**, 5515.

Breslow, R.; Czarnik, A.W.; Lauer, M.; Leppkes, R.; Winkler, J.; Zimmerman, S., *J.Am.Chem.Soc.* (1986) **108**, 1969.

pyridoxal which can be recovered and chemically recycled by reductive amination to the amine. Modifications of this system led to the pyridoxamines **188b-d** which similarly incorporate the critical diamine side-chain substituent. It was found that **188c** and **188d** give the best rate accelerations while, **188b** seems to have too short a tether to facilitate the crucial proton transfer. The promising results with **188** is quite interesting and should provide a working model to develop a more practical system. The major limitation to the practical use of **188a** is the long multi-step (ca 18 steps) syntheses and attendant resolution required for its preparation; the newer systems 188c,d being somewhat simpler to prepare.

SCHEME 43

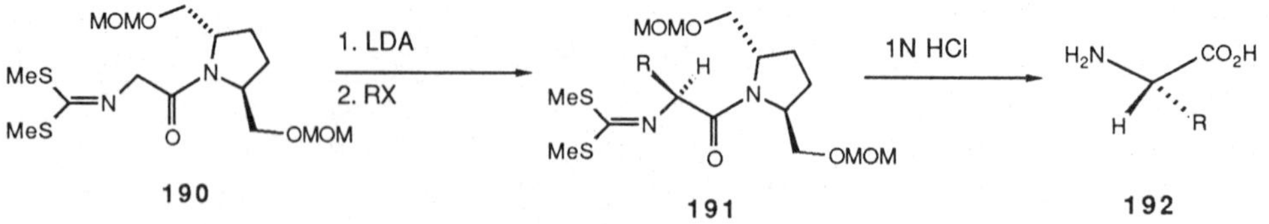

Ikegami, S.; Uchiyama, H.; Hayama, T.; Katsuki, T.; Yamaguchi, M., *Tetrahedron* (1988) **44**, 5333.
Ikegami, S.; Hayama, T. ; Katsuki, T.; Yamaguchi, M., *Tetrahedron Lett.* (1986) **27**, 3403.

RX	%de(191)	%YIELD (191)	%ee (192)	%YIELD (192)
MeI	98	84	95 (S)	97
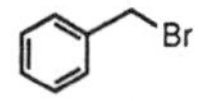 Br (benzyl)	98	85	97 (S)	92
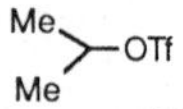 Me₂CH-OTf	98	51	- (S)	81
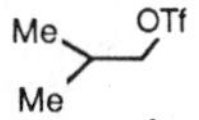 Me₂CHCH₂-OTf	97	68	97 (S)	96
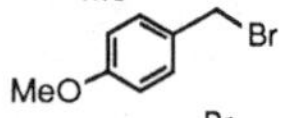 MeO-C₆H₄-CH₂Br	96	91	95 (S)	88
H-C≡C-CH₂Br	98	95	98 (S)	51

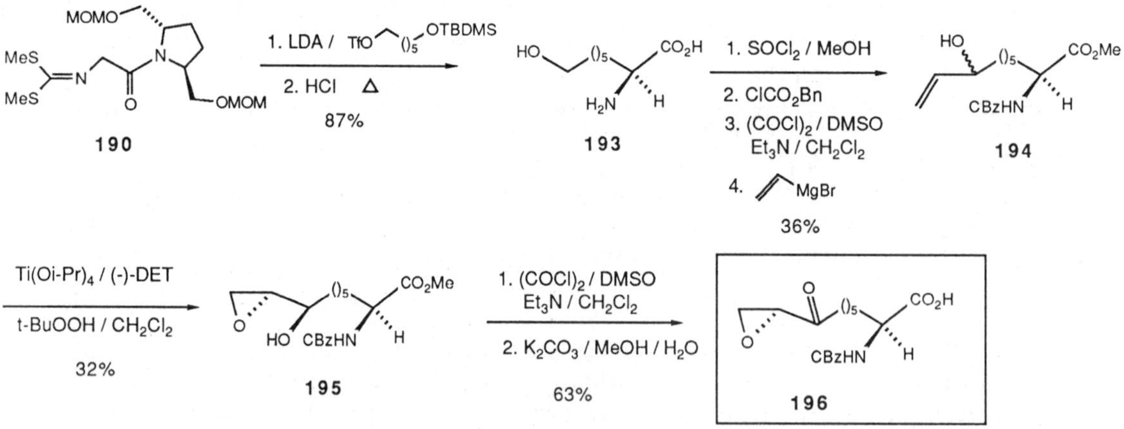

Katsuki and co-workers [45] have recently devised a potentially quite useful glycine enolate equivalent as shown in Scheme 43. Glycine amide **190** was found to undergo smooth and highly stereoselective enolate alkylation with active electrophiles affording the adducts **191**. Simple acid hydrolysis and ion-exchange chromatography furnished the free amino acids **192** in excellent chemical yields and high ee's. To further demonstrate the utility of the method, the authors prepared the protected amino acid **196** (Aoe) a component of the physiologically active peptide chlamydocin. Enolate alkylation of **190** with 6-*t*-butyldimethylsiloxy-1-hexyl triflate and acid hydrolysis furnished the amino acid **193**. Protection, followed by oxidation to the ω-aldehyde and vinyl Grignard homologation gave the allylic alcohol **194**. Sharpless epoxidation effected a kinetic resolution to **195** which was subsequently oxidized and hydrolyzed to the N-CBz derivative of Aoe (**196**). The authors note that attempts to remove the carbobenzoxy group were unsuccessful due to the lability of **196**.

An ingenious and practical glycine enolate equivalent has recently been developed by Evans and Weber [46] as a means of constructing β-hydroxy-α-amino acids via aldol bond constructions. The first report of this new method was elegantly demonstrated with an asymmetric synthesis of the unusual C9 amino acid MeBmt (**202**), found in the immunosuppressive peptide cyclosporine. As shown in Scheme 44, the optically active chloroacetate derivative is converted into the key, crystalline isothiocyanate **198** by reaction with sodium azide followed by reduction and thiocyanate formation. Of a variety of metal enolates examined, it was found that the stannous triflate mediated aldolization gave both good yields and high levels of stereoselectivity. The initially formed aldol adducts undergo spontaneous cyclization on the isothiocyanate to form the thiono urethanes (**205**) which are isolated by silica gel chromatography. A variety of aldehydes smoothly undergo this reaction to give predominantly the *syn*-aldol adducts (Table). For the specific case of MeBmt, The requisite aldehyde **199** was synthesized by stereoselective methylation of the Evans amide enolate **203** followed by reduction and Swern oxidation of **204** to **199**. Condensation of **199** with the tin enolate of **198** furnished **200** as a 94 : 6 ratio of diastereomers. Hydrolysis of the chiral auxilliary (which can be recovered), N-methylation and complete hydrolysis gave MeBmt (**202**) in optically active form. It should not be underemphasized that the high-yielding conversion of the aldol adducts to the simple, yet difficult mono-N-methylated amino acids is an important and greatly needed synthetic transformation. The success of this reaction is due to the facile double alkylation of the thiono urethane to the corresponding N,S-dimethyl oxazolidine salt that readily hydrolyzes to the free N-methyl amino acid. Full experimental details accompany this work.

Another example of the above methodology was reported from the same laboratories [47] in the context of a total synthesis of the antifungal cyclic hexapeptide echinocandin D (**210**); the complete synthesis of this natural product is outlined in the last chapter. As shown in Scheme 45, condensation of the tin enolate of **198** with *p*-(benzyloxy)phenylacetaldehyde affords the expected heterocycle **207** in 72% yield. Removal of the chiral auxilliary, t-BOC protection and

SCHEME 44

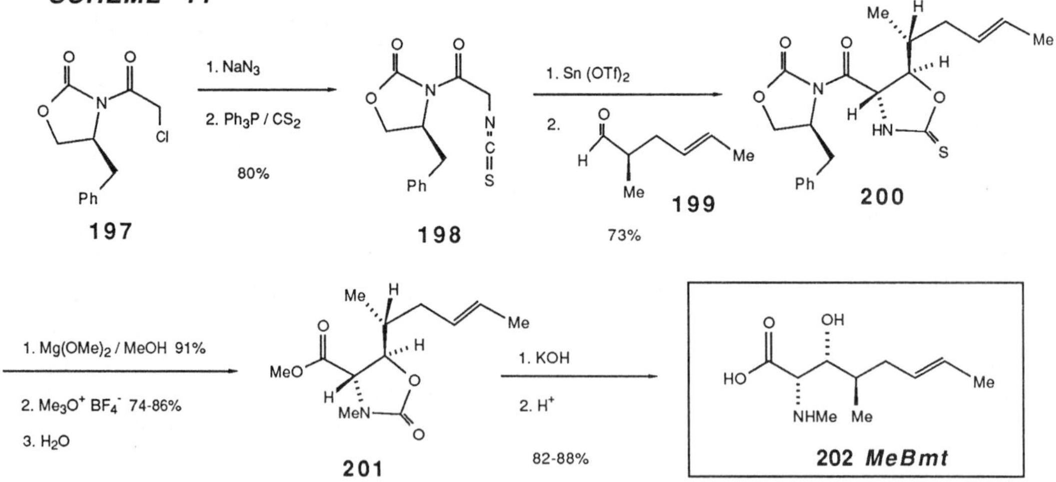

Evans,D.A.; Weber, A.E., *J.Am.Chem.Soc.* (1986) **108**, 6757.

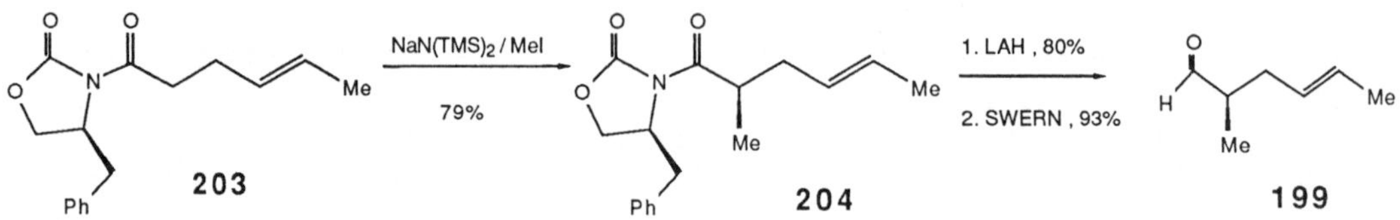

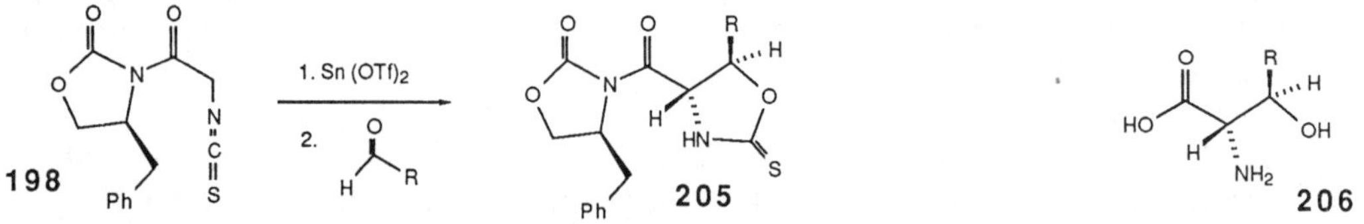

RCHO	RATIO	YIELD % (205)	AMINO ACID (206)
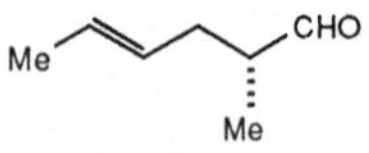 Me⌇⌇CHO (Me)	94:6	73	70%
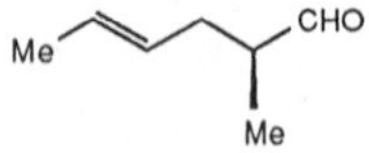 Me⌇⌇CHO (Me)	97:3	71	67%
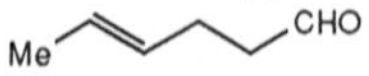 Me⌇⌇CHO	93:7	81	70%
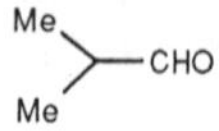 Me₂CH–CHO	99:1	92	69%
MeCHO	91:9	75	—
PhCHO	99:1	91	—

SCHEME 45

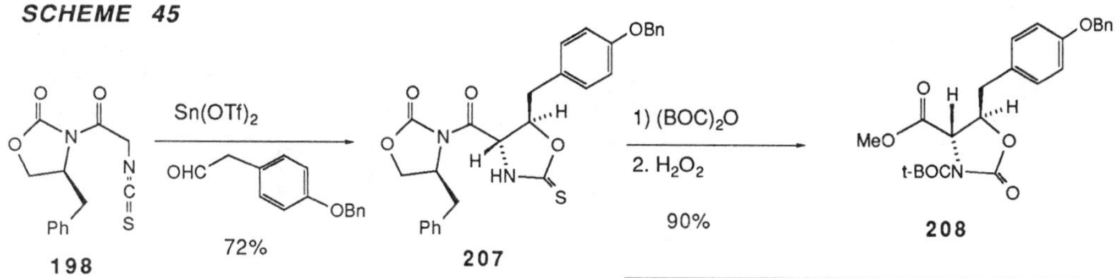

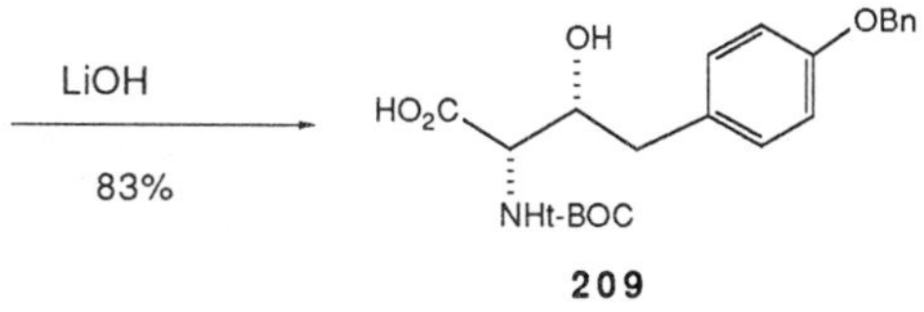

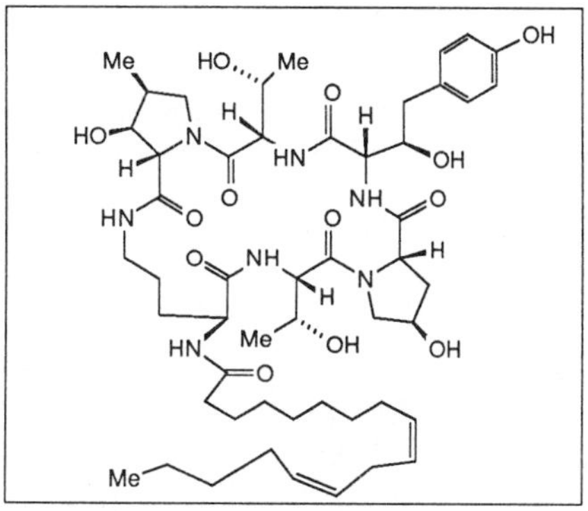

Evans, D.A.; Weber, A.E., *J.Am.Chem.Soc.* (1987)**109**, 7151.

210, ECHINOCANDIN D

conversion of the thiono urethane to the urethane is effected with hydrogen peroxide. Finally, hydrolysis of the methyl ester and the cyclic urethane furnishes the acid **209** which is subsequently coupled and incorporated into echinocandin.

Genet and associates[48] have initiated developing the asymmetric enolate allylation of an achiral glycine Schiff base with optically active pi-allyl-palladium complexes. As illustrated in Scheme 46, the benzophenone imine of glycine methyl ester (**211**) is treated with LDA followed by allylation with allyl acetate and a palladium catalyst carrying an optically active phosphine ligand. Of a large variety of chiral ligands examined, the best results were obtained with the (+)-DIOP ligand. After hydrolysis of the initial allylation adduct, allylglycine methyl ester (**212**) was obtained in 60% yield and 57% ee. In a related paper[49], the authors demonstrate a variety of *racemic* allylations of **213** employing the same basic chemistry without the optically active ligands. In all, seven allylated glycine derivatives (**214**) were reported. Although the results of this preliminary study are not readily useful, this type of approach, particularly if an efficient and general catalytic alkylation method is developed may prove to be of significant industrial potential.

In a related study, Duhamel and co-workers[50] studied the methylation of a similar glycine ketimine with optically active methyl sulfates derived from sugars. Glycine methyl ester is converted to two series of

SCHEME 46

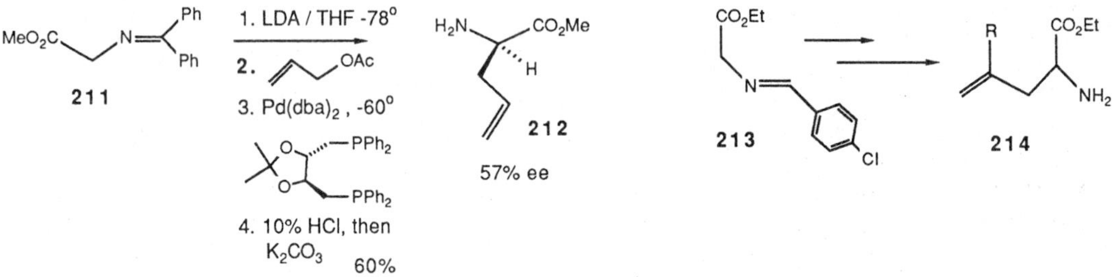

Genet, J.P.; Feroud, D.; Juge, S.; Montes, J.R., *Tetrahedron Lett.* (1986) **27**, 4573.
Genet, J-P.; Juge, S.; Montes, J.R.; Gaudin, J-M., *J.Chem.Soc.Chem.Comm.* (1988) 718.

SCHEME 47

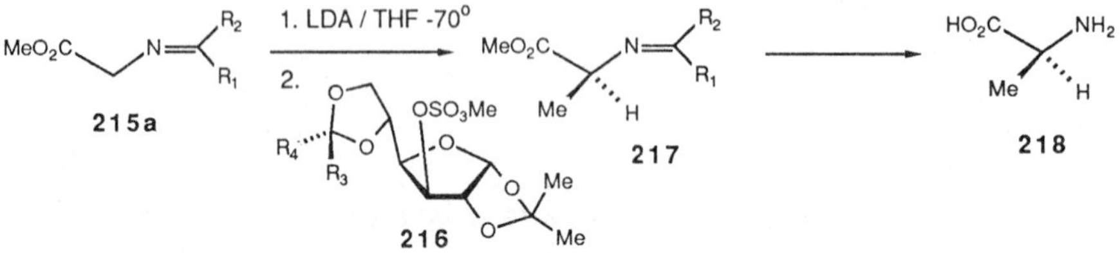

Duhamel, P.; Eddine, J.J.; Valnot, J-Y. *Tetrahedron Lett.* (1987) **28**, 3801.
Duhamel, L.; Duhamel, P.; Fouquay, S.; Eddine, J.J.; Peschard, O.; Plaquevent, J-C.; Ravard, A.; Solliard, R.;
Valnot, J-Y.; Vincens, H., *Tetrahedron* (1988) **44**, 5495.

R_1	R_2	R_3	R_4	YIELD (%217)	%ee (218)
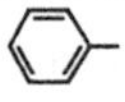 ⬡–	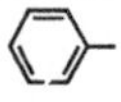 –⬡	Me	Me	60	40
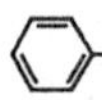 ⬡–	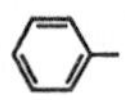 –⬡	Me	t-Bu		56
⬡(t-Bu)–	t-Bu	Me	Me		66
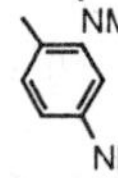 ⬡(NMe₂)–	t-Bu	Me	t-Bu		76

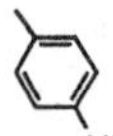

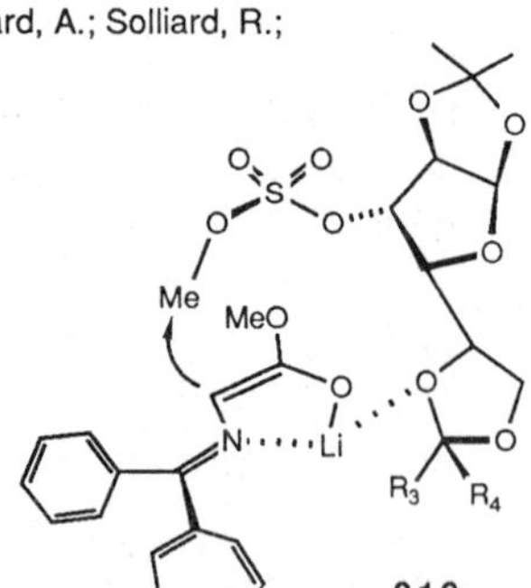

ketimines (**215**), metallated and treated with a variety of methyl sulfates
216 furnishing the methylated derivatives **217**. The yield is only
reported in one of the better cases and the %ees range from 40-76% in the
best cases (Scheme 47, Table). The authors postulate that the internally
chelated lithium enolate coordinates to one of the ketal oxygens to deliver
the methyl group selectively from one face (shown, **219**). Although quite
interesting from the point of view of asymmetric synthesis, the rather
exotic and specialized alkylating reagents (**216**) are unlikely to be of
utility.

The above examples of Schiff base alkylation focused on the employment of *acyclic* systems. The wide range of success in inducing high asymmetric induction are common to the entire area of acyclic stereoselection. Some of the notable and exceptional achievements in attaining stereocontrol in these systems will undoubtedly provide a solid foundation for the further refinement and development of practical glycine enolates for asymmetric amino acid synthesis. The general simplicity of preparing the acyclic starting materials is an inherent advantage of this general strategy that may now be compared to some *cyclic* Schiff base derivatives.

Schollkopf and co-workers[51] found that optically active imidazolidinones could be prepared from alanine or phenylalanine isonitriles (**220**) and (S)-1-phenethylamine. The amide **221** is cyclized and immediately metallated to furnish the lithio-derivative **222** (Scheme 48) that undergoes stereoselective alkylation to furnish the α,α-disubstituted derivatives **223**. The %de in the alkylations was generally quite high as were the yields of alkylation. Hydrolysis of the imidazolidinone to the corresponding amino acid is accomplished with ethanolic KOH followed by N-acetylation to the isolated products **224**. Presumably, the basic conditions required to cleave the heterocycle preclude the analogous development of a glycine system due to the danger of racemizing the mono-substituted imidazolidinones. Following the preliminary communication[51], a full paper describing the alanine system was published with complete experimental details.

In a related approach, Schollkopf [53] reported the preparation of optically active 3,6-dihydro-2H-1,4-oxazine-2-ones from a variety of optically active α-hydroxyacids and phenyl glycine (Scheme 49). The heterocycle **227** is converted into the mono-lactim ether **228** that undergoes enolate alkylation with several activated electrophiles furnishing the homologated systems **229** in excellent yields. Unlike the imidazolidinone system (**223**) above, the oxazinone can be cleaved with mild acid treatment to furnish the α-alkylated phenylglycine derivatives **230**. The chiral auxilliary **231** must be separated from the amino acid, but may be recovered and recycled. In all cases, the asymmetric induction with benzyl bromide gives the best results, regardless of the nature directing group R_1. It is possible that a pi-stacking interaction in the transition state as was proposed above by McIntosh[41] may be responsible for the superior induction with the benzyl electrophiles. A strictly analogous system was reported [54] utilizing 2-furylglycine (**233**) as the amino acid template. Following the exact same route as that above, the α-alkylated-2-furyl glycines (**237**) are obtained via alkylation of the oxazinone **235**. It is surprising that a corresponding glycyl system

SCHEME 48

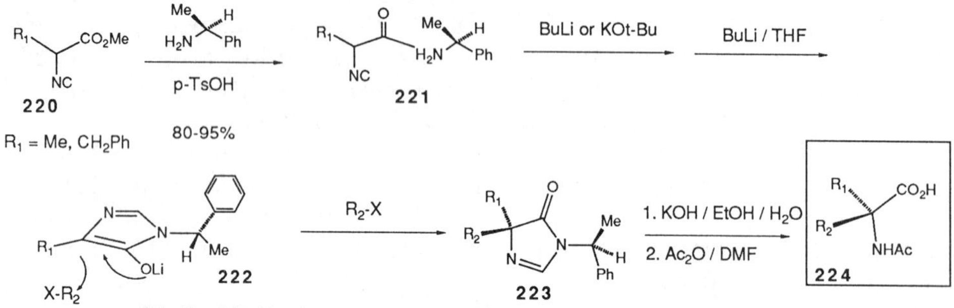

R_1—CH(CO$_2$Me)(NC) **220** R_1 = Me, CH$_2$Ph   + H$_2$N—CH(Me)(Ph) $\xrightarrow[\text{80-95\%}]{\text{p-TsOH}}$ **221** $\xrightarrow{\text{BuLi or KOt-Bu}}$ $\xrightarrow{\text{BuLi / THF}}$

222 $\xrightarrow{R_2\text{-X}}$ **223** $\xrightarrow[\text{2. Ac}_2\text{O / DMF}]{\text{1. KOH / EtOH / H}_2\text{O}}$ R_1R_2C(CO$_2$H)(NHAc) **224**

Schollkopf, U.; Hausberg, H.H.; Hoppe,I.; Segal, M.; Reiter,U. *Angew.Chem.Int.Ed.Engl.* (1978) **17**, 117.

R_1	R_2	ALKYLATION YIELD	%de	ABSOLUTE CONFIG.
Me	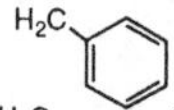 H$_2$C—benzyl	94	>90-100	S
Me	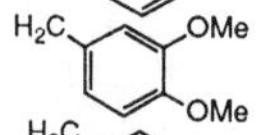 H$_2$C—(OMe,OMe)phenyl	90	76-80	S
Me	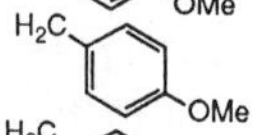 H$_2$C—(OMe)phenyl	90	72	S
Me	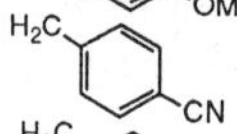 H$_2$C—(CN)phenyl	86	>95	S
Me	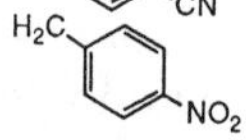 H$_2$C—(NO$_2$)phenyl	84	>95	S
Me	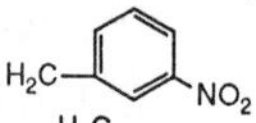 H$_2$C—(NO$_2$)phenyl	80	>95	S
Me	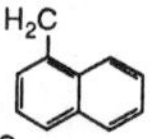 H$_2$C—naphthyl	84	>95	S
Me	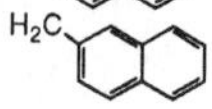 H$_2$C—naphthyl	83	>95	S
Me	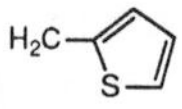 H$_2$C—thienyl	87	>95	S
Me	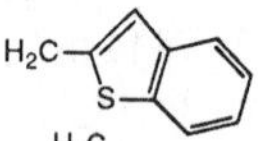 H$_2$C—benzothienyl	80	>95	S
Me	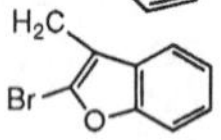 H$_2$C—(Br)benzofuranyl	88	>95	S
Me	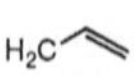 H$_2$C—CH=CH$_2$	66	~17	S
Et	CH$_2$Ph	85	95	S
Me	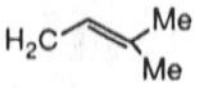 H$_2$C—CH=C(Me)Me	74	34	S
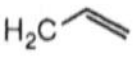 H$_2$C=CH—	CH$_2$Ph	75	90	S
Me	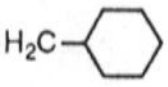 H$_2$C—cyclohexyl	20	~35	S
n-C$_3$H$_7$	CH$_2$Ph	77	90	S
i-C$_3$H$_7$	CH$_2$Ph	78	95	R
p-BrBnCH$_2$Ph	CH$_2$Ph	72	100	R
CH$_2$Ph	Me	89	20	R
	i-C$_3$H$_7$	62	65	S
CH$_2$Ph	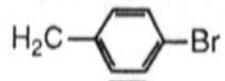 H$_2$C—(Br)phenyl	87	100	S
i-C$_4$H$_9$	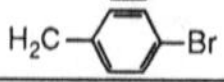 H$_2$C—(Br)phenyl	67	95	S

SCHEME 49

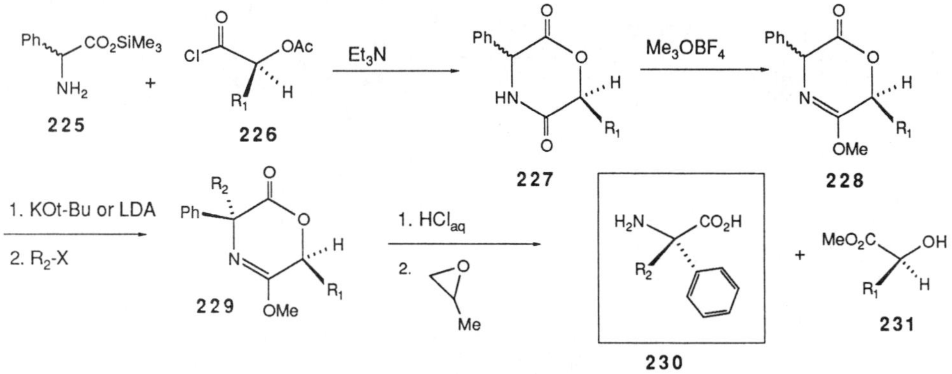

Hartwig, W.; Schollkopf, U., *Liebigs Ann.Chem.* (1982) 1952.

R_1	R_2	ALKYLATION YIELD	%de	ABSOLUTE CONFIG.
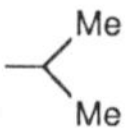 Me / Me (isopropyl)	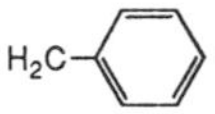 H_2C–Ph (benzyl)	90	>95	S
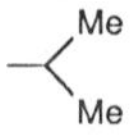 Me / Me (isopropyl)	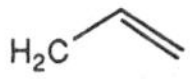 H_2C–CH=CH$_2$ (allyl)	95	83	S
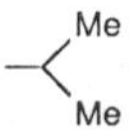 Me / Me (isopropyl)	Me	95	60	S
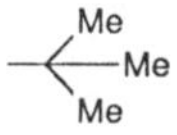 Me / Me–Me (tert-butyl)	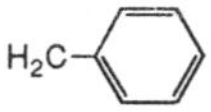 H_2C–Ph (benzyl)	90	>95	S
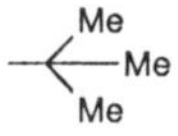 Me / Me–Me (tert-butyl)	Me	95	90	S
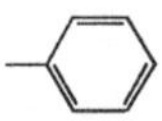 cyclohexyl	Me	85	77	S
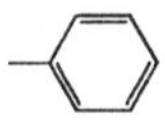 methylcyclohexyl	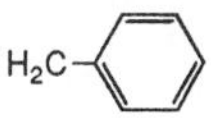 H_2C–Ph (benzyl)	91	>95	S
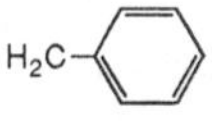 H_2C–cyclohexyl	Me	86	49	S

SOAA—C

SCHEME 50

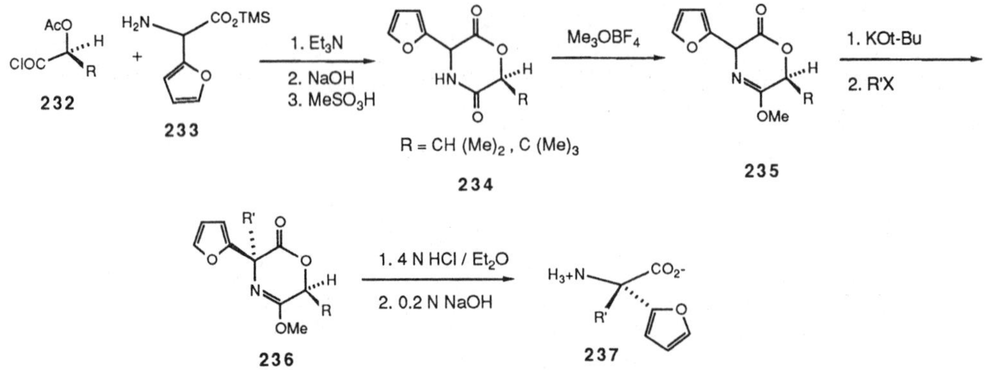

Schollkopf, U.; Scheuer, R., *Liebigs Ann. Chem.* (1984) 939.

R_1	R_2	YIELD	CONFIG.	%de
i-Pr	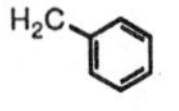 H_2C—(phenyl)	91	S	>95
i-Pr	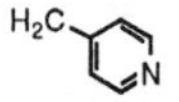 H_2C—(pyridyl, N)	85	S	>95
i-Pr	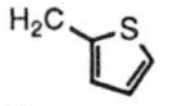 H_2C—(thienyl, S)	92	S	>95
i-Pr	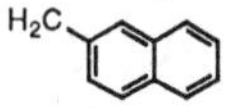 H_2C—(naphthyl)	91	S	>95
i-Pr	 H_2C—(allyl)	93	S	70
i-Pr	CH_3	92	S	58
i-Pr	C_2H_5	83	S	61
i-Pr	$CH(CH_3)_2$	81	S	62
i-Pr	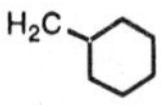 H_2C—(cyclohexyl)	75	S	62
t-Bu	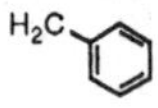 H_2C—(phenyl)	92	S	>95
t-Bu	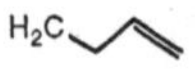 H_2C—(allyl)	85	S	>95
t-Bu	CH_3	92	S	>95

analogous to **228** and **235** has not been reported since, the relatively mild acid hydrolysis conditions after alkylation should not racemize most mono-substituted amino acids. The oxazinone concept provides a nice complement to the bis-lactim ethers developed extensively in the same laboratories.

SCHEME 51

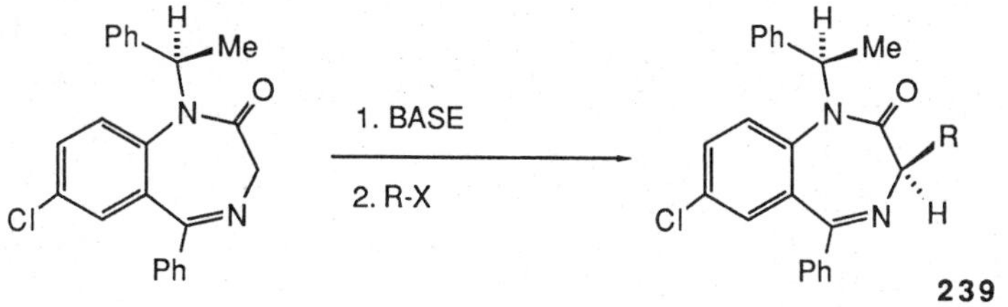

3-31 % de

(EtOCOCl / LDA/THF 25° = 85% de)

Decorte, E.; Toso, R.; Sega, A.; Sunjic, V.; Ruzic-Toros, Z.; Kojic-Prodic,B.; Bresciani-Pahor,N.; Nardin, G.; Randaccio, L., *Helv.Chim.Acta.* (1981) **64**, 1145

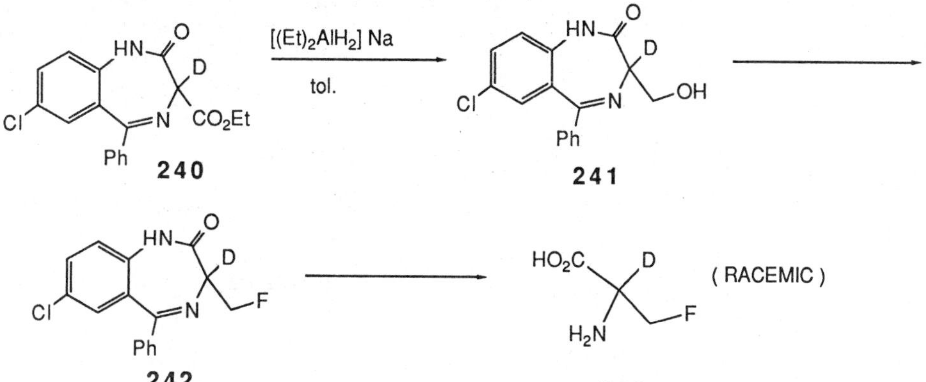

Sunjic and collaborators [55] discovered that an optically active benzodiazepinone (**238**, Scheme 51) underwent diastereoselective alkylation reactions to afford the adducts **239**. The %de was generally quite low (3-31%) with most alkylating reagents; the notable exception being 85% de with ethylchloroformate. Cleavage of the heterocycle to the optically active amino acids was not commented on. A racemic synthesis of α-deuterio-β-fluoroalanine (**243**) was reported from the deuteriated adduct **240**. These studies parallel the earlier reports[51] of the imidazolidinone alkylations by Schollkopf (Scheme 48).

Hayashi, Ito and co-workers [56] have discovered an extremely interesting catalytic asymmetric aldol condensation between aldehydes and methyl isocyanoacetate employing an optically active gold(I)complex. As shown in Scheme 52, the ferrocenyl ligand (**244**) is admixed with a

SCHEME 52

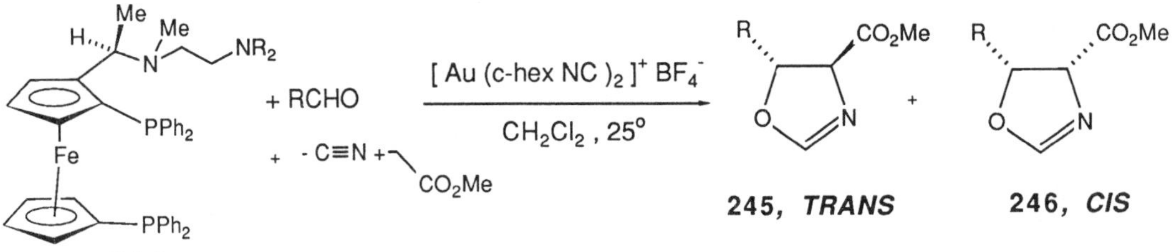

245, TRANS   **246, CIS**

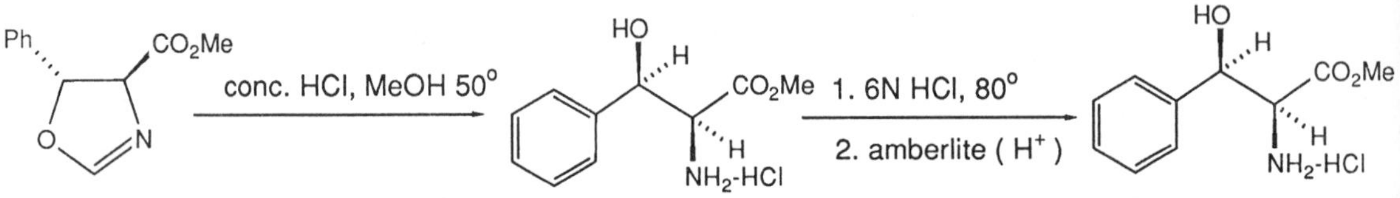

247   **248**   **249**

Ito, Y.; Sawamura, M.; Hayashi, T., *J.Am.Chem.Soc.* (1986) **108**, 6405.

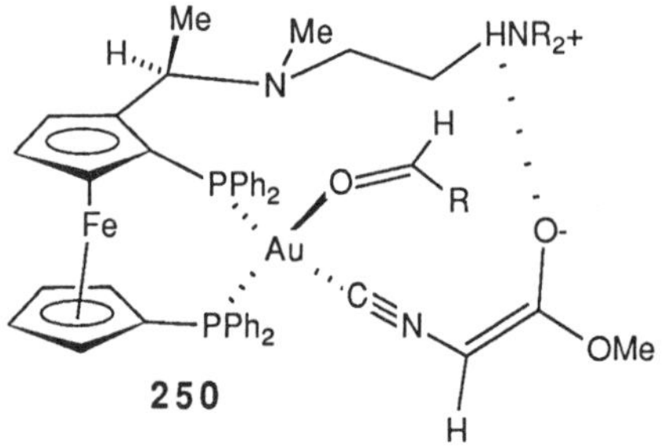

Ligand: a : R=Et
           b : R=Me

250

ALDEHYDE	LIGAND	YIELD	TRANS/CIS	%ee TRANS	%ee CIS
PhCHO	a	98	89/11	96 (4S,5R)	49 (4R, 5R)
	b	91	90/10	94	4 (4S , 5S)
Me⁀⁀CHO (Me)	a	83	81/19	84	52 (4R, 5R)
	b	97	80/20	87	0
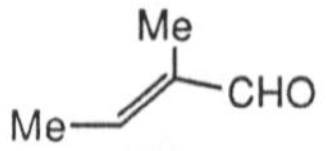 Me⁀CHO (Me)	b	89	91/9	95	31
MeCHO	a	100	84/16	72	44
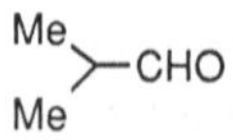 Me₂CH—CHO	a	99	98/2	92	
cyclohexyl—CHO	a	95	97/3	90	
	b	96	98/2	81	
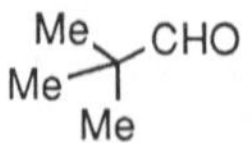 Me₃C—CHO	a	100	100/0	97	

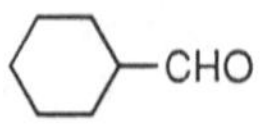

SCHEME 53

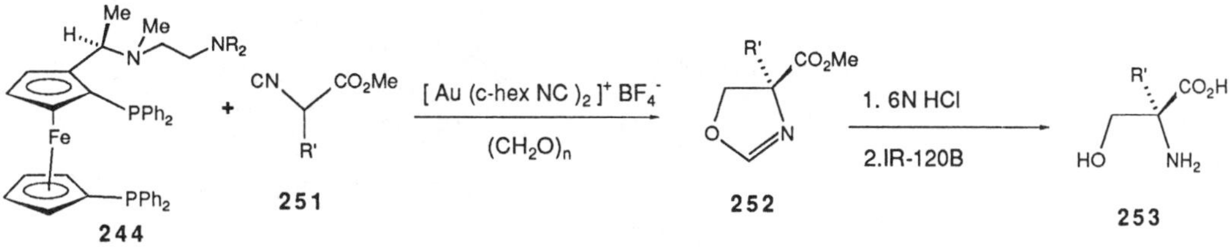

Ito, Y.; Sawamura, M.; Shirakawa, E.; Hayashizaki, K.; Hayashi, T., *Tetrahedron Lett.* (1988) **29**, 235.

NR$_2$	R'	YIELD (% 252)	%ee
—NMe$_2$	H	99	52
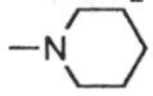 —N⟨⟩	H	89	44
—NMe$_2$	Me	100	64
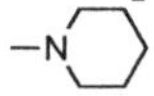 —N⟨⟩	Me	95	63
—NMe$_2$	Et	89	70
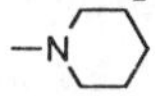 —N⟨⟩	Et	89	66
—NMe$_2$	i-Pr	99	71
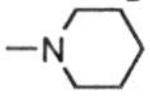 —N⟨⟩	i-Pr	96	81
—NMe$_2$	Ph	75	67

gold(I) salt and methyl isocyanoacetate in methylene chloride at room temperature, followed by addition of the aldehyde. After reaction at ambient temperature for 20 hours, the oxazolines **245** and **246** were isolated by bulb-to-bulb distillation. The reaction shows selectivity for the *trans* -isomers (Table) with enantiomeric excesses ranging as high as 97%. The *cis*-isomers, on the other hand are obtained in mediocre or poor optical purity. After the initial bulb-to-bulb distillation, the *cis*- and *trans*-isomers are separated by column chromatography and their respective optical purities determined by NMR using a chiral shift reagent. The authors rationalize the stereoselectivity by invoking the complex **250** as the reactive intermediate. The use of gold as opposed to silver or copper catalysts was shown to be essential and was attributed to the high affinity of gold for the phosphines. In addition , the authors speculate that there is no coordination of the amine to gold which they propose to be critical for coordinating the isocyanoacetate in the transition state of the aldol reaction. In one case, **247** was converted in high (unspecified) yield to the *threo*-β-phenylserine (**249**) derivative by hydrolysis with

Asymmetric Derivatization of Glycine

SCHEME 54

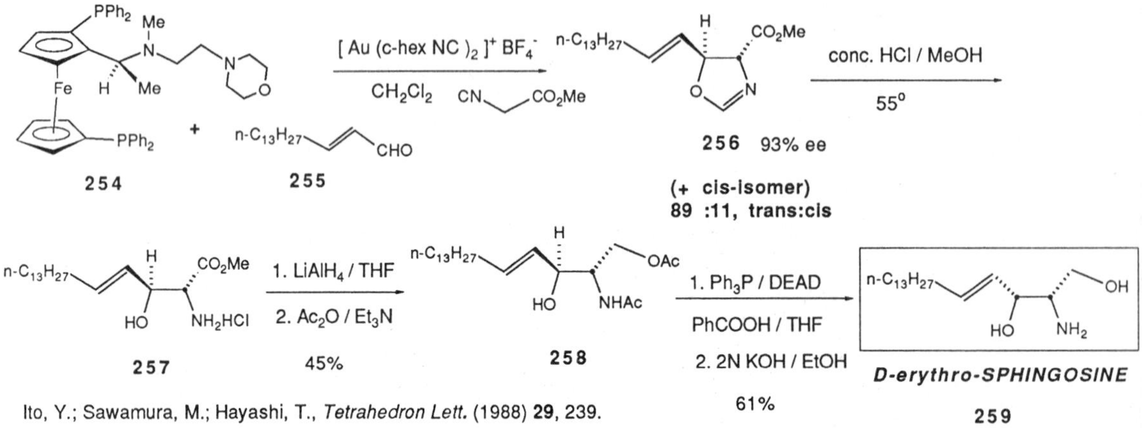

Ito, Y.; Sawamura, M.; Hayashi, T., *Tetrahedron Lett.* (1988) **29**, 239.

concentrated HCl at 50° (to **248**) and subsequent hydrolysis of the methyl ester and ion-exchange chromatography. An additional paper[56b] on the "fine-tuning" of the chiral amine appendage has also appeared that employs either morpholine or piperidine in place of the diethyl- and dimethylamino groups. It should be noted that the chemical yields for these reactions are excellent.

In a subsequent communication[57], the same group reported on the condensation of formaldehyde with substituted methyl isocyanoacetates (**251**) using the same catalyst **244** (Scheme 53). As above, the aldol condensation proceeds in a stereoselective manner with incipient trapping of the initial aldol on the isocyanate to furnish the oxazolines **252** as the isolated products. The chemical yields for the formation of **252** are excellent, but the %enantiomeric excess is somewhat modest ranging from 44-81%. Hydrolysis of the oxazoline is performed with 6N HCl followed by ion-exchange chromatographic elution of the α-alkylated serine derivatives **253**.

An elegant demonstration[58] of the catalytic asymmetric aldol methodology was achieved in the synthesis of D-*erythro*-sphingosine (**259**) as shown in Scheme 54. Asymmetric aldolization of aldehyde **255** with catalyst **254** furnished the oxazoline **256** as an 89 : 11 *trans : cis* mixture in 93% ee. Hydrolysis to the methyl ester (**257**) followed by reduction , acylation and Mitsunobu inversion of stereochemistry at C-3 provided, after removal of the acetyl groups, D-erythro-sphingosine **259**.

SCHEME 55

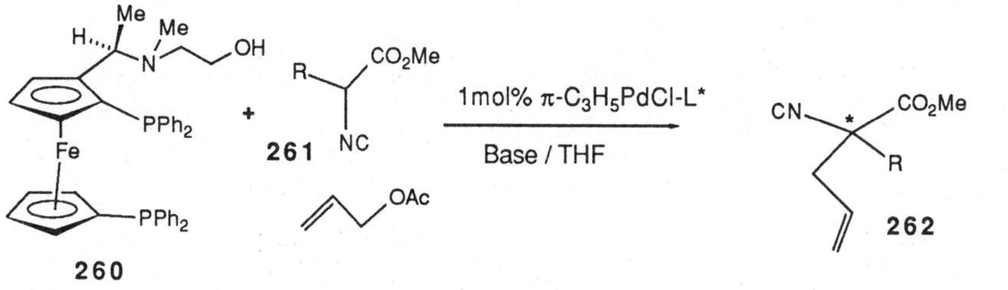

Ito, Y.; Sawamura, M.; Matsuoka, M.; Matsumoto, Y.; Hayashi, T., *Tetrahedron Lett.* (1987) **28**, 4849.

R	BASE	ADDITIVE	YIELD %	%ee
Ph	DBU	ZnBr$_2$	55	39
Ph	DBU	ZnCl$_2$	63	37
Ph	DBU	ZnCl$_2$	54	36
Ph	DBU	-	22	7
Ph	K$_2$CO$_3$	ZnCl$_2$	28	24
Ph	K$_2$CO$_3$	-	76	0
Me	DBU	ZnBr$_2$	35	20

The major drawbacks to the above methodology are the rather exotic catalyst employed (which must be synthesized) , the expense of gold and the variable stereoselectivities that seem quite substrate-dependent. On the other hand, the catalyst can be recovered and reused, although a cycling efficiency has not yet been reported. This work is highly significant in that it opens an entire new area for further development which in principle, holds great promise.

As a first attempt to extend the utility of the chiral ferrocenyl catalysts in asymmetric amino acid synthesis, the asymmetric allylation of substituted methyl isocyanoacetates was examined [59] using a derived palladium π-allyl complex of **260** (Scheme 55). Although the %ee reached a maximum of only 39%, this work illustrates that additional potential for employing chiral catalysts in the construction of amino acids is quite promising. Neither the absolute configurations of the product isocyano esters (**262**) nor conditions to convert these substances to the amino acids was determined.

SCHEME 56

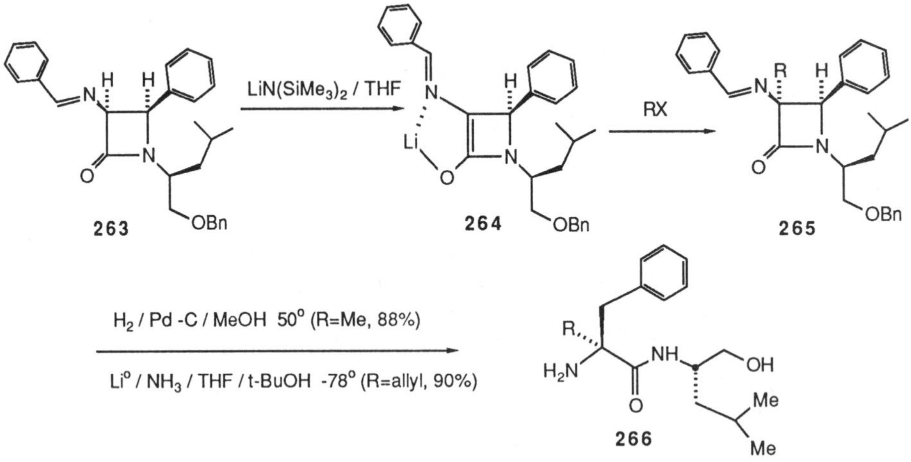

Ojima, I.; Chen, H-J.C.; Nakahashi, K., *J.Am.Chem.Soc.* (1988) **110**, 278.

SCHEME 57

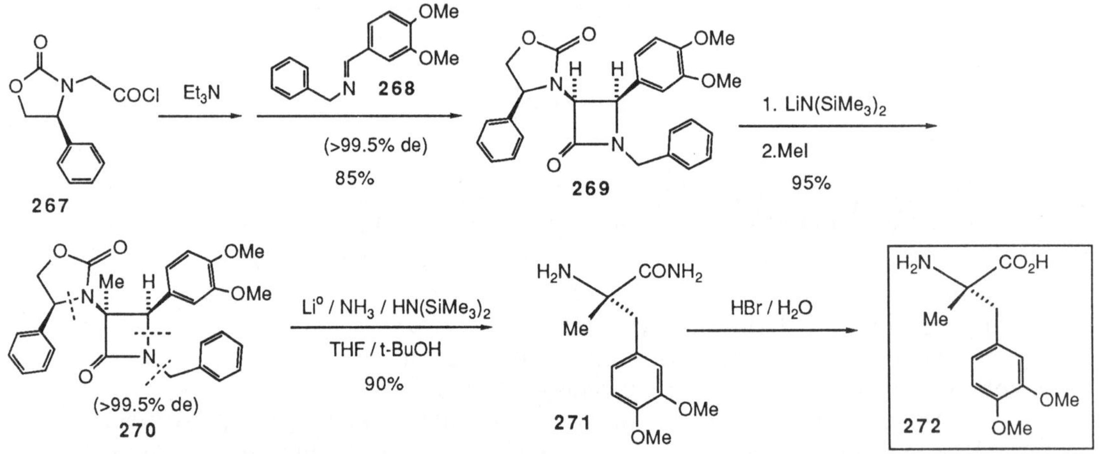

Ojima, I.; Chen, H-J.C.; Qiu, X., *Tetrahedron* (1988) **44**, 5307.
Ojima, I.; Chen, H-J.C.; Nakahashi, K., *J.Am.Chem.Soc.* (1988) **110**, 278.
Ojima,I.; Shimizu, N.; Qiu,X.; Chen, H-j.C.; Nakahashi, K., *Bull.Chem.Soc.Fr.* (1987), 649.

(S)- α – METHYL-DOPA

CONFIG (273)	R	CONFIG (274)	%de (275)	CONFIG (275)
S	Me	S	>98	3S,4R
S	Me	R	>98	3S,4R
R	i-Pr	R	>98	3R,4S
R	i-Pr	S	>98	3R,4S
R	H₂C⟨phenyl⟩	S	>98	3R,4S
R	H₂C⟨⟩SMe	S	>98	3R,4S

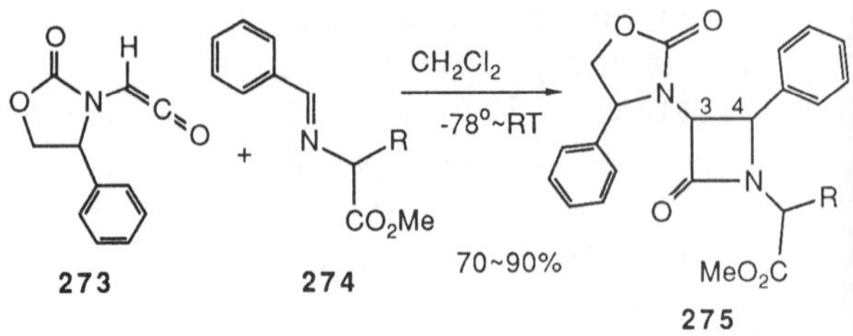

SCHEME 58

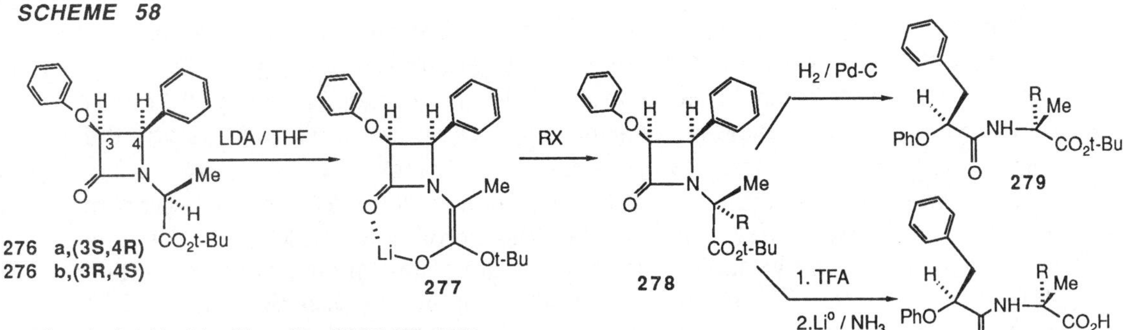

276 a,(3S,4R)
276 b,(3R,4S)

Ojima, I.; Qui, X., *J.Am.Chem.Soc.* (1987) **109**, 6537.

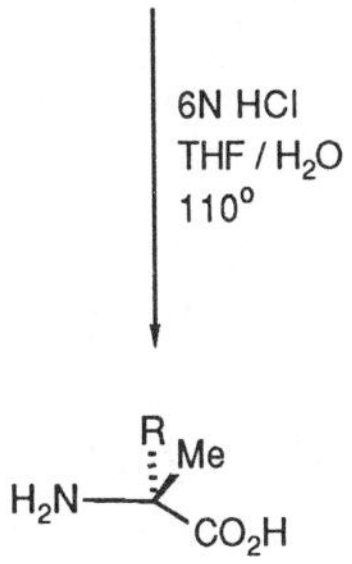

RX	SUBSTRATE (276)	YIELD %(278)	%de	CONFIG.
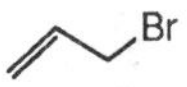 (allyl)Br	a	95	>98	R
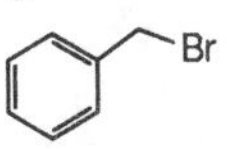 (benzyl)Br	a	96	>98	R
EtBr	b	95	>98	R
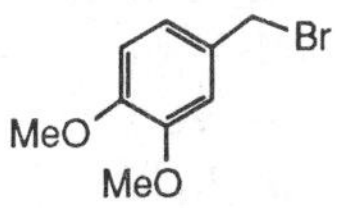 MeO, MeO (benzyl)Br	b	95	93	R

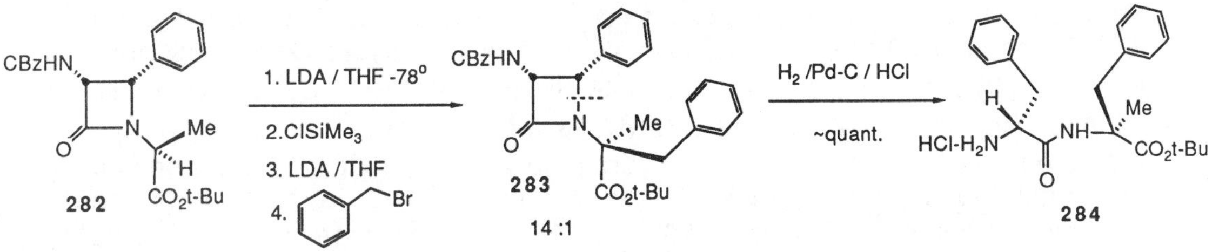

Ojima and associates[60] have developed a clever and potentially quite useful way to make various optically active amino acids and peptide derivatives through stereoselective alkylations of β-lactams. The overall strategy is predicated on the use of 4-phenyl-substituted β-lactams which can be reductively cleaved at the N-1/C-4 (benzylic) bond to produce the corresponding phenylalanine-containing amino acid or peptide. As shown in Scheme 56, the benzylidene Schiff base of a 3-amino-4-phenyl β-lactam (**263**) is metallated with lithium hexamethyldisilazane to give the internally chelated enolate **264**. Alkylation occurs in a stereoselective manner *anti* -to the 4-phenyl substituent providing **265**. Hydrogenolysis in the case of R=Me or Birch reduction in the case of R=allyl provides the α-alkylated phenyl alanine peptides of leucinol (**266**)

SOAA—C*

in high yield as virtually optically pure substances. Utilizing a related strategy, the asymmetric synthesis of (S)-α-methyl-DOPA (**272**) was accomplished as shown in Scheme 57. The appropriate β-lactam **269** was prepared by the classical ketene/imine cyclocondensation using the elegant asymmetric technology developed by Evans[61]. Thus, condensation of the Evans ketene precursor **267** with benzylidene **268** gave β-lactam **269** in high yield and >99.5% de. Stereoselective alkylation of this substance with methyl iodide as above, gave the methylated derivative **270** in >99.5% de. Birch reduction of this compound provided amide **271** which upon hydrolysis to the acid gave (S)-α-methyl-DOPA, an effective antihypertensive drug. To demonstrate the potentiality of the approach, a series of optically active β-lactams (**275**) of the same basic substitution pattern were prepared according to the Evans protocol (Table); the %de in every case exceeded 98%. The obligate 4-aryl-substituted β-lactam required for the pentultimate reductive β-lactam cleavage confines this methodology to the synthesis of α-alkylated phenylalanine derivatives

In a related series of investigations, Ojima and Qiu [62] reported on the stereoselective alkylations of amino acid residues incorporated at N-1 of the β-lactam ring as shown in Scheme 58. In this system, the derived enolate of **276** is proposed to react from the seven-membered chelate **277** directing alkylation from the side *anti* -to the C-4 phenyl moiety. Reductive cleavage can be performed to access either the t-butyl ester **279** or the acid **280**; either derivative can then be hydrolyzed completely to the α-alkylated amino acid **281** with 6N HCl at 110º. By applying this concept to the 3-amino-substituted β-lactam **282**, stereoselective alkylation (14 : 1) followed by hydrogenolysis gives the corresponding phenylalanine dipeptide **284**. This approach is more versatile than the systems described in Schemes 56 and 57 since, the type of α-alkylated amino acid is determined by the type of amino acid incorporated in the starting β-lactam. Additional studies employing these concepts for making di-, tri-and higher peptides will be presented in the chapter on asymmetric hydrogenation.

An acyclic Schiff base system that employs the stereocontrol elements of a cyclic system has been reported on by Mukaiyama and associates [63]. The optically active ketoacid **285** is condensed with glycine *t*-butyl ester to furnish **286** in good yield (Scheme 59). Formation of the magnesium alkoxide **287** followed by enolate formation and aldolization gives the β-hydroxy adducts in good but not excellent %ee's. Cleavage of the chiral auxilliary and protection of the nitrogen as a t-BOC urethane provides the carbinols **289**. It is assumed that the magnesium

SCHEME 59

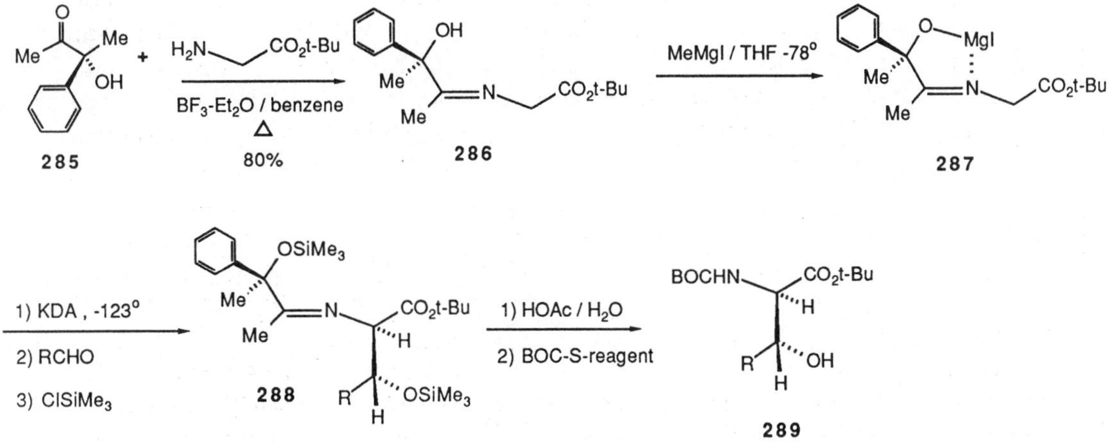

Nakatsuka, T.; Miwa, T.; Mukaiyama, T., *Chem.Lett.* (1981) 279.

ALDEHYDE	YIELD (% 289)	THREO : ERYTHRO	%ee
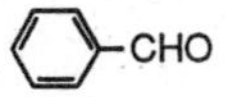 ⬡-CHO	67	76 : 24	64
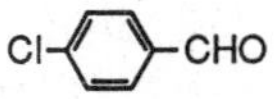 Cl-⬡-CHO	51	75 : 25	71
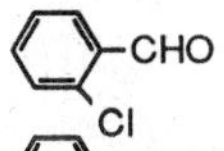 ⬡-CHO (Cl)	51	69 : 31	75
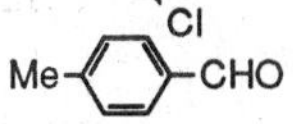 Me-⬡-CHO	62	81 : 19	62
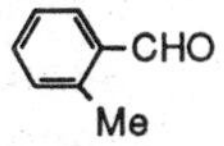 ⬡-CHO (Me)	67	92 : 8	70
 Me⌒CHO (Me)	46	58 : 42	43

chelate provides a rigid, cyclic transition state for aldol reaction. It is unlikely that this aldolization template can effectively compete with more extensively studied systems, such as those developed by Schollkopf, Hayashi, Evans and Seebach.

Husson and co-workers [64] have found that phenylglycinol serves as an interesting chiral auxilliary to prepare optically active α-amino nitriles; reasonable precursors to amino acids. The oxazolidine **291** is prepared from **290**, formaldehyde and KCN in excellent yield. Enolate formation with LDA and alkylation furnishes the adducts **292** in good yields and modest diastereoselectivity. A second alkylation sequence provides the α,α-disubstituted systems **293**. Conditions to cleave these

SCHEME 60

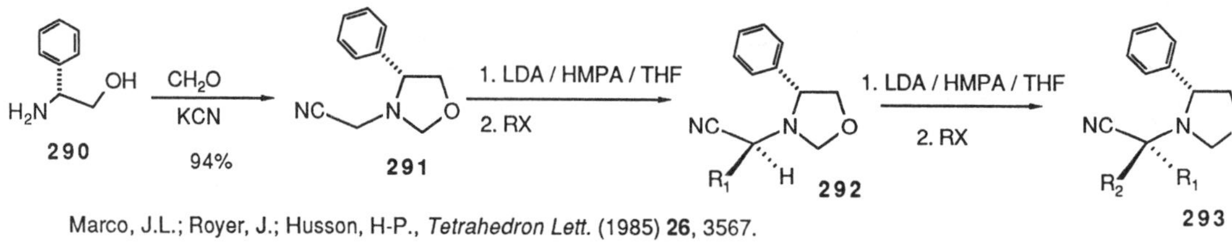

Marco, J.L.; Royer, J.; Husson, H-P., *Tetrahedron Lett.* (1985) **26**, 3567.

R$_1$	YIELD (%)	%de(292)	R$_1$	R$_2$	YIELD (%, 293)	%de(293)	CONFIG.
Me	55	38	Me	Et	70	64	S
Et	60	50	Et	Me	68	36	R
n-Pr	65	62	Me	H$_2$C-C$_6$H$_5$	73	52	S
H$_2$C-CH=CH$_2$	51	44			68	40	S
H$_2$C-C$_6$H$_5$	65	68	Me	H$_2$C-C$_6$H$_3$(OMe)$_2$	72	50	S

adducts to the corresponding amino acids were reported [65] by the same authors in syntheses of 1-amino-1-cyclopropane carboxylic acids (Scheme 61). Condensation of **291** with epibromohydrin gave four diastereomeric adducts **294**. Hydrolysis of the nitrile with hot sodium hydroxide followed by dilute acidic cleavage of the aminal and esterification afforded the *trans* -product **295** and the *cis* - product which suffers spontaneous lactonization to **296**. The phenylglycinol auxilliary is destructively removed by catalytic hydrogenolysis and tosylated to afford the separable substances **297** and **298**. Lithium dimethyl cuprate addition to **297** provides **299** which is detosylated by Birch reduction to *allo*-coronamic acid (**300**, 30% ee). Similarly, **295** and **296** can be deprotected to the corresponding free amines by catalytic hydrogenation. The relatively modest %ee's again reflect the difficulties associated with acyclic and non-chelation controlled enolate reactions.

An interesting glycine equivalent has been investigated by Kolb and Barth [66] (Schemes 63 and 64). Prolinol (**301**) is formylated , methylated and acetalized to give **303** (SDMP). Condensation of various amino acid methyl esters with SDMP provides the formamidines **304**. Enolate alkylation provides the α,α-disubstituted species **305** which can be cleaved to the free amino acids **306**. In a related series, propargyl amine is condensed with SDMP and silylated to furnish **307**. Double enolate alkylation furnishes **308** which can be cleaved with hydrazine-hydrate to the optically active 1,1-disubstituted alkynes **309**. Sodium metal in methanol effects desilylation to **310** which can be oxidized with

SCHEME 61

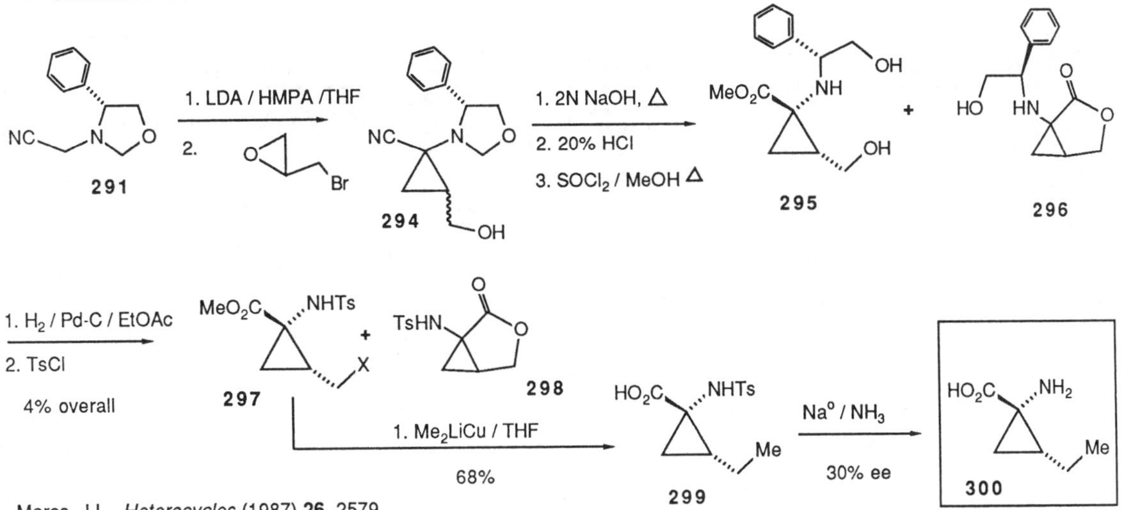

NC — **291**

1. LDA / HMPA /THF
2. (epoxide) Br

294 OH

1. 2N NaOH, Δ
2. 20% HCl
3. SOCl₂ / MeOH Δ

MeO₂C NH OH **295** + HO HN O **296**

1. H₂ / Pd-C / EtOAc
2. TsCl

4% overall

MeO₂C NHTs **297** X + TsHN O **298**

1. Me₂LiCu / THF

68%

HO₂C NHTs Me **299**

Na° / NH₃

30% ee

HO₂C NH₂ Me **300**

ALLOCORONAMIC ACID

Marco, J.L., *Heterocycles* (1987) **26**, 2579

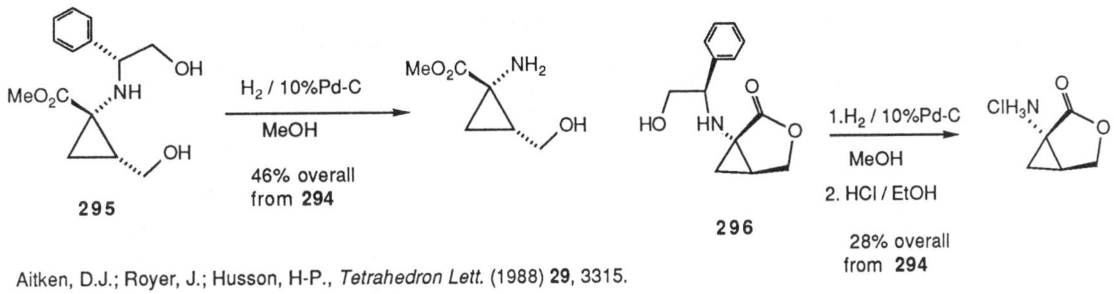

MeO₂C NH OH **295**

H₂ / 10%Pd-C
MeOH

46% overall
from **294**

MeO₂C NH₂ OH

HO HN O **296**

1. H₂ / 10%Pd-C
MeOH
2. HCl / EtOH

28% overall
from **294**

ClH₃N O

Aitken, D.J.; Royer, J.; Husson, H-P., *Tetrahedron Lett.* (1988) **29**, 3315.

ruthenium and periodate to the amino acids **306**. The chemical yields for the alkylations are generally good, but the %ee's are rather dissappointing; the notable exception being the synthesis of α-methyl phenylalanine (84% ee). The employment of the formamidine moiety for activating proton abstraction α- to nitrogen for asymmetric synthesis has been extensively developed by Meyers [67]; the above methodology being a specific application to amino acid synthesis. Further development of this strategy to improve the %ee's of the alkylations should prove to be useful.

SCHEME 62

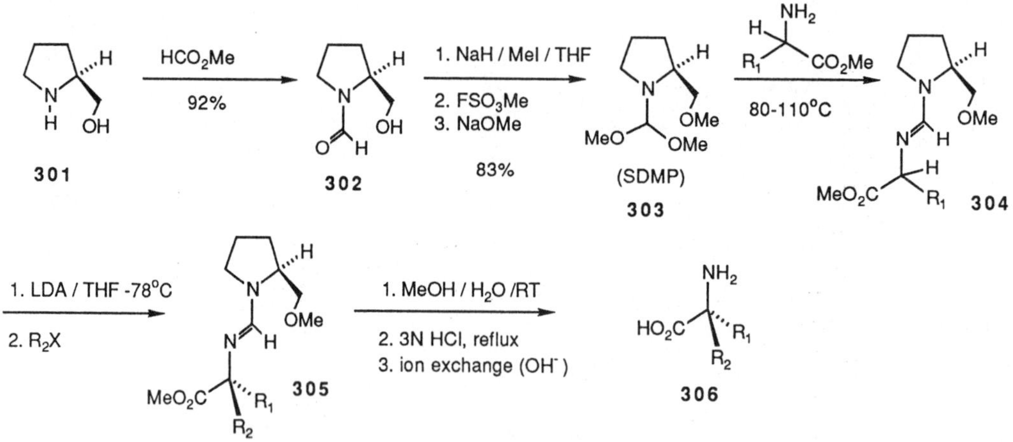

Kolb, M.; Barth, J. *Liebigs Ann. Chem.* (1983) 1668; Kolb, M.; Barth, J., *Tetrahedron Lett.* (1979) 2999.

R_1	R_2	% (304)	%(305)	%(306)	%ee (306)	CONFIG.
Me	CHF_2	66	36	59		
	Et		80	76		
	$n\text{-}C_4H_9$		78	79		R
	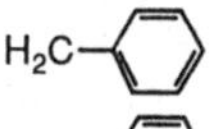 $H_2C\text{-}$ (benzyl)		60	68	51	R
	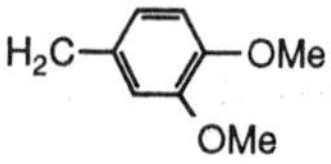 $H_2C\text{-}$ (dimethoxybenzyl, OMe/OMe)		80	31	30	R
	$H_2CH_2CCO_2Me$		43			
$CH_2CH(Me)_2$	Me	89	46	50		
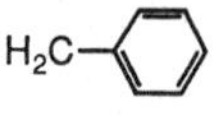 $H_2C\text{-}$ (benzyl)	Me		68	78	15	S
	Et		81	53		
	CHF_2		81	58		
	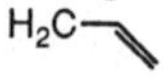 $H_2C\text{-}$ (allyl)		54	(unstable)		

SCHEME 63

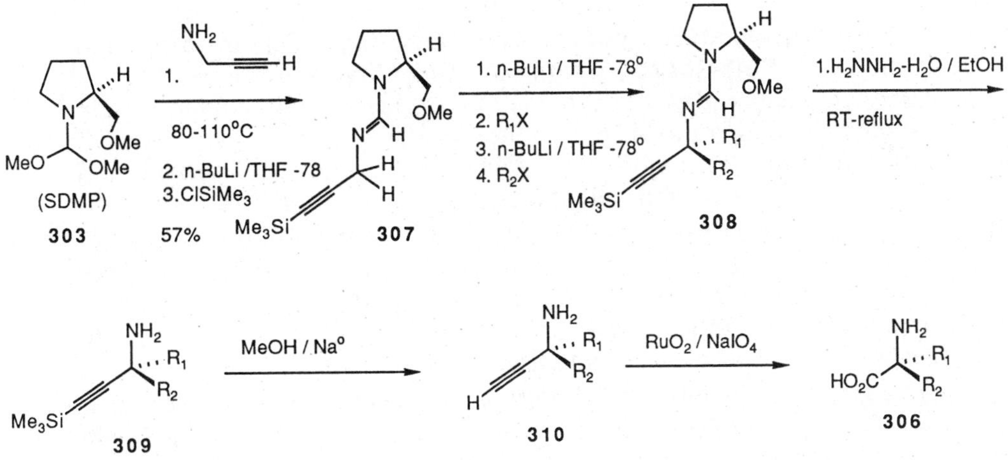

Kolb, M.; Barth, J. *Liebigs Ann. Chem.* (1983) 1668; Kolb, M.; Barth, J., *Angew.Chem.Int.Ed.Engl.* (1980) 19, 725.

R_1	R_2	% (308)	%(309)	%(310)	%ee (306)	CONFIG.
	Me	84				
H	n-C_4H_9	85	56	90	9	S
	$H_2C-C_6H_5$	66	77	94	15	S
Me	Et	76	71	96		S
	n-C_4H_9	73	86	96		S
	$H_2C-C_6H_5$	75	84	90	84	S
n-C_4H_9	Me	77	64	100		R
$H_2C-C_6H_5$	Me	79	84	97	67	R

3. Enolate Functionalization of Imidazolidinones, Oxazolidinones, and Oxazolines : Self-Reproduction of Chirality

Seebach and associates have made an extensive contribution to the practical asymmetric synthesis of amino acids via the formation of various cyclic aminals of numerous amino acids and the subsequent enolate alkylation of such systems with a high degree of diastereoselectivity. This work, which will be detailed below has been summarized and reviewed by the authors [68]. It is noteworthy that virtually all of the chemistry reported by these authors has been published with full experimental details. The first system reported on was the enolate alkylation of the bicyclic aminal (**307**) prepared from L-proline and pivaldehyde (Scheme 64) [69]. Azeotropic removal of water in pentane in the presence of TFA gives rise to a single diastereomer in high yield with the relative configuration shown. This material can be prepared on a large scale and stored indefinitely. Treatment of **307** with LDA in THF at dry ice temperature affords the enolate **310** which suffers highly diastereoselective alkylation with *retention* of configuration. Since the initial stereogenic center in proline controls the relative stereochemistry in the formation of the aminal stereogenic center and is subsequently planarized in the enolate reaction, the authors have coined the term 'self reproduction of chirality' or a self-immolative process to describe this interesting stereochemical outcome. The authors speculate that the enolate (**310**) adopts a conformation that places the aminal methine in a pseudo-axial disposition and hindering electrophilic approach *anti*-to the *t*-butyl group presumed to be pseudo-equatorially disposed. The instability of enolate **310** above -30º C has precluded a rigorous x-ray structural study. As shown in the table, a wide range of electrophiles participate in highly diastereoselective functionalizations of this system in good to excellent yields and virtually complete diastereoselectivity. In cases where a second stereogenic center is produced, such as in the aldol reactions, the diastereomer ratios are also good to excellent.

Hydrolysis of the homologated heterocycles (**308**) directly to the free amino acids turns out to be most difficult typically requiring 15-48% HBr at ambient to reflux temperatures. The bulkier the α-R residue, the more drastic are the conditions required to cleave the aminal. This is in marked contrast to the starting material **307** which is reported to readily hydrolyze upon exposure to moisture. A manifestation of this problem is the paucity of free amino acids reported in this account; only α-methyl- and α-benzyl proline being reported. Fortunately, the authors have found that several richly nucleophilic reagents such as lithium amides and alkyl lithium reagents open the aminal of **308** to produce the corresponding amides and ketones, respectively. Hydride reduction of **312** opens the

SCHEME 64

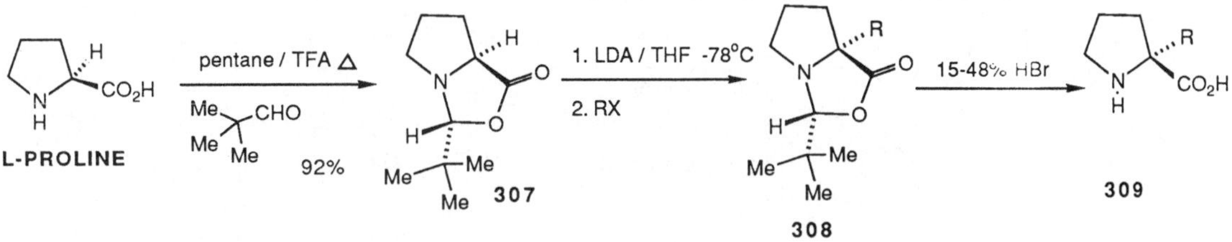

Seebach, D.; Naef, R., *Helv.Chim.Acta* (1981) **64**, 2704
Seebach, D.; Boes, M.; Naef, R.; Bernd Schweizer, W., *J.Am.Chem.Soc.* (1983) **105**, 5390.

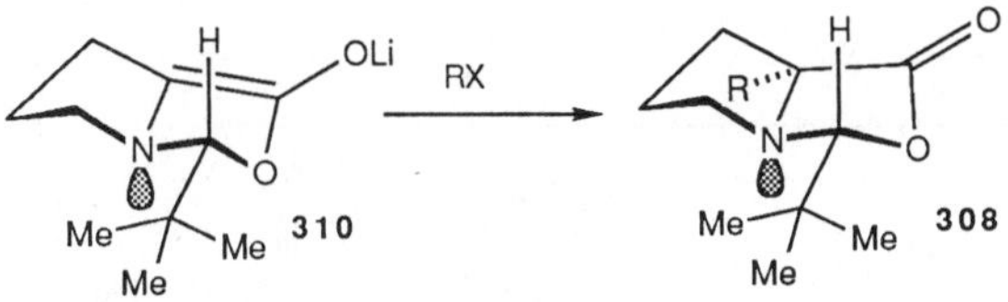

ELECTROPHILE	YIELD (% 308)	RATIO	%de	AMINO ACID(%309)	%ee
MeI	93		>99	80%	>99
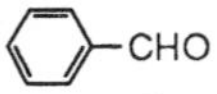 ⬡—CHO	73	95:5	>99		
 allyl—Br	87		>99		
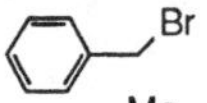 benzyl—Br	91		>99	77%	>99
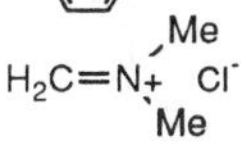 H₂C=N+(Me)(Me) Cl⁻	56		>99		
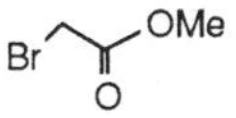 Br—CH₂—CO—OMe	40		>99		
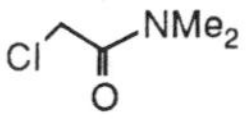 Cl—CH₂—CO—NMe₂	70		>99		
 ⬡··Cr(CO)₃	30		>99		
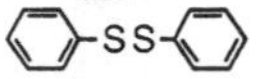 ⬡—S-S—⬡	84		>99		
MeCHO	88	94:6	>99		
acetone	94		>99		
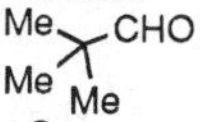 Me₂(Me)C—CHO	85	6:4	>99		
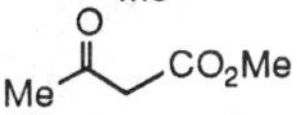 Me—CO—CH₂—CO₂Me	62	2:1	>99		
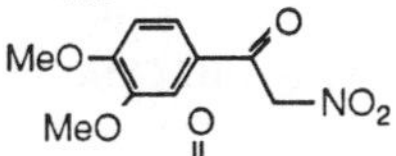 (MeO)₂C₆H₃—CO—CH₂—NO₂	79		>99		
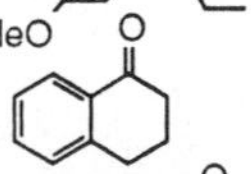 tetralone	70	7:3	>99		
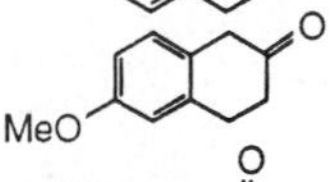 MeO-tetralone	67	7:3	>99		
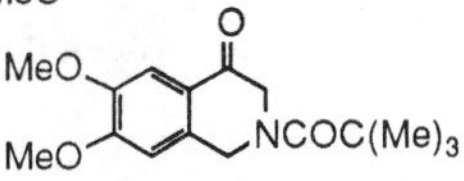 (MeO)₂—isoquinolinone—NCOC(Me)₃	91		>99		

ELECTROPHILE	YIELD (%308)	RATIO	%de
cyclohexenone	85 (1,2 adduct with DMPU) 50 (1,2 adduct) 21 (1,4 adduct)	95:5	>99
dihydropyridinone (NBn)	50 (1,4 adduct)	>95:5	>99
Ac$_2$O	80		>99
MeO—(—CO$_2$Me) MeO	80		>99
MeOCOCl	85		>99

308 → (Nuc) → 311

312 → (LiAlH$_4$ / THF, 87%) → 313

NUCLEOPHILE	R	R'	YIELD (311)
LiNMe$_2$	H$_2$C—(Ph)	-NMe$_2$	43%
LiNHMe	H$_2$C—(Ph)	-NHMe	90%
MeLi	H$_2$C—(Ph)	-Me	88%
(Ph)-Li	H$_2$C—(Ph)	-(Ph)	70%
MeO—(Ph)—NHLi	-H$_2$C—	-HN—(Ph)—OMe	88%

314 → (MeO—(Ph)—CH$_2$NHLi, 88%) → 315 → → → 316, (-)- BREVIANAMIDE B

317 → (pivaldehyde) → 318 (racemic)

319   ‖→   320

aminal, but the incipient imine is also cleanly reduced to the N-alkyl amino alcohol **313**. For the purpose of preparing amino acids and derivatives, the most useful protocol seems to be the lithium amide opening to the amides. The author of this monograph can attest to the usefulness of this approach [70] as the sequence from **314** to **315** proceeds in high yield on a multi-gram scale; the latter has been successfully employed in a total synthesis of (-)-brevianamide B (**316**).

Since the product amino acids (**309**) are α,α-disubstituted and generally not prone to racemization, the somewhat brutal hydrolysis conditions should be tolerable in many cases.

In considering the extension of this methodology to the homologues azetidinecarboxylic acid (**317**) and pipecolic acid (**319**), the authors note that (S)-**317** afforded the bicyclic system **318**, but as a racemate; pipecolic acid gave no bicyclic adduct (**320**) with pivaldehyde despite numerous attempts. The subtle and special stereoelectronic factors that allow the acetalization to occur so readily and without attendant racemization for the bicyclo [3.3.0] system are not readily apparent.

SCHEME 65

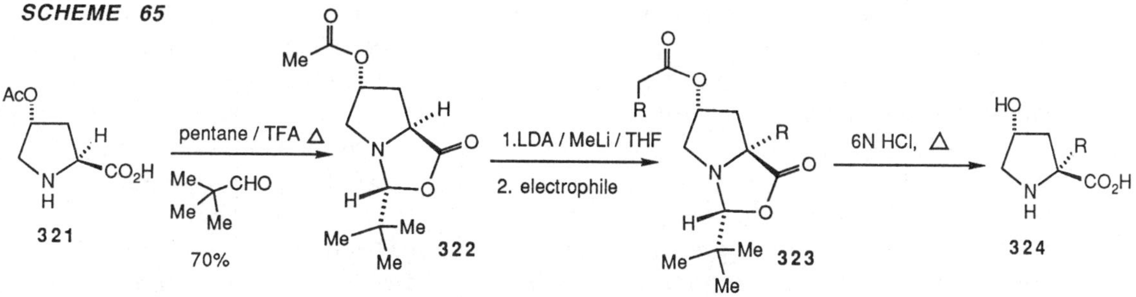

Weber, T.; Seebach, D., *Helv.Chim.Acta* (1985) **68**, 155.

ELECTROPHILE	R	YIELD (%323)	%ds	YIELD (%324)	%ds
MeI	Me	32	>95	29	>95
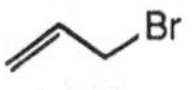 allyl Br	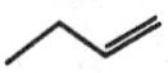 allyl	23	>95	20	>95
acetone	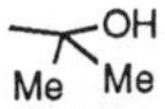 C(Me)(Me)OH	76	>95		
CH₃CHO	CH(Me)(H)OH	74	65		
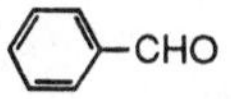 Ph-CHO	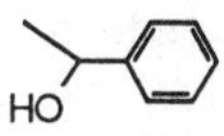 CH(OH)Ph	50	81		
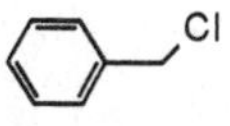 benzyl Cl	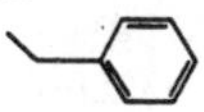 CH₂Ph			37	>95

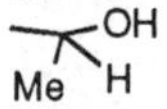

The closely realted 4-hydroxyproline system [71] is detailed in Scheme 65. It is noteworthy that the acetate protecting group of the template **322** competes for the base and thus, consumes a full molar equivalent of the electrophile furnishing the di-functionalized derivatives **323** which are similarly processed to the corresponding α-substituted-4-hydroxyproline derivatives **324**. Cleavage of the aminal and modified acetate is effected with refluxing 6N HCl; ion exchange purification of the HCl-salt affords the pure zwitterions in modest chemical yields and excellent %ds.

Another bicyclo[3.3.0] aminal was prepared from the cyclic hemithio aminal **325** derived from L-cysteine [72]. The pivaldehyde adduct **326** is formed in good yield under the standard conditions. Enolization with LDA or lithium *t*-butoxide at -100°C followed by aldol condensation provided the carbinols **327** in good to excellent diastereoselectivities. The authors note that the enolate starts to suffer elimination of thiolate (*retro 5-endo*-trig) above -60°C which requires enolate generation and electrophilic quench below -75°C; warming the reaction above -60°C subsequently results in good yields of the aldol products and very little elimination product. The diastereoselectivity of C-C bond formation at the α-center of the amino acid moiety is >99%de; the diastereoselectivity at the carbinol center is very good ranging from 65-96%ds. The relative configuration (**327**) is assumed to be that shown by correlation of the benzaldehyde adduct to α-methyl-β-phenylserine.

Useful additions to this general repertoire were the successive developments of threonine[73] and serine [74] templates for homologation of the α-carbons; these are shown in Schemes 67 and 68. Condensation of threonine methyl ester with metyl iminobenzoate furnished the *trans*-isomer **332**; the corresponding *cis*-isomer **336** is obtained from n-benzoyl threonine methyl ester and thionyl chloride. Enolate formation with LDA in THF at dry ice temperature followed by addition of various electrophiles affords the adducts **333** and **337**, respectively. The electrophile approaches the enolate from the face *anti*- to the threonine methyl residue. The free amino acids (**334** and **338**) are obtained in high yield by hydrolysis of the oxazolines with refluxing 6N HCl and ion exchange dehydrochlorination. The serine templates **340** and **341** were prepared by acetalization with pivaldehyde and N-formylation; the major diastereomer (**340**) is separated from the minor component by crystallization. Enolate formation and homologation with several electrophiles demonstrate excellent diastereoselectivity; the electrophiles attack virtually exclusively *anti*- to the *t*-butyl residue. Improvements in yield are realized by addition of HMPT or DMPU. The

SCHEME 66

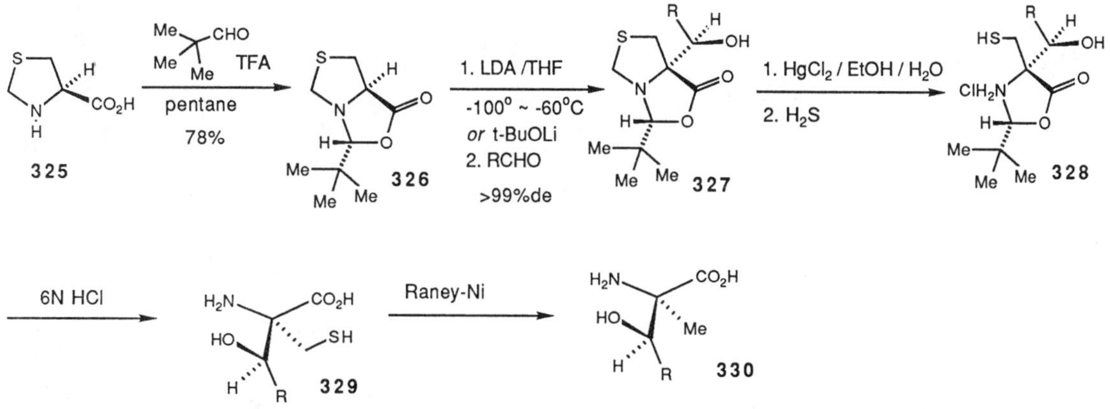

Seebach, D.; Weber, T., *Tetrahedron Lett.* (1983) **24**, 3315.
Seebach, D.; Weber, T., *Helv.Chim.Acta* (1984) **67**, 1650.

ALDEHYDE	R	YIELD (%327)	%Diastereoselectivity
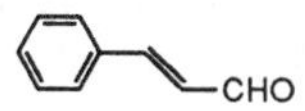 (cinnamaldehyde)	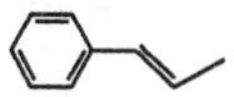 (styryl)	48	89
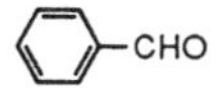 (benzaldehyde)	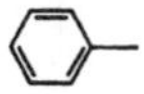 (phenyl)	64	92
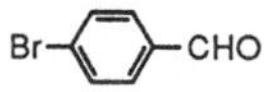 Br-(benzaldehyde)	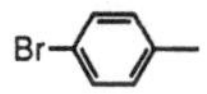 Br-	65	82
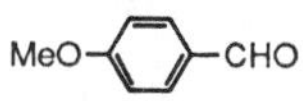 MeO-(benzaldehyde)	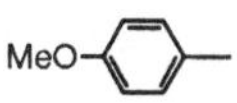 MeO-	68	96
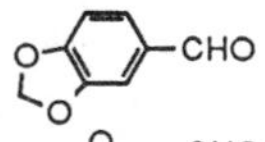 (methylenedioxybenzaldehyde)	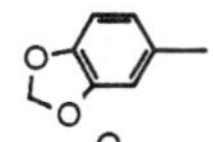	48	96
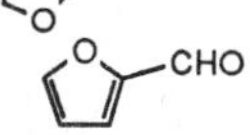 (furaldehyde)	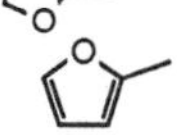	45	88
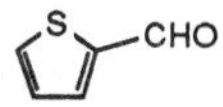 (thiophene-CHO)	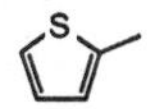	44	94
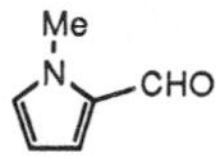 (N-Me-pyrrole-CHO)	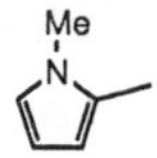	55	90
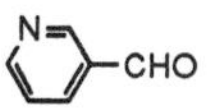 (pyridine-CHO)	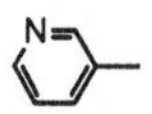	65	65

aminal is removed by refluxing 6N HCl to afford the α-functionalized serine derivatives **(343)**; the α-methyl substance being reported in 93% yield.

SCHEME 67

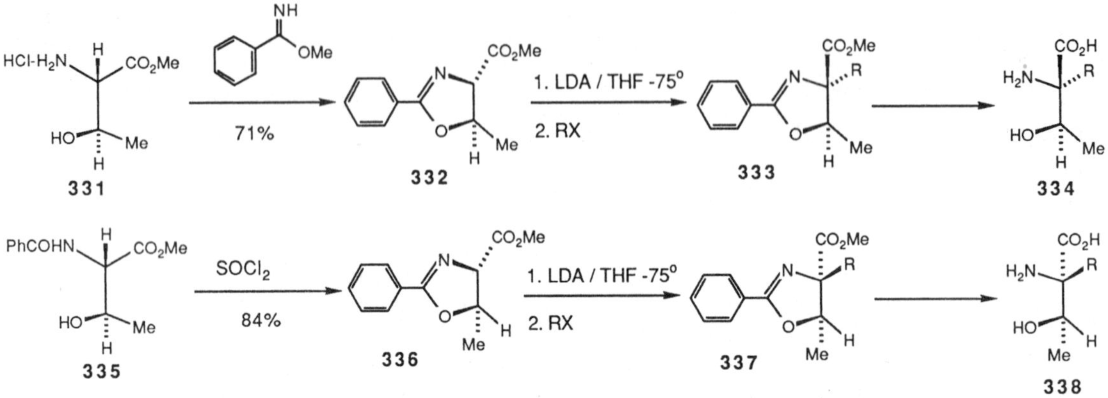

Seebach, D.; Aebi, J.D., *Tetrahedron Lett.* (1983) **24**, 3311.
Seebach, D.; Aebi, J.D.; Gander-Coquoz, M.; Naef, R. *Helv.Chim.Acta* (1987) **70**, 1194.

ELECTROPHILE	R	YIELD(%)	%ds
MeI	-Me	94 (**333**)	93
		93 (**337**)	95
EtI	-Et	94(**337**)	95
		91(**333**)	95
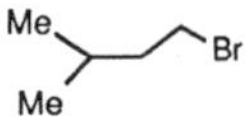 Me₂CHCH₂Br	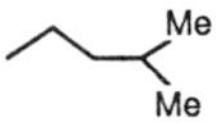	62(**333**)	>98
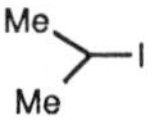 Me₂CHI	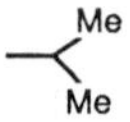	85(**333**)	>98
 allyl Br		96(**337**)	>98
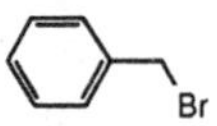 benzyl Br	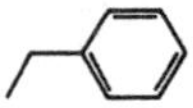	93(**337**)	>98
acetone	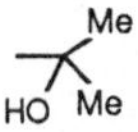	68(**337**)	>95
MeCHO	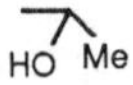	97(**333**)	60
PhCHO		- (**333**)	70
		91(**337**)	70
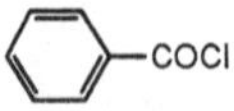 PhCOCl	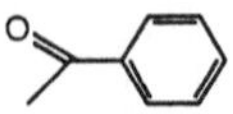	83(**333**)	>95

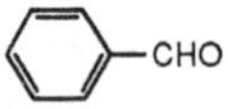

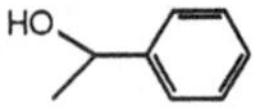

SCHEME 68

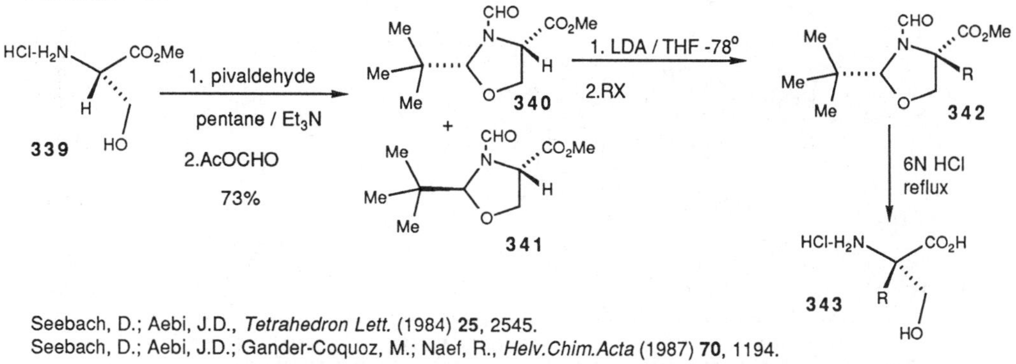

Seebach, D.; Aebi, J.D., *Tetrahedron Lett.* (1984) **25**, 2545.
Seebach, D.; Aebi, J.D.; Gander-Coquoz, M.; Naef, R., *Helv.Chim.Acta* (1987) **70**, 1194.

ELECTROPHILE	R	YIELD (%342)	%ds
MeI	Me	68	>98
EtI	Et	62	>98
 allyl Br	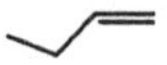 allyl	57	>98
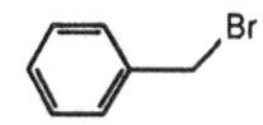 benzyl Br	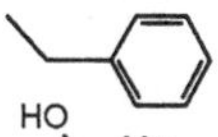 CH2Ph	52	97
acetone	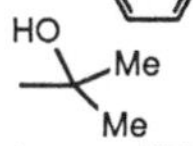 C(OH)Me2	58	>95
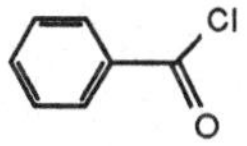 PhCOCl	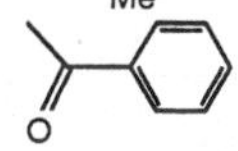 PhCO	70	>95

The above investigations culminated in the development of a general and practical approach to preparing a wide variety of novel amino acids employing optically active oxazolidinone and imidazolidinone derivatives of several amino acids (Schemes 69-89). The general approach [75] for the imidazolidinones is illustrated in Scheme 69. Amino acid methyl or ethyl esters (**344**) are treated with concentrated methyl amine to form the corresponding N-methyl amides. Imine formation with pivaldehyde by azeotropic removal of water provides the pivaloyl imines **345**. Treatment of the latter with methanolic HCl followed by acylation with benzoyl chloride effects stereoselective ring closure to the *anti*-imidazolidinones **346** as the major product. The corresponding *syn*-isomers **347** are available by treatment of **345** with benzoic anhydride at 130°C. The following amino acids were converted into the imidazolidinones: alanine; methionine; valine; phenylglycine; phenylalanine; aspartic acid and glutamic acid[76] (lysine and ornithine appear in Scheme 85). Enolate homologation of these derivatives proceeds in generally good yields and excellent diastereoselectivities (Scheme 70, Table). Unfortunately in

SCHEME 69

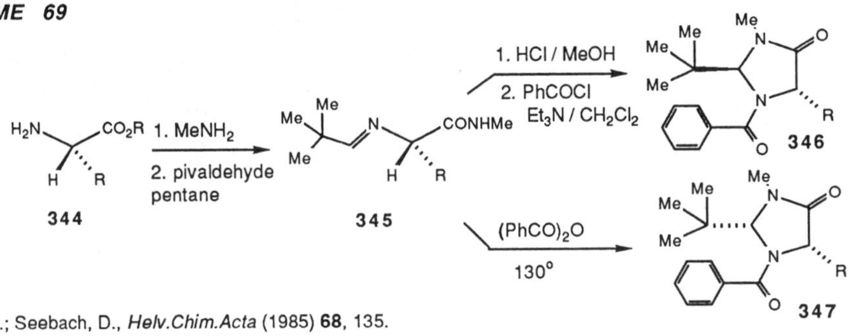

Naef, R.; Seebach, D., *Helv.Chim.Acta* (1985) **68**, 135.

AMINO ACID	R	YIELD (%346)	(%347)
Alanine	Me	78	13
Methionine	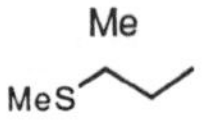 MeS⌇	64	56
Valine	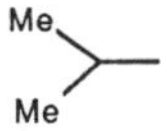 Me / Me (isopropyl)		78
Phenylglycine	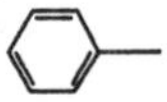 (phenyl)	79	
Phenylalanine	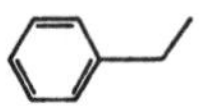 (benzyl)		57
Aspartic Acid	CH$_2$COOH	31	7
Glutamic Acid	CH$_2$CH$_2$COOH	38	

SCHEME 70

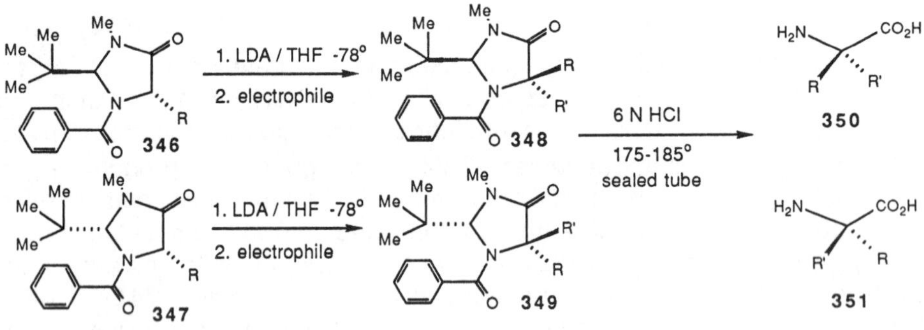

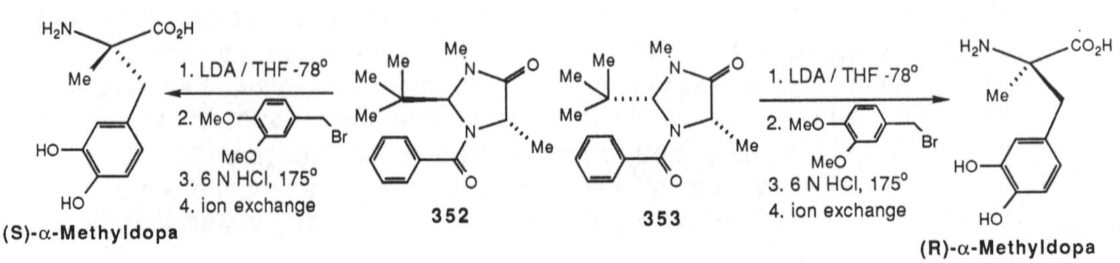

Seebach, D.; Aebi, J.D.; Naef, R.; Weber, T., *Helv.Chim.Acta* (1985) **68**, 144.
Calderari, G.; Seebach, D., *Helv.Chim.Acta* (1985) **68**, 1592.
Aebi, J.D.; Seebach, D. *Helv.Chim.Acta* (1985) **68**, 1507.

SUBSTRATE	R	ELECTROPHILE	R'	YIELD% (348 / 349)	%ds	AMINO ACID% (350 / 351)
347	Me	EtI	Et	90	>95	
346		Me‚—CH=CH—NO₂ (Me, NO₂)	—CH(Me)—NO₂	62	>90	
		Br—CH₂—C₆H₅	Me, —CH₂—C₆H₅ (Et)	73	>95	
347	Me₂CH (i-Pr)	MeI	Me	70	>95	95
		EtI	Et	81	>99	
		Br—CH₂—CH=CH₂ (allyl)	—CH₂—CH=CH₂	89	>99	
346	MeS—CH₂—CH₂—	MeI	Me	66	>95	72
347		EtI	Et	55-84	>95	
346		i-PrI	i-Pr	50	93	
347	C₆H₅—	EtI	Et	80	>95	
347	HOOCCH₂	Br—CH₂—C₆H₅, MeI	Me, —CH₂—C₆H₅	56	>95	94.6
				46.5	>95	
346	HOOCCH₂CH₂	Br—CH₂—C₆H₅, MeI	Me, —CH₂—C₆H₅	76.9	>95	90.1
				41.2	>95	77

these systems, the final hydrolysis to the amino acids again requires rather severe conditions, typically 6N HCl at 175-185°C in a sealed tube. In most cases, the α,α-disubstituted amino acids (**350 / 351**) which are not racemization-prone can withstand these conditions as several examples in the Table testify (72-95% yields). Certainly, various acid-labile substituents would be precluded under such conditions. An illustration of the methodology was nicely applied to the asymmetric syntheses of (R)- and (S)-α-methyldopa from **353** and **352**, respectively.

Prompted by the highly stereoselective alkylations of the imidazolidinones derived from the amino acids detailed above, these authors readily recognized the tremendous potential of preparing the corresponding optically active glycine derivatives [77]. Three separate approaches to the preparation of the requisite glycine templates (**356** and **357**) was examined: 1) resolution of the racemic substance **355** via the derived mandelate salts (Scheme 71); 2) oxidative degradation of the optically active serine system **359** (Scheme 72) via the acid **360**; and 3) oxidative degradation of the methionine-derived vinyl imidazolidinone **362** (Scheme 73). In the final analysis, the resolution (Scheme 71) proves

SCHEME 71

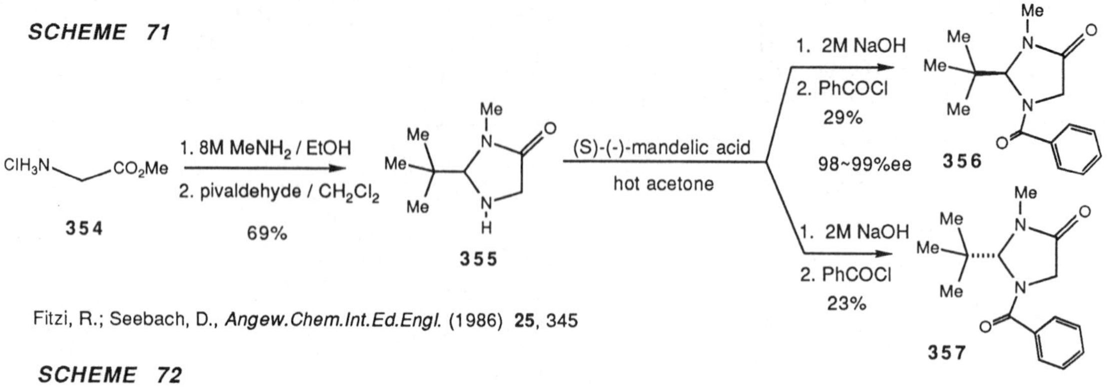

Fitzi, R.; Seebach, D., *Angew.Chem.Int.Ed.Engl.* (1986) **25**, 345

SCHEME 72

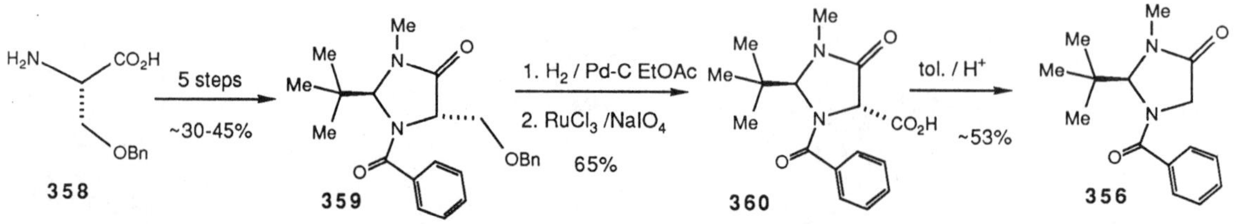

Seebach, D.; Miller, D.D.; Muller, S.; Weber, T., *Helv.Chim.Acta.* (1985) **68**, 949.

SCHEME 73

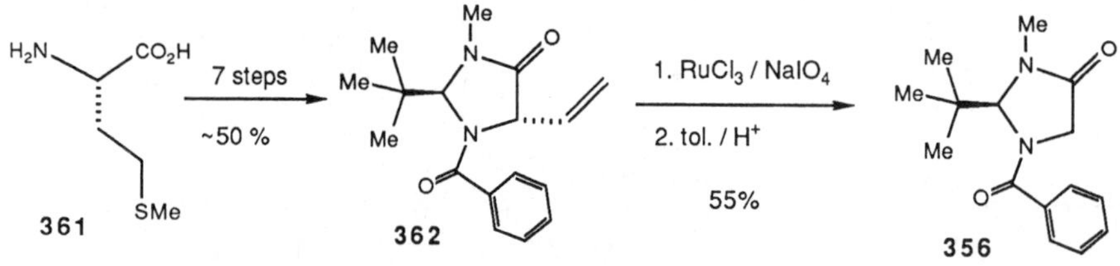

Seebach, D., *Modern Synthetic Methods* (1986) **4**, 128

to be the most practical since, both antipodes become available and the relative costs of the reagents makes this most cost-effective method. These compounds have also been resolved[78] by preparative high-pressure liquid chromatography on a chiral silica gel column. The corresponding N-CBz and N-t-BOC substances shown below (**A-D**) in each enantiomorphic form are now commercially available; they are however, more than three times as expensive than the glycine templates of Schollkopf (**10**) or of Williams (**459 / 462**).

SCHEME 74

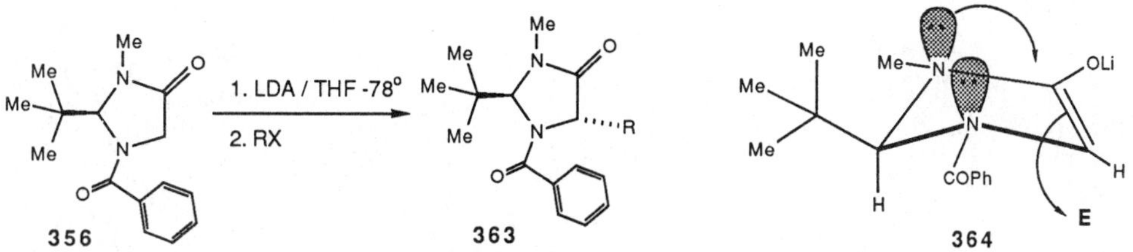

356 → 363, 364

Seebach, D.; Juaristi, E.; Miller, D.D.; Schickli, C.; Weber, T., *Helv.Chim.Acta.* (1987) **70**, 237

ELECTROPHILE	R	YIELD (%363)	%ds
MeI	-Me	90	95
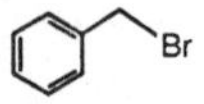 (benzyl bromide)	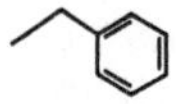 (ethylbenzene)	83	>95
n-BuI	-n-Bu	89	>95
i-PrI	-i-Pr	27	>95
acetone	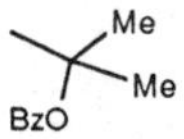 Me / Me (BzO)	89	>95

SCHEME 75

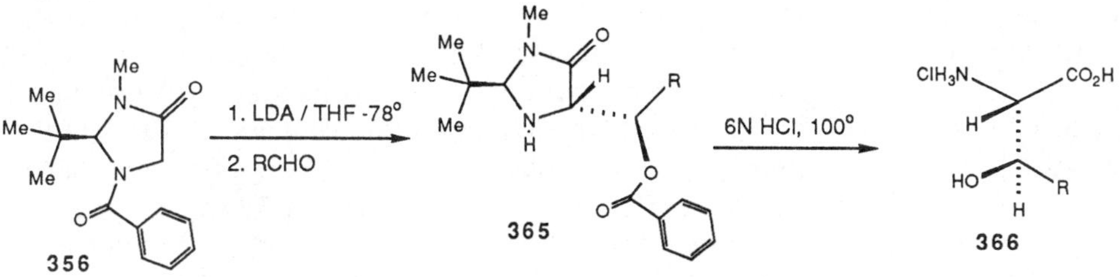

356 → 365 → 366

ALDEHYDE	YIELD (%365)	%ds	AMINO ACID (% 366)
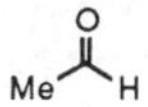 Me—CHO	75	86	>98
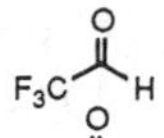 F3C—CHO	41	63	>98
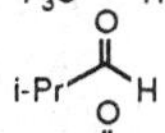 i-Pr—CHO	79	95	68
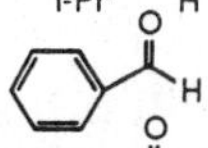 (benzaldehyde)	85	92	54
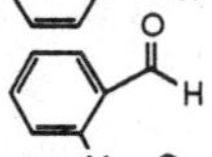 (o-methylbenzaldehyde)	81	88	
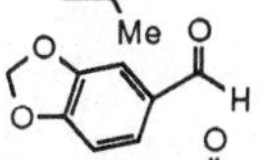 (methylenedioxybenzaldehyde)	77	96	
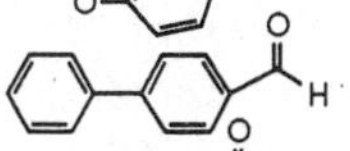 (biphenyl aldehyde)	79	93	
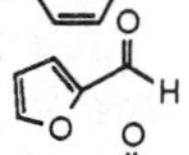 (furfural)	78	86	
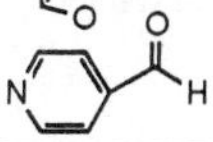 (pyridine carbaldehyde)	71	89	81

Enolate functionalization of the benzoylated glycine amide **356** proceeds with a high degree of stereoselectivity (Scheme 74)[79]. The authors propose that the enolate adopts a conformation (**364**) that places the *t*-butyl group in a pseudoequatorial position and both nitrogen atoms are pyramidal with the lone pairs pseudoaxial. The combined steric effect of the *t*-butyl group and the *anti*-stereoelectronic effect of the 'enamine' moiety would favor the observed selectivity of electrophilic attack *anti*- to the *t*-butyl group. The condensation with acetone again revealed that the benzoyl group migrates to the incipient alkoxide; a phenomena that is general with the condensations of numerous aldehydes (Scheme 75). These reactions proceed with high *threo*-selectivity furnishing the β-hydroxy-α-amino acids **366** after hydrolysis in boiling 6N HCl. A salient illustration of this methodology is reported in an approach to the unusual cyclosporine amino acid MeBmt (**202**) as shown in Scheme 76. Utilizing **356**, aldol condensation with the C_7-aldehyde furnishes **367** in very good yield and 93% ds. N-Methylation furnishes **368** which is further hydrolyzed to the N-methyl amide **369**; attempts to effect the complete hydrolysis to MeBmt were however, unsuccessful.

Aldol condensations of the methionine system **370** provides the adducts **371** in good yields and excellent diastereoselectivity (Scheme 77). When the hydrolyses of these substances are performed in 6N HCl at 150°C in a sealed tube, an unusual S-demethylation accompanies ring closure to the tetrahydrothiophenes **372**. The corresponding oxazolidinone **374** was prepared in the usual way and condensed with acetaldehyde to afford the acid-sensitive *threo*-aldol in 78% yield (Scheme 78). Raney-Nickel desulfurization (to preclude tetrahydrothiophene formation) and hydrolysis afforded the known α-ethyl-*allo*-threonine **375**.

An alternate use of the methionine-derived imidazolidinone **370** was demonstrated [80] with access to vinyl glycine derivatives as shown in Scheme 79. Oxidative elimination as described above provides **362** which is treated with base and functionalized with carbonyl derivatives and alkylating reagents (Schme 80). The ambident nature of the enolate derived from **362** is evident in the aldolizations where significant

SCHEME 76

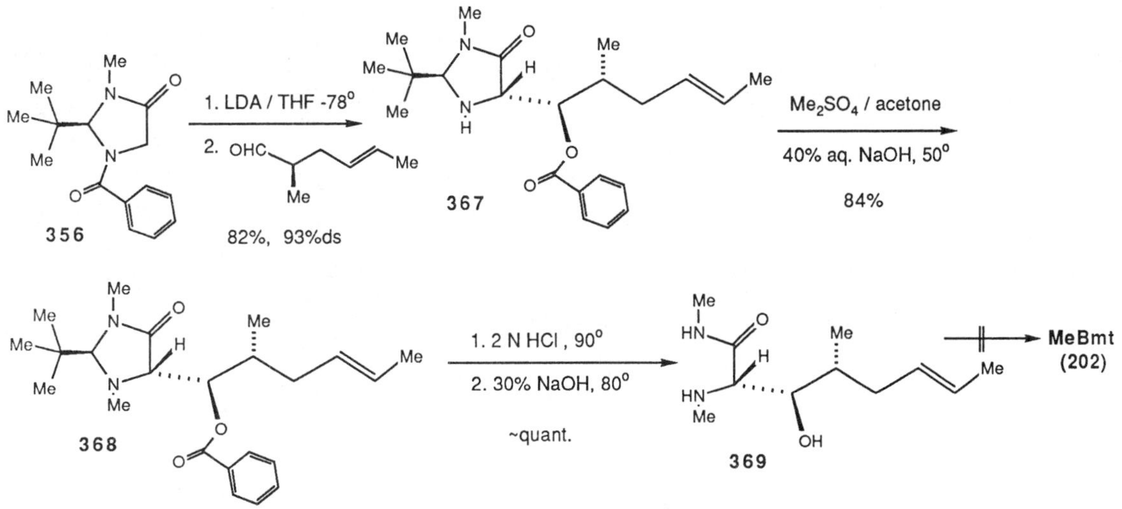

356 → 1. LDA / THF -78° 2. OHC Me → **367** 82%, 93%ds → Me₂SO₄ / acetone 40% aq. NaOH, 50° 84%

368 → 1. 2 N HCl, 90° 2. 30% NaOH, 80° ~quant. → **369** → MeBmt (202)

Seebach, D.; Juaristi, E.; Miller, D.D.; Schickli, C.; Weber, T., *Helv.Chim.Acta.* (1987) **70**, 237.

SCHEME 77

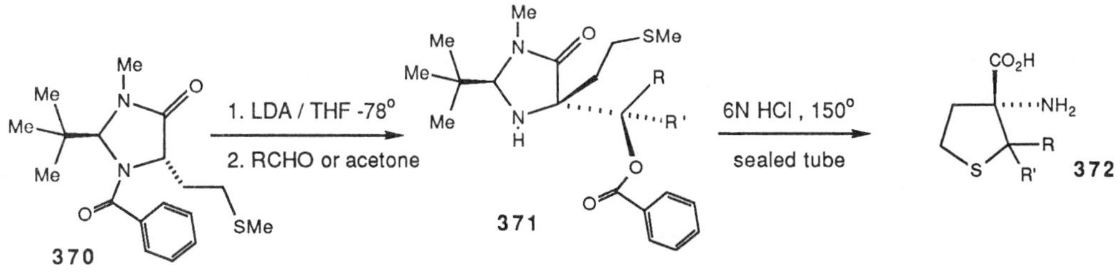

370 → 1. LDA / THF -78° 2. RCHO or acetone → **371** → 6N HCl, 150° sealed tube → **372**

ELECTROPHILE	YIELD (% 371)	%ds	AMINO ACID (% 372)
MeCHO	73	>95	30
⬡-CHO	75	>95	70
acetone	71	>95	53

SCHEME 78

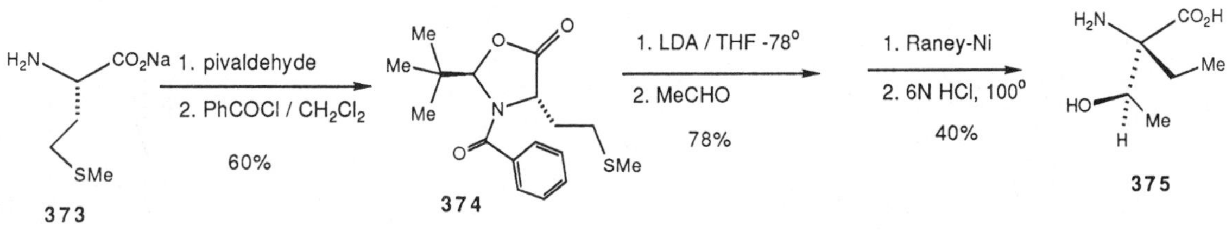

373 → 1. pivaldehyde 2. PhCOCl / CH₂Cl₂ 60% → **374** → 1. LDA / THF -78° 2. MeCHO 78% → 1. Raney-Ni 2. 6N HCl, 100° 40% → **375**

SCHEME 79

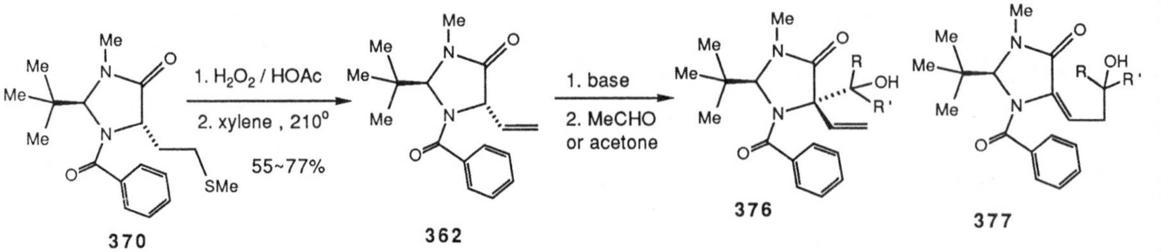

Seebach, D.; Juaristi, E.; Miller, D.D.; Schickli, C.; Weber, T., *Helv.Chim.Acta.* (1987) **70**, 237.

ELECTROPHILE	BASE	RATIO (376 : 377)	YIELD%	%ds
acetone	LDA	5 : 95	35	>95
MeCHO	LDA	3 : 2	40	>90
MeCHO	LDA / MgBr$_2$	>95 : 5	55	>90
MeCHO	LDA / Ti(NMe$_2$)$_4$	>95 : 5	48	>90

SCHEME 80

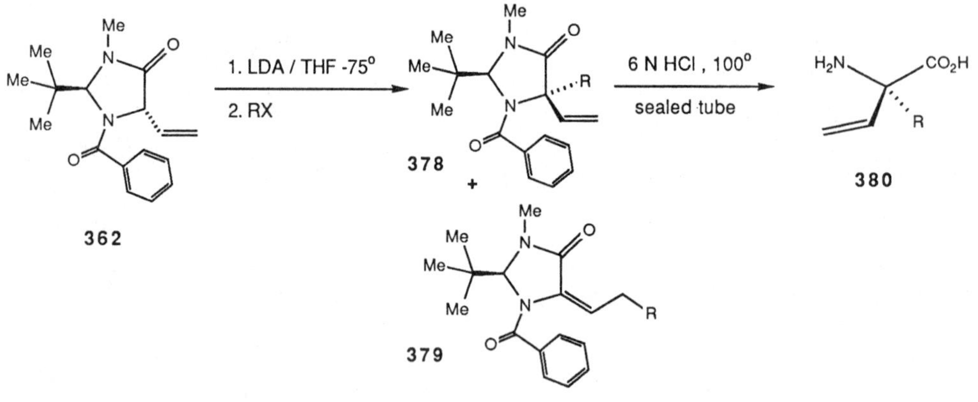

Weber, T.; Aeschimann, R.; Maetzke, T.; Seebach, D., *Helv.Chim.Acta.* (1986) **69**, 1365

ELECTROPHILE	RATIO(378 : 379)	YIELD %	%ds	AMINO ACID (% 380)
MeI	>95 : 5	61	89	37
EtI	>95 : 5	62	90	63
⟍—Br	>95 : 5	77	79	
⬡—Br	>95 : 5	75	87	

SCHEME 81

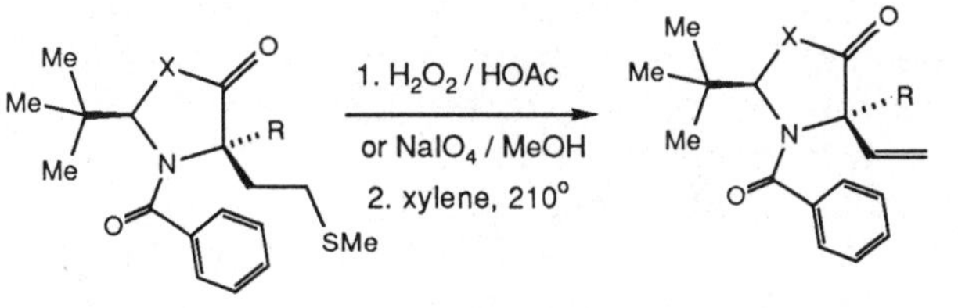

X	R	YIELD (% 382)
-NMe	Me	62
-NMe	Et	75
O	H	13(2S,4R) 26(2S,4S)
O	Me	75

Weber, T.; Aeschimann, R.; Maetzke, T.; Seebach, D., *Helv.Chim.Acta.* (1986) **69**, 1365.

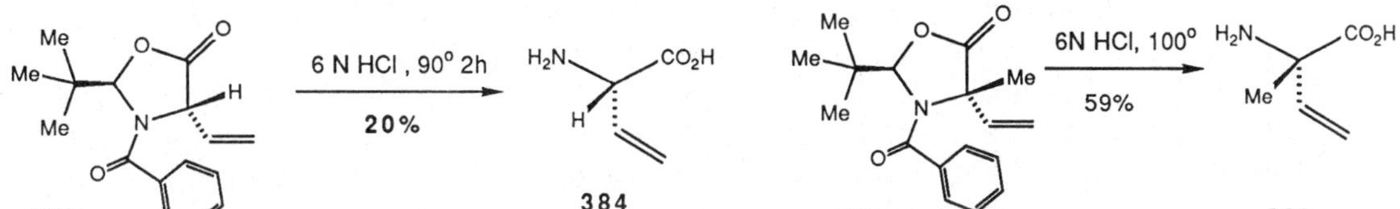

SCHEME 82

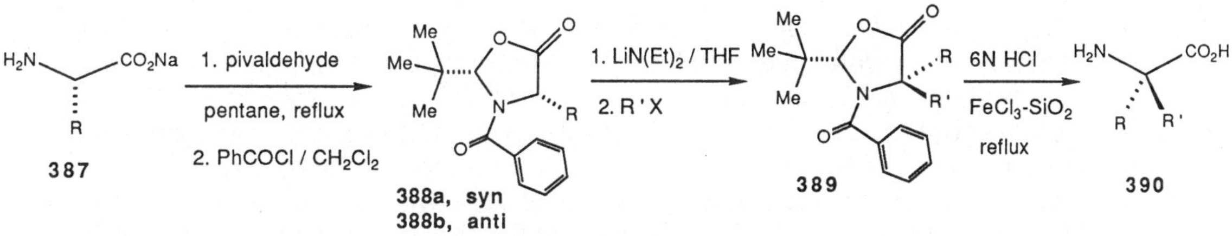

Seebach, D.; Fadel, A., *Helv.Chim.Acta.* (1985) **68**, 1243.

R	YIELD(% 388)	ELECTROPHILE	YIELD(% 389)	%ds	AMINO ACID (% 390)
Me -	69(syn) 11.5(anti)	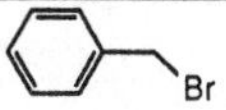 ⬡—Br	93	>98	95
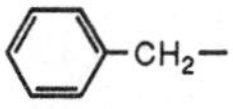 ⬡—CH₂—	92%(syn)	MeI 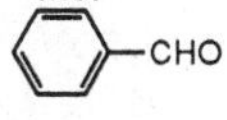 ⬡—CHO	88 96	>98 >95	
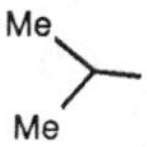 Me₂CH (isopropyl)	67.5(syn) 17.3(anti)	MeI MeI 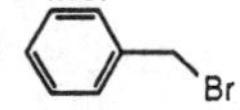 ⬡—Br	53 75 40	>99 >85 >93	
MeSCH₂CH₂-	70(syn) ~14(anti)	MeI	69	>95	

amounts of the 1,4-adducts (**377**) are produced under the standard conditions (LDA/THF). The authors found that the corresponding magnesium and titanium enolates were much more selective for the desired 1,2-adducts **376**. The alkylations on the other hand, were very selective for the 1,2-adducts **378** with ratios of >95:5 for **378:379** being typical. The diastereoselectivity of these reactions was not as high as related cases discussed above and may reflect a departure from conformation **364** due to more extensive delocalization of this enolate. Hydrolysis of the adducts **378** in hot 6N HCl provided the α-alkylated vinyl glycines **380**.

An alternate, related   strategy[80] to access the α-alkylated vinyl glycine derivatives was developed that simply involves a shift in the timing of the oxidative elimination (Scheme 81). The α-alkylated methionine derivatives **381** can be oxidized and heated to afford the corresponding vinylic substances **382**. Hydrolysis of the unsubstituted oxazolidinone (**383**) to vinyl glycine itself (**384**) was accomplished, albeit in low yield; little, if any racemization accompanies this transformation but an exact %ee was not reported. The α-methyl vinylglycine (**386**) is produced in good yield under the same conditions.

A full account[81] describing the preparation of the previously elusive oxazolidinones was published in 1985 and embraced the optically active oxazolidinones of alanine, phenylalanine, valine and methionine (Scheme 82). The general protocol involves reaction of the sodium salt of the free amino acid with pivaldehyde to furnish the corresponding imine. The imine is then cyclized in the presence of the acylating reagent (benzoyl chloride in the present case) to give a *syn/anti* mixture of the desired oxazolidinones **388**. In all cases the *syn*-isomer (**388a**) is the major product that must be separated from the minor *anti*-isomer (**388b**). It was found that LDA was not a suitable base for deprotonation of **388**; curiously, LDEA (lithium diethyl amide) proved to give very good results in enolate generation and derivatization. The α-functionalized substances **389** are hydrolyzed under somewhat milder conditions (6N HCl, reflux temperature in the presence of $FeCl_3$-SiO_2) than the imidazolidinones discussed above.

During the extensive development of this concept by Seebach, et.al., a Merck group[82] reported on   related oxazolidinones[83] derived from benzaldehyde or 2,4-dichlorobenzaldehyde and phenylalanine or alanine as shown in Scheme 83. In all cases, the *syn*-isomers (**392a/ 401**) were formed as the major products. Enolate formation with potassium hexamethyldisilyl amide and alkylation with either methyl iodide or benzyl bromide provided the adducts **396, 397** and **402**, respectively. In all cases the electrophile approached from the face *anti*- to the aromatic acetal moiety. The α-methyl phenylalanine derivatives **398 / 399** are

SCHEME 83

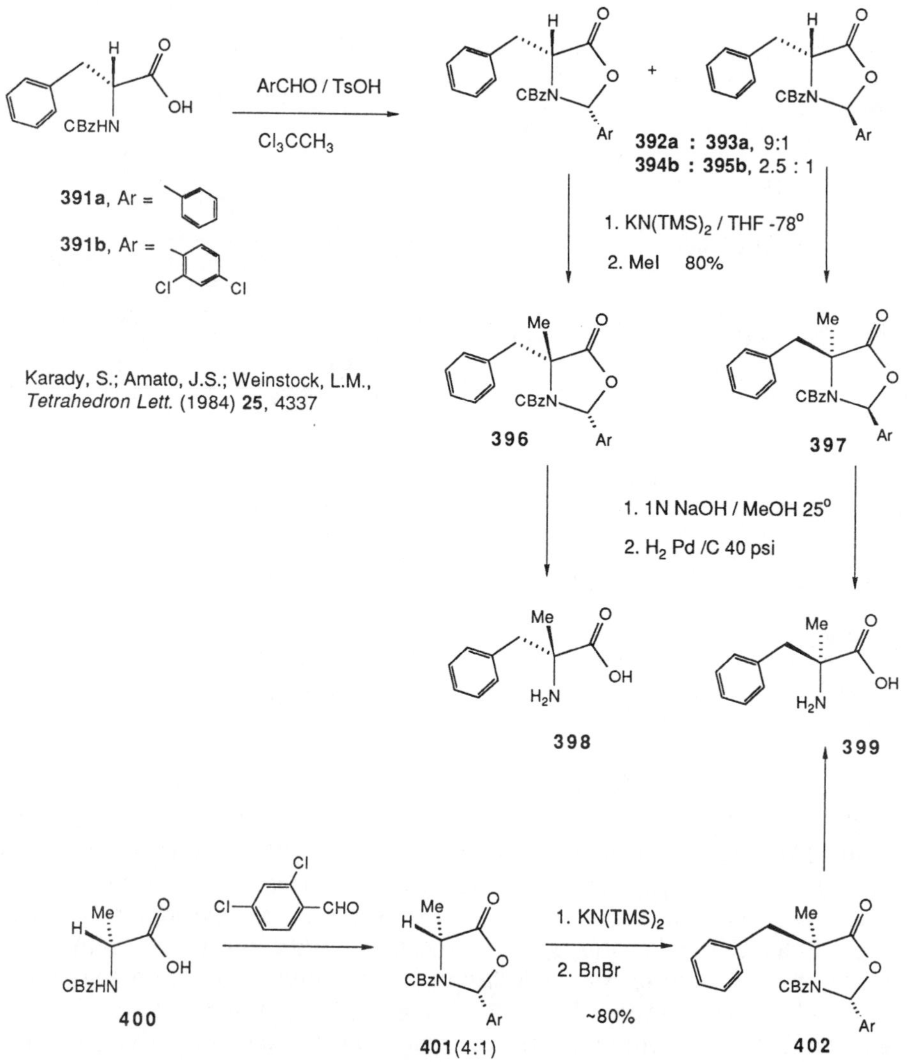

391a, Ar =

391b, Ar =

Karady, S.; Amato, J.S.; Weinstock, L.M.,
Tetrahedron Lett. (1984) **25**, 4337

392a : 393a, 9:1
394b : 395b, 2.5 : 1

1. KN(TMS)$_2$ / THF -78°
2. MeI   80%

396   **397**

1. 1N NaOH / MeOH 25°
2. H$_2$ Pd /C 40 psi

398   **399**

1. KN(TMS)$_2$
2. BnBr
~80%

400   **401**(4:1)   **402**

produced by treatment with methanolic sodium hydroxide and catalytic hydrogenation to furnish the amino acids. This mild deblocking protocol offers several complimentary advantages to the more drastic acidic hydrolysis of the pivaldehyde acetals. It seems reasonable that the direct hydrogenolysis of **396, 397** or **402** would yield the corresponding N-benzyl amino acids[83], which should be further reducible to the free amino acids; no comment on this attractive possibility is made. Full experimental details have not yet been published for this potentially very useful method.

SCHEME 84

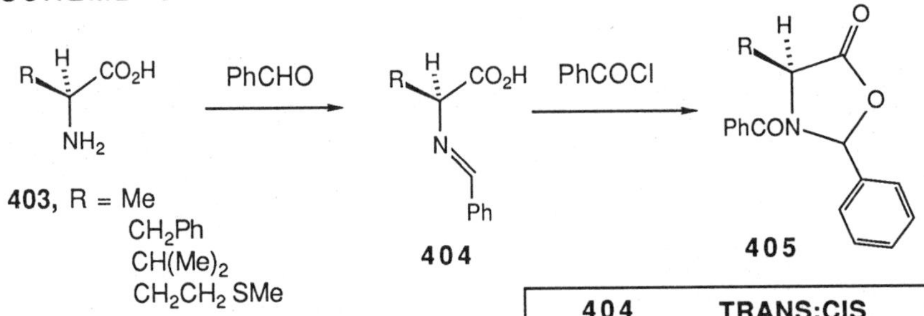

403, R = Me
       CH$_2$Ph
       CH(Me)$_2$
       CH$_2$CH$_2$SMe

Fadel, A.; Salaun, J.,
Tetrahedron Lett. (1987) **28**, 2243.

404	TRANS:CIS	405(%)
R = Me	7.5:1	94%
CH$_2$Ph	3:1	88%
CH(Me)$_2$	4.5:1	92%
CH$_2$CH$_2$SMe	4:1	93%

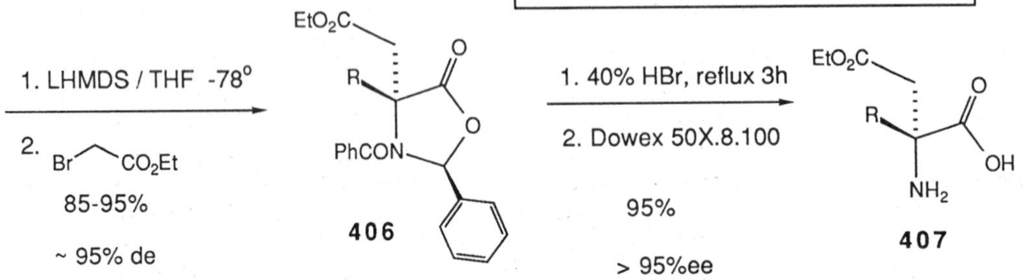

1. LHMDS / THF -78°

2. Br⌒CO$_2$Et

85-95%

~ 95% de

1. 40% HBr, reflux 3h

2. Dowex 50X.8.100

95%

> 95%ee

Fadel and Salaun [84] extended this strategy to embrace the phenyl oxazolidinones (**405**) of alanine, phenylalanine, valine and methionine (Scheme 84). By pre-forming the benzylidine (**404**) and effecting the concomitant acylation/cyclization, the *anti-* isomers (**405**) are formed as the major products; this contrasts to the pivaldehyde examples shown in Scheme 82 which prefer the *syn-* isomers. Enolate formation with lithium hexamethyldisilyl amide and alkylation with ethyl bromoacetate affords the *anti*-alkylated adducts **406** in excellent yields and excellent diastereoselectivity. Removal of both the acetal and benzoyl groups is accomplished with 40% HBr at reflux. The α-alkylated β-ethyl aspartates **407** are obtained in >95% ee.

Very recently, direct extension of the Seebach system to the difficult basic amino acids lysine, ornithine and tryptophan was reported (Schemes 85-87). The ω-N-CBz methyl esters of lysine (**408**) or ornithine (**409**) are converted into the N-methyl imidazolidinones **410** in the usual way. Due to the additional acidic proton of the urethane, these substances are converted into their *dianions* by treatment with two equivalents of

SCHEME 85

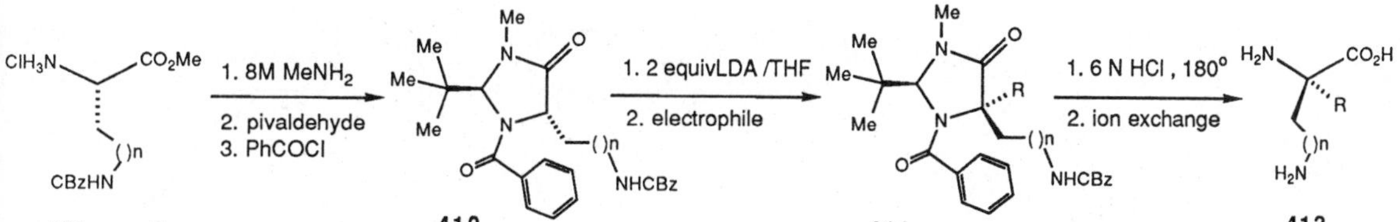

408, n = 3
409, n = 2

Gander-Coquoz, M.; Seebach, D., *Helv.Chim.Acta* (1988) **71**, 224.

n	ELECTROPHILE	YIELD (% 411)	%ds	AMINO ACID (% 412)
4	MeI	59	93	66.3
4	(benzyl bromide) Br	51	85	83
4	MeCHO	51	>80	
4	(phenyl)—CHO	46	85	
3	MeI	34	90	~quant.

SCHEME 86

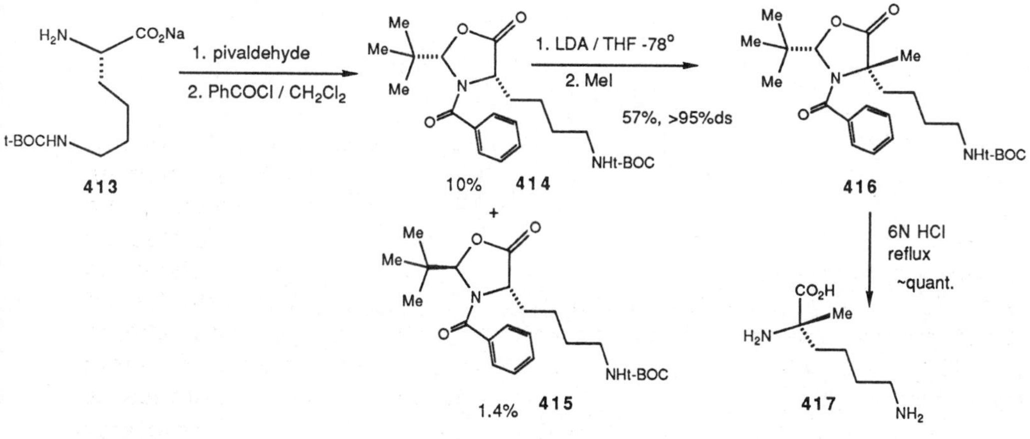

413

10% 414
+
1.4% 415

416

6N HCl
reflux

~quant.

417

LDA and then quenched with an electrophile. The diastereoselectivities range from >80- 93% ds and the yields are moderate. Cleavage of the N-CBz group and the heterocycle are achieved with 6N HCl at 180°C in a sealed tube, furnishing the α-functionalized amino acids **412**. The corresponding oxazolidinone of lysine (**414**, Scheme 86) is prepared from

SCHEME 87

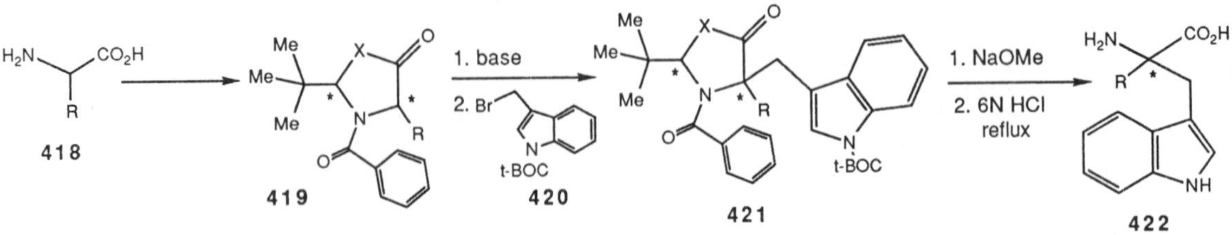

Gander-Coquoz, M.; Seebach, D., *Helv.Chim.Acta* (1988)**71**, 224.

R	X	CONFIG.(419)	BASE	YIELD (%421)	%ds	AMINO ACID (% 422)
H	NMe	2S	LDA	69	>90	
Me	NMe	2R,5S	LDA	63	>95	
H	O	rac.	LDEA	34	>95	
Me	O	2R,4S	LDA	35		
Me	O	2R,4S	LDA/HMPT	65		87
MeSCH₂CH₂	O	2R,4S	LDA	12		

SCHEME 88

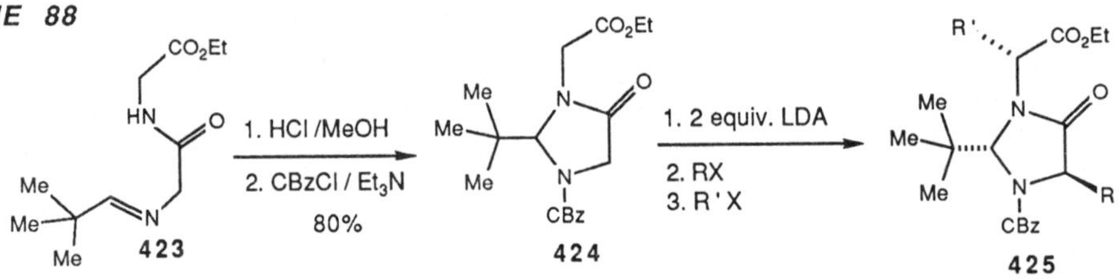

Polt, R.; Seebach, D., *Helv.Chim.Acta* (1987) **70**, 1930.

RX	R'X	YIELD (% 425)	Diastereomeric ratio
MeI	HOH	77	50 : 1
MeI	MeI	81	7 : 1
EtI	HOH	61	50 : 1
EtI	EtI	47	11 : 1
PhCH₂Br	MeI	75	8 : 1
PhCH₂Br	PhCH₂Br	87	25 : 1

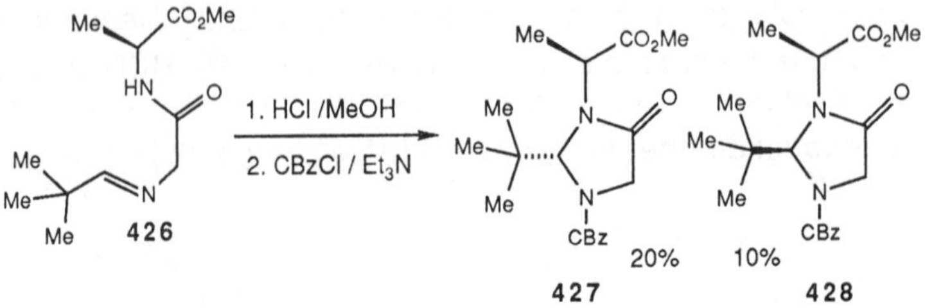

the ε-N-t-BOC lysine (**413**) and alkylated as above with methyl iodide in good yield and >95% ds to afford, after acidic deprotection, α-methyllysine **417**, a known enzyme inhibitor.

The tryptophan derivatives (**422**, Scheme 87) are accessible by alkylation of various oxazolidinones or imidazolidinones (**419**) with the labile 3-bromomethyl-N-t-BOC indole **420** to afford the adducts **421**. Base treatment followed by hydrolysis in boiling 6N HCl provides the optically active derivatized tryptophan derivatives **422** (only the α-methyl case of **422** is reported).

In a very interesting, preliminary study, Seebach and Polt[86] have extended these useful heterocycles to a direct asymmetric synthesis of dipeptides as shown in Schemes 88 and 89. Gly-Gly ethyl ester is condensed with pivaldehyde to furnish **423**. Treatment with methanolic HCl followed by concomitant acylation/cyclization with benzyl chloroformate provides the racemic imidazolidinone **424** in 80% yield. Treatment of this substance with two equivalents of LDA followed by sequential addition of two electrophilic species reveals that the dianion alkylates sequentially in the imidazolidinone ring and then the exocyclic position. The diastereoselectivity of this process is surprisingly high giving ratios up to 50:1. The authors note that treatment of **424** with one molar equivalent of LDA followed by methyl iodide quench gives rise to the same dimethylated product (**425**, R=R'= Me) and unreacted **424**. Thus the dianion seems to be favored over the monoanion under normal conditions using LDA. The diastereoselectivity of alkylation at the imidazolidinone moiety is virtually 100% ds; the diastereomer ratios refer to the epimeric ratios at the exocyclic position. Encouraged by these stereoselective alkylations, the Gly-Ala imine **426** was converted into the imidazolidinones **427** / **428** and the Ala-Gly imine **429** was converted into **430** / **431**; the yields in each case were modest. An elegant demonstration of stereoselective alkylations is detailed where the (+)- and (-)- dimethyl alkylation products (**432**) are produced from (+)-**427** and (-)-**430**, respectively. It is fascinating that the alkylation of (+)-**427** *inverts* the stereogenic center at the exocyclic position (7:1 ratio). Hydrolysis of the methyl ester furnished the acid **433** that could either be hydrogenated and hydrolyzed to D-Ala-D-Ala (**434**) or condensed with L-Ala and processed to the tripeptide **436**. It seems obvious that great potential exists in this methodology to access a wide variety of oligopeptide problems; forthcoming reports on these extensions are anticipated from the Zurich laboratories.

The extensive use of stereoselective enolate functionalizations of the Seebach heterocycles provides a useful and practical approach to the asymmetric synthesis of non-proteinogenic amino acids. The recent

SCHEME 89

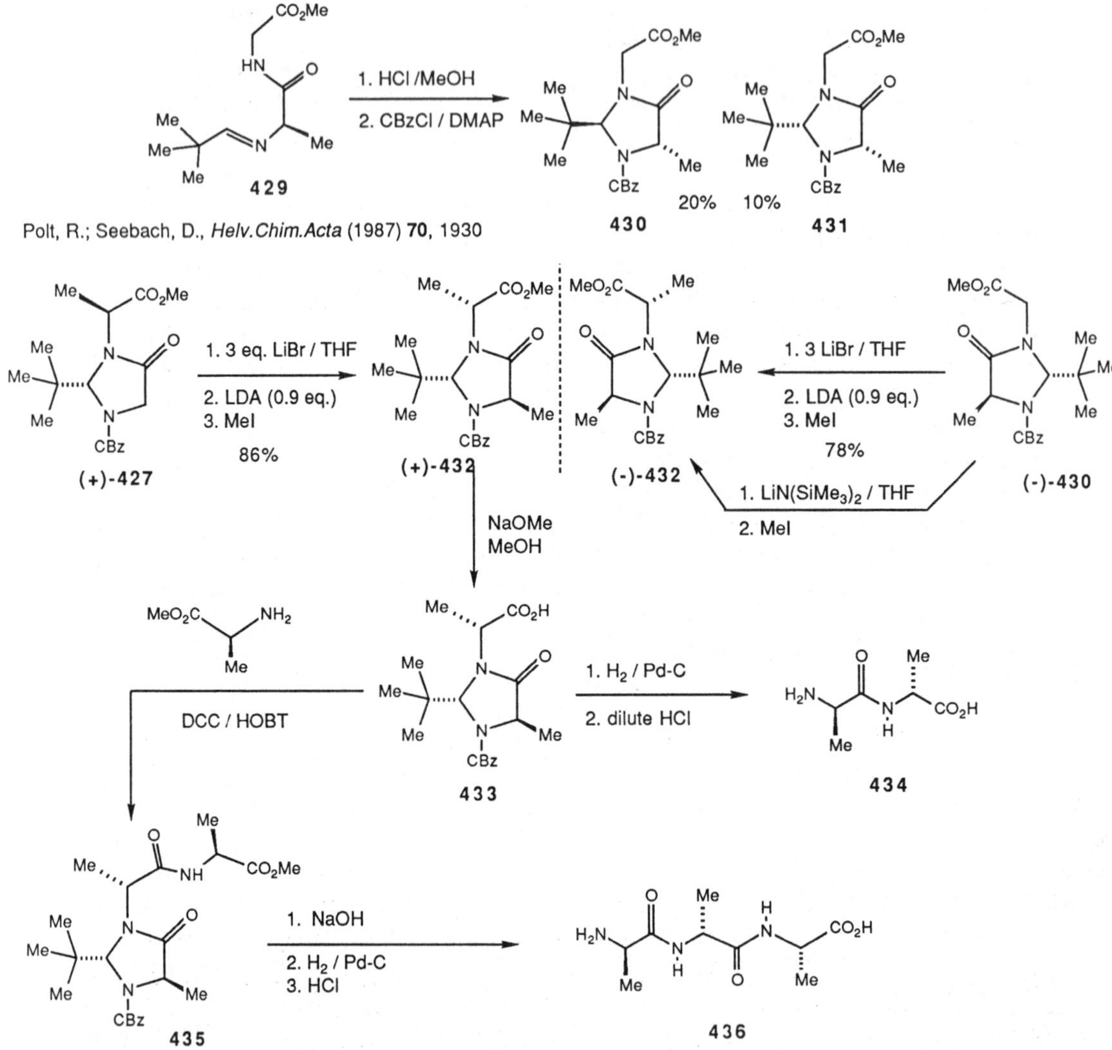

Polt, R.; Seebach, D., *Helv.Chim.Acta* (1987) **70**, 1930

commercial availability of two pairs of enantiomorphic imidazolidinones, the thorough reporting of full experimental detail and the overall simplicity of the method will undoubtedly encourage many chemists to try their hand at this chemistry. The only inherent limitation that appears obvious are the somewhat harsh acidic conditions required for the final deblocking to the final amino acids. For α-R groups that are inherently acid stable, this method should become particularly useful for preparing α,α-disubstituted amino acids.

4. Achiral Glycine Enolates

In the context of the aldol methodology touched on above, two syntheses of MeBmt, the unusual amino acid constituent of cyclosporine are detailed in Schemes 90 and 91. As shown in Scheme 90, Rich[87] and co-workers condensed the achiral glycinate **437** with the racemic aldehyde **438**. The authors note that the usual difficulties associated with the N-methylation of a derivatized amino acid are obviated by the employment of **437**. A nearly 1:1 mixture of cyclic urethanes are produced resulting from incipient alkoxide attack on the t-BOC group. The acids **439 / 440** are then separated and resolved by the use of (+)-ephidrine to afford **202** (MeBmt). Several other MeBmt analogs (**441**) were prepared by the same protocol; full experimental details accompany this work.

In an analogous approach that concomitantly appeared, Schmidt and Siegel [88] prepared the optically active aldehyde **440** by the use of the Helmchen system **442**. Aldol condensation of **444** with N,N,O-*tris*-(trimethylsilyl)glycine (**445**) provided a 75:25 mixture of **446 : 447** which were separated by medium pressure chromatography. The major isomer was shown to possess the correct relative stereochemistry at the α-carbon but the incorrect stereochemistry at the β-carbon which was

SCHEME 90

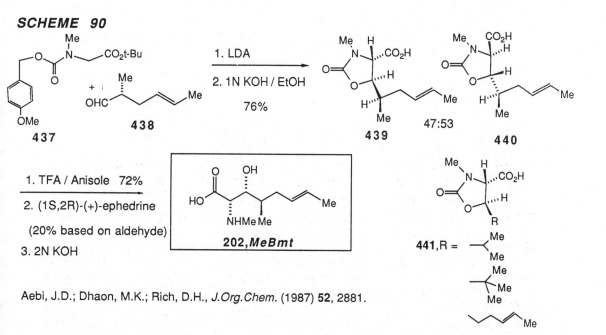

Aebi, J.D.; Dhaon, M.K.; Rich, D.H., *J.Org.Chem.* (1987) **52**, 2881.

Asymmetric Derivatization of Glycine

SCHEME 91

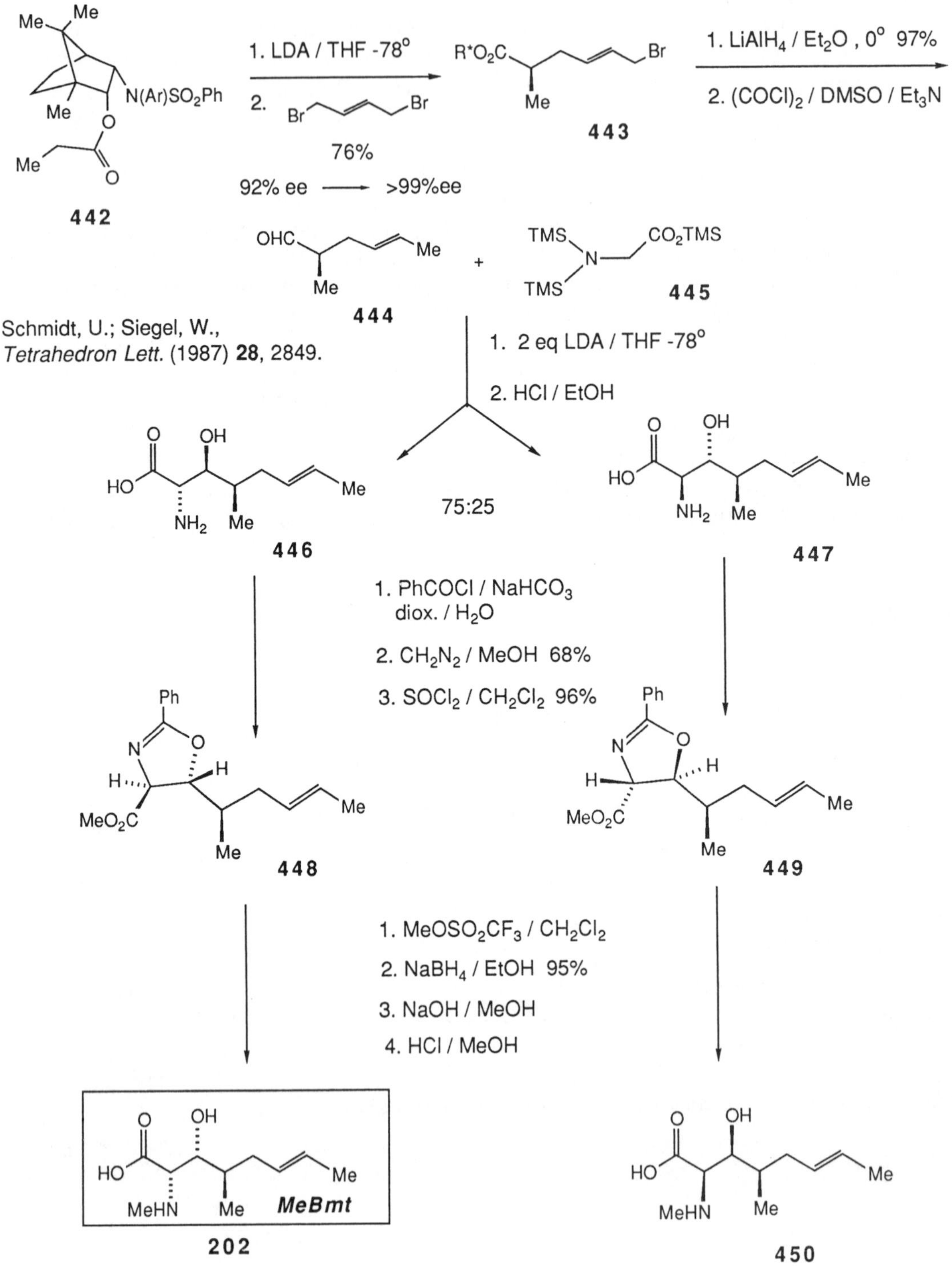

1. LDA / THF -78°

2. Br~~~~~Br
   76%
   92% ee ⟶ >99%ee

442

1. LiAlH$_4$ / Et$_2$O , 0° 97%

2. (COCl)$_2$ / DMSO / Et$_3$N

443

444 + **445**

Schmidt, U.; Siegel, W.,
Tetrahedron Lett. (1987) **28**, 2849.

1. 2 eq LDA / THF -78°

2. HCl / EtOH

75:25

446   **447**

1. PhCOCl / NaHCO$_3$
   diox. / H$_2$O

2. CH$_2$N$_2$ / MeOH 68%

3. SOCl$_2$ / CH$_2$Cl$_2$ 96%

448   **449**

1. MeOSO$_2$CF$_3$ / CH$_2$Cl$_2$

2. NaBH$_4$ / EtOH 95%

3. NaOH / MeOH

4. HCl / MeOH

MeBmt

202   **450**

subsequently inverted through the oxazoline **448**. N-Methylation with 'magic methyl' followed by reduction and hydrolysis of the oxazoline provided **202** (MeBmt); the diastereomer **450** was produced by the same procedure. The above two examples serve to illustrate that access to optically active amino acids via achiral glycine condensations with either an optically active homologation reagent to effect kinetic resolution or by classical resolution techniques should still be seriously considered in the repertoire of the synthetic amino acid chemist.

5. Enolate Derivatization of Optically Active Oxazinones.

A landmark paper on asymmetric synthesis by Kagan [89] and associates in 1968 provided a solid framework from which to design glycinates based on optically active 2,3,5,6-tetrahydro-4H-oxazin-2-ones (termed oxazinones herein for brevity); this contribution will be discussed in the chapter dealing with hydrogenations. The key feature of this work is the placement of phenyl substituents benzylic to the amino acid nitrogen and oxygen atoms that permit a reductive cleavage of the heterocycle under mild conditions. Two very recent *enolate* applications of this heterocyclic system have appeared in a preliminary form.

Dellaria and Santarsiero [90] have prepared the 5-phenyl-2,3,5,6-tetrahydro-4H-oxazin-2-ones **452-454** from phenylglycinol [91] as shown in Scheme 92. The labile system **452** must be immediately acylated or benzylated to preclude piperazinedione formation. The authors found that the base and solvent were critical in effecting efficient enolate generation and alkylation. It was found that sodium hexamethyldisilyl amide in THF/DME or DME gave good to excellent yields of alkylation adducts **455** with the kinetic diastereoselectivities typically in excess of 200 : 1. These workers also demonstrated that the *anti-* isomers (**455**) are thermodynamically more stable than the *syn-* isomers, but the equilibrium ratios were only between 1.3~3.4 : 1. These observations strongly indicate that the enolate derived from **453** is selectively attacked from the more accessible face *anti-* to the C-5 phenyl substituent. Curiously, the N-benzyl substrate affords the *opposite* sense of diastereoselectivity furnishing the *syn-* product **458** as the major component (*syn : anti* = 17 : 1). Conversion of **455** to the corresponding amino acid ethyl esters is accomplished by acidic hydrolysis of the lactone to the ethyl ester-HCl **456** and subsequent catalytic hydrogenation to the ethyl ester-HCl **457**. The % ee's of **457** exceed 99%.

Williams and Im [92] examined the enolate alkylation of the 5,6-diphenyl-2,3,5,6-tetrahydro-4H-oxazin-2-ones **459** and **462** as shown in Scheme 93. These systems have been much more extensively examined as *electrophilic* glycinates and their preparation will be detailed thoroughly

Asymmetric Derivatization of Glycine

SCHEME 92

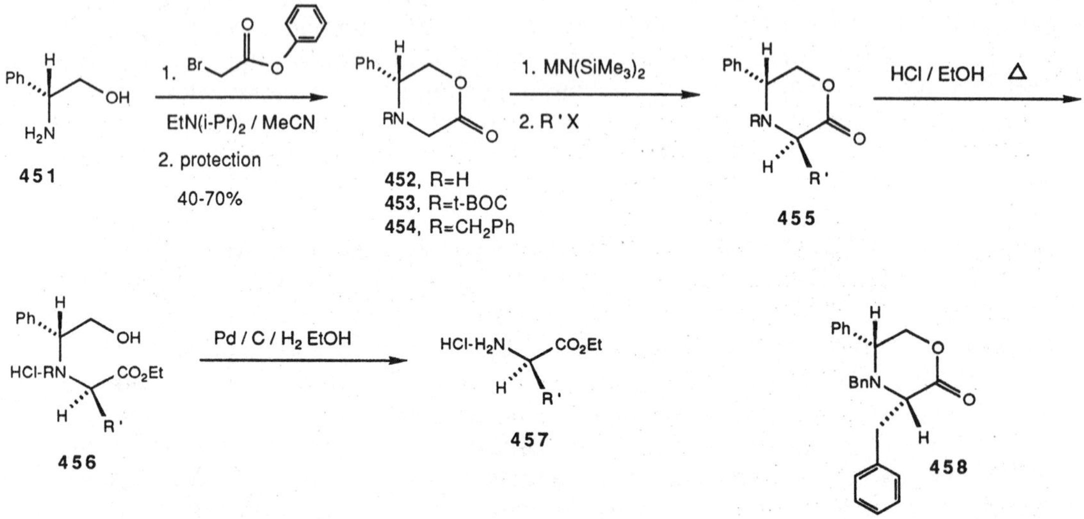

451 → 1. (phenyl bromoacetate), EtN(i-Pr)$_2$ / MeCN 2. protection 40-70% → 452, R=H; 453, R=t-BOC; 454, R=CH$_2$Ph → 1. MN(SiMe$_3$)$_2$ 2. R'X → 455 → HCl / EtOH Δ

456 → Pd / C / H$_2$ EtOH → 457

458

Dellaria, J.F.; Santarsiero, B.D, *Tetrahedron Lett.* (1988) **29**, 6079.

R'X	R	BASE / SOLVENT	YIELD (455)	KINETIC RATIO	AMINO ACID %
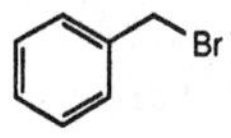 (benzyl bromide)	t-BOC	NaN(SiMe$_3$)$_2$ / THF : DME (1:1)	78	>200 : 1	61
MeI			85	>200 : 1	
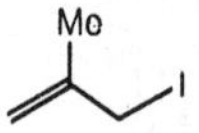 (2-methylallyl iodide)			86	>200 : 1	
 (allyl bromide)			90	>200 : 1	
n-butyl iodide			0	-	
n-butyl iodide		NaN(SiMe$_3$)$_2$ / DME	78	83 : 1	92
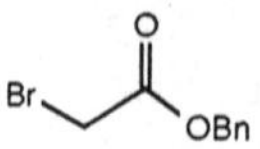 (benzyl bromoacetate)			59	3.5 : 1	
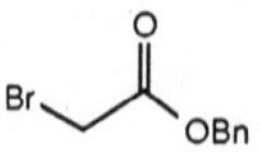 (benzyl bromoacetate)		LiN(SiMe$_3$)$_2$ / THF	65	18.5 : 1	

SCHEME 93

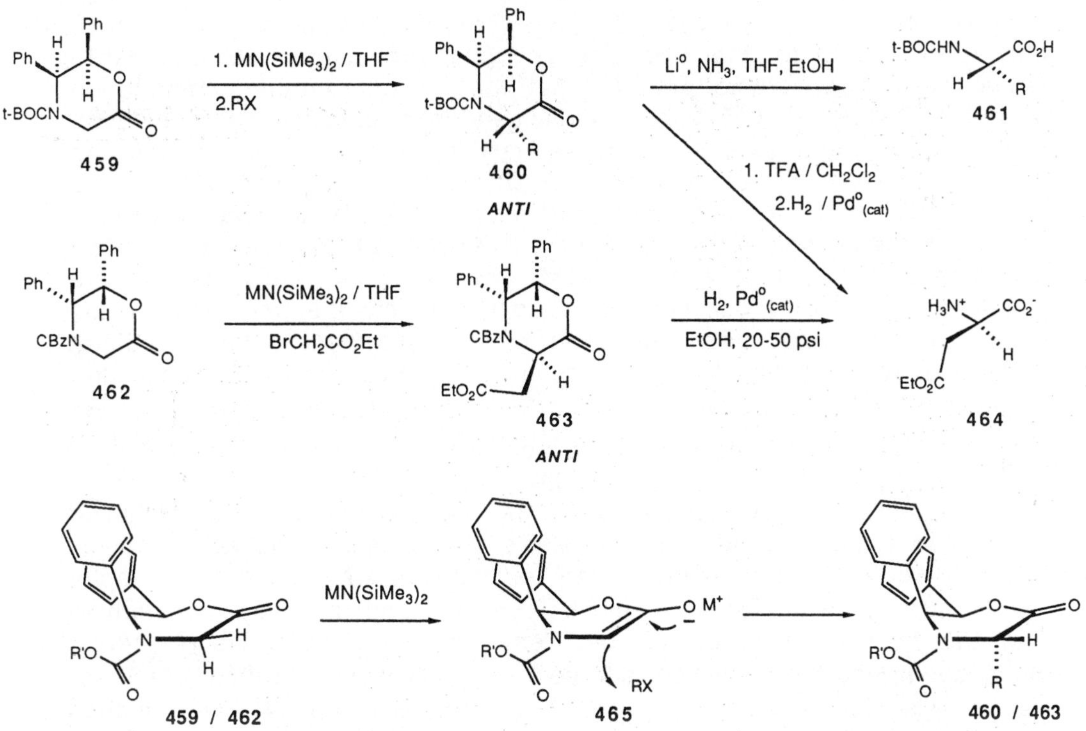

Williams, R.M.; Im, M-N., *Tetrahedron Lett.* (1988) **29**, 6075

LACTONE SUBSTRATE / YIELD		RX	AMINO ACID DERIVATIVE YIELD		%ee
459	462		461	464	
91		MeI	54		97.2
48 (60)		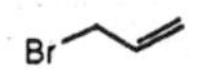 Br⌇⟍⟍ (allyl)	50-70		98
68 (85)		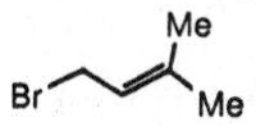 Br⌇⟍⟍=C(Me)Me	52		100
70 (77)		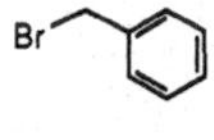 Br⌇benzyl		76	98.2
	61 (77)	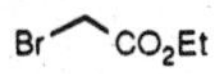 Br⌇CO₂Et		71	95.9
56		 I⌇⟍⟍=			
48		I⌇⟍⟍⟍I			
42		I⌇⟍⟍⟍OBn			
68		Br⌇⟍≡SiMe₃			

in the next section. As in the above mono-phenyl system, the enolates derived from these heterocycles display excellent selectivity for kinetic alkylation *anti-* to the two phenyl rings providing **460** and **463**, respectively. The authors propose that the enolate adopts conformation **465** that places the C-5 phenyl substituent in a pseudo-axial orientation leaving the bottom face sterically accessible to electrophilic attack. As will be discussed in depth in the next section, the oxazinones **460** can be directly converted into the correspnding N-t-BOC amino acids (**461**) by dissolving metal reduction; alternatively, the BOC group can be removed and then catalytic hydrogenation affords the free amino acids (**464**). The N-CBz adducts (**463**) can be directly converted into the free amino acids by either dissolving metal reduction or catalytic hydrogenation. Since these oxazinones incorporate the 'second' phenyl ring originally used in the Kagan[89] system, this method has the significant advantage over the monophenyl systems (**453**) of being able to directly access the free, or t-BOC-protected amino acids without the additional hydrolytic manipulation of the lactone. As shown in the Table, the alkylation products are obtained in good to excellent yields and the %ee's of the final amino acid products are ~96-100%. In a very preliminary example, it was found that the 3'-iodopropane product **466** (Scheme 94) could be quantitatively transformed into the bicyclic system **467** by treatment with trimethylsilyl iodide. Subsequent enolate alkylation afforded the α-alkylated proline precursor **468** in 76% yield and >99% de (see **314 / 315**, Scheme 64). This method would also seem to hold promise for preparing α,α-disubstituted amino acids as the preliminary example **469** to **470** illustrates.

SCHEME 94

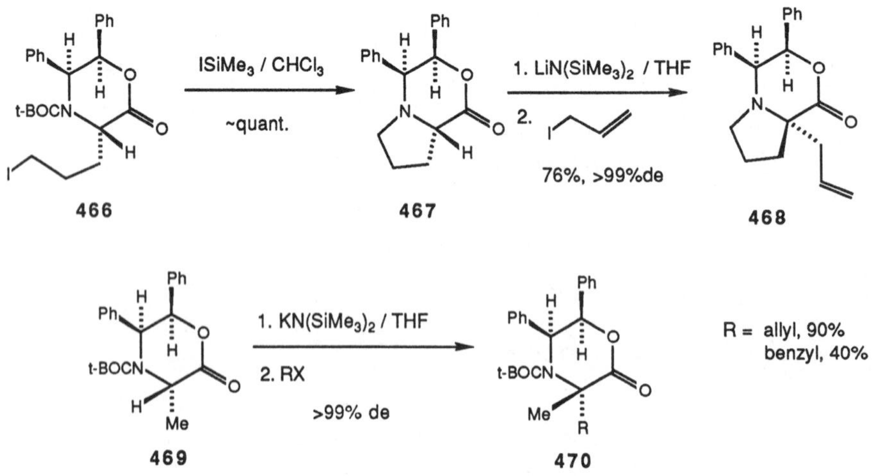

6. Enolate Derivatization of Transition Metal Complexes

An interesting asymmetric synthesis of *erythro*-β-hydroxy-L-histidine (**476**) , a constituent of bleomycin, was recently reported by Ohno, et. al.[93]; the method derives from extensive related work by Belokon, et.al., described below. The optically active copper complex **472** is prepared from the Gly-D-Phe peptide **471**, pyruvic acid and copper acetate. Reaction of this complex with aldehyde **473** in water at room temperature afforded a 3 : 1 mixture (50% ee) of the desired substance **476** along with the enantiomorph in 49% yield. Co-crystallization of this material with D-tartaric acid afforded optically pure **476**. The authors postulate that the phenyl ring of the phenylalnine moiety shields one enantiotopic face of the Gly residue . The relative stereochemistry of the aldol condensation is controlled by interaction of the developing negative charge on the aldehyde oxygen with the pyruvimine moiety (**474**) to afford the aldol adduct **475**. Addition of H_2S precipitates the copper; the peptide is hydrolyzed and the amino acids are separated by chromatography. Presumably, this method is limited to electrophiles that can coordinate to the copper as in the present case.

SCHEME 95

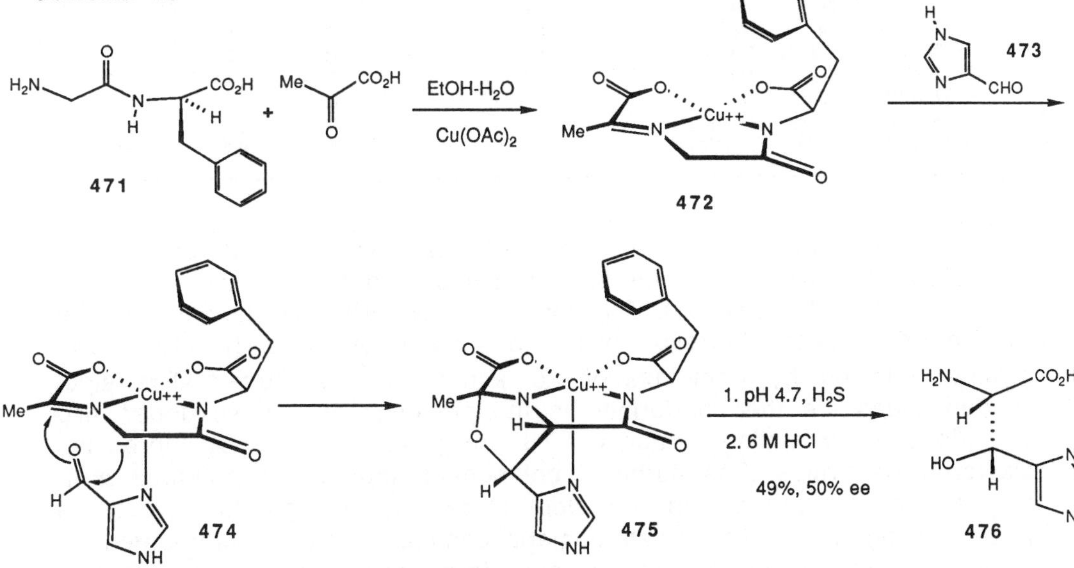

Owa, T.; Otsuka, M.; Ohno, M., *Chem. Lett.* (1988) 83.

SCHEME 96

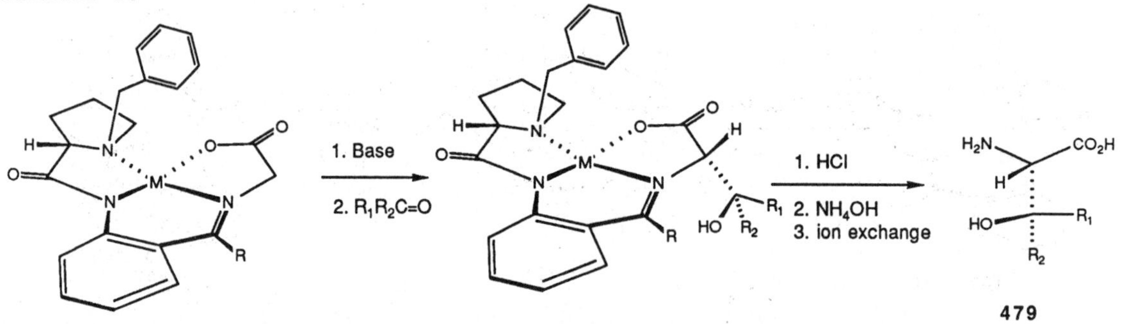

Belokon, Y.N.; Bulychev, A.G.; Vitt, S.V.; Struchkov, Y.T.; Batsanov, A.S.; Timfeeva, T.V.; Tsryapkin, V.A.; Ryzhov, M.G.; Lysova, L.A.; Bakhmutov, V.I.; Belikov, V.M., *J.Am.Chem.Soc.* (1985) **107**, 4252.

R	$R_1R_2C=O$	M	BASE	AMINO ACID (%479)	CONFIG.	%ee	*threo : allo*
Me	HCHO	Ni	Et₃N	75-82	S	96	-
Me	HCHO	Ni	NaOMe	66-77	R	87-89	-
Ph	HCHO	Ni	Et₃N	75	S	83	-
Ph	HCHO	Ni	NaOMe	95	R	88	-
Ph	MeCHO	Ni	NaOMe	72	R	84	20:1
Ph	MeCHO	Ni	Et₃N	32	S	78	2:1
Ph	Me₂CO	Ni	NaOMe	54	R	72	-
Ph	Me₂CO	Ni	NaOMe	56	R	98	-
Me	Me₂CO	Cu	NaOMe	55	R	70	-
Me	PhCHO	Cu	NaOMe	67	R	74	50:1
Me	PhCHO	Ni	NaOMe	67	R	82	34:1
Ph	PhCHO	Ni	NaOMe	67	R	82	34:1
Ph	PhCHO	Cu	NaOMe	59	R	80	50:1

Belokon, et.al., [94] have extensively developed an interesting and potentially quite useful glycine enolate complex based on the Ni or Cu complexes **477** (Scheme 96). The complex is prepared by standard peptide coupling of N-benzyl-L-proline with *o*-aminoacetophenone (R=Me) and glycine followed by complexation to either Ni^{+2} or Cu^{+2}. The aldol condensations are then performed in the presence of a base under mild conditions to give the adducts **478** in good yields and impressive diastereoselectivities. The carbonyl component approaches primarily *anti*-to the N-benzyl group and the aldol stereoselection greatly favors the *threo*- configuration. The diastereomeric complexes can be separated by chromatography in stereochemically pure form, and hydrolyzed to the free amino acids. The chiral complex can be recovered in good to excellent

SCHEME 97

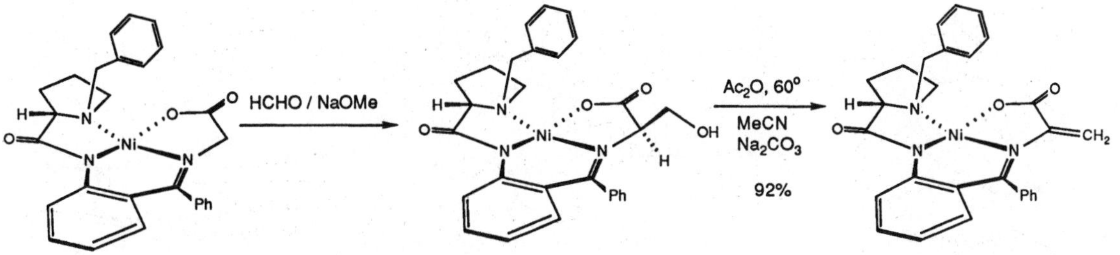

Belokon, Y.N.; Sagyan, A.S.; Djamgaryan, S.M.; Bakhmutov, V.I.; Belikov,, V.M., *Tetrahedron* (1988) **44**, 5507

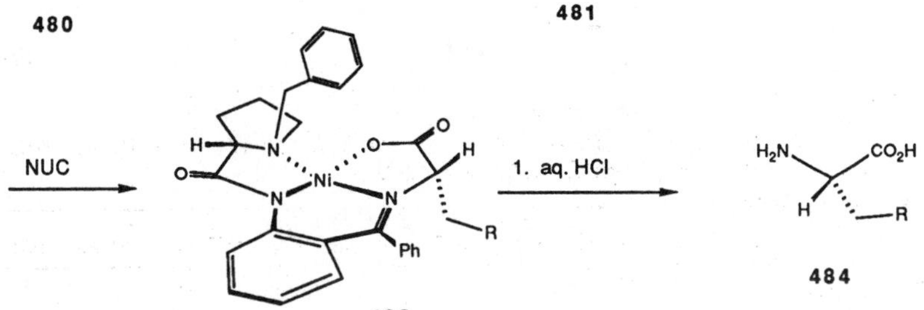

NUCLEOPHILE	SOLVENT	DIASTEREOMER RATIO (%)	R	AMINO ACID (% 484)	%ee
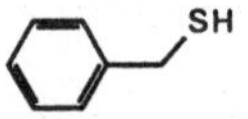 ⬡—CH₂SH	MeCN	90 : 3	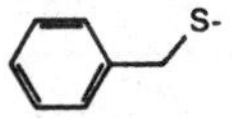 ⬡—CH₂S-	90	98
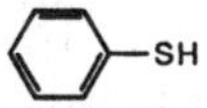 ⬡—SH	MeCN	93 : 4.8	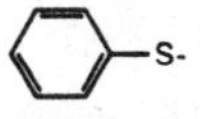 ⬡—S-	92	98
 EtO₂C⌇CO₂Et	MeCN	90 : 8.1	-CH₂CO₂H	80	80
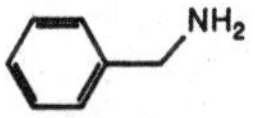 ⬡—CH₂NH₂	MeCN	90 : 5	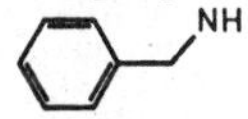 ⬡—CH₂NH	75	>90
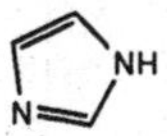 imidazole NH	MeCN	90 : 5	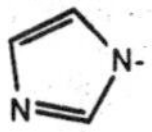 imidazol-N-	85	>90
Me₂NH	MeCN	85 : 5	Me₂N-	93	70
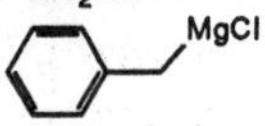 ⬡—CH₂MgCl	THF	55	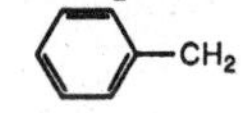 ⬡—CH₂	82	50
MeOH	MeOH	90 : 2	MeO	90	90

SCHEME 98

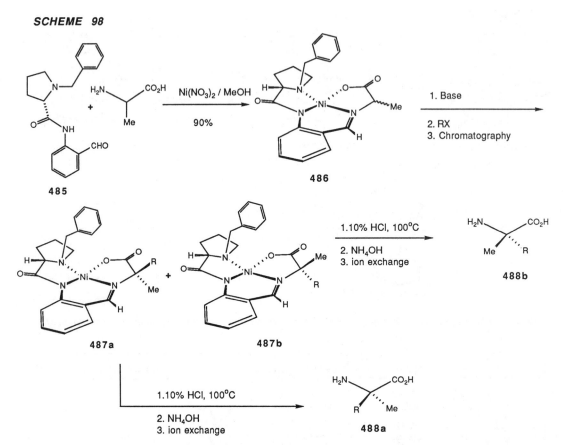

Belokon, Y.N.; Chernoglazova, N.I.; Kochetkov, C.A.; Garbalinskaya, N.S.; Belikov, V.M., J.C.S.Chem.Comm. (1985) 171.

RX	Base/ RATIO (%487b:487a)		AMINO ACID (% 488)
	n-BuLi	10%NaOH/ PTC	
MeI	92	-	77
⟨benzyl bromide⟩ Br	51 : 40	63 : 31	45 / 35
⟨allyl bromide⟩ Br	56 : 33	62 : 22	49 / 27

yields (60-90%). While the %ee's are modest in many cases, a few salient examples (96-98%ee) clearly demonstrate that this methodology has great potential.

In a related study,[95] the dehydroalanine substrate **482** was prepared from the formaldehyde adduct (**481**) derived from **480**. This material proved to be a reactive Michael acceptor for a variety of nucleophiles. The thermodynamically most stable *anti-* isomers **483** are formed and hydrolyzed as above to the free amino acids (**484**). The protected cysteine derivatives are the outstanding examples being produced in 98% ee.

Alkylations [96] of the Ala complex **486** (Scheme 98) with reactive alkylating reagents furnished the corresponding α-alkylated alanine derivatives (**488**) after hydrolysis. The %de's in these reactions were not very good, but separation of the pure diastereomeric complexes (**487a/b**) by silica gel chromatography followed by hydrolysis gave enantiomerically pure amino acids (**488**).

Several other 'enolate-type' of systems will be covered in the final section of this chapter that include deracemization methods.

B. GLYCINE CATIONS

The pioneering work of Ben-ishai[97a] describing racemic electrophilic glycine equivalents primarily for Friedel-Crafts type of C-C bond-forming reactions has been recently translated into several asymmetric versions. The first four reports appeared concomitantly by four independent groups in 1985.

Schollkopf and co-workers[97b] found that chlorination of the enolate derived from the Val-Gly bis-lactim ether **10** with hexachloroethane provided the labile chloride **489** as a 94 : 6, *cis : trans* mixture (Scheme 99) . The authors found that **489** was labile to elimination of HCl forming the corresponding aromatic 2,5-dimethoxy-3-isopropylpiperazine; thus immediate condensation of **489** with three esters of sodium malonate provided the malonates **490** in *ca* 65% yield. The malonate anions attacked the electrophilic glycine carbon of **489** with *inversion* giving the *trans-* isomer as the major product. Hydrolysis with 0.25N HCl provided the corresponding methyl; isopropyl; and *tert*-butyl esters (**491**) of β-carboxy aspartic acid methyl ester in good yields and >95% ee.

Extending this system [98] to Friedel-Crafts couplings, the chloride **489** was condensed with various electron-rich aromatic compounds in the presence of tin(IV)chloride in methylene chloride giving the *anti-*adducts **492** in good yields. Hydrolysis of these substances in dilute hydrochloric acid gave the arylglycines (**493**) in good yields and 84~>95%ee. The well-

SCHEME 99

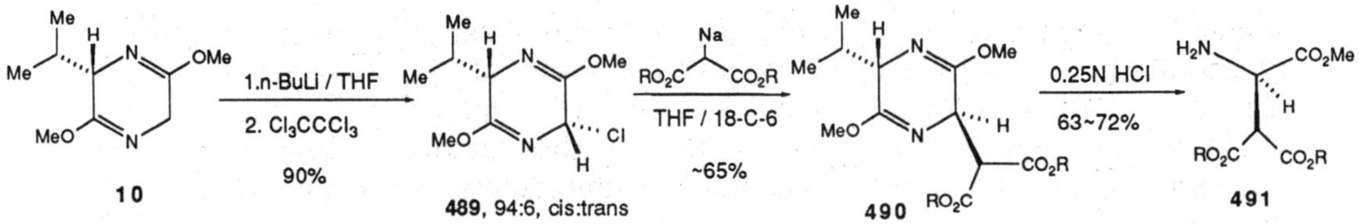

Reagents/conditions (Scheme 99):
- 10 → 1. n-BuLi / THF; 2. Cl₃CCCl₃ — 90% → **489**, 94:6, cis:trans
- **489** + RO₂C–CH(Na)–CO₂R, THF / 18-C-6, ~65% → **490**
- **490** → 0.25N HCl, 63~72% → **491**

R = Me, i-Pr, t-Bu
>95% ee

Schollkopf, U.; Neubauer, H-J.; Hauptreif, M., *Angew. Chem. Int. Ed. Engl.* (1985) **24**, 1066.

SCHEME 100

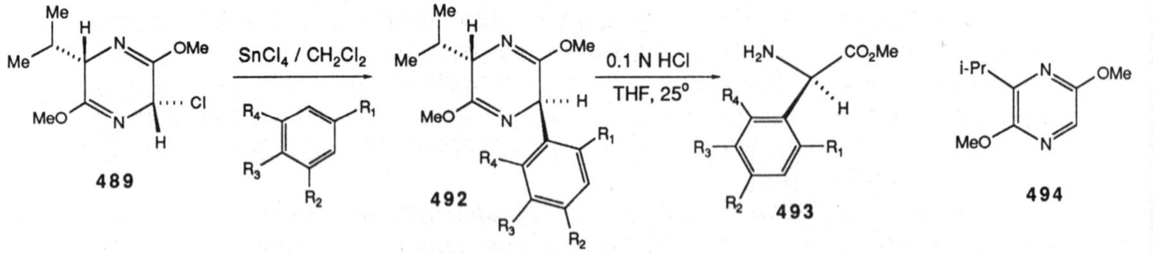

Reagents/conditions (Scheme 100):
- **489** + arene (R₁–R₄), SnCl₄ / CH₂Cl₂ → **492**
- **492** → 0.1 N HCl, THF, 25° → **493**
- **494**

Schollkopf, U.; Gruttner, S.; Anderskewitz, R.; Egert, E.; Dyrbusch, M., *Angew. Chem. Int.Ed. Engl.* (1987) **26**, 683.

R₁	R₂	R₃	R₄	Friedel-Crafts Yield	HYDROLYSIS(%)	%ee
OEt	H	OEt	H	65	89	>95
OMe	OMe	H	OMe	67	60	>95
H	OMe	H	H	} 62 (11.5 : 1 ; p:o)	69	~84
OMe	H	H	H			

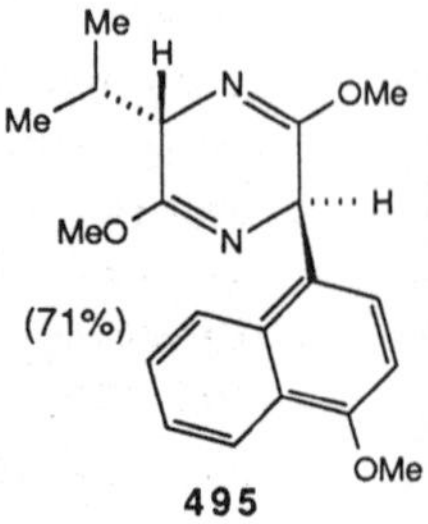

(71%)

495

SCHEME 101

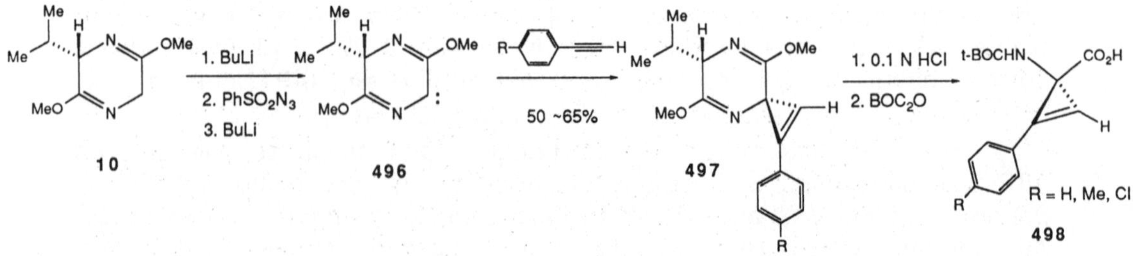

Reagents/conditions (Scheme 101):
- 10 → 1. BuLi; 2. PhSO₂N₃; 3. BuLi → **496**
- **496** + R–C₆H₄–C≡C–H, 50 ~65% → **497**
- **497** → 1. 0.1 N HCl; 2. BOC₂O → **498**

R = H, Me, Cl
498

Schollkopf, U.; Hupfeld, B.; Kuper, S.; Egert, E.; Dyrbusch, M., *Angew. Chem. Int.Ed.Engl.* (1988) **27**, 433 known propensity of arylglycines to racemize illustrates that the present methodology is a potentially powerful means to access these difficult amino acids in optically active form. The major drawback to the electrophilic bis-lactim ether **489**, is the lability and unwelcome aromatization to **494** which accompanies the desired coupling products.

In a very interesting, preliminary study, Schollkopf, et.al.,[99] generated the electrophilic carbene species **496** by diazotization of lithio- **10**. Reaction of this species with three aryl acetylenes gave the spiro-cyclopropenes **497** in good yield. The addition occurs stereospecifically giving the *anti*-isomers as evidenced by single crystal x-ray analysis of **497** (R=H). Hydrolysis in the usual way affords the novel 1-amino-2-arylcyclopropene-1-carboxylic acids (>95% ee) ; subsequent protection of the nitrogen as the t-BOC derivatives provides **498**. The use of glycine carbene equivalents has been little, if at all studied for the synthesis of novel amino acids; the present work should stimulate additional studies in this area.

Enders and Steglich, et.al. [100] have employed the N-benzoyl α-bromo glycine esters **499** in asymmetric enamine additions to the chiral imines **500** (Scheme 102). The diastereoselectivities favor the *anti*- adducts **502**. While the diastereoselectivities were high, the %ee's were low (27-67%ee); optically pure isomers could however, be obtained by careful crystallization. When a second chiral auxilliary was added , as in the case of **503+504**, the %de=%ee= >98%. This study did not comment on the possibility of converting the adducts **502,505** or **506** to the corresponding amino acids. This is expected to be rather difficult as the hindered esters might require somewhat brutal conditions potentially resulting in racemization.

Yamamoto, et.al., [101] have studied the additions of various allylic organometallic reagents to the chiral imines **507** (Scheme 103). In contrast to the Enders/Steglich system above, these workers first examined attaching the chiral auxilliary to the glycine nitrogen atom. It was found that allyl-9-BBN reagents gave the best yields and highest diastereoselectivities, whereas the corresponding zinc, magnesium and titanium reagents gave poor results. However, there was a wide range of %ee's (54-96%) that seemed to worsen with substitution on the allyl organometallic reagent. The addition of benzyl zinc bromide to **510** furnished a 74 : 26 ratio of **511 : 512**. The relative stereochemistry of the major isomer was determined to be D- by reductive cleavage of the chiral benzyl amine moiety to phenylalanine butyl ester. Heteroatom-substituted allyl organometallics (**517**) gave the adducts **518 / 519** with a preference for the Cram-*threo* isomers (all four diastereomers were produced in each case). As in the case of the Enders/Steglich

Asymmetric Derivatization of Glycine

SCHEME 102

Kober, R.; Papadopoulos, K.; Miltz, W.; Enders, D.; Steglich, W., *Tetrahedron* (1985) **41**, 1693.

R	X	%500	%502
Me	CH₂	87	80
	S		91
Et	CH₂	86	83
	CH₂	89	89
	CH₂	81	93
	S		92
	CH₂	87	84
	S		95
	CH₂	95	80

SCHEME 103

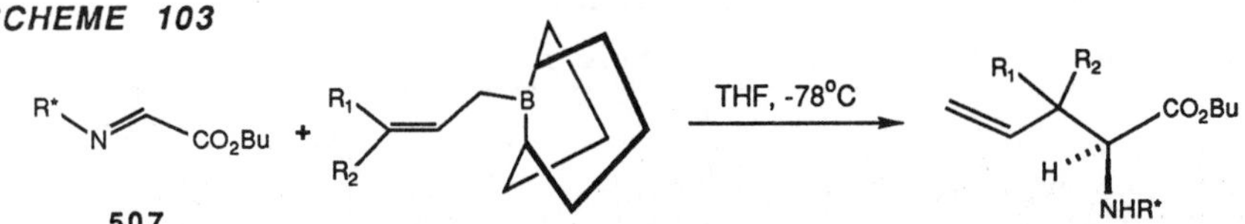

Yamamoto, Y.; Ito, W.; Maruyama, K., *J.Chem.Soc.Chem.Comm.* (1985) 1131
Yamamoto, Y.; Ito, W., *Tetrahedron* (1988) **44**, 5415.

R*	R₁	R₂	YIELD (% 509)	%ee	erythro : threo
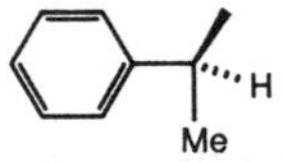 (1-phenylethyl)	H	H	92	92	
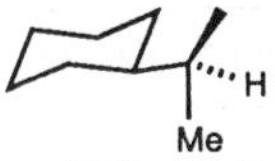 (1-cyclohexylethyl)	H	H	94	96	
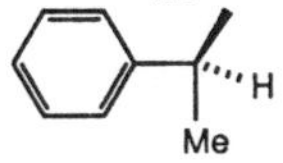 (1-phenylethyl)	Me	Me	33	54	
Me—⟨C₆H₄⟩—SO₂	Me	H	75		85 : 15

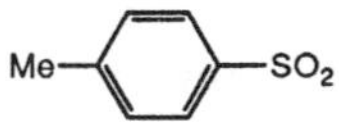

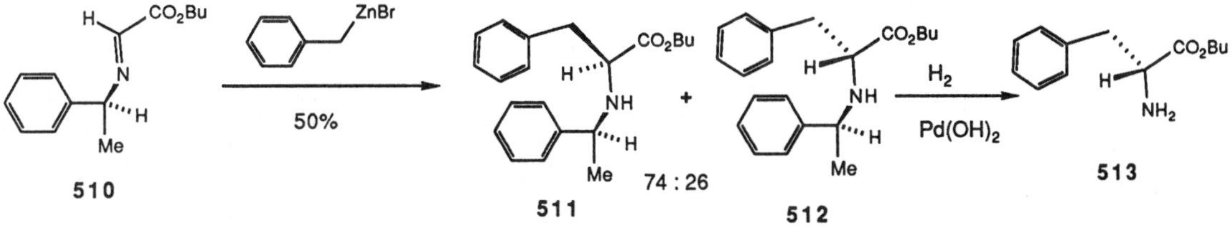

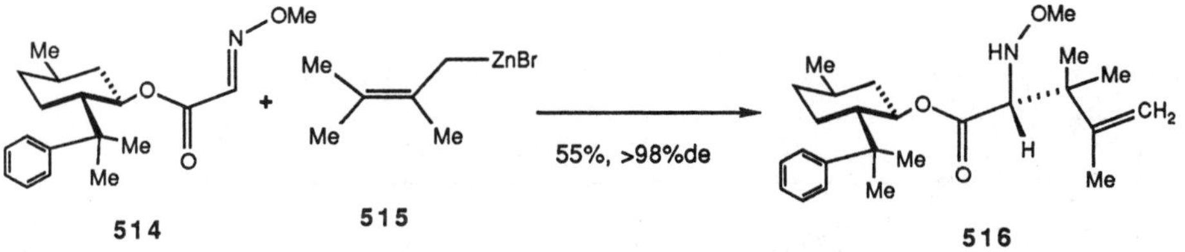

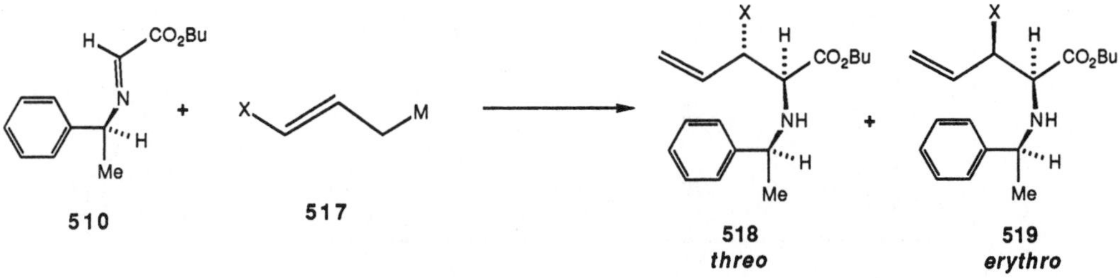

X	M	YIELD (%518+519)	*Threo : Erythro*
OMe	ZnBr	67	6 : 1
OMe	Ti(Oi-Pr)$_3$	20	2 : 1
OMe	AlEt$_3$Li	14	2 : 1
OPh	ZnBr	54	1 : 1
OPh	AlEt$_3$Li	21	2 : 1
SMe	ZnBr	50	1 : 1
Sme	B(OMe)$_2$	42	1 : 1

systems, the 8-phenylmenthyl oximino ester **514** gave excellent diastereoselectivity with trimethallyl zinc bromide to afford **516** with >98% de.

SCHEME 104

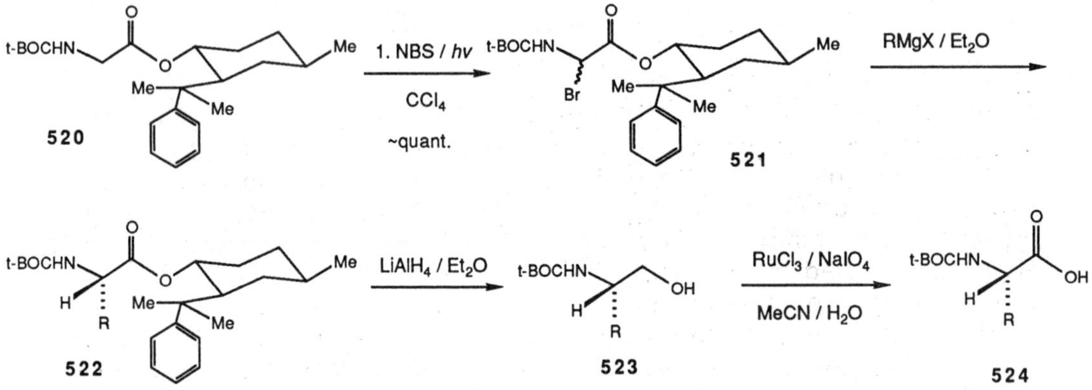

Ermert, P.; Meyer, J.; Stucki, C.; Schneebeli, J.; Obrecht, J-P., *Tetrahedron Lett.* (1988) **29**, 1265.

R	%522	%523	%524	%ee
Me	71	75	46	92
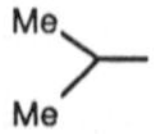 Me$_2$CH	54	81	58	89
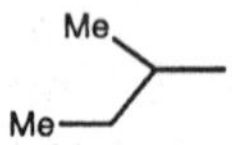 Me-CH(Me)-CH$_2$	65	88	70	95
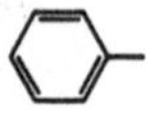 Ph	78	72	78	82

SCHEME 105

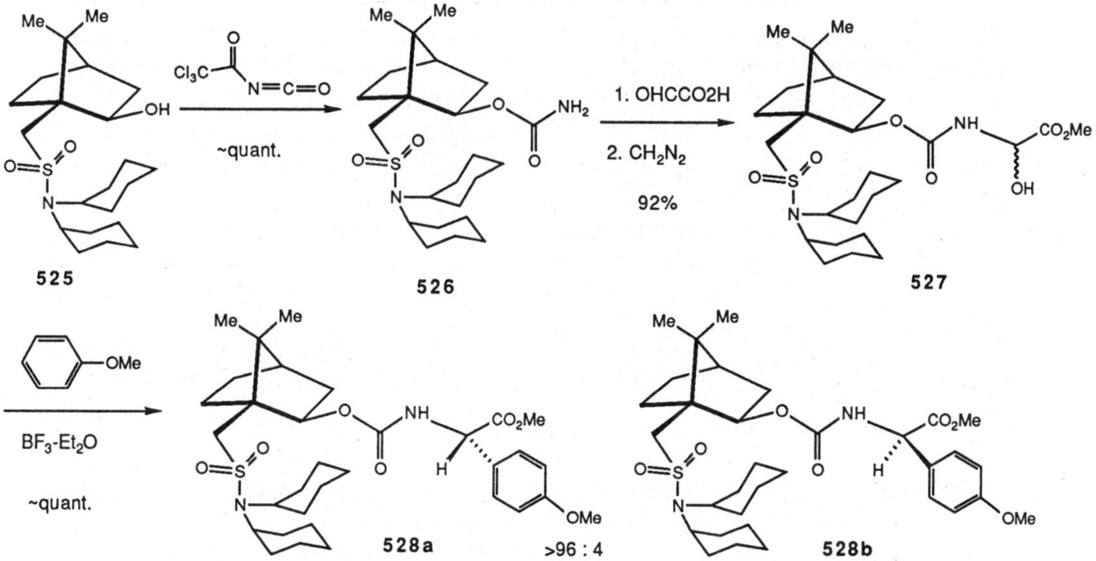

Harding, K.E.; Davis, C.S., *Tetrahedron Lett.* (1988) **29**, 1891.

Another example of the use of an 8-phenylmenthyl ester was recently reported by Obrecht, et.al. [102] (Scheme 104). Bromination of **520** with NBS provided a 1:1 diastereomeric mixture of **521** that was directly used without further purification. Grignard additions proceeded from the expected less hindered face on the incipient imine resulting from Grignard-mediated elimination of HBr to afford the adducts **522**. As alluded to above, the authors note that all attempts to remove the phenylmenthyl group by hydrolysis or transesterification without racemization were unsuccessful. This problem mandated a clumsy reductive removal of the the esters with LAH affording the amino alcohols (**523**) that were subsequently oxidized with ruthenium to the corresponding N-t-BOC amino acids (**524**). The %ee's are generally high ranging from 82-95% ee. The redox manipulation of this system will certainly limit the scope of α-R groups that can be tolerated.

Harding and Davis [103] have recently employed the Oppolzer camphor sultam **525** to access an optically active electrophilic glycinate. Conversion to the urethane **526** proceeds in quantitative yield followed by condensation with gloxylic acid and esterification furnishes **527** as a mixture of diastereomers. Reaction of this substance with anisole and boron trifluoride etherate proceeds with excellent stereoselectivity providing a >96 : 4 ratio of **528a** : **528b** in essentially quantitative yield. As with the related hindered esters noted above, the authors comment that hydrolytic removal of the chiral auxilliary to provide the amino acid without racemization is problematical.

The most extensively studied optically active electrophilic glycine template has been developed by Williams, et.al. [104-109]. The glycine templates **459/462** are prepared as shown in Scheme 106. Inexpensive benzoin is converted into the oxime and stereospecifically hydrogenated to the racemic *erythro*-amino alcohols **530**; these are subsequently resolved through the agency of the derived L-glutamate salts according to Tishler, et.al [104b] on a large scale providing each optical isomer **530a** and **530b** of >98%ee. These amino alcohols have recently become commercially available. Each isomer is then separately alkylated with ethyl bromoacetate; acylated with either benzylchloroformate or di-*t*-butyldicarbonate and; finally lactonized with catalytic *p*-TsOH in hot benzene or toluene to afford the crystalline lactones **459 / 462** in ~65% overall yield from the amino alcohols. The entire sequence from benzoin is accomplished without any chromatographic separations. All four isomers **459a / 459b; 462a / 462b** as well as the corresponding racemic substances are now commercially available. As in the Seebach glycine systems (Scheme 71) , the benefit of the resolution is manifested in providing acces to either D- or L-configured amino acids in a predictable manner.

Bromination of these oxazinones with NBS in refluxing carbon tetrachloride proceeds in essentially quantitative yield. The experimental protocol simply involves cooling the solution after the reaction is complete (ca. 1~1 1/2 h, reflux) precipitating the insoluble succinimide which is filtered off leaving, after evaporation, a white amorphous powder of the corresponding bromide (**531**). The bromide is reported to be unstable to silica gel purification and is used directly (crude) for the subsequent coupling reactions. The relative stereochemistry of the bromide is *anti-* and only a single diastereomer is produced in the reaction. The corresponding chloride can be obtained by chlorination with *t*-butyl hypochlorite but apparently offers few advantages over the bromides. Reaction of **531** with various organometallic reagents in the presence of zinc chloride results in displacement of the halogen providing the homologated oxazinones **532**. In most cases, the relative stereochemistry of the coupling reactions proceeds with net *retention* providing *anti-***532**.

The authors speculate that the zinc(II) salt coordinates to the halogen ultimately providing the reactive iminium species **534**. Since the phenyl rings are *cis*, the sterically least encumbered approach is from the face *anti-* to the two phenyl substituents (shown). Depending on which type of BOC protecting group is employed, two different types of reductive protocol have been devised. In the case of the N-CBz systems, either catalytic hydrogenation on a Pd[o] catalyst or dissolving metal reduction

SCHEME 106

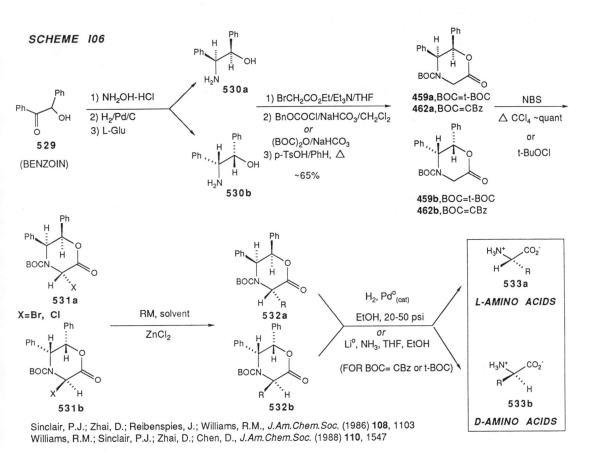

Sinclair, P.J.; Zhai, D.; Reibenspies, J.; Williams, R.M., *J.Am.Chem.Soc.* (1986) **108**, 1103
Williams, R.M.; Sinclair, P.J.; Zhai, D.; Chen, D., *J.Am.Chem.Soc.* (1988) **110**, 1547

BOC GROUP	NUCLEOPHILE	REACTION CONDITIONS	YIELD of 532	DEPROTECTION METHOD	YIELD of AMINO ACID	%ee
CBz	OSiMe₂t-Bu / OEt	ZnCl₂ /THF 25°C	74%	H₂/PdCl₂ (cat) EtOH, 20psi	85% ETHYLASPARTATE	> 96%
	SiMe₃	ZnCl₂ /THF 25°C	66%	H₂/PdCl₂ (cat) EtOH, 20psi	93% NORVALINE	>98%
	SiMe₃	ZnCl₂ /THF 25°C	66%	Li° /NH₃ /EtOH	90% ALLYLGLYCINE	>91%
	H₃CZnCl	THF/ -78°C	46%	H₂/PdCl₂ (cat) EtOH, 20psi	100% ALANINE	>96%
	Bu₂Cu (CN) Li	THF/ Et₂O -78°C	48%	H₂/PdCl₂ (cat) EtOH, 20psi	52% NORLEUCINE	>99%
	OSiMe₃	ZnCl₂ (cat) CH₃CN / 25°C	72%	H₂/PdCl₂ (cat) EtOH, 40psi	91% HOMOPHENYLALANINE	>96%
	MeO— OSiMe₃	ZnCl₂/ THF MeCN	72%	H₂/PdCl₂ (cat) EtOH, 40psi	94% p-MeO-HOMOPHENYL-ALANINE	>98%
	SiMe₃	ZnCl₂ /THF 25°C	82%	H₂/PdCl₂ (cat) EtOH, 20psi	91% CYCLOPENTYLGLYCINE	>96%
	SiMe₃	ZnCl₂ /THF 25°C	82%	Li° /NH₃ /EtOH	94% CYCLOPENTENYLGLYCINE	>96%
	(furan)	ZnCl₂ /THF 25°C	64%	H₂/PdCl₂ (cat) EtOH, 20psi	H₃+N H CO₂⁻ / O H 89%	>96%
	(furan)—CH₃	ZnCl₂ /THF 25°C	66%	H₂/PdCl₂ (cat) EtOH, 20psi	89% DIHYDROFURANOMYCIN	not determ.
t-BOC	SiMe₃	ZnCl₂ /THF 25°C	63%	Li° /NH₃ /EtOH	70% N-t-BOCALLYLGLYCINE	>98%
	SiMe₃	ZnCl₂ /THF 25°C	59%	Li° /NH₃ /EtOH	70% N-t-BOC-CYCLOPENTENYLGLYCINE	>95%

directly provides the free zwitterionic amino acids **533**. In the corresponding N-t-BOC systems, dissolving metal reduction directly provides the N-t-BOC-protected amino acids **535**. This is the only *direct* asymmetric synthesis of N-t-BOC-protected α–amino acids reported to date. The Table lists the results of surveying a variety of coupling reactions with **531**, the coupling conditions, the reduction method, the amino acid produced in each case and the %ee. The chemical yields for **532** reflect the two-step conversion of the oxazinones **459 / 462** into the bromides **531** and hence to **532**. The authors note that with richly basic reagents, such as methyl zinc chloride or the cuprate, relatively modest yields result when compared to the 'neutral' , electron-rich nucleophiles. This is due to a competing reduction of the halides **531** back to the starting oxazinones (**459 / 462**) and is interpreted as involving an electron transfer radical-radical coupling mechanism in these cases. The %ee's are excellent, typically exceeding 96%ee. Full experimental details accompany this work.

Several illustrations of the complementary utility of these systems to the enolate-based aprroaches are provided in Schemes 107-109. The crystalline allylated substance **536** was osmylated to provide the γ-butyrolactones **538** as a 1:1 diastereomeric mixture in 78% yield. The initially formed diol **537** spontaneously rearranges to the thermodynamically more stable γ-butyrolactone isomer under the reaction conditions. Reductive removal of the chiral auxilliary, followed by acylation with benzyl chloroformate provided **539** and **540** which were separated by silica gel chromatography. Isomer **540** was previously converted into the unusual β–lactam antibiotic clavalanine (**541**) by a Hoffmann-La Roche group; the Roche synthesis of **540** is a multi-step preparation from D-xylose. Although the oxidation of **536** proceeds without stereocontrol, the brevity of the approach remains an attractive element of functionalizing a derivatized oxazinone such as **536**.

Scheme 108 details an extremely short and convenient synthesis of chiral glycine derivatives [106]. The bromide **531** is simply reduced with tritium carrier gas on Pd⁰ at 1 atmosphere in tritiated water / THF. Ion-exchange isolation provided **542** in ~31-38% chemical yields that was 88~93% optically pure and had a specific activity of 0.78 Ci/mmol. The authors note [106b,c] that the material obtained by this procedure contained less than 1% of the corresponding di-tritiated material **543** which is found as a major contaminant in the classical enzymatic exchanges with serine hydroxymethyl transferase and glutamic-pyruvic transaminase. The contamination of **543** produced by the enzymatic exchange protocol precluded the use of this material for the Alberta group[106b,c] who were studying the stereospecific abstraction of the α-

SCHEME 107

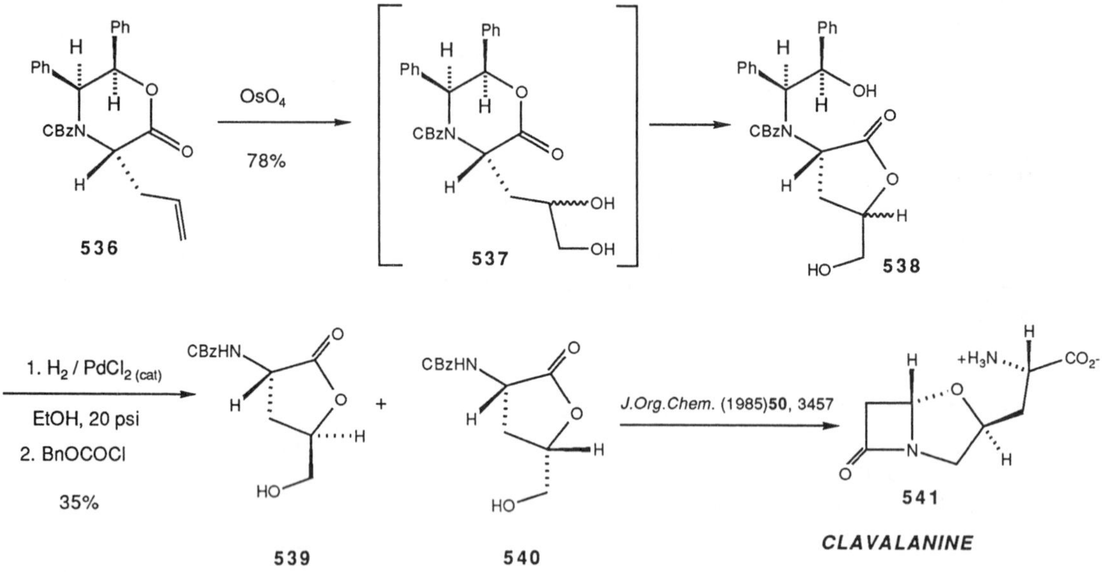

536 → OsO$_4$ 78% → **537** → **538**

1. H$_2$ / PdCl$_{2 (cat)}$
EtOH, 20 psi
2. BnOCOCl
35%

539 + **540**

J.Org.Chem. (1985) 50, 3457 → **541** CLAVALANINE

SCHEME 108

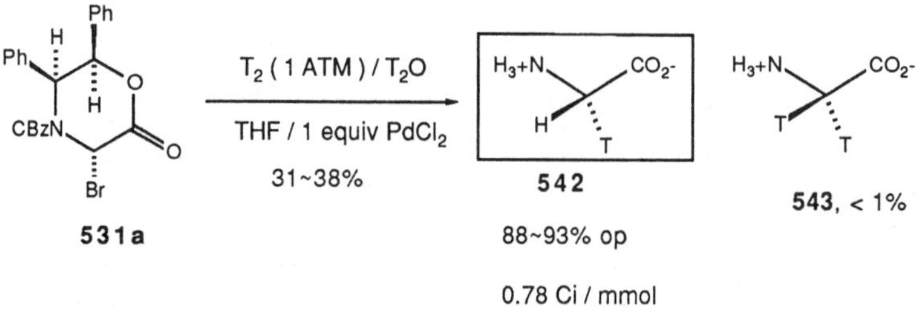

531a → T$_2$ (1 ATM) / T$_2$O, THF / 1 equiv PdCl$_2$, 31~38% → **542** 88~93% op, 0.78 Ci / mmol + **543**, < 1%

Williams, R.M.; Zhai, D.; Sinclair, P.J., *J.Org.Chem.* (1986) **51**, 5021
Ramer, S.E.; Cheng, H.; Palcic, M.M.; Vederas, J.C., *J.Am.Chem.Soc.* (1988) **110**, 8526.

SCHEME 109

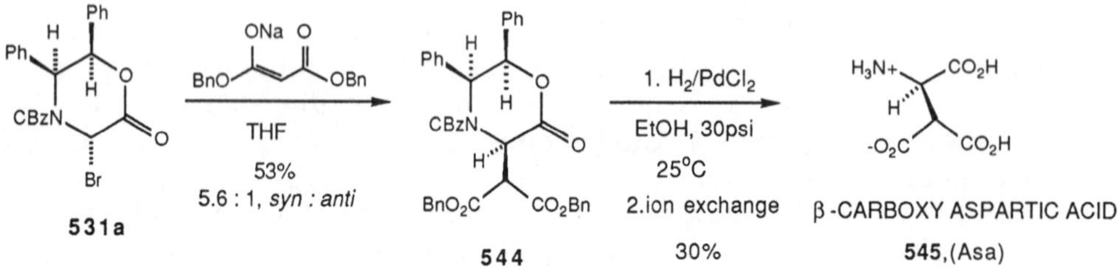

531a → (ONa, BnO, OBn), THF, 53%, 5.6 : 1, *syn : anti* → **544** → 1. H$_2$/PdCl$_2$, EtOH, 30psi, 25°C, 2.ion exchange, 30% → β-CARBOXY ASPARTIC ACID **545**,(Asa)

Williams, R.M.; Sinclair, P.J.; Zhai, W., *J.Am.Chem.Soc.* (1988) **110**, 482 methine protons of terminal glycine amides by the enzyme peptidyl α-amidating monooxygenase (PAM). The corresponding α-deuterio glycines were obtained [106a] by hydrogenating **531a** or **531b** on a Pd[0] catalyst in D_2O giving material of 84-90% isotopic purity, 77-82%ee and 54% yield. The authors note that if D_2O is not employed, the Pd catalyst exchanges protons from the solvent giving material of reduced *isotopic* purity. The Alberta group note that at 1 atm pressure which is required for tritium reactions, a full molar equivalent of $PdCl_2$ must be employed to obtain reasonable yields. On the other hand, the deuterium reduction can be carried out with catalytic amounts of $PdCl_2$ at 20-40 psi. The primary advantage of this synthesis is the overall simplicity (two steps from commercially available **459**), the high isotopic purity and the introduction of the isotopic atom in the last step. This contrasts to alternate chemical syntheses of chiral glycine that are often multi-step and carry the isotope through many manipulations.

Scheme 109 details an asymmetric synthesis [107] of the recently discovered amino acid β-carboxy aspartic acid (Asa, **545**) that was obtained from ribosomal protein hydrolysates by Koch, et.al. Asa is a notoriously unstable amino acid that is sensitive to decarboxylation under acidic conditions and eliminiation of ammonia under basic conditions. The inherent lability of Asa is sufficiently problematical that the harsh conditions employed in conventional peptide-sequencing techniques (resulting in production of Asp in most cases) have limited the detection of Asa in natural systems. Coupling of **531a** with sodium dibenzyl malonate in THF furnished in 53% yield, the malonate **544** as a 5.6 : 1, *syn : anti* mixture of diastereomers that was separated by chromatography. As will be discussed below, the highly nucleophilic malonate anion suffers primarily S_N2 displacement of the bromide providing *syn*-**544** as the major product. Reduction of all five benzylic residues over a Pd[0] catalyst provided Asa (**545**) as a 4 : 1 mixture with Asp. These were easily separated by acidic ion exchange chromatography which alllows Asa (pKa = 0.8) to pass freely off the column. The small amount of Asp produced in the reduction presumably results from the small amount of HCl produced from the $PdCl_2$ and reflects the sensitive nature of this difficult amino acid. The pure Asa is obtained in 30% yield from **544** and >98%ee. This was the first optically active sample of Asa obtained since, the natural material was never isolated in sufficient quantity to collect the usual data. The %ee was determined by decarboxylation to Asp and Mosher amide formation.

These workers have devised a straightforward and practical experimental protocol for directly preparing the zwitterionic amino acids in a pure form as shown in Scheme 110. In the case of saturated 'R' groups (ie., stable to catalytic hydrogenation), the Kagan-type[89] reductive

SCHEME 110

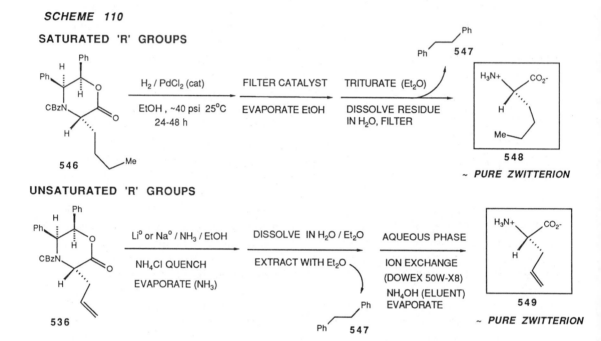

SATURATED 'R' GROUPS

546 → H$_2$ / PdCl$_2$ (cat), EtOH, ~40 psi 25°C, 24-48 h → FILTER CATALYST, EVAPORATE EtOH → TRITURATE (Et$_2$O), DISSOLVE RESIDUE IN H$_2$O, FILTER → 548 ~ *PURE ZWITTERION* (547)

UNSATURATED 'R' GROUPS

536 → Li° or Na° / NH$_3$ / EtOH, NH$_4$Cl QUENCH, EVAPORATE (NH$_3$) → DISSOLVE IN H$_2$O / Et$_2$O, EXTRACT WITH Et$_2$O → AQUEOUS PHASE, ION EXCHANGE (DOWEX 50W-X8), NH$_4$OH (ELUENT), EVAPORATE → 549 ~ *PURE ZWITTERION* (547)

cleavage is typically chosen. The substrate (eg., **546**) is dissolved in ethanol and hydrogenated over a catalytic amount of PdCl$_2$ at room temperature at 20-50 psi for 1-2 days in a glass hydrogenation bottle. The reaction is purged with nitrogen, the catalyst filtered off and the ethanol evaporated leaving an oily residue. This residue is then triturated several times with ether or pentane to remove the bibenzyl (**547**) produced in the reductive cleavage of the chiral auxilliary. The remaining water-soluble residue becomes solid during the trituration and is subsequently dissolved in water and filtered through cotton and concentrated to afford essentially pure amino acid (eg., **548**). A small amount of HCl from the PdCl$_2$ accompanies the crude amino acid that may be easily removed by exposure to a quick ion exchange filtration. For most applications where the amino acid will be transformed into an ester or urethane for peptide coupling, the crude materials are sufficiently pure to utilize directly without further purification.

In the case of unsaturated or hydrogenolizable 'R' groups, the dissolving metal protocol is employed. The substrate (eg., **536**) is dissolved in liquid ammonia containing ethanol at -33°C and either lithium or sodium metal is added to the reaction mixture. The metal is added until the blue color persists for *ca.* 1 min. and then quenched with solid ammonium chloride. The ammonia is allowed to evaporate and the ethanolic residue partitioned between water and ether in a separatory funnel. Extraction with ether again removes the bibenzyl (**547**) leaving the pure amino acid in the aqueous phase. The aqueous phase is then filtered through an ion exchange resin affording essentially pure zwitterionic unsaturated amino acids (eg., **549**). In the case of the dissolving metal reduction of the t-BOC substances (**532** to **535**) the same protocol is followed, except that the aqueous phase is acidified to pH = 3 after the organic extraction of **547** and extracted with ethyl acetate, etc. to afford the pure t-BOC amino acids (**535**). The authors point out that, although the chiral auxilliary is sacrificed in the final reductions, this system offers an important advantage over numerous other amino acid syntheses that require expensive, time-consuming chromatographic separations, recovery and 'recycling' of chiral auxilliaries (rarely done in practice) and hydrolyses of esters, etc. to obtain the amino acids themselves. In the present case the chiral auxilliaries are polar, water-soluble substances that, even if it were possible to recover , would require a difficult separation from the products. Thus, the destruction of the chiral auxilliary in this case turns out to be a significant *advantage* since the final processing converts the chiral auxilliary into an inocuous substance of greatly different solubility properties than the amino acids or t-BOC amino acids and is easily removed by trituration or extraction. The raw cost of the amino alcohols **530** of course, preclude the application of this chemistry to large , multi-kilo industrial scale syntheses. As with virtually all of the (non-catalytic) wholly 'organic' amino acid syntheses, this system is most appropriate for the basic research chemist who needs rapid and predictable access to a large number of structurally diverse amino acids in optically active form.

A very useful new coupling reaction has been developed by the same group [108, 109] as shown in Scheme 111. The oxazinones are brominated in the usual way and then condensed with trialkyl tin acetylides in the presence of zinc chloride in warm carbon tetrachloride to afford the crystalline alkynes **550** as single diastereomers (*anti*). Dissolving metal reduction directly provides exclusively the E-vinyl glycine derivatives

SCHEME 111

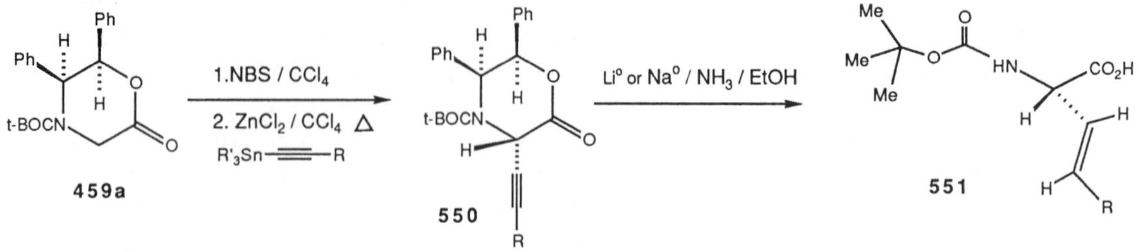

Williams, R.M., Zhai, W., Tetrahedron (1988) 44, 5425
Zhai, D.; Zhai, W., Williams, R.M., *J.Am.Chem.Soc.* (1988) **110**, 2501.

R	R'	BOC	% 550 [a]	% 551 via Na°	% ee via Na°	% 551 via Li°	% ee via Li°
-CH$_3$	n-Bu	t-BOC	99	79	64	18-80	80~>98
-n-C$_3$H$_7$	n-Bu		70	80	61	20	72
-n-C$_6$H$_{13}$	n-Bu		61	74	56	16	65
-(CH$_2$)$_2$OSiMe$_2$t-Bu	n-Bu		71	71	68	-	-
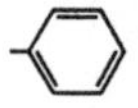 ⬡	n-Bu		99				
-SiMe$_3$	Me		69				
-CH$_2$OSiMe$_2$t-Bu	n-Bu		99				
-(CH$_2$)$_2$OSiMe$_2$t-Bu	Me	CBz	74				
-n-C$_6$H$_{13}$	n-Bu		53				
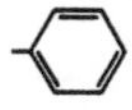 ⬡	n-Bu		55				
Me	n-Bu		70				

a. Yields are for the two-step conversion of **459** or **462** to **531** to **550**

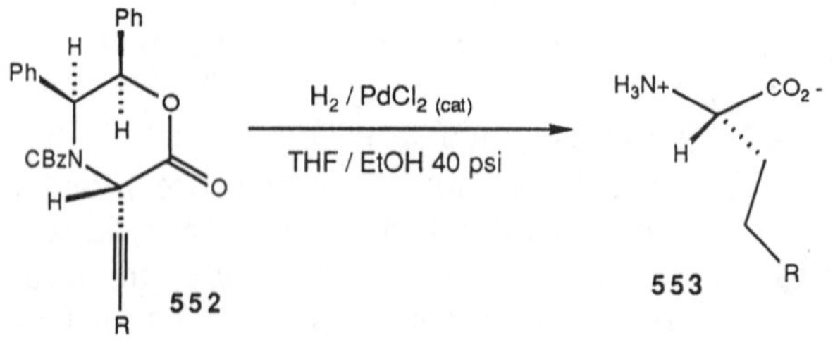

R = Ph, 57%, 94.4%ee
C$_6$H$_{13}$, 68%, >98%ee

551 in good chemical yields and good to excellent %ee's. The authors note that the partial racemization that occurs in several cases must either attend the reduction step , the subsequent work-up or %ee determination (involving removal of the BOC group and Mosher amide formation) since, the alkynes **550** are stereochemically pure (>99% de and >99% ee). This method provides the first stereocontrolled, asymmetric synthesis of γ-substituted vinyl amino acids[110]. It is also possible to effect complete saturation of the alkynes (**552**) by catalytic hydrogenolysis to the free amino acids **553**.

To study the mechanism of coupling to the bromides, a solvent/Lewis acid study[105] was undertaken as shown in Schemes 112-114. When **531** is condensed with the ketene silyl acetal of ethyl acetate in methylene chloride using zinc chloride as the Lewis acid, the *syn* isomer (**555**) is produced as the major product to the extent of at least 45 : 1, *syn : anti* (**555 : 554**). When a more polar solvent is used, such as THF, the selectivity decreases giving as little as 14 : 1, *syn : anti*. When a very powerful Lewis acid (AgOTf) is used, the ratio decreases to 2 : 1, *syn : anti*. These results indicate that in the non-polar solvent methylene chloride and with a weak Lewis acid zinc chloride, the electron-rich ketene silyl acetal effects a clean S_N2 displacement of the bromide. When the conditions are changed to encourage formation of the iminium species **534** (more polar solvent, strong halophile), more of the S_N1 product (*anti*) begins to appear. When a slightly less electron-rich nucleophile is employed such as the silyl enol ether of acetophenone (Scheme 113) a predominance of the *syn-* isomer (**557**) is produced but in a poorer ratio (3.4 : 1, *syn : anti*) than the above system under identical conditions. As the reaction conditions are changed toward a more polar solvent and a more powerful Lewis acid, the selectivity completely *reverses* giving as high as a 24.5 : 1, *anti : syn* ratio of **556 : 557**. When an electron-releasing substituent is added to increase the nucleophilicity of the silyl enol ether (Scheme 114), intermediate behaviour (expressed as selectivity) between the above two extremes is displayed as expected. The authors caution workers to be cognizant of the relative nucleophilicity of the specific reagent under consideration since the reaction conditions can be modified to favor either the S_N2 or S_N1 pathways. The best selectivities are obtained with very weak 'neutral' carbon nucleophiles, such as the allylic silanes and the tin acetylides which typically give exclusive formation of the S_N1 (*anti*) products.

The *syn-* and *anti*-oxazinones from a given coupling reaction are readily distinguishable by examination of the Δ,δ of the benzylic methine protons (Ha and Hb) in the lactone ring in the proton NMR. The *anti*-isomers consistently display a larger Δδ for Ha/Hb than the corresponding

SCHEME 112

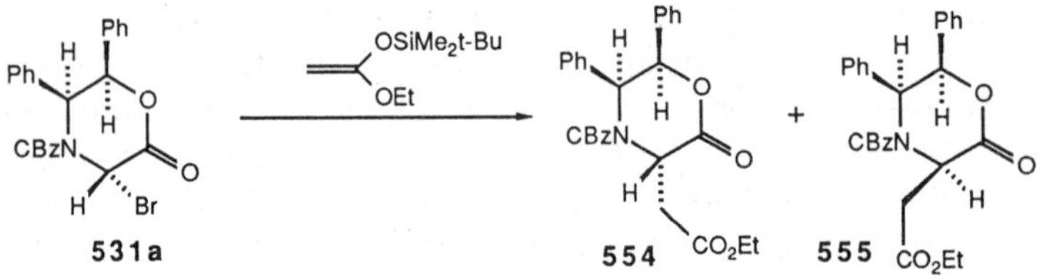

SOLVENT	LEWIS ACID	ANTI		SYN
THF	AgOTf	1	:	2
THF	ZnCl$_2$	1	:	14-45
CH$_2$Cl$_2$	ZnCl$_2$	1	:	45

SCHEME 113

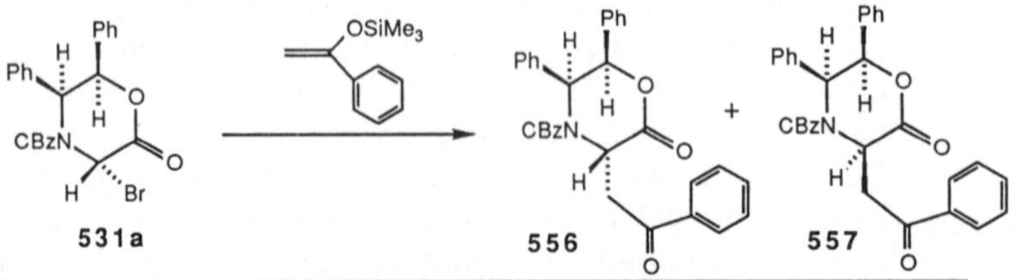

SOLVENT	LEWIS ACID	ANTI		SYN
THF	AgOTf	24.5	:	1
MeCN	ZnCl$_2$	14.5	:	1
THF	ZnCl$_2$	7	:	1
CH$_2$Cl$_2$	ZnCl$_2$	1	:	3.4

SCHEME 114

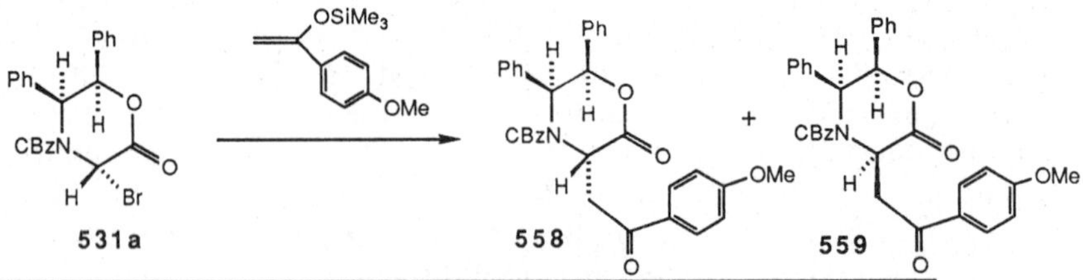

SOLVENT	LEWIS ACID	ANTI		SYN
THF	AgOTf	5.9	:	1
MeCN	ZnCl$_2$	2.9	:	1
THF	ZnCl$_2$	1	:	1.6
CH$_2$Cl$_2$	ZnCl$_2$	1	:	11.2

syn-isomers. The authors also note that the NMR spectra of these substances must be recorded at ~398 K to obviate the line-broadening induced by slow conformational exchange of the urethane moiety.

SYN

$\Delta\delta H_a$, H_b~0.6~0.7 ppm

ANTI

$\Delta\delta H_a$, H_b~0.9~1.1 ppm

Furthermore, the *anti-* isomers have consistently proven to be crystalline substances and the corresponding *syn*-isomers prove to be oily. The authors note that if the selectivity of a given coupling reaction is modest, the general difference in the physical properties of the two diastereomers permits simple crystallization of the *anti*-isomers (generally the major products) consistently resulting in >96~98% de of the homologated lactones and a correspondingly high %ee for the final amino acids.

In a very recent series of investigations, the Colorado State group has broadened the utility of the oxazinone templates to include Wittig-type homologations and asymmetric 1,3-dipolar cycloadditions as shown in Schemes 115 and 116. Arbuzov reaction of **531** with trimethyl phosphite cleanly produces the Steglich-type [111] phosphonate **560** as a single, crystalline diastereomer. This material smoothly condenses with formaldehyde furnishing the dehydro-Ala lactone **561** in high yield. The corresponding di-deuterio substance **561**-d2 is similarly produced with *para*-formaldehyde-d2. Cyclopropanation of **561**-d2 with trimethylsulfoxonium methylide provides essentially a single diastereomeric cyclopropane **563**; dissolving metal reduction furnishes the t-BOC-protected, stereospecifically labeled 1-amino-1-cyclopropane carboxylic acid **564** in 50% yield. Alternatively, condensation of **560** with acetaldehyde provides a single geometric isomer **562** that can be similarly cyclopropanated in a stereocontrolled manner and reduced to *nor*-coronamic acid. These interesting substances should prove to be valuable Michael acceptors for preparing a variety of difficult β-substituted amino acids.

Removal of the t-BOC group from **459** with either trimethylsilyl iodide or TFA furnishes the secondary amine **565**. Reaction of this substance with an aldehyde (eg., benzaldehyde) in the presence of acid and dimethylmaleate furnishes the bicyclic substances **566** and **567** in 70%

SCHEME 115

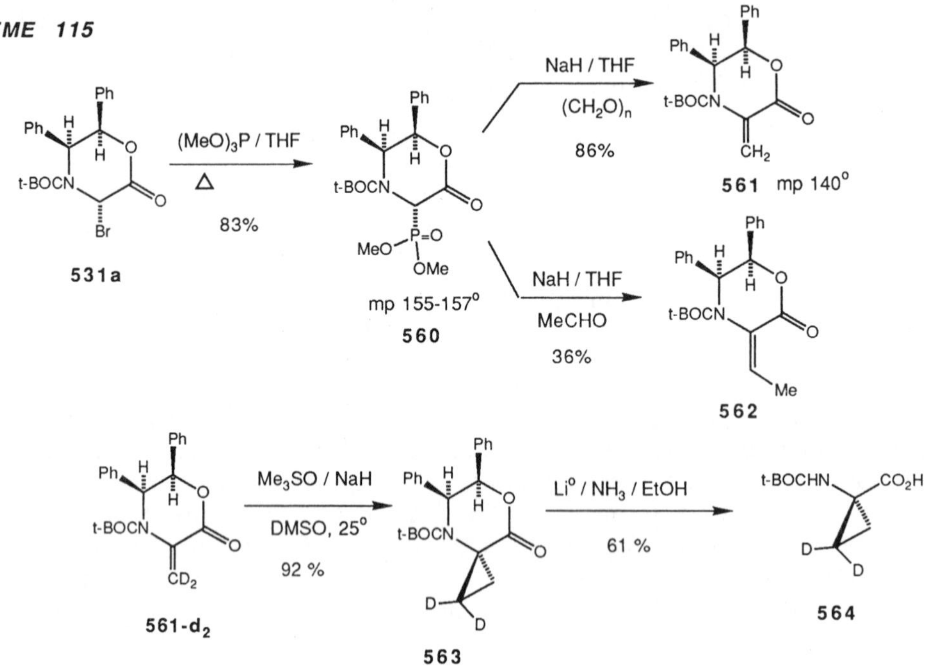

SCHEME 116

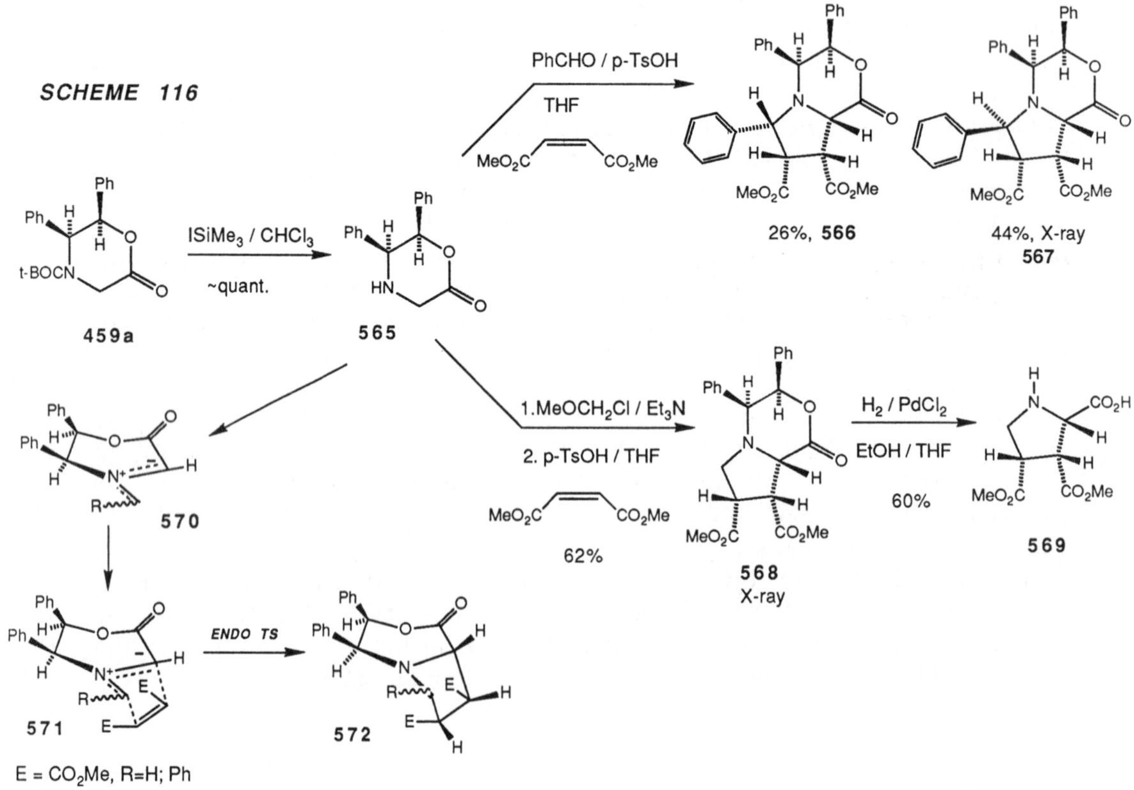

E = CO₂Me, R=H; Ph

combined yield. An X-ray stereostructure for the major isomer (**567**) secured the relative stereochemistry and NOE experiments on **566** secured the relative stereochemistry shown. Alternatively, condensation of **565** with chloromethyl methyl ether furnishes the labile hemi-aminal that is directly treated with acid and dimethylmaleate as above to furnish a single stereoisomeric adduct **568**; again, an X-ray structure firmly secured the relative stereochemistry. Reduction of **568** affords the 3,4-di-carbomethoxy proline (**569**) in quantitative yield. The relative stereochemistries of the dipolar cycloaddition adducts indicates that the ylide **570** that is generated in-situ, suffers exclusively *endo* attack by the olefin from the least hindered face (**571**, shown). The epimeric mixture in the case of **566/567** merely reflects the E : Z ratio of the incipient ylides ; complete *endo*-selectivity is observed in both systems. This asymmetric version of the well-known 1,3-dipolar cycloadditions of amino acid derivatives should find numerous applications in the preparation of optically active, substituted pyrrolidines and prolines.

These workers have also recently demonstrated that the bromides **531** undergo stereoselective reactions with aryl cuprates and electron-rich aromatic compounds under Friedel-Crafts conditions (Scheme 117). Unlike all of the homologated substances discussed above, the adducts **573** are subject to reductive cleavage at the Ar-C-N linkage and therefore requires an alternate protocol for removing the chiral auxilliary in accessing the difficult, racemization-prone α-aryl glycines. Employing the method of Weineges[112] used on a related oxazinone, the lactones **573** are treated with p-TsOH or trimethylsilyl iodide to remove the t-BOC group and then hydrolyzed in aqueous HCl to afford the hydroxy acids **574**. Treatment of these substances with sodium *meta*-periodate at pH 3 leads to clean *oxidative* cleavage of two molar equivalents of benzaldehyde from **574** without destroying the Ar-C-N linkage. The corresponding α-aryl glycine derivatives **575** are obtained in good overall yields and good to excellent %ee's.

Belokon, et.al.[113], have examined the electrophilic properties of the bromides **576** obtained from the above-described Nickel complex **480**. Bromination of **480** with bromine in isoprpyl alcohol affords a diastereomeric mixture (ratio = 2 : 1) of **576**. Nucleophilic additions to the purified bromide, results in the adducts **577**; the *anti*-isomers are formed with excellent diastereoselectivity (Table) in accordance with the corresponding enolate reactions discussed earlier. In the coupling with diethyl malonate anion, the corresponding L-Asp (**578**) is obtained in 50% yield and 80% ee by aqueous HCl treatment and recovery of **480** (92% yield).

SCHEME 117

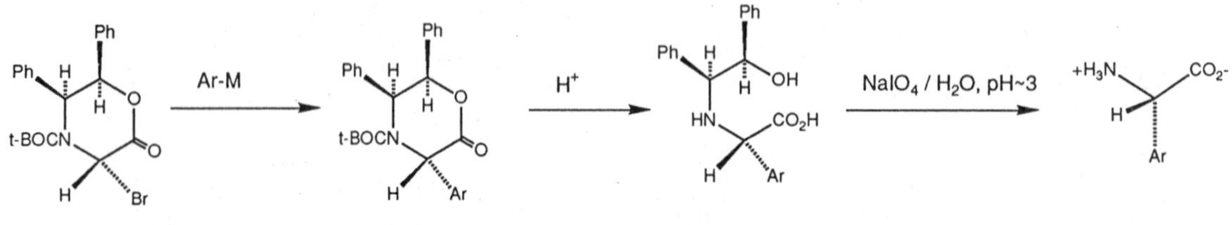

Ar-M	SOLVENT / CONDITIONS	%573	%574	%575	%ee (575)
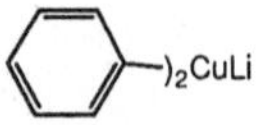 (phenyl—)$_2$CuLi	Et$_2$O / THF, -78° C	45	71	60	82
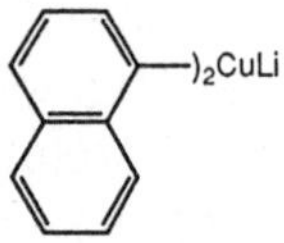 (naphthyl—)$_2$CuLi	Et$_2$O / THF, -78° C	43	67	65	94
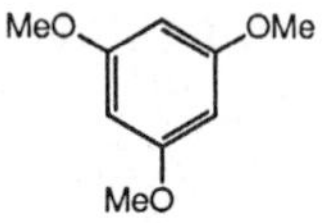 (3,5-dimethoxyphenyl)	THF / ZnCl$_2$ 25° C	83	~quant.	70	91
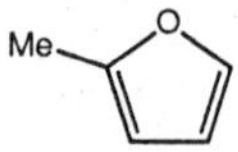 (5-methylfuran)	THF / ZnCl$_2$ 25° C	68	86	85	93
 (furan)	THF / ZnCl$_2$ 25° C 4A	43		26 (3 steps)	90

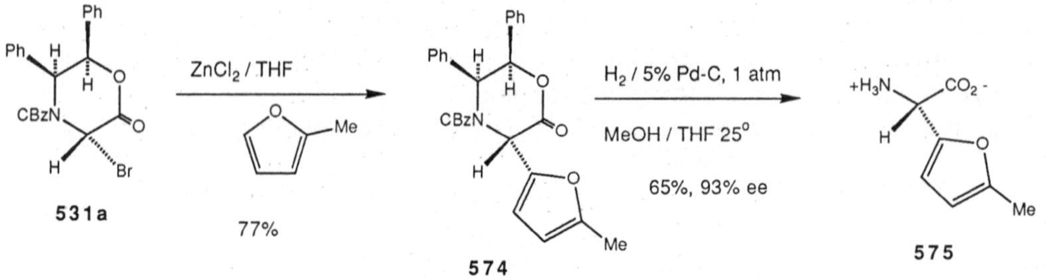

SCHEME 118

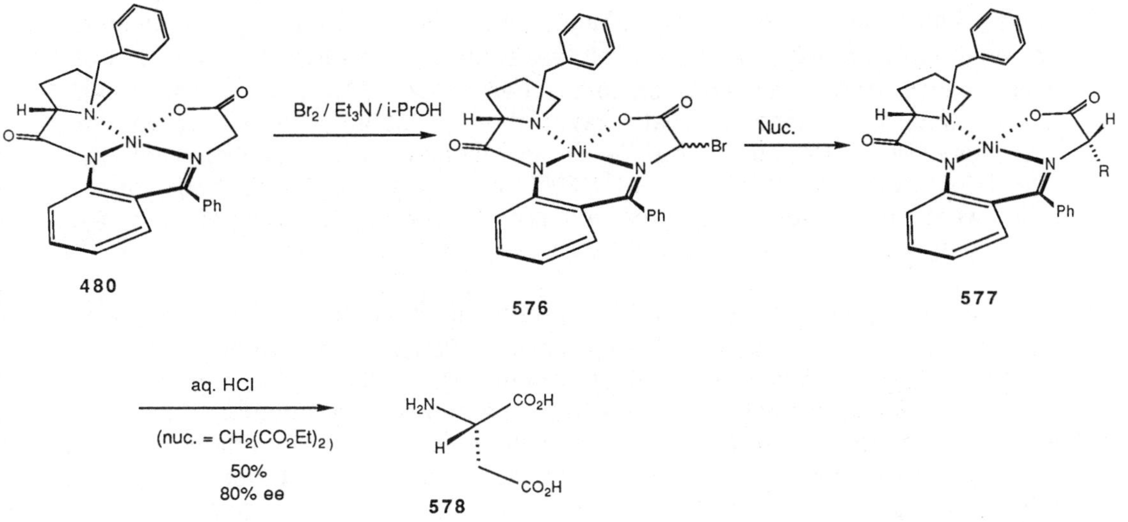

480 → 576 → 577

Reagents: Br$_2$ / Et$_3$N / i-PrOH; Nuc.

aq. HCl
(nuc. = CH$_2$(CO$_2$Et)$_2$)
50%
80% ee

578

Belokon, Y.N.; Popkov, A.N.; Chernoglazova, N.I.; Saporovskaya, M.B.; Bakhmutov, V.I.; Belikov, V.M., *J.Chem.Soc.Chem.Comm.* (1988) 1336.

NUCLEOPHILE	CONDITIONS	DIASTEREOMER RATIO (577)	YIELD (% 577)
NaOMe	MeOH, rt	98 : 2	>90
NaOPh	MeCN	91 : 9	60
Me$_2$NH	MeCN, rt	99 : 1	>90
CH$_2$(CO$_2$Et)$_2$	MeCN, KOt-Bu, rt	9 : 1	>90
n-BuLi	THF, -70		
n-BuZnCl	THF, -70	4 : 1	13

Mukaiyama and associates [114] examined the condensations of tin(II) enolates with the optically active Schiff base **580** (Scheme 119). The relative stereochemistry of the adducts (**581**) was ascertained by closure to the β-lactams **582**, reduction of the esters , reductive removal of the chiral benzyl group and acylation to the benzoates **584** and **585**. The Table shows that the reactions greatly favor the cis isomers (**584**). While the chemical yields are good, the %ee's are modest. These selectivities are consistent with the related chiral Schiff base couplings described above. Somewhat better %ee's (77-84% ee) are obtained with the phenylglycine-derived imine **586**, but the *cis : trans* ratios are not as high (75 : 25; 67 : 33) .

A daring electrophilic glycine system has recently been investigated by Easton, and co-workers [115] as shown in Scheme 120. Several glycine-containing dipeptides are brominated with NBS and coupled with several nucleophiles. Methanolysis, allylation, malonate condensation and tri-butyl tin deuteride reductions are examined for replacement of the bromide. The yields are generally modest and the diastereoselectivities are generally very poor ranging from *ca.* 3 : 1 to 1 : 1 in most cases determined. Being acyclic systems, it is expected that good stereocontrol will not be easy to obtain since the stereogenic center is three atoms away from the reacting center and is subject to conformational motion. It is significant that only the glycine moiety is oxidized in the bromination which should provide a provocative framework from which to more carefully investigate the general strategy of modification of peptides.

SCHEME 119

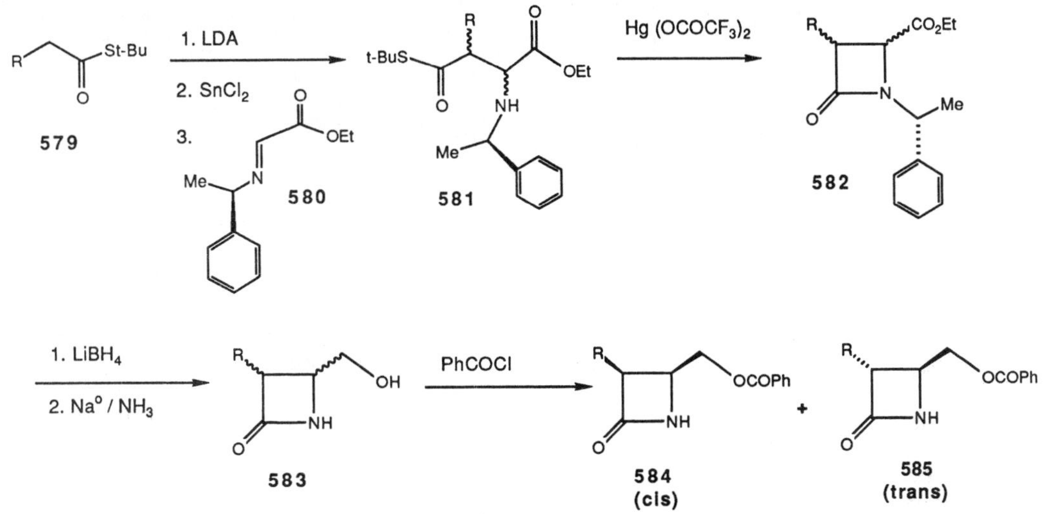

Yamada, T.; Suzuki, H.; Mukaiyama, T., *Chem. Lett.* (1987) 293.

R	Yield (% 581)	cis : trans	%ee
Me	78	91 : 9	70
Et	78	95 : 5	72
i-Pr	79	95 : 5	71

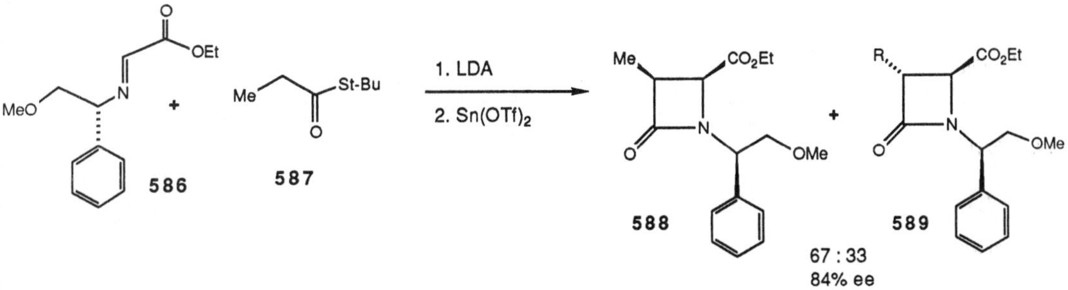

67 : 33
84% ee

SOAA—E*

SCHEME 120

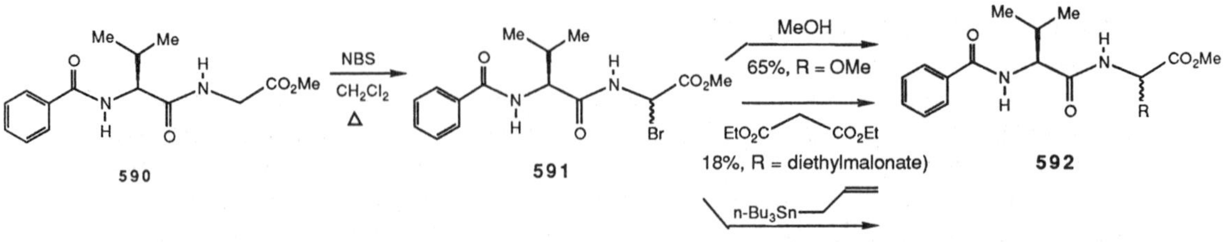

Easton, C.J.; Scharfbilling, I.M.; Tan, E.W., *Tetrahedron Lett.* (1988) **29**, 1565.   (R=allyl) 3 : 1

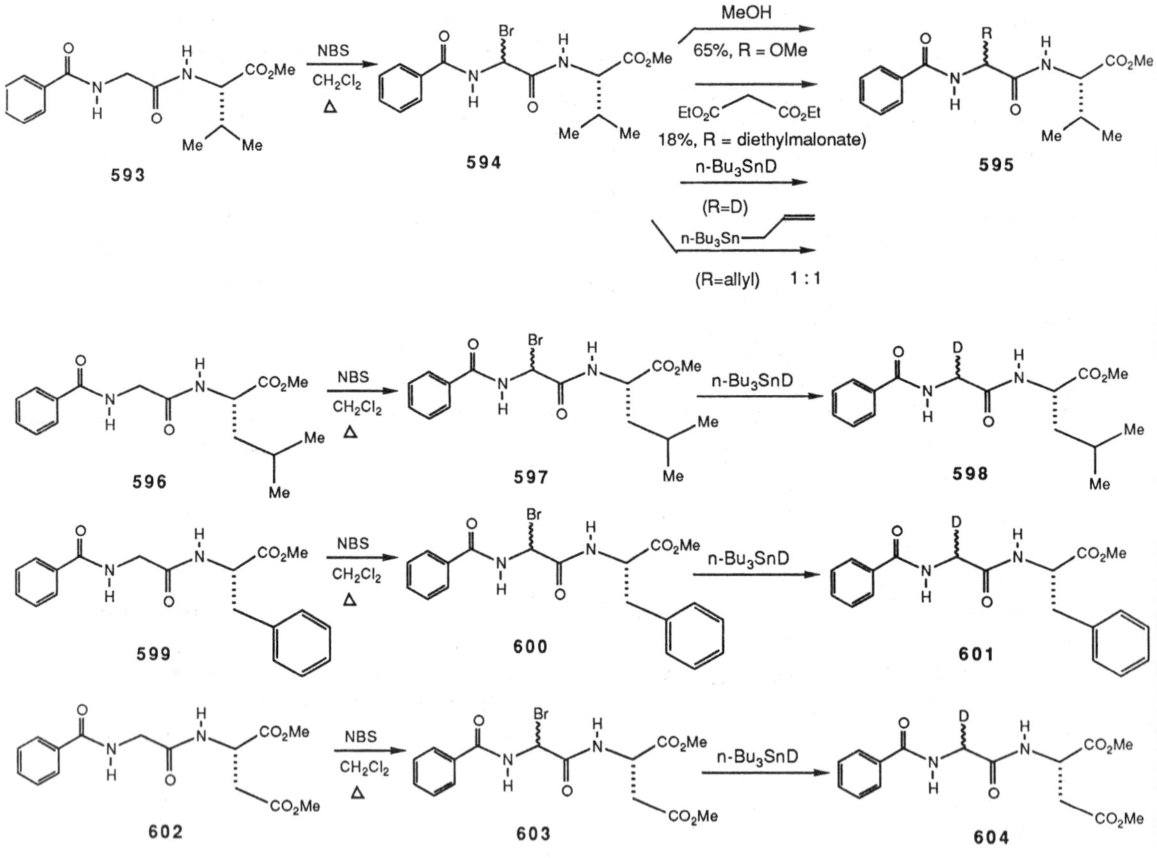

(R=allyl) 1 : 1

C. MISCELLANEOUS METHODS

1. Retroracemization

Several reports have appeared dealing with the retroracemization or deracemization of racemic amino acids. While there is no truly general *chemical* method that produces very high levels of enantiomeric excess, the following examples serve to illustrate a promising area that may become increasingly significant. The *enzymatic* kinetic resolutions which effectively perform the same task will be discussed in Chapter 7.

Belokon, and associates [116] have examined the Cu(II) complexation of **605** to various racemic amino acids (**606**) under thermodynamic equilibration in the presence of base (Scheme 121). As observed above for these types of complexes, the *anti-* complex **608** is more stable due to steric interactions of the R$_2$ group with the N-benzyl residue of the proline moiety. After reaction with sodium methoxide for 1 hour at ambient temperature, the complex is decomposed with 1 N HCl; the optically enriched amino acid (**609**) is separated from the chiral auxilliary (**605**) which is generally recovered in high yield without loss of optical integrity. The optical purities are generally low with 55% being the best example (valine).

An interesting system for the preparation of chiral glycine has been examined by Bosnich and co-workers [117] as shown in Scheme 122. Pyridine-2-carboxylic acid (**610**) is condensed with glycine ethyl ester to furnish the amide **611**. This substance is complexed to the C-2 symmetric triamine **612** with dicobalt hexacarbonyl to form the complex **613**. The enantiotopic hydrogens are then exchanged selectively with deuterium oxide at pH=11.2 and the reaction quenched with DCl. Ion-exchange separation and acid/base manipulation of the final product furnishes the optically active glycine derivative **614**. The authors note that the enantiomeric excess increases with the extent of reaction providing 88% ee at 50.5% deuteration and extrapolated to ~100% ee at 72% deuteration. The practical problem with this approach, is that although the %ee can become very high at extended reaction time, the percent of the *dideuterio* glycine also increases with the extent of reaction. While this method is extremely interesting from a mechanistic standpoint, the high levels of contamination by the dideuterio species renders this chemistry impractical for synthetic preparations of optically active mono-deuterio glycine.

Duhamel and associates[118-123] have extensively studied the enantioselective protonation of lithium enolates of amino acid Schiff bases as shown in Scheme 123. Treatment of the benzylidene derivatives **615** with LDA furnishes the enolates **616**; subsequent protonation with various tartrates (**617**) provides the optically enriched substances **618**.

SCHEME 121

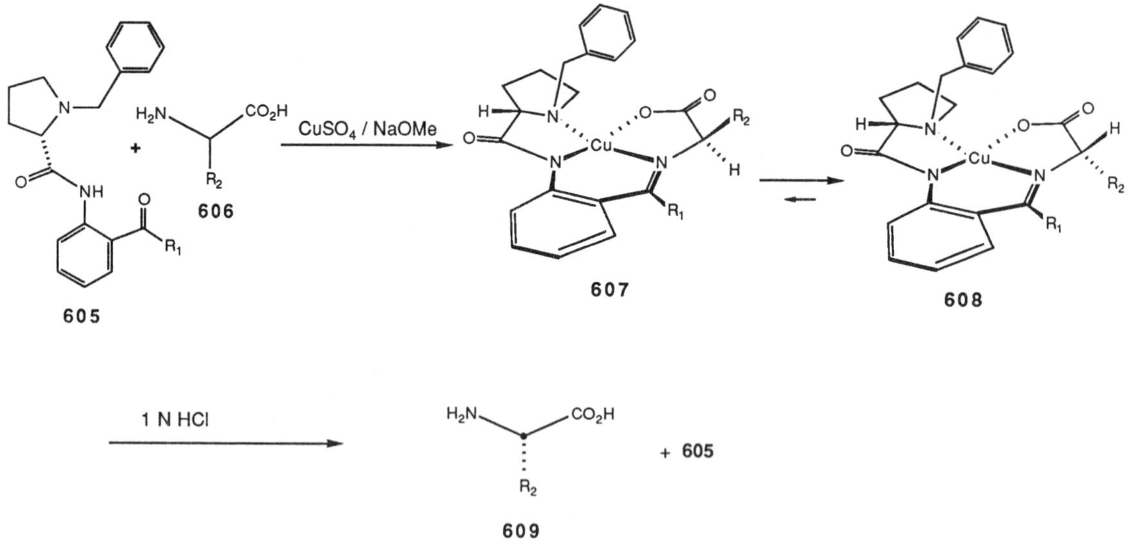

Belokon, Y.N.; Zel'ter, I.E.; Bakhmutov,V.I.; Saporovskaya, M.B.; Ryzhov, M.G.; Yanovsky, A.I.; Struchkov, Y.T.; Belikov, V.M., *J.Am.Chem.Soc.* (1983) **105**, 2010.

R₁	R₂	ENANTIOMERIC PURITY (609)
H	Me	0
H	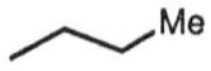 (propyl)—Me	12% (S)
H	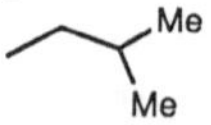 (sec-butyl)—Me	22% (S)
H	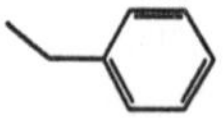 (benzyl ethyl)	42% (S)
H	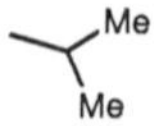 isopropyl (Me, Me)	54% (S)
H	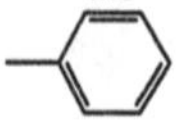 (benzyl)	35% (S)
Me	Me	36% (S)
Me	Me	33% (S)

SCHEME 122

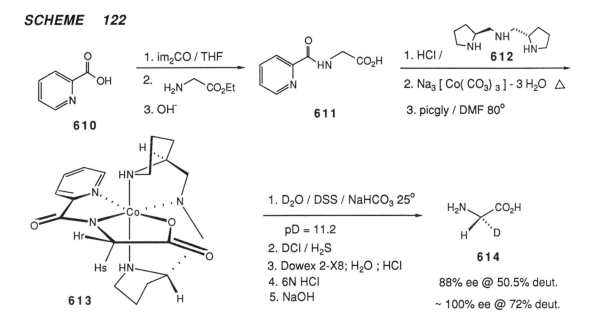

Dukuzovic, Z.; Roberts, N.K.; Sawyer, J.F.; Whelan, J.; Bosnich, B., *J.Am.Chem.Soc.* (1986) **108**, 2034.

The chemical yields are rather good, but the enantiomeric excess is generally modest; the best example is 62% ee for the phenyl glycine case employing the 1-adamantyl tartrate.

These workers have also examined combining the enantioselective protonation with a chiral amide base (**620**, Scheme 124). The authors hypothesize that the lithium enolate (i.e., **621**) is coordinated to the amine employed as the base. The additional chirality of the optically active ligand (L*) improves the %ee up to a maximum of 70% ee in the phenyl glycine case.

In a related study, Duhamel and co-workers have also examined the asymmetric carboxylation of the α-lithioamine **626** employing a chiral amide base (**625**, Scheme 125). This disconnection for the synthesis of amino acids is a surprisingly rarely studied approach. Numerous new methods for the asymmetric carbanionic C-C functionalization α- to nitrogen have recently become available and may portend a future area of investigation. Carboxylation of the anion **626** with various chloroformates

SCHEME 123

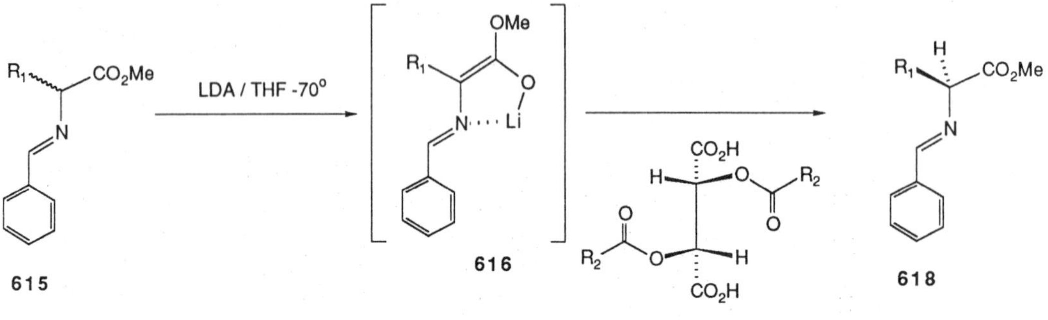

Duhamel, L.; Plaquevent, J-C., *J.Am.Chem.Soc.* (1978) **100**, 7415.
Duhamel, L.; Plaquevent, J-C., *Bull.Soc.Chim.Fr.* (1982) II-75

R_1	R_2	S : R RATIO (618)	YIELD (%618)
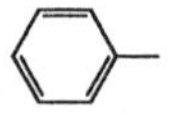 (phenyl)	t-Bu	79 : 21	85
	t-BuCH$_2$	59 : 41	84
	t-BuCH$_2$CH$_2$	70 : 30	82
	Ph	57 : 43	80
	PhCH$_2$	55 : 45	81
	PhCH$_2$CH$_2$	54 : 46	83
	(Me)$_2$CBr	70 : 30	85
	cyclohexyl	71 : 29	85
	β-styryl	65 : 35	86
	1-adamantyl	**81 : 19**	**79**
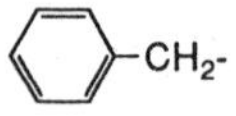 (benzyl) -CH$_2$-	t-Bu	63 : 37	65
	t-BuCH$_2$	54 : 46	64
	t-BuCH$_2$CH$_2$	59 : 41	65
	Ph	**67 : 33**	**60**
	PhCH$_2$	57 : 43	61
	PhCH$_2$CH$_2$	51 : 49	64
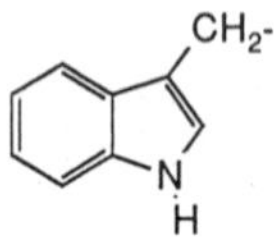 (indol-3-yl)-CH$_2$-	**t-Bu**	**65 : 35**	**95**
	t-BuCH$_2$	55 : 45	93
	t-BuCH$_2$CH$_2$	57 : 43	92
	Ph	56 : 44	90
	PhCH$_2$	52 : 48	95
	PhCH$_2$CH$_2$	53 : 47	95
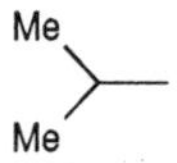 (isopropyl, Me$_2$CH)	t-Bu	67 : 33	82
	Ph	**69 : 31**	**76**
	cyclohexyl	67 : 33	80

SCHEME 124

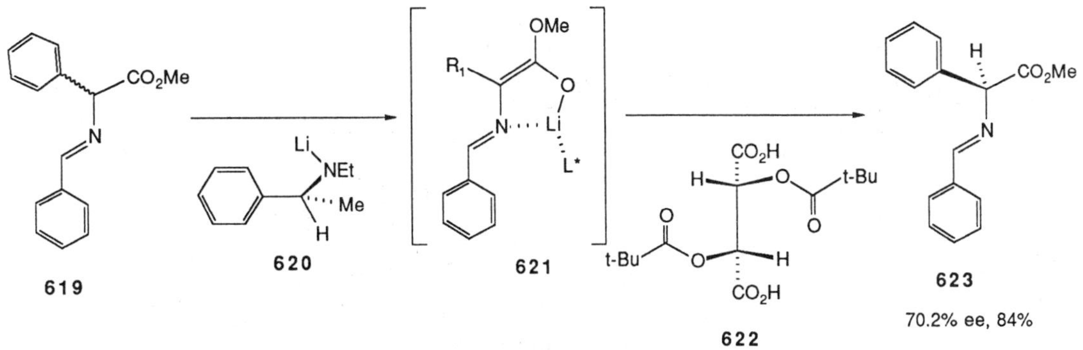

Duhamel, L.; Plaquevent, J-C., *Bull.Soc.Chim.Fr.* (1982) II-75
Duhamel, L.; Plaquevent, J-C., *Tetrahedron Lett.* (1980), **21**, 2521
Duhamel, L.; Fouquay, S.; Plaquevent, J-C., *Tetrahedron Lett.* (1986), **27**, 4975
Duhamel, L.; Duhamel, P.; Launay, J-C.; Plaquevent, *Bull.Soc.Chim.Fr.* (1984) II-421

SCHEME 125

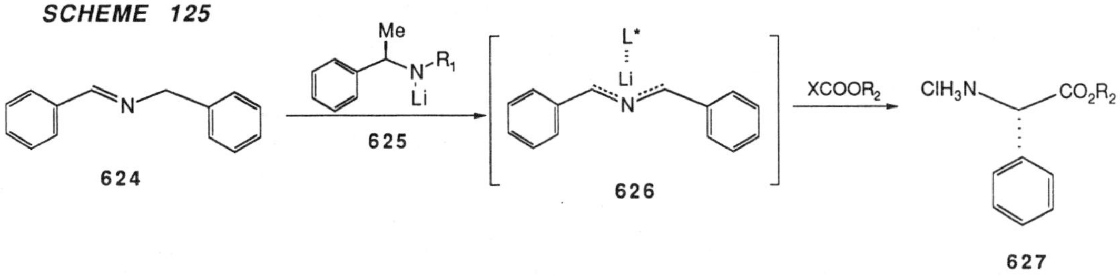

Duhamel, L.; Duhamel, P.; Fouquay, S.; Eddine, J.J.; Peschard, O.; Plaquevent, J-C.; Ravard, A.;
Solliard, R.; Valnot, J-Y.; Vincens, H., *Tetrahedron* (1988) **44**, 5506.

R₁	R₂	X	YIELD (%627)	%ee
Et	Me	MeO	40	0
Et	Me	Cl	58	35 (S)
Pr	Me	Cl	60	32 (S)
Et	Et	Cl	56	41 (S)
Pr	Et	Cl	55	34 (S)
Et	Bu	Cl	40	40 (S)

or dimethyl carbonate furnishes the esters **627** after acidic removal of the benzylidene. The %ee's are low ranging from 0-41%ee. It would seem rather promising to examine the asymmetric carboxylation of the elegant chiral formamidine systems developed by Meyers [67].

2.Chiral Aziridines

The aziridine or azirine moiety is rarely found in nature; two interesting examples are the natural azirines azirinomycin (**628**), an antibiotic obtained from *Streptomyces aureus* [124] and the recently discovered cytotoxic marine natural product dysidazirine, isolated[125] from the marine sponge *Dysidea fragilis*. These unusual structures provide a new and interesting challenge to the synthetic chemist but also contain potential for designing new types of biologically active amino acids.

628, AZIRINOMYCIN   **629, DYSIDAZIRINE**

Seebach and associates[126] have recently examined the diastereoselective alkylations of the optically active aziridines **632** and **633** prepared from (S)-phenethylamine (**630**) and 2,3-dibromopropionate (**631**, Scheme 126). The two diastereomers are separated by chromatography and separately examined for enolate alkylations. Compound **632** displayed highly diastereoselective reactions furnishing **634** with overall net *retention* of stereochemistry. In contrast, the other isomer **633** gave mixtures of **634** and **635**; net *retention* being the predominant pathway. The authors conclude that the conformation of the lithium carbanion derived from **633** is more sterically congested than in the case of **632**; partial equilibration of lithio-**633** to the more stable lithio-**632** is invoked to explain the relatively poor selectivity displayed in the case of **633**. Related N-benzyl and N-*tert*-butyl aziridine carboxylates obtained from serine are also found to undergo stereoselective alkylation reactions giving optically active products. These species are all proposed to possess pyramidal carbanion structures with the non-bonded electron pairs on nitrogen and carbon occupying *anti*-relationships on the three-membered ring.

SCHEME 126

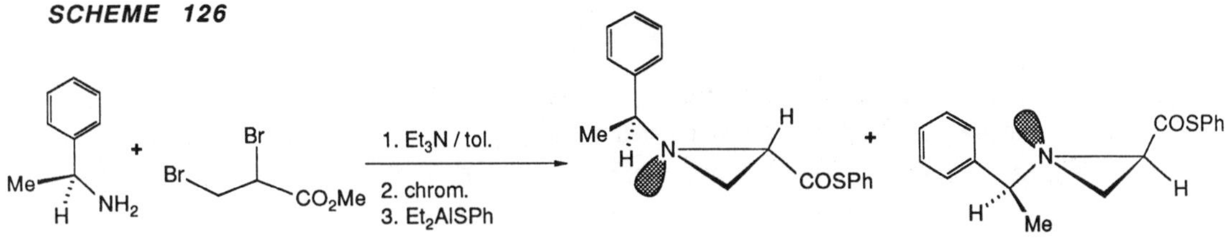

Haner, R.; Olano, B.; Seebach, D., *Helv.Chim.Acta.* (1987) **70**, 1676

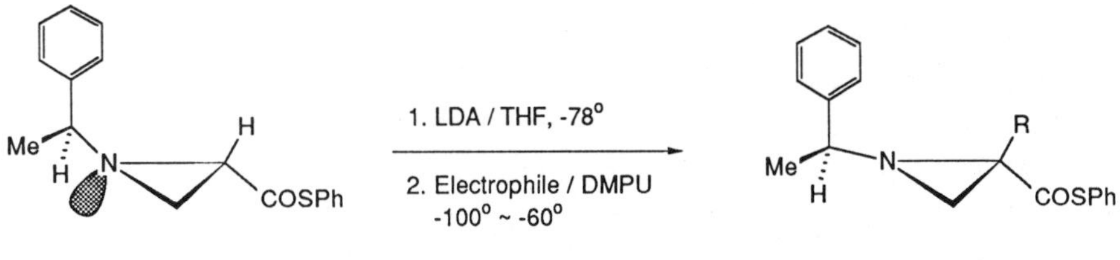

ELECTROPHILE	R	YIELD (% 634)
MeOD	D	55
MeI	Me	62
EtI	Et	51
 ⟶⟶—Br	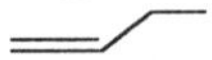 ⟶⟶	59
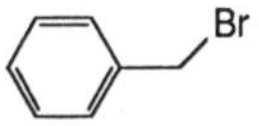 (benzyl)—Br	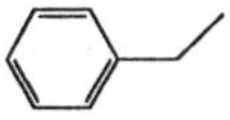 (phenethyl)	60
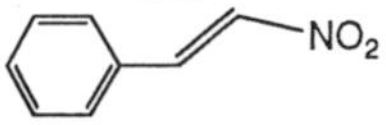 (styryl)—NO₂	 (benzyl-CH-CH₂)—NO₂	79

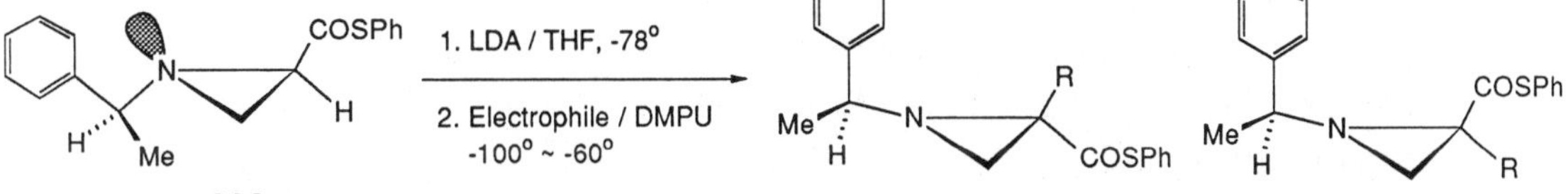

ELECTROPHILE	R	YIELD (% 634 + 635)	RATIO (634 : 635)
MeOD	D	~60	1 : 2
MeI	Me	51	1 : 4
PhCH₂Br	PhCH₂	57	1 : 2

References   Chapter   1

1. For summaries of the bis-lactim ether method, see: a) Schollkopf, U., *Tetrahedron* (1983) **39**, 2085; b) Schollkopf, U., *Pure & Appl. Chem.* (1983) **55**, 1799; c) Schollkopf, U.,*Topics Curr. Chem.* (1983) **109**, 65.

2. a) Schollkopf, U.; Hartwig, W.; Groth, U., *Angew.Chem. Int.Ed.Engl.* (1979) **18**, 863; b) Schollkopf, U.; Hartwig, W.; Groth, U.; Westphalen, K-O., *Liebigs Ann.Chem.* (1981) 696.

3. Schollkopf, U.; Hartwig, W.; Groth, U., *Angew.Chem.Int.Ed.Engl.* (1980) **19**, 212.

4. Schollkopf, U.; Groth, U.; Hartwig, W., *Liebigs Ann.Chem.* (1981) 2407.

5. Schollkopf, U.; Groth, U.; Deng, C., Angew.Chem.Int.Ed.Engl. (1981) 20, 798.

6. Schollkopf, U.; Hartwig, W.; Pospischil, K-H.; Kehne, H., *Synthesis* (1981) 966.

7. Schollkopf, U.; Groth, U.; Westphalen,K-O.; Deng, C., *Synthesis* (1981) 969.

8. Schollkopf, U.; Neubauer, H-J., *Synthesis* (1982) 861.

9. Schollkopf, U.; Busse, U.; Kilger, R.; Lehr, P., *Synthesis* (1984) 271.

10. Jiang, Y.; Schollkopf, U.; Groth, U., *Scientia Sinica B* (1984) 27, 566.

11. Schollkopf, U.; Groth, U., *Angew.Chem.Int.Ed.Engl.*, (1981) **20**, 977.

12. Schollkopf, U.; Nozulak, J.; Groth, U., *Tetrahedron* (1984) **40**, 1409.

13. Groth, U.; Schollkopf,U.; Chiang,Y-C., *Synthesis* (1982) 864.

14. For reviews of vinyl glycine derivatives and other unsaturated amino acids as suicide inhibitors, see: a) Walsh, C., *Enzymatic Reaction Mechanisms* (1979) W.H. Freeman, San Francisco; b) Walsh, C., *Tetrahedron* (1982) **38**, 871.

15. Nozulak, J.; Schollkopf, U., *Synthesis* (1982) 866.

16. Schollkopf, U.; Nozulak, J.; Groth, U., *Synthesis* (1982) 868.

17. Groth, U.; Chiang, Y-C.; Schollkopf, U., *Liebigs Ann.Chem.* (1982) 1756.

18. Groth, U.; Schollkopf, U., *Synthesis* (1983) 37.

19. Schollkopf, U.; Nozulak,J.; Grauert, M., *Synthesis* (1985) 55.

20. Grauert, M.; Schollkopf, U., *Liebigs Ann.Chem.* (1985) 1817.

21. Groth, U.; Schollkopf, U., *Synthesis* (1983) 673.

22. Schollkopf, U.; Bardenhagen, J., *Liebigs Ann.Chem.* (1987) 393.

23. Neubauer, H-J.; Baeza, J.; Freer, J.; Schollkopf, U., *Liebigs Ann.Chem.* (1985), 1508.

24. Gull, R.; Schollkopf, U., *Synthesis* (1985) 1052.

25. Schollkopf, U.; Pettig,D.; Busse, U., *Synthesis* (1986) 737.

26. Schollkopf, U.; Schroder, J., *Liebigs Ann.Chem.* (1988) 87.

27. Schollkopf, U.; Kuhnle, W.; Egert, E.; Dyrbusch, M., *Angew.Chem.Int.Ed.Engl.* (1987) **26**, 480.

28. Pettig, D.; Schollkopf, U., *Synthesis* (1988) 173.

29. Schollkopf, U.; Pettig, D.; Schulze, E.; Klinge, M.; Egert, E.; Benecke, B.; Noltemeyer, M., *Angew.Chem.Int.Ed.Engl.* (1988) **27**, 1194.

30. Schollkopf, U.; Tiller, T.; Bardenhagen, J., *Tetrahedron* (1988) **44**, 5293.

31. Schollkopf, U.; Westphalen, K-O.; Schroder, J.; Horn, K., *Liebigs Ann.Chem.* (1988) 781.

32. Schollkopf, U.; Hinrichs, R.; Lonsky, R., *Angew.Chem. Int. Ed. Engl.* (1987) **26**, 143.

33. Schollkopf, U.; Lonsky, R., *Synthesis* (1983) 675.

34. Schollkopf, U.; Lonsky, R.; Lehr, P., *Liebigs Ann. Chem.* (1985) 413.

35. Subramanian, P.K.; Woodard, R.W., *J. Org. Chem.* (1987) **52**, 15.

36. Holler, T.P.; Spaltenstein, A.; Turner, E.; Klevit, R.E.; Shapiro, B.M.; Hopkins, P.B., *J. Org. Chem.* (1987) **52**, 4421.

37. Baldwin, J.E.; Adlington, R.M.; Robinson, N.G., *Tetrahedron Lett.* (1988) **29**, 375.

38. Hartwig, W.; Born, L., *J.Org.Chem.* (1987) **52**, 4352.

39. Schollkopf, U.; Tolle,R.; Egert, E.; Nieger, M., *Liebigs Ann.Chem.* (1987) 399.

40. For some early (and low %ee)Schiff base-derived asymmetric syntheses of amino acids, see: a)from (-)-menthone : Oguri, T.; Shioiri, T.; Yamada, S-i., *Chem.Pharm.Bull.* (1977)**25**, 2287 and references cited therein;b) from α-phenethylamines : Harada, K.; Tamura, M.; Suzuki, S., *Bull.Chem.Soc. Jpn.* (1978) 2171; c) with chiral bases : Yamashita, T.; Mitsui, H.; Watanabe, H.; Nakamura, N., *Bull.Chem.Soc.Jpn.* (1982) **55**, 961; d)Yamada, S-I.; Oguri, T.; Shioiri, T., *J .C.S.Chem.Comm.* (1976) 136.; e) Oguri, T.; Kawai,N.; Shioiri, T.; Yamada, S-I., *Chem. Pharm. Bull.* (1978) **26**, 803;f) a subsequent report on α,α-disubstitution from the hydroxypinanones appeared : Bajgrowicz, J.A.; Cossec, B.; Pigiere, Ch.; Jacquier, R.; Viallefont, Ph., *Tetrahedron Lett.* (1983) **24**, 3721; El Achqar, A.; Boumzebra, M.; Roumestant, M-L.; Viallefont, P., *Tetrahedron* (1988) **44**, 5319; El Achqar, A.; Roumestant, M-L.; Viallefont, P., *Tetrahedron Lett.* (1988) **29**, 2441; g) asymmetric synthesis (95% ee) of Aoe (see **196**) utilizing the hydroxypinanone auxilliary : Jacquier, R.; Lazaro,R.;Raniriseheno,   H.; Viallefont, P., *Tetrahedron Lett.*, (1984) **25**, 5525; h) Casella, L.; Jommi, G.; Montanari,   S.; Sisti, M., *Tetrahedron Lett.* (1988) **29**, 2067.

41. a)McIntosh, J.M.; Mishra, P., *Can.J.Chem.* (1986) **64**, 726.;b)McIntosh, J.M.; Leavitt, R.K., *Tetrahedron Lett.* (1986)**27**, 3839; c) McIntosh, J.M.; Leavitt, R.K.; Mishra, P.; Cassidy, K.C.; Drake, J.E.; Chadha, R., *J.Org.Chem.* (1988)**53**, 1947.

42. Jiang, Y.; Liu, G.; Deng, R.; Wu, S., *The Third International Kyoto Conference on New Aspects of Organic Chemistry*, November, 1985 Kyoto, Japan, Abstr.#O-29.

43. Kuzuhara,H.; Watanabe, N.; Ando, M., *J.C.S.Chem.Comm.* (1987) 95.

44. a)Breslow, R.; Czarnik,A.W.; Lauer, M.; Leppkes, R.; Winkler, J.; Zimmerman, S., *J.Am.Chem.Soc.* (1986) **108**, 1969. ; b) Tabushi, I.; Kuroda, Y.; Yamada, M.; Higashimura, H.; Breslow, R., *J.Am.Chem.Soc.* (1985) **107**, 5545; c) Breslow, R.; Chmielewski, J.; Foley, D.; Johnson, B.; Kumabe, N.; Varney, M.; Mehra, R., *Tetrahedron* (1988) **44**, 5515.

45.   a)Ikegami, S.; Hayama, T.; Katsuki, T.; Yamaguchi, M., *Tetrahedron Lett.* (1986) **27**, 3403; b)Ikegami, S.; Uchiyama, H.; Hayama, T.; Katsuki, T.; Yamaguchi, M., *Tetrahedron* (1988) **44**, 5333.

46.   Evans,D.A.; Weber, A.E., *J.Am.Chem.Soc.* (1986) **108**, 6757.

47.   Evans, D.A.; Weber, A.E., *J.Am.Chem.Soc.* (1987) **109**, 7151.

48.   a)Genet, J.P.; Ferroud, D.; Juge, S.; Montes, J.R., *Tetrahedron Lett.* (1986) **27**, 4573; b) Genet, J-P.; Juge, S.; Montes, J.R.; Gaudin, J-M., *J.Chem.Soc.Chem.Comm.* (1988)718.

49.   Ferroud,D.; Genet, J.P.; Kiolle, R., *Tetrahedron Lett.* (1986) **27**, 23.

50.   a)Duhamel, P.; Eddine, J.J.; Valnot, J-Y., *Tetrahedron Lett.* (1987) **28**, 3801; b) Duhamel, P.; Eddine, J.J.; Valnot, J-Y., *Tetrahedron Lett.* (1984) **25**, 2355; c) Duhamel, P.; Valnot, J-Y.; Eddine, J.J., *Tetrahedron Lett.* (1982) **23**, 2863; d) Duhamel, L.; Duhamel, P.; Fouquay, S.; Eddine, J.J.; Peschard, O.; Plaquevent, J-C.; Ravard, A.; Solliard, R.; Valnot, J-Y.; Vincens, H., *Tetrahedron* (1988) **44**, 5495.

51.   Schollkopf, U.; Hausberg, H.H.; Hoppe, I.; Segal, M.; Reiter, U., *Angew. Chem. Int.Ed. Engl.* (1978) **17**, 117.

52.   Schollkopf, U.; Hausberg, H.H.; Segal, M.; Reiter, U.; Hoppe, I.; Saenger, W.; Lindner, K., *Liebigs Ann. Chem.* (1981) 439.

53.   Hartwig, W.; Schollkopf, U., *Liebigs Ann. Chem.* (1982) 1952.

54.   Schollkopf, U.; Scheuer, R., *Liebigs Ann. Chem.* (1984) 939.

55.   Decorte, E.; Toso, R.; Sega, A.; Sunjic, V.; Ruzic-Toros, Z.; Kojic-Prodic, B.; Bresciani-Pahor, N.; Nardin, G.; Randaccio, L., *Helv. Chim.Acta.* (1981) **64**,1145.

56.   a) Ito, Y.; Sawamura, M.; Hayashi, T., *J.Am.Chem.Soc.* (1986) **108**, 6405; b) Ito, Y.; Sawamura, M.; Hayashi, T., *Tetrahedron Lett.* (1987) **28**, 6215.

57.   Ito, Y.; Sawamura, M.; Shirakawa, E.; Hayashizaki, K.; Hayashi, T., *Tetrahedron Lett.* (1988) **29**, 235.

58.   Ito, Y.; Sawamura, M.; Hayashi, T., *Tetrahedron Lett.* (1988) **29**, 239.

59.   Ito, Y.; Sawamura, M.; Matsuoka, M.; Matsumoto, Y.; Hayashi, T., *Tetrahedron Lett.* (1987) **28**, 4849.

60.   a) Ojima, I.; Chen, H-J.C.; Nakahashi, K., *J.Am.Chem.Soc.* (1988) **110**, 278; b) Ojima, I.; Shimizu, N.; Qiu, X.; Chen, H-J.C.; Nakahashi, K., *Bull.Chem.Soc.Fr.*, (1987) 649.

61.   Evans, D.A.; Sjogren, E.B., *Tetrahedron Lett.* (1985) **26**, 3783.

62.   Ojima, I.; Qiu, X., *J.Am.Chem.Soc.* (1987) **109**, 6537.

63.   Nakatsuka, T.; Miwa, T.; Mukaiyama, T., Chem.Lett. (1981) 279.

64.   Marco, J.L.; Royer, J.; Husson, H-P., *Tetrahedron Lett.* (1985) **26**, 3567.

65.   a) Marco, J.L., *Heterocycles* (1987) **26**, 2579; b)Aitken, D.J.; Royer, J.; Husson, H-P., *Tetrahedron Lett.* (1988) **29**, 3315.

66. a) Kolb, M.; Barth, J., *Liebigs Ann.Chem.* (1983) 1668; b)Kolb, M.; Barth, J., *Tetrahedron Lett.* (1979) 2999; c) Kolb, M.; Barth, J., *Angew.Chem.Int.Ed.Engl.* (1980) **19**, 725.

67. For a review of chiral formamidine use in asymmetric synthesis, see:Meyers, A.I., *Aldrichimica Acta* (1985) **18**, 59.

68. An extensive review of Seebachs' amino acid methodology has appeared: Seebach, D.; Imwinkelried, R.; Weber, T., *Modern Synthetic Methods* (1986) **4**, 128, Springer-Verlag, Heidelberg.

69. a)Seebach, D.; Naef, R., *Helv.Chim.Acta* (1981) **64**, 2704; b) Seebach, D.; Boes, M.; Naef, R.; Bernd Schweizer, W., *J.Am.Chem.Soc.* (1983) **105**, 5390.

70. Williams, R.M.; Glinka, T.; Kwast, E., *J.Am.Chem.Soc.* (1988) **110**, 5927.

71. Weber, T.; Seebach, D., *Helv.Chim.Acta* (1985) **68**, 155.

72. a) Seebach, D.; Weber, T., *Tetrahedron Lett.* (1983) **24**, 3315; b) Seebach, D.; Weber, T., *Helv.Chim.Acta* (1984) **67**, 1650.

73. a)Seebach, D.; Aebi, J.D., *Tetrahedron Lett.* (1983) **24**, 3311; b) Seebach, D.; Aebi, J.D.; Gander-Coquoz, M.; Naef, R., *Helv.Chim.Acta* (1987) **70**, 1194.

74. a) Seebach, D.; Aebi, J.D., Tetrahedron Lett. (1984) 25, 2545; b) ref. 73b.

75. a)Naef, R.; Seebach, D., *Helv.Chim.Acta* (1985) **68**, 135; b) Seebach, D.; Aebi, J.D.; Naef, R.; Weber, T., *Helv.Chim.Acta* (1985) **68**, 144; c) Calderari, G.; Seebach, D.; *Helv.Chim.Acta* (1985) **68**, 1592; d) Aebi, J.D.; Seebach, D., *Helv.Chim.Acta* (1985) **68**, 1507.

76. A preliminary report of stereoselective α- and β-alkylation of Aspartic acid was reported: Seebach, D.; Wasmuth, D., *Angew.Chem.Int.Ed.Engl.* (1981) **20**, 971.

77. a) Fitzi, R.; Seebach, D., *Angew.Chem.Int.Ed.Engl.* (1986) **25**, 345; b) Seebach, D.; Miller, D.D.; Muller, S.; Weber, T., *Helv.Chim.Acta* (1985) **68**, 949; c) ref. 68.

78. Seebach, D.; Muller, S.G.; Gysel, U.; Zimmermann, J., *Helv. Chim.Acta* (1988) **71**, 1303.

79. Seebach, D.; Juaristi, E.; Miller, D.D.; Schickli, C.; Weber, T., *Helv.Chim.Acta* (1987) **70**, 237.

80. Weber, T.; Aeschimann, R.; Maetzke, T.; Seebach, D., *Helv.Chim.Acta* (1986) **69**, 1365.

81. Seebach, D.; Fadel, A., *Helv.Chim.Acta* (1985) **68**, 1243.

82. Karady, S.; Amato, J.S.; Weinstock, L.M., *Tetrahedron Lett.* (1984) **25**, 4337.

83. An original description of preparing N-alkylated FMOC amino acids through reduction of various oxazolidinones of amino acids was reported: Freidinger, R.M.; Hinkle, J.S.; Perlow, D.S.; Arison, B.H., *J.Org.Chem.* (1983) **48**, 77.

84. Fadel, A.; Salaun, J., *Tetrahedron Lett.* (1987) **28**, 2243.

85. Gander-Coquoz, M.; Seebach, D., *Helv.Chim.Acta* (1988) **71**, 224.

86. Polt, R.; Seebach, D., *Helv.Chim.Acta* (1987) **70**, 1930.

87. Aebi, J.D.; Dhaon, M.K.; Rich, D.H., *J.Org.Chem.* (1987) **52**, 2881.

88. Schmidt, U.; Siegel, W., *Tetrahedron Lett.* (1987) **28**, 2849.

89. Vigneron, J.P.; Kagan, H.; Horeau, A., *Tetrahedron Lett.* (1968) 5681.

90. a)Dellaria, J.F.; Santarsiero, B.D., *Tetrahedron Lett.* (1988) **29**, 6079; b) Dellaria, J.F.; Santarsiero, B.D., *J.Org.Chem.* (1989) **54**, 0000.

91. Oxazinone **453** was prepared by a slightly different procedure and investigated as an electrophilic glycinate, see: Sinclair, P.J., Ph.D. Thesis, Colorado State University (1987).

92. Williams, R.M.; Im, M-N, *Tetrahedron Lett.* (1988) **29**, 6075.

93. Owa, T.; Otsuka, M.; Ohno, M., *Chemistry .Lett.* (1988) 83.

94. Belokon, Y.N.; Bulychev, A.G.; Vitt, S.V.; Struchkov, Y.T.; Batsanov, A.S.; Timfeeva, T.V.; Tsryapkin, V.A.; Ryzhov, M.G.; Lysova, L.A.; Bakhmutov, V.I.; Belikov, V.M., *J.Am.Chem.Soc.* (1985) **107**, 4252.

95. Belokon, Y.N.; Sagyan, A.S.; Djamgaryan, S.M.; Bakhmutov, V.I.; Belikov, V.M., *Tetrahedron* (1988) **44**, 5507.

96. Belokon, Y.N.; Chernoglazova, N.I.; Kochetkov, C.A.; Garbalinskaya, N.S.; Belikov, V.M., *J.Chem.Soc. Chem.Comm.* (1985) 171.

97. a) For leading references, see: Ben-Ishai, D.; Sataty, I.; Bernstein, Z., *Tetrahedron* (1976) **32**, 1571; Bernstein, Z.; Ben-Ishai, D., *Tetrahedron* (1977) **33**, 881; Zoller, U.; Ben-Ishai, D., *Tetrahedron* (1975) **31**, 863, and earlier refernces cited therein; b)Schollkopf, U.; Neubauer, H-J.; Hauptreif, M., *Angew.Chem.Int.Ed.Engl.* (1985) **24**, 1066.

98. Schollkopf, U.; Gruttner, S.; Anderskewitz, R.; Egert, E.; Dyrbusch, M.,*Angew.Chem.Int.Ed.Engl.* (1987) **26**, 683.

99. Schollkopf, U.; Hupfeld, B.; Kuper, S.; Egert, E.; Dyrbusch, M., *Angew.Chem.Int.Ed.Engl.* (1988) **27**, 433.

100. Kober, R.; Papadopoulos, K.; Miltz, W.; Enders, D.; Steglich, W., *Tetrahedron* (1985) **41**, 1693. For a more recent study involving synthesis of racemic amino acids with the related N-acylimino acetates, see: Bretschneider, T.; Miltz, W.; Munster, P.; Steglich, W., *Tetrahedron* (1988) **44**, 5403.

101. a)Yamamoto, Y.; Ito, W.; Maruyama, K., *J.Chem.Soc.Chem.Comm.* (1985) 1131.; b) Yamamoto, Y.; Ito, W., *Tetrahedron* (1988)**44**, 5415.

102. Ermert, P.; Meyer, J.; Stucki, C.; Schneebeli, J.; Obrecht, J-P., *Tetrahedron Lett.* (1988) **29**, 1265.

103. Harding, K.E.; Davis, C.S., *Tetrahedron Lett.* (1988) **29**, 1891.

104. a)Sinclair, P.J.; Zhai, D.; Reibenspies, J.; Williams, R.M., *J.Am.Chem.Soc.* (1986) **108**, 1103; b) Weijlard, J.; Pfister,K.; Swanezy, E.F.; Robinson, C.A.; Tishler, M., *J.Am.Chem.Soc.* (1951) **73**, 1216.

105. Williams, R.M.; Sinclair, P.J.; Zhai, D.; Chen, D., *J.Am.Chem.Soc.* (1988) **110**, 1547.

106. a) Williams, R.M.; Zhai, D.; Sinclair, P.J., *J.Org.Chem.* (1986) **51**, 5021; b) Ramer, S.E.; Cheng, H.; Palcic, M.M.; Vederas, J.C., *J.Am.Chem.Soc.* (1988) **110**, 8526; c) Ramer, S.E., Ph.D. Thesis, The University of Alberta, Edmonton (1988).

107. Williams, R.M.; Sinclair, P.J.; Zhai, W.,*J.Am.Chem.Soc.* (1988) **110**, 482.

108. Zhai, D.; Zhai, W.; Williams, R.M., *J.Am.Chem.Soc.* (1988) **110**, 2501.

109. Williams, R.M.; Zhai, W., *Tetrahedron* (1988) **44**, 5425.

110. For related preparations of racemic vinyl amino acids from an electrophilic glycine derivative, see: Castelhano, A.L.; Horne, S.; Taylor, G.J.; Billedeau, R.; Krantz, A., *Tetrahedron* (1988) **44**, 5451 and references cited therein.

111. Kober, R.; Steglich, W., *Liebigs Ann.Chem* (1983) 599.

112. Weineges, K.; Brachmann, H.; Stahnecker, P.; Rodewald, H.; Nixdorf, M.; Imgartinger, H., *Liebigs Ann.Chem.* (1985) 366.

113. Belokon, Y.N.; Popkov, A.N.; Chernoglazova, N.I.; Saporovskaya, M.B.; Bakhmutov, V.I.; Belikov, V.M.; *J.Chem.Soc.Chem.Comm.* (1988) 1336.

114. Yamada, T.; Suzuki, H.; Mukaiyama, T., *Chemistry Lett.* (1987) 293.

115. Easton, C.J.; Scharfbilling, I.M.; Tan, E.W., *Tetrahedron Lett.* (1988) **29**, 1565.

116. Belokon, Y.N.; Zel'ter, I.E.; Bakhmutov, V.I.; Saporovskaya, M.B.; Ryzhov, M.G.; Yanovsky, A.I.; Struchkov, Y.T.; Belikov, V.M., *J.Am.Chem.Soc.* (1983) **105**, 2010.

117. a)Dukuzovic, Z.; Roberts, N.K.; Sawyer, J.F.; Whelan, J.; Bosnich, B., *J.Am.Chem.Soc.* (1986) **108**, 2034; b) for a related series of studies for enantioselective H/D exchange via Cobalt(III) chelates, see: Keyes, W.E.; Legg, J.I., *J.Am.Chem.Soc.* (1976) **98**, 4970, and references cited therein.

118. Duhamel, L.; Plaquevent, J-C., *J.Am.Chem.Soc.* (1978) **100**, 7415.

119. Duhamel, L.; Plaquevent, J-C., *Bull.Chim.Soc.Fr.* (1982) II-75.

120. Duhamel, L.; Plaquevent, J-C., *Tetrahedron Lett.* (1980) **21**, 2521.

121. Duhamel, L.; Fouquay, S.; Plaquevent, J-C., *Tetrahedron Lett.* (1986) **27**, 4975.

122. Duhamel, L.; Duhamel, P.; Launay, J-C.,; Plaquevent, J-C., *Bull.Chim.Soc.Fr.* (1984) II-421.

123. Duhamel, L.; Duhamel, P.; Fouquay, S.; Eddine, J.J.; Peschard, O.; Plaquevent, J-C.; Ravard, A.; Solliard, R.; Valnot, J-Y.; Vincens, H., *Tetrahedron* (1988) **44**, 5506.

124. Miller, T.W.; Tristam, E.W.; Wolf, F.J., *J.Antibiotics* (1971) **24**, 48.

125. Molinski, T.F.; Ireland, C.M., *J.Org.Chem.* (1988) **53**, 2103.

126. a) Haner, R.; Olano, B.; Seebach, D., *Helv.Chim.Acta.* (1987) **70**, 1676; b) Seebach, D.; Haner, R., *Chemistry Lett.* (1987) 49.

CHAPTER 2

HOMOLOGATION OF THE β–CARBON

Sources of electrophilic , nucleophilic and radical-based 'serine' or 'alanine' templates as homologation reagents, is a potentially very useful approach for making a variety of amino acids that nicely complements the more abundantly studied glycine templates discussed in Chapter 1. These β-homologation reagents are just recently being recognized as useful additions to the synthetic amino acid chemists' arsenal. L- and D-Serine have provided the basic raw materials for most of these approaches; cysteine and aspartic acid also providing potential activating functionality for bond-formation at the β-carbon. L-Aspartic acid and L-cysteine are the least expensive of these starting materials and are inexpensive enough to be seriously considered for industrial scale applications. L-Serine is somewhat more costly, but is certainly inexpensive enough for most research applications. The corresponding D-isomers of all three amino acids are substantially more costly (see cost comparison in the *Preface*). Since this is a relatively new area, this chapter is necessarily small dealing with only the most useful, and practical methods. The related homologations at the γ-carbon resulting from very similar methionine-, homoserine- , aspartic- and glutamic acid-derived templates will also be covered in this chapter.

A. β-CARBON HOMOLOGATIONS

A series of very recent papers from Vederas and associates [1-3] have detailed the preparation and ring-opening reactions of β-lactones derived from serine. Full experimental details accompany these excellent contributions. These workers have ingeniously exploited the stereoelectronic parameters of the β-lactone to obviate the β-elimination reactions that often attends activation of the β-position for displacement. The direct ring-opening elimination by enolization of the β-lactone is formally a *retro-4-endo-trig* [4] process and is therefore, generally disfavored on stereoelectronic grounds. As shown in Scheme 1, N-CBZ (**1a**)

SCHEME 1

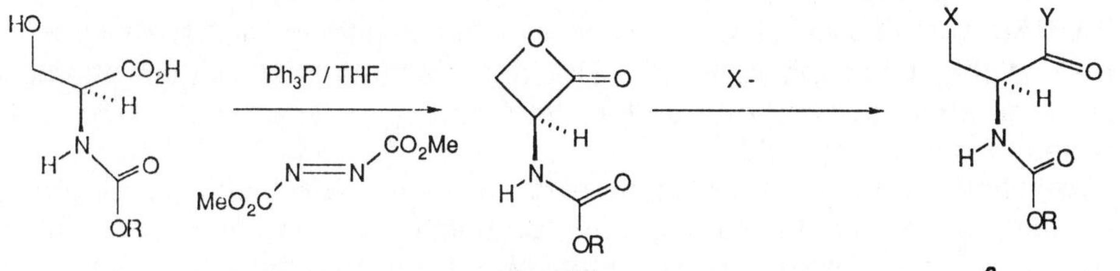

1a, R = CH$_2$Ph
1b, R = C(Me)$_3$

2a, 60%
2b, 72%

3

Arnold, L.D.; Kalantar, T.H.; Vederas, J.C., *J.Am.Chem.Soc.* (1985) **107**, 7105.

β-LACTONE (2a/2b)	NUCLEOPHILE (X⁻)	X (3)	Y (3)	CONFIG.	YIELD (%3)
2a	NH$_{3(g)}$ / MeCN, 0°	-OH	NH$_2$	L	77
2a	NH$_{3(g)}$ / THF, 0°	NH$_3^+$	O-	D	75
2b	NH$_{3(g)}$ / THF, 0°	NH$_3^+$	O-	L	79
2a	N(Me)$_3$ / THF, 0°	N(Me)$_3^+$	O-	L	100
2a	HSCH$_2$CH$_2$NH$_3^+$Cl⁻	-⁺NH$_2$CH$_2$CH$_2$SH	O-	D	76
2a	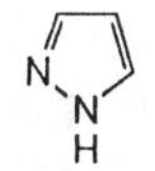 pyrazole (N-H)	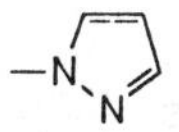 —N-pyrazolyl	OH	D	71
2a	NaOAc	OAc	OH	L	97
2a	NaOMe	OH	OMe	DL	88
2a	PhCH$_2$SNa	SCH$_2$Ph	OH	D	78
2a	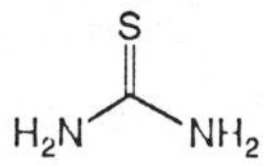 thiourea (H$_2$N-C(=S)-NH$_2$)	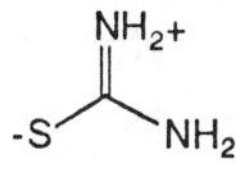 -S-C(=NH$_2^+$)-NH$_2$	O-	DL	56
2a	MgCl$_2$	Cl	OH	D	94
2a	MgBr$_2$	Br	OH	L	99

and N-t-BOC (**1b**) serine are cyclized to the corresponding β-lactones **2a** (60%) and **2b** (72%) , respectively with dimethyl azodicarboxylate (DMAD) and triphenyl phosphine. The reaction is performed by pre-forming the Ph₃P-DMAD complex to allow generation of the β-lactone in the absence of the free, highly nucleophilic Ph₃P. The authors note that earlier attempts to perform this cyclization under 'standard' Mitsunobu conditions proceeded in only 1.4% yield.

Reaction of these substances with a variety of 'soft' nucleophiles gives good to excellent yields of the products (**3**) resulting from alkyl-oxygen cleavage (Table). Only relatively 'hard' nucleophiles such as ammonia or methoxide give significant amounts of acyl-oxygen cleavage. Even in these difficult cases, alteration of the solvent from acetonitrile to tetrahydrofuran inverts the acyl-oxygen cleavage to the desired alkyl-oxygen cleavage. Most conveniently, the CBz and t-BOC protecting groups are maintained in the products, making this a highly attractive and practical β-homologation template.

A complementary β-lactone system[2] that *directly* provides access to the corresponding free amino acids (**5**) is depicted in Scheme 2. The t-BOC group of **2b** is removed with acid to form the corresponding amine salts **4a** and **4b**; the tosylate salt being reported as a crystalline, stable substance amenable to storage. These interesting compounds readily add even very weak nucleophiles (such as trifluoroacetate) at the β-carbon directly forming the free amino acids. The authors note that some functionality introduced in the above (acylated) versions are labile to the conditions required to remove the nitrogen protecting groups (such as azido); the complementary salts (**4**) allow the preparation of even relatively labile amino acids in a very simple manner. The yields are generally excellent and the diversity of nucleophiles (S,O,N, C, halogen) that provide the desired β-substituted products is impressive. It is also significant that some of the addition reactions (eg., sulfur nucleophiles) *can be performed in water* near pH 5. These additions take place despite the relative lability of the parent salt **4** in water ($t_{1/2}$ ~ 2.5h in unbuffered water). The examples with the amine-thiols, such as cysteine and β-mercaptoethylamine should be compared to the additions to the acylated species (**2**) above. The addition of β-mercaptoethylamine to **2** provides the product of nitrogen attack whereas the *same* nucleophile attacks the salt **4** with the *reverse* chemoselectivity giving the corresponding sulfide. These beautifully complementary methods should provide convenient access to a wide variety of interesting β-substituted α-amino acids in optically pure form.

SCHEME 2

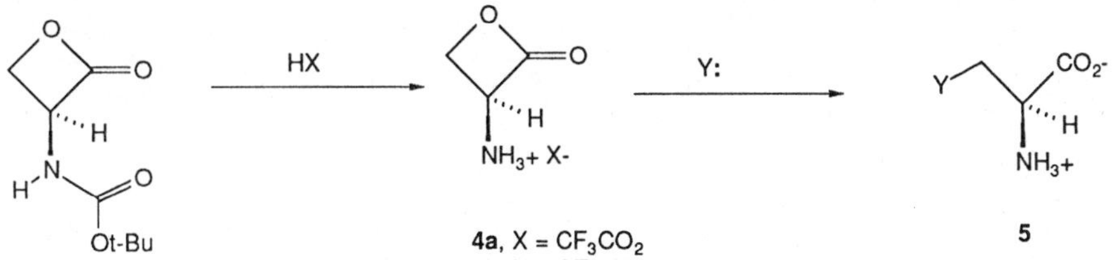

2b

4a, X = CF$_3$CO$_2$
4b, X = OTs

5

Arnold, L.D.; May, R.G.; Vederas, J.C., *J.Am.Chem.Soc.* (1988) **110**, 2237.

NUCLEOPHILIC REAGENT (Y:)	CONDITIONS	Y	YIELD (%5)
L-cysteine	pH 5.5	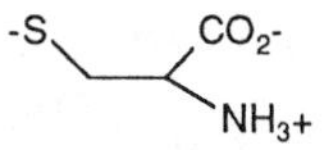	93
H$_2$NCH$_2$CH$_2$SH-HCl	pH 5.5	-SCH$_2$CH$_2$NH$_2$	85
LiSH	MeCN / THF	-SH	88
Na$_2$S$_2$O$_3$	pH 5.0	-SSO$_3$Na$^+$	83
Me$_2$S	TsOH / TFA	-S$^+$Me$_2$	88
CF$_3$CO$_2^-$	TFA	-OOCF$_3$	87
K$_2$HPO$_4$	18-crown-6 / DMF	-OPO$_3$H$_2$	87
conc. HCl	30 min.	-Cl	92
n-Bu$_4$NCN	DMF, -10°	-CN	84
NaN$_3$	DMF	-N$_3$	96
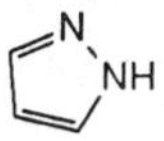	DMF	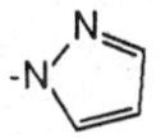	77

SCHEME 3

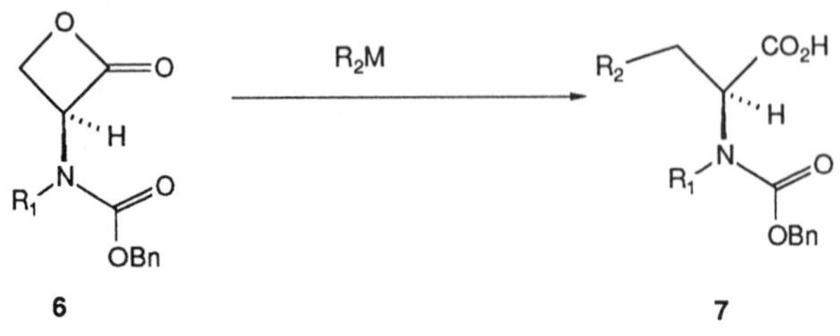

Arnold, L.D.; Drover, J.C.G.; Vederas, J.C., *J.Am.Chem.Soc.* (1987) **109**, 4649.

R$_1$	REAGENT (R$_2$M)	R$_2$	YIELD (%7)	%DECREASE in ee
H	MeLi / CuCN	Me	47	1.7
Bn	MeLi / CuBr-SMe$_2$	Me	70	2.4
Bn	MeLi / CuCN	Me	72	17.5
Bn	MeLi / CuCN	Me	92	1.0
H	*n*-BuLi / CuCN	*n*-Bu	62	0
Bn	*n*-BuLi / CuCN	*n*-Bu	76	11.7
H	*i*-PrMgCl / CuBr-SMe$_2$	*i*-Pr	44	<0.5
Bn	*i*-PrMgCl / CuBr-SMe$_2$	*i*-Pr	83	0
Bn	*sec*-BuLi / CuCN	*sec*-Bu	76	n.d.
H	*t*-BuLi / MeLi / CuCN	*t*-Bu	48	0
Bn	*t*-BuLi / CuCN	*t*-Bu	38	5.6
Bn	*t*-BuLi / CuBr-SMe$_2$	*t*-Bu	51	0
H	H$_2$C=CHMgCl / CuBr-SMe$_2$	H$_2$C=CH-	47	0
Bn	H$_2$C=CHLi / CuCN	H$_2$C=CH-	56	27.2
H	PhMgCl / CuBr-SMe$_2$	Ph	55	0
H	PhLi / CuCN	Ph	46	67.4
Bn	PhMgBr / CuBr-SMe$_2$	Ph	60	3.3
Bn	PhLi / CuCN	Ph	25	4.7
Bn	PhLi / CuBr-SMe$_2$	Ph	36	14.2

In extending these substrates to C-C bond-forming homologations, these workers [3] have examined the addition of various cuprates to the serine β-lactone **6** as shown in Scheme 3. A variety of lithium and Grignard-derived cuprates add regiospecifically to the β-methylene of the mono-and di-protected systems **6**. The authors were concerned that the mono-protected systems (**6**, R_1 =H) carrying the acidic N-H proton , would suffer side reactions resulting from base-mediated deprotonation and lead to azlactone (oxazoline) or oxazolinone formation by rupture of the acyl-oxygen and/or alkyl-oxygen bonds, respectively. This does indeed seem to occur as evidenced by the formation of N-CBz-serine after work-up using conditions that do *not* hydrolyze the β-lactone; an optimal temperature range of -23⁰ ~ -15⁰C minimizes this process and maximizes the substitution reactions. The chemical yields of the products resulting from the mono-substituted system (**6**, R_1=H) are somewhat lower than those with the di-protected (**6**, R_1=Bn) substrates. In accordance with this hypothesis, the di-protected substrates are reported to exhibit none of the N-CBz serine side-product reactions that the mono-protected system displays; this is manifested as better chemical yields. However, the loss of optical purity in the di-protected systems is significant in some instances, but is negligible for the mono-protected substrate. In spite of the somewhat moderate yields, these results provide a practical and convenient method for introducing carbon functionality at the β-position.

The nucleophilic ring-opening of serine-derived aziridines has been recognized for some time, but has recently been gaining increased attention. An elegant and creative example has recently been reported by Baldwin and associates [5] as shown in Scheme 4. Conversion of the previously known N-trityl aziridine **8** into the activated *para*-nitrophenyl amide **9** is accomplished in high yield. This substance is condensed with a stabilized Wiitig reagent in refluxing toluene to afford the potentially versatile ylide **10** in 49% yield. Subsequent Wittig reaction with *para*-formaldehyde in hot benzene provided the γ-methylene glutamic acid derivative **11** in excellent yield. Similarly, reaction of **10** with acetaldehyde provided the unsaturated homologue **13** as a 12 : 1 mixture of geometric isomers in good yield. Hydrolytic cleavage of the amide and esters of **11** and **13** in hot 3N hydrochloric acid afforded the corresponding free amino acids **12** and **14**, respectively. This protocol allows for the difficult, simultaneous β- and γ-homologation of an electrophilic alanine equivalent in two simple operations from readily available starting materials.

SCHEME 4

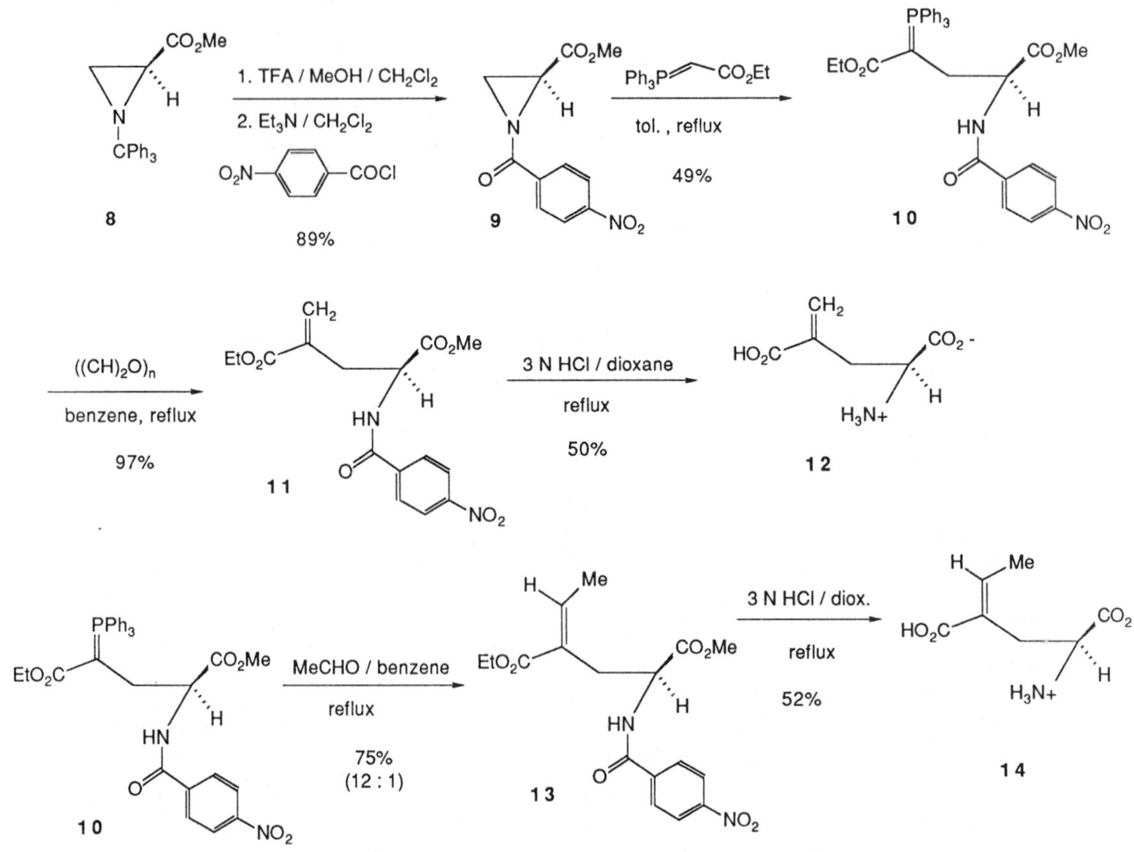

Baldwin, J.E.; Adlington, R.M.; Robinson, N.G., *J.Chem.Soc. Chem.Comm.* (1987) 153.

An interesting synthesis of the 1-carbon-extended homologues of the substances just described has been devised by Baldwin and associates 6 as illustrated in Scheme 5. The β-iodoalanine derivatives **15** and **18** were condensed with the methacryl tin reagents **16** and **19**, respectively under free radical conditions furnishing the adducts **17** and **20** in good yields. The utilization of free radical-based glycine and alanine equivalents has been studied relatively little for the synthesis of optically active amino acids. The great deal of recent research devoted to free radical C-C bond-forming reactions promises to find increasing utility in the synthesis of amino acids; the present case elegantly attests to this.

SCHEME 5

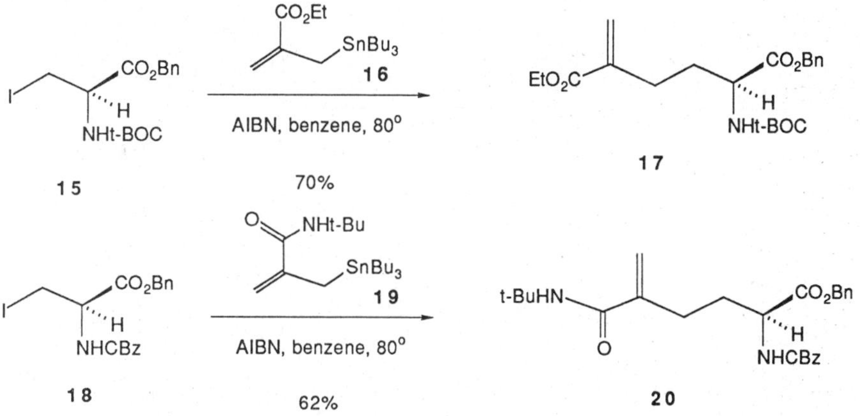

Baldwin, J.E.; Adlington, R.M.; Birch, D.J.; Crawford, J.A.; Sweeney, J.B., *J.Chem.Soc.Chem.Comm.* (1986) 1339.

A clever and efficient use of Millers' β-lactam synthesis [7] from serine has recently been exploited by Baldwin and co-workers[8] in a total synthesis of the potent L-glutamate agonist quisqualic acid (**25**, Scheme 6). The strategy developed economizes on the oxidation state of the hydroxamic acid inherent in the Miller β-lactam protocol and employs an interesting rearrangement of the N-hydroxy β-lactam to an isoxazolidinone that exposes the nitrogen atom to acylation. The conversion of β-hydroxy-α-amino acids into the corresponding optically active β-lactams by the intramolecular Mitsunobu reaction is now a routine and reliable method to install nitrogenous functionality at the β-position. The N-hydroxy β-lactam **21** was cleanly rearranged into the isoxazolidin-5-one **22** by treatment with catalytic lithium ethanethiolate. This substance was condensed with ethoxycarbonyl isocyanate providing the urea **23** which was immediately ring-opened with alkalai to furnish the salt **24**. Removal of the t-BOC group with trifluoroacetic acid directly provided L-quisqualic acid in 89% overall yield from **22**. The authors note that application of the same procedure to D-serine provided D-quisqualic acid; the optical purity of both enantiomers was confirmed by reductive conversion to **26**.

Effenberger and Weber[9] have studied the Friedel-Crafts-type alkylation reactions of protected threonine triflates as shown in Scheme 7. N-Phthalimido threonine methyl ester is converted into the corresponding triflate **27** and condensed with refluxing benzene in the presence of trifluoroacetic acid to afford a mixture of products **28-30**.

Homologation of the β-Carbon

SCHEME 6

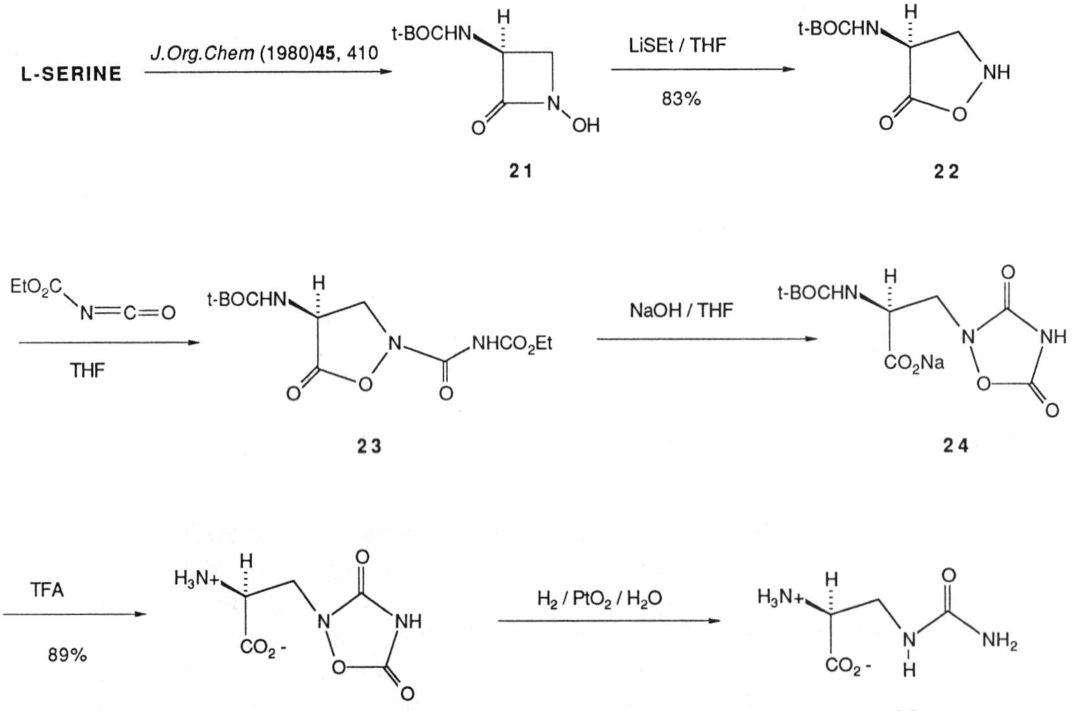

Baldwin, J.E.; Adlington, R.M.; Birch, D.J., *J.Chem.Soc. Chem.Comm.* (1985) 256.

SCHEME 7

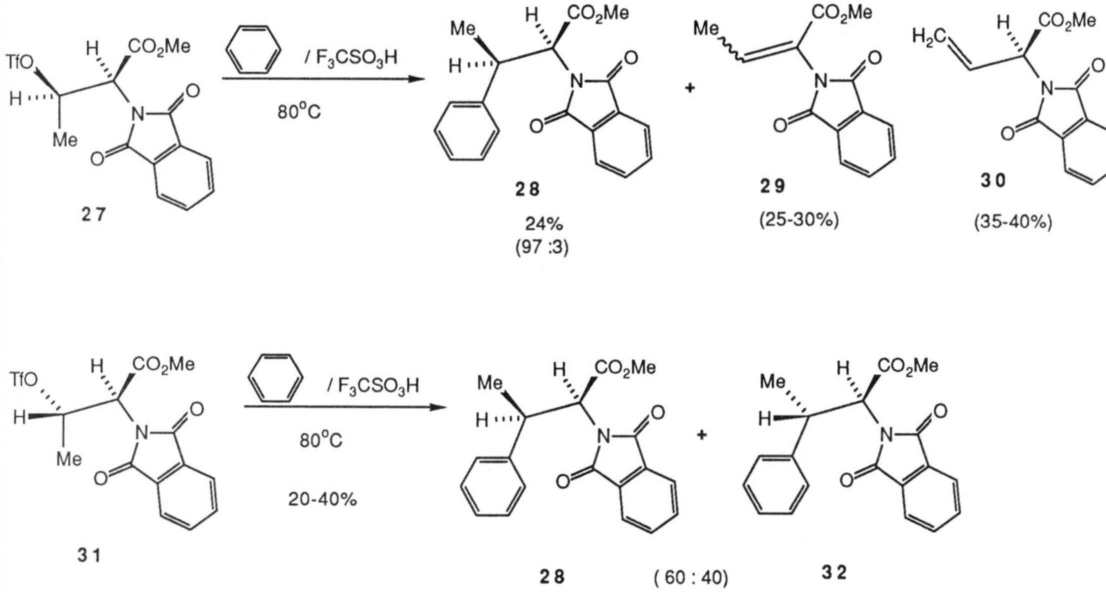

28
24%
(97 :3)

29
(25-30%)

30
(35-40%)

27

31

28   (60 : 40)   32

Effenberger, F.;; Weber. T., *Angew.Chem.Int.Ed.Engl.* (1987) **26**, 142.

The desired β-methyl phenylalanine is obtained in low yield but with a high level of diastereochemical integrity (97 : 3, **28** : **32**). In contrast, the *allo*-threonine system **31** gives a similar mixture of substitution and elimination products, but the desired β-methyl phenylalnine derivative is obtained as a 60 : 40 mixture of **28** : **32**. The authors attribute this to differences in the conformational stabilities of putative carbocations formed in the reactions. The present method is of interest from a mechanistic standpoint, but the high level of elimination products that attend the substitutions will certainly limit the practical utility of the procedure. The Vederas β-lactone protocol [1-3] should be contrasted to this and related acyclic β-activated substrates where β-elimination is often a competing reaction that can preclude the synthetic utility of such approaches.

Sasaki and associates [10] have devised an unambiguous protocol for activating the β-position for substitution and obviating the elimination problem by changing the oxidation state of the carboxyl group as shown in Scheme 8. The method also economizes on the use of L-serine to access both the L- and D-configured templates; D serine being substantially more costly a starting material than the L-isomer. N-t-BOC carbomethoxy L-

SCHEME 8

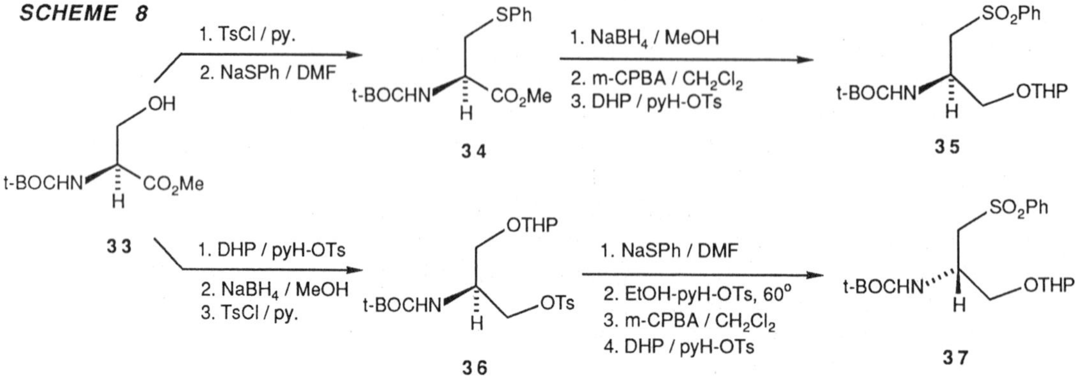

SCHEME 9

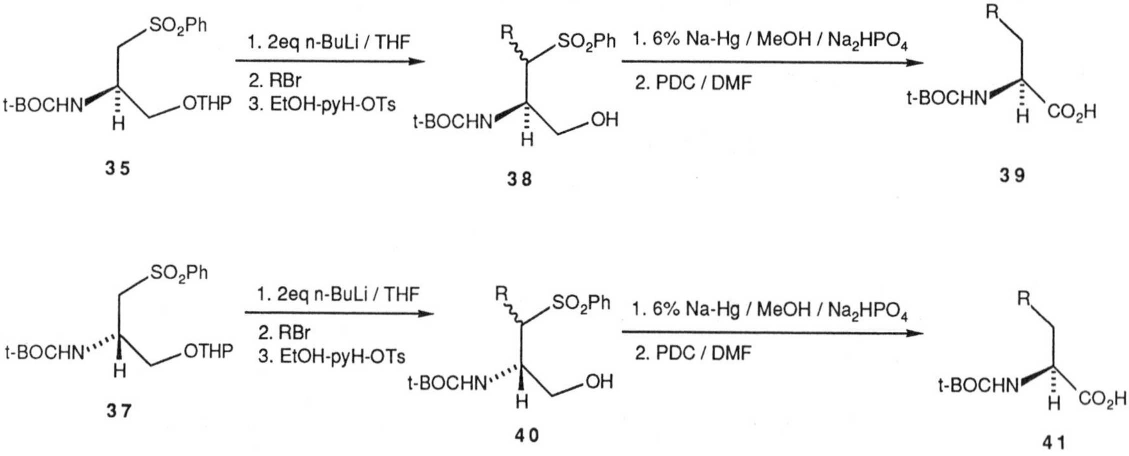

Sasaki, N.A.; Hashimoto, C.; Potier, P., *Tetrahedron Lett.* (1987) **28**, 6069.

RBr	YIELD (%39)
EtBr	71
i-PrBr	66

serine (**33**) is tosylated and converted into the phenyl sulfide **34**. Reduction of the carboxyl group, oxidation to the sulfone and THP protection affords the L-template **35**. By altering the ordering of steps, processing in a similar fashion provides the corresponding D-template **37**. These substances can be metallated and alkylated to afford the substituted sulfones **38** after removal of the THP ether. Reductive removal of the sulfur and oxidation of the alcohol with pyridinium dichromate affords the t-BOC-protected L-amino acids **39**. The same protocol when applied to **37** provides the corresponding D-isomers **41**. The advantages of this work are in providing access to both L- and D-configured amino acids as well as maintaining the t-BOC protecting group throughout. The rather clumsy redox manipulation will certainly limit the types of functionality that may be tolerated.

SCHEME 10

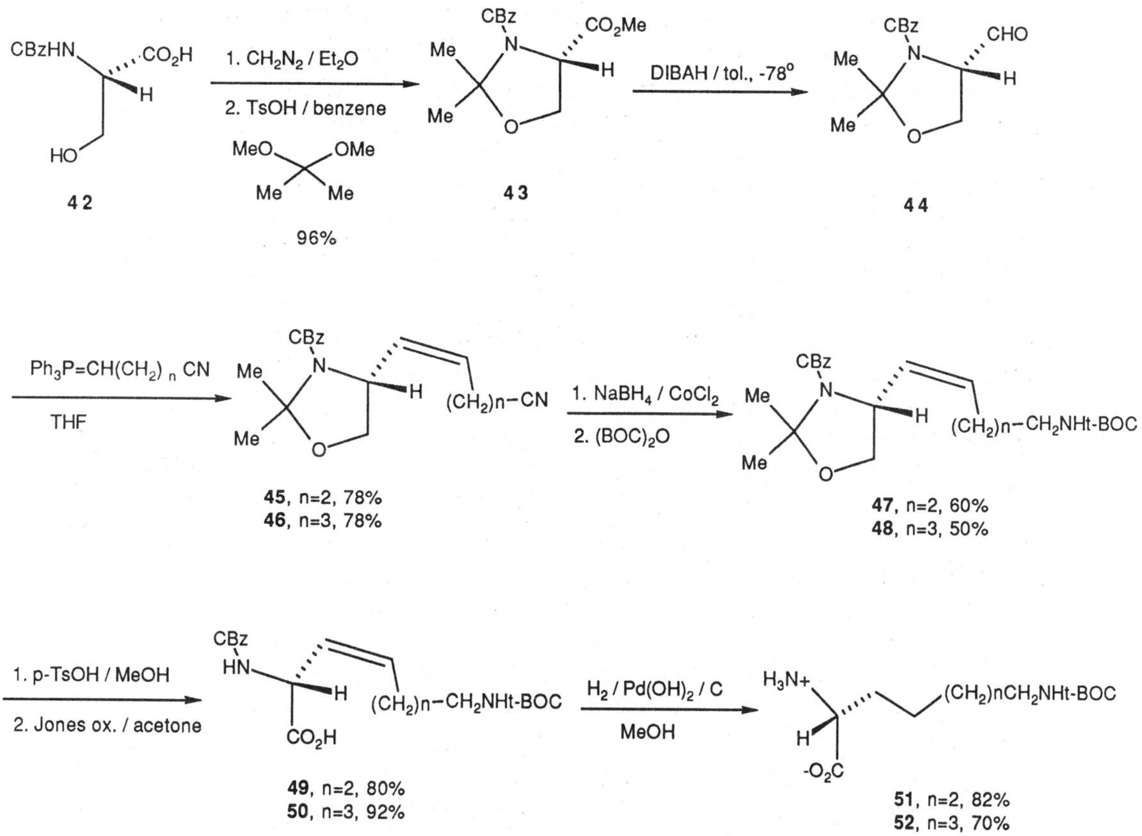

Beaulieu, P.L.; Schiller, P.W., *Tetrahedron Lett.* (1988) **29**, 2019.

An increasingly useful strategy that is being recognized is the manipulation of protected serinal[11] systems for homologation to unusual amino acids. As shown in Scheme 10, N-CBz L-serine (**42**) is esterified and ketalized to provide **43** in excellent yield[12]. Reduction to the key aldehyde is achieved in 70% overall yield from **42**. Wittig reaction of this substance with non-stabilized ylides proceeds in good yields to afford the Z-olefins **45 / 46**. Reduction of the nitrile and protection affords **47 / 48** which are subjected to de-ketalization and Jones oxidation to the difficult Z-vinyl glycine derivatives **49 / 50** in good overall yields. These interesting substances are subsequently hydrogenated to afford the differentially protected D-α,ω-diamino alkanoic acids **51 / 52**. Presumably, the valuable substance **44** should provide a general entry to synthesizing a variety of difficult, racemization-prone β,γ-unsaturated amino acids for which very few good syntheses have been developed. Forthcoming reports from Beaulieu and Schiller [12] addressing this vital chemistry is expected.

Garner and Park [13] have independently developed an interesting protocol to install β,γ-unsaturation via the related N-t-BOC serinal system **53** (Scheme 11). Addition of vinyl magnesium bromide to **53** affords a 6 : 1, *erythro : threo* mixture of carbinols **54**. These substances were then converted into the urethanes **55** and subjected to a palladium-mediated allylic rearrangement to **56**. De-ketalization to **57** followed by permanganate oxidation afforded the incipient diol aldehyde **58** which was isolated as the lactol **59** as a 2.5 : 1 mixture of diastereomers. Oxidation with N-bromo urea afforded the lactone **60** which was hydrolyzed in quantitative yield to afford 5-O-carbamoylpolyoxamic acid (**61**) the unusual amino acid constituent of the polyoxins. Full experimental details accompany this work as well as the paper[11a] describing the preparation of the serinal template **53**. The primary difficulty that has previously been encountered with serinal equivalents is the great propensity of the labile α-amino aldehyde to suffer partial or complete racemization. The configurational stability of **53** (obtained in 93-95% ee) is reported as being quite acceptable; additional utility of the relatively new substances **44** and **53** for a wide range of synthetic objectives is expected.

Rapoport and associates[14] have examined the conversion of the L-serine carboxyl group into the corresponding ketones as shown in Scheme 12; full experimental details accompany this work. N-Phenylsulfonyl L-serine (**62**) is treated with a variety of organometallic reagents to afford the corresponding ketones **63**. The yields are good but the reactions require 4-6 molar equivalents of the organometallic reagent. Processing of these keto-adducts to D-α-amino acids was accomplished in several ways. In two instances, the ketone was transformed into the dithioketals

SCHEME 11

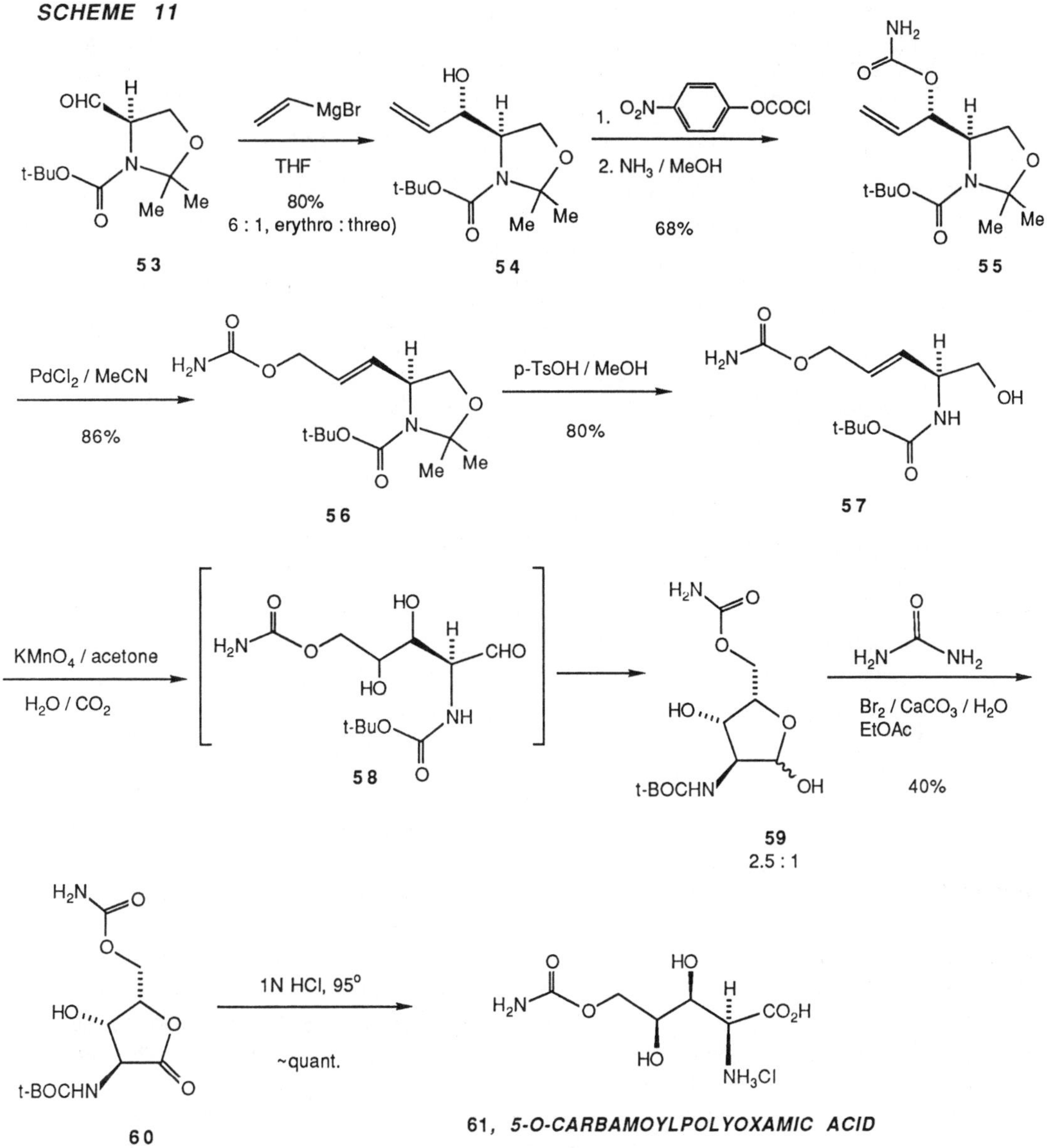

Garner, P.; Park, J.M., *J.Org.Chem.* (1988) **53**, 2979.

64 and subsequently reduced to the methylene derivatives **65** with Raney-Nickel. Alternatively, triethylsilane reduction of **66** proceeded in 81% yield. The authors note that catalytic reductions proved troublesome. The methyl ketone **67** was reduced to the corresponding secondary alcohols **68** and **69** as shown in Scheme 14. Sodium borohydride reduction provided a 7 : 3 ratio of **68** : **69** whereas, L-selectride gave a 1 : 99 ratio. The authors were presumably hoping for improved selectivity for the formation of **68** which serves as a precursor for the rare and extremely expensive D-*allo*-threonine (>US$400.00/gram) . The derivatized alcohols (**65**) were oxidized to the D-amino acids by treatment with oxygen on a platinum catalyst at 55-60°C (**70**) followed by removal of the N-phenylsulfonyl group with refluxing 48% HBr in the presence of phenol affording the D-amino acids **71**. D-*Allo*-threonine **72** was obtained by dissolving metal reduction of the N-phenylsulfonyl moiety in very good yield. The platinum oxidation of the 4-hydroxybutyl derivative (**65**) concomitantly effected oxidation of both primary alcohols giving the corresponding diacid. In this way, four representative α-amino acids were synthesized: D-norleucine, D-α-aminopimelic acid, D-DOPA, and D-*allo*-threonine. The authors also report the synthesis and utility of N-phenylsulfonyl prolyl chloride (**73**, Scheme 16) as a useful derivatizing agent for the determination of optical purity. The amino acid methyl ester is acylated with **73** and examined by HPLC for diastereomeric purity. This reagent can be used as an alternative to the more costly MTPACl (Moshers' acid chloride) for the precise determination of %ees'. The present methodology offers a straightforward entry into preparing the generally costly D-amino acids from L-serine. As with related approaches requiring redox manipulations, limitations on the functionality tolerable must be carefully considered.

A series of reports by Nakajima, et.al. [15-18] have examined the nucleophilic ring-opening reactions of serine-derived aziridines ; Scheme 17 illustrates the general method to prepare the requisite aziridines. O-Benzyl-L-serine benzene sulfonate (**75**) is tritylated and cyclized to the aziridine **77** via the tosylate in 55% overall yield. Trifluoroacetic acid removes the trityl group providing the key aziridine **78**. This substance can be acylated, for example with N-CBz-Gly (**79**) or hydrolyzed to the free base **80**. Schemes 18, 19 and 20 illustrate the ring-opening reactions with amines, alcohols and thiols, respectively. Two series of aziridines are employed; that outlined above from serine and the corresponding system derived from threonine. The reactions from the threonine systems are particularly significant since the stereogenic center at the β-position is maintained as *threo* in the final products by the double inversion protocol. The aziridines share the same stereoelectronic protection to β-elimination-related side reactions as the Vederas β-lactones; β-

SCHEME 12

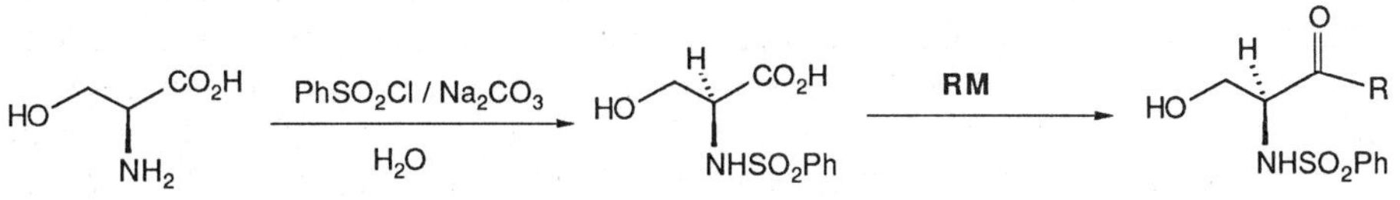

Maurer, P.J.; Takahata, H.; Rapoport, H., *J.Am.Chem.Soc.* (1984) **106**, 1095.

RM	R (63)	YIELD (%63)
n–BuLi / n-PrMgBr	 ⌒Me	78
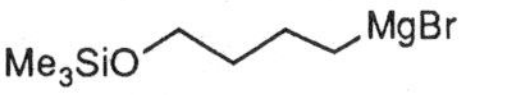 Me₃SiO⌒⌒⌒MgBr	 ⌒⌒OH	55
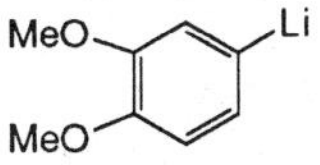 MeO⌬Li / MeO	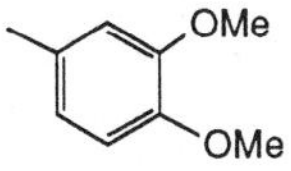 ⌬OMe / OMe	83
MeLi / TMEDA	Me	60

SCHEME 13

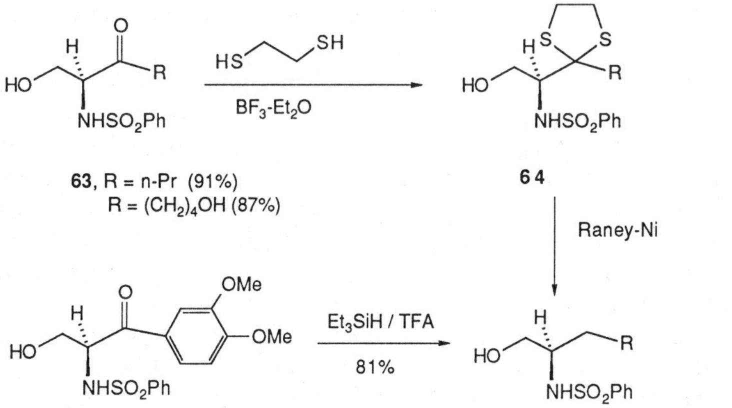

63, R = n-Pr (91%)
R = (CH₂)₄OH (87%)

64

Raney-Ni

66

Et₃SiH / TFA
81%

65, R = n-Pr (83%)
R = (CH₂)₄OH (83%)

SCHEME 14

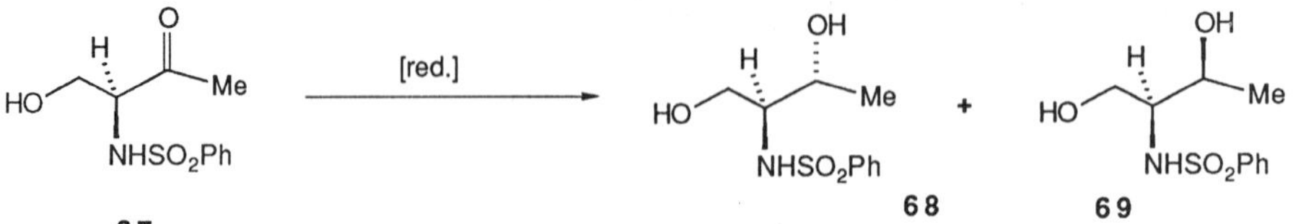

68   69

NaBH₄ / EtOH   7 : 3

L-selectride   1 : 99

SCHEME 15

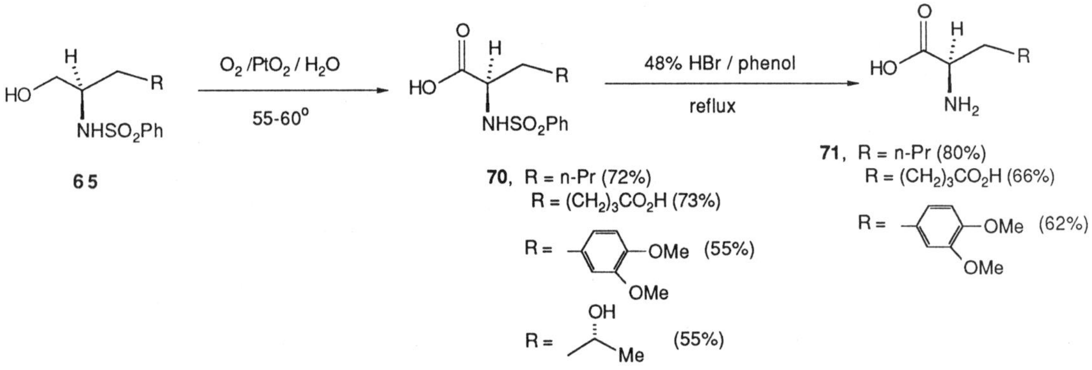

65

O₂ /PtO₂ / H₂O

55-60°

70, R = n-Pr (72%)
     R = (CH₂)₃CO₂H (73%)

R = ——⟨benzene⟩—OMe (55%)
         OMe

R = ⟨structure⟩ (55%)
      OH / Me

48% HBr / phenol

reflux

71, R = n-Pr (80%)
     R = (CH₂)₃CO₂H (66%)

R = ——⟨benzene⟩—OMe (62%)
         OMe

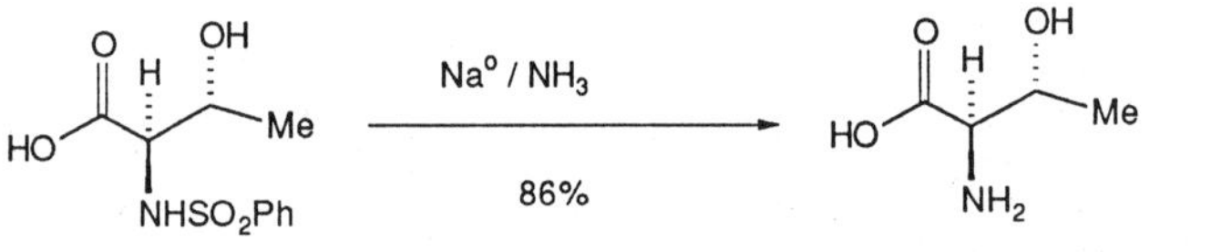

70

Na° / NH₃

86%

72, *D-ALLOTHREONINE*

SCHEME 16

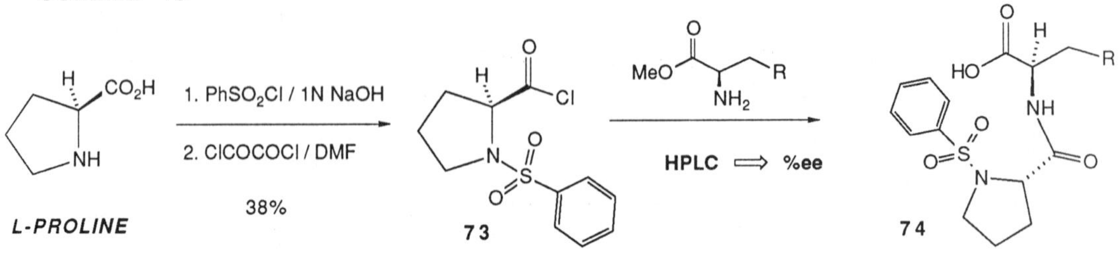

L-PROLINE

1. PhSO₂Cl / 1N NaOH

2. ClCOCOCl / DMF

38%

73

MeO₂C—⟨structure⟩—R
         NH₂

HPLC ⟹ %ee

74

SCHEME 17

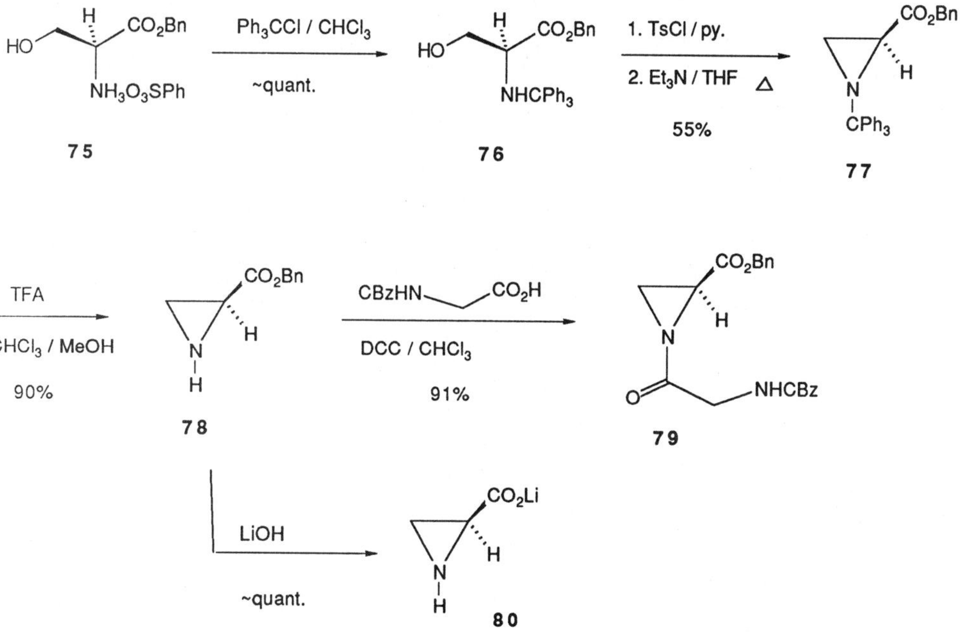

Nakajima, K.; Takai, F.; Tanaka, T.; Okawa, K., *Bull.Chem.Soc. Jpn.* (1978) **51**, 1577.

elimination in these cases would formally constitute a disfavored[4] *retro-3-endo-trig* process. Since the acylated versions of these systems readily participate in nucleophilic ring-opening processes, it would seem promising to examine the corresponding N-t-BOC substrates under reaction conditions commensurate with the t-BOC residue. Complete experimental details accompany these reports.

These workers have also prepared all stereoisomers of the interesting cysteine adducts **91** (Scheme 20). Using a similar protocol, Parry and Naidu[19] prepared (S)-(2-carboxypropyl)-L-cysteine in optically active form and elucidated the absolute stereochemistry of the natural amino acid.

Homologation of the β-Carbon

SCHEME 18

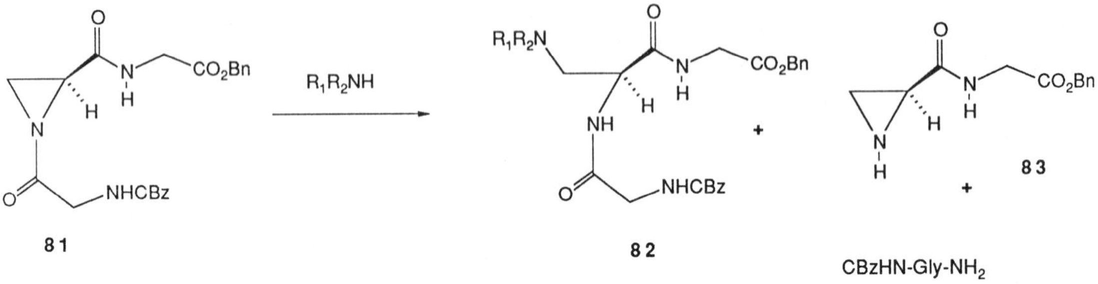

Nakajima, K.; Tanaka, T.; Morita, K.; Okawa, K., *Bull.Chem.Soc.Jpn.* (1980) **53**, 283.

AMINE	R_1	R_2	%82	%83	%84
NH_3	H	H	0	93	94.5
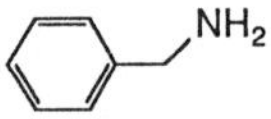 (benzyl)—NH_2	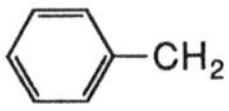 (phenyl)—CH_2	H	0	92	95
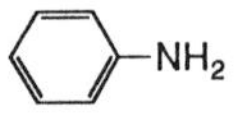 (phenyl)—NH_2	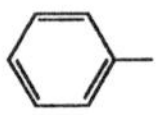 (phenyl)—	H	52	19	11.5
Et_2NH	Et	Et	88.5	0	0

SCHEME 19

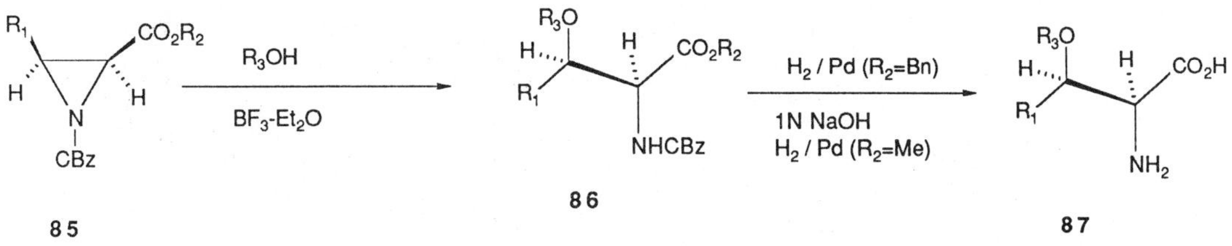

Nakajima, K.; Neya, M.; Yamada, S.; Okawa, K., *Bull. Chem. Soc. Jpn.* (1982) **55**, 3049.

R_1	R_2 (86)	R_3	YIELD (% 86)	YIELD (%87)
H	Bn	Me	64	
H	Bn	i-Pr	95	87
H	Bn	Bu	73	
H	Bn	t-Bu	58	100
H	Bn	n-Hex.	57	82
H	Bn	Bn	100	
H	Bn	Ph	27	
Me	Me	Me	98	
Me	Me	i-Pr	98	86
Me	Me	Bu	92	75
Me	Me	t-Bu	94	82
Me	Me	n-Hex.	92	83
Me	Me	Bn	98	
Me	Me	Ph	57	

SCHEME 20

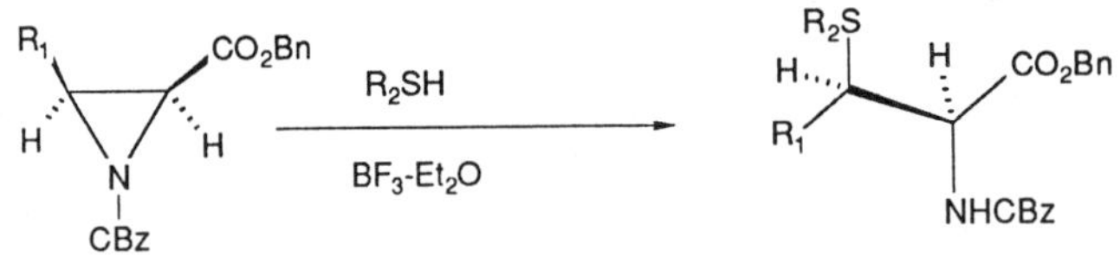

Nakajima, K.; Oda, H.; Okawa, K., *Bull.Chem.Soc.Jpn.* (1983) **56**, 520.
Nakajima, K.; Okawa, K., *Bull.Chem.Soc. Jpn.* (1983) **56**, 1565.

R₁	R₂	YIELD (%89)
Me	H	86
i-Pr	H	82
i-Bu	H	74
t-Bu	H	90
c-hex.	H	81
Ph	H	78
Bn	H	72
Me	Me	92
i-Pr	Me	73
i-Bu	Me	69
t-Bu	Me	53
c-hex.	Me	52
Ph	Me	82
Bn	Me	68

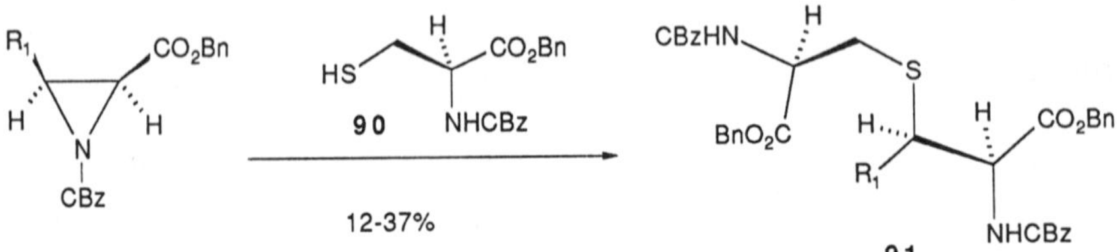

12-37%

B. γ-CARBON HOMOLOGATIONS

L-Glutamic acid is the least expensive amino acid and is produced on an enormous scale, primarily for the food industry. It is somewhat surprising that more use has not been made of this abundant, inexpensive substance as a γ-homologation template for amino acid synthesis. While this section is certainly not comprehensive with respect to the large number of papers that employ glutamic (and aspartic) acid for semi-synthetic manipulations, an attempt has been made to cover the basic strategies for manipulation of the γ-position that have recently emerged.

Baldwin and associates[20] have examined the γ-enolate reactions of a carefully protected L-glutamic acid system as shown in Scheme 21. To block competing enolization at the α-position, L-Glu is converted into the γ-methyl ester (**92**) and then sequentially esterified and tritylated to afford the key substance **94** in 35% overall yield. The bulky *tert*-butyl ester and trityl groups create a very sterically congested environment near the α-methine proton making this site kinetically inaccessible. Treatment with a sterically hindered base (lithium isopropyl cyclohexylamide) followed by aldol condensation provided the adducts **95** with no apparent loss of stereochemical integrity at the α-position. The carbinols are obtained as mixtures of diastereomers at the newly created stereogenic center. The yields vary considerably from case to case, but the low cost of the starting materials renders the present approach quite practical even in the lowest-yielding instances.

Optically active homoserine should serve as an excellent substrate for γ-homologations; the extremely high cost of L-homoserine (>US$30/gram; the D-isomer is not readily available) has, however severely limited these applications. Two recent syntheses of optically active homoserine systems have appeared; one from methionine and the other from aspartic acid. In the first approach, Baldwin and Flinn [21] have devised a very simple procedure illustrated in Scheme 22. L-Methionine is treated with methyl bromide producing the corresponding sulfonium salt which is subsequently solvolyzed with hot aqueous sodium bicarbonate. These reaction conditions provide an aqueous solution of homoserine that was directly acylated and esterified to the labile derivative **96**. The authors note that **96** is unstable to chromatography giving the corresponding γ-lactone; this is a well-known tendency of homoserine esters that has additionally hampered utilization of this amino acid. Direct treatment of crude **96** with pyridinium chlorochromate provided the key aldehyde **97** in 40% overall yield from methionine. Wittig

Homologation of the β-Carbon

SCHEME 21

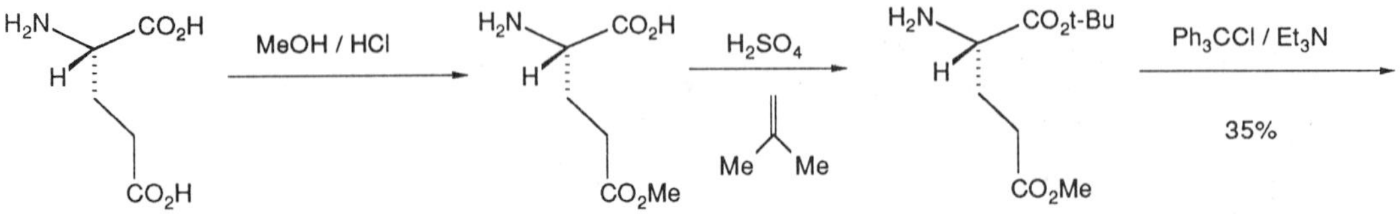

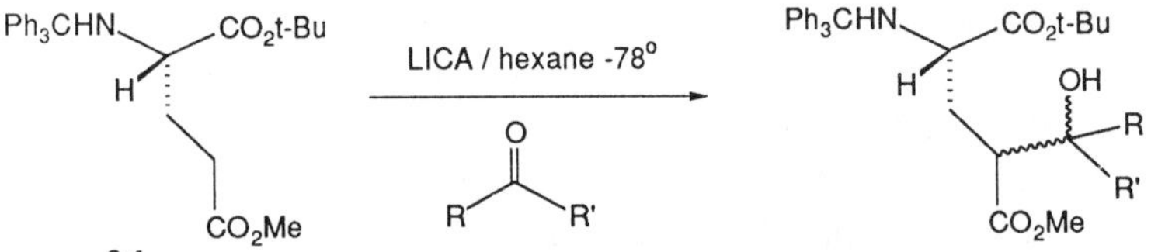

Baldwin, J.E.; North, M.; Flinn, A.; Moloney, M.G., *J.Chem.Soc.Chem.Comm.* (1988) 828.

ELECTROPHILE	R	R'	SOLVENT	YIELD (%95)
EtCHO	Et	H	THF	68
			hexane	68
PhCHO	Ph	H	THF	10
			hexane	35
acetone	Me	Me	THF	20
			hexane	40
H$_2$CO	H	H	THF	17
			hexane	30
Me$_2$CHCHO	Me$_2$CH	H	THF	25
			hexane	50
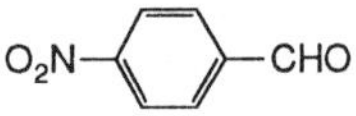 O$_2$N—⬡—CHO	p-NO$_2$Ph	H	hexane	95

condensation of **97** with allylidenetriphenylphosphorane provides the dieneamino acid **98** as 1 : 1, E : Z mixture. Similarly, formylmethylenetriphenylphosphorane provides **99** as the E-isomer in good yield. Stereoselective access to E-**98** via Wittig homologation of **99** with methylenetriphenylphosphorane is accomplished in 25% yield (unoptimized). Removal of the t-BOC and *para*-methoxybenzyl protecting groups from **98** provides L-2-amino hept-4,6-dienoic acid **100**. The unsaturated aldehyde **99** was envisaged to serve as a precursor to the natural product bulgecine. Reduction of the aldehyde with sodium borohydride furnished the E-allylic alcohol **101** in good yield. Sharpless

SCHEME 22

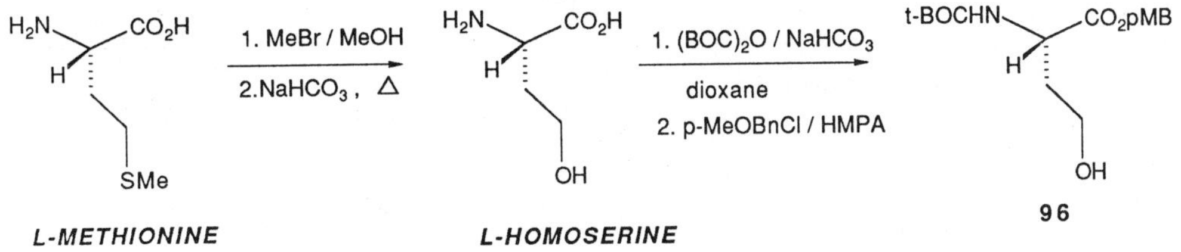

H2N
CO2H
H
SMe
L-METHIONINE
1. MeBr / MeOH
2.NaHCO3 , Δ
H2N
CO2H
H
OH
L-HOMOSERINE
1. (BOC)2O / NaHCO3
dioxane
2. p-MeOBnCl / HMPA
t-BOCHN
CO2pMB
H
OH
96

Baldwin, J.E.; Flinn, A., *Tetrahedron Lett.* (1987) **28**, 3605.

pMB = *para*-methoxybenzyl

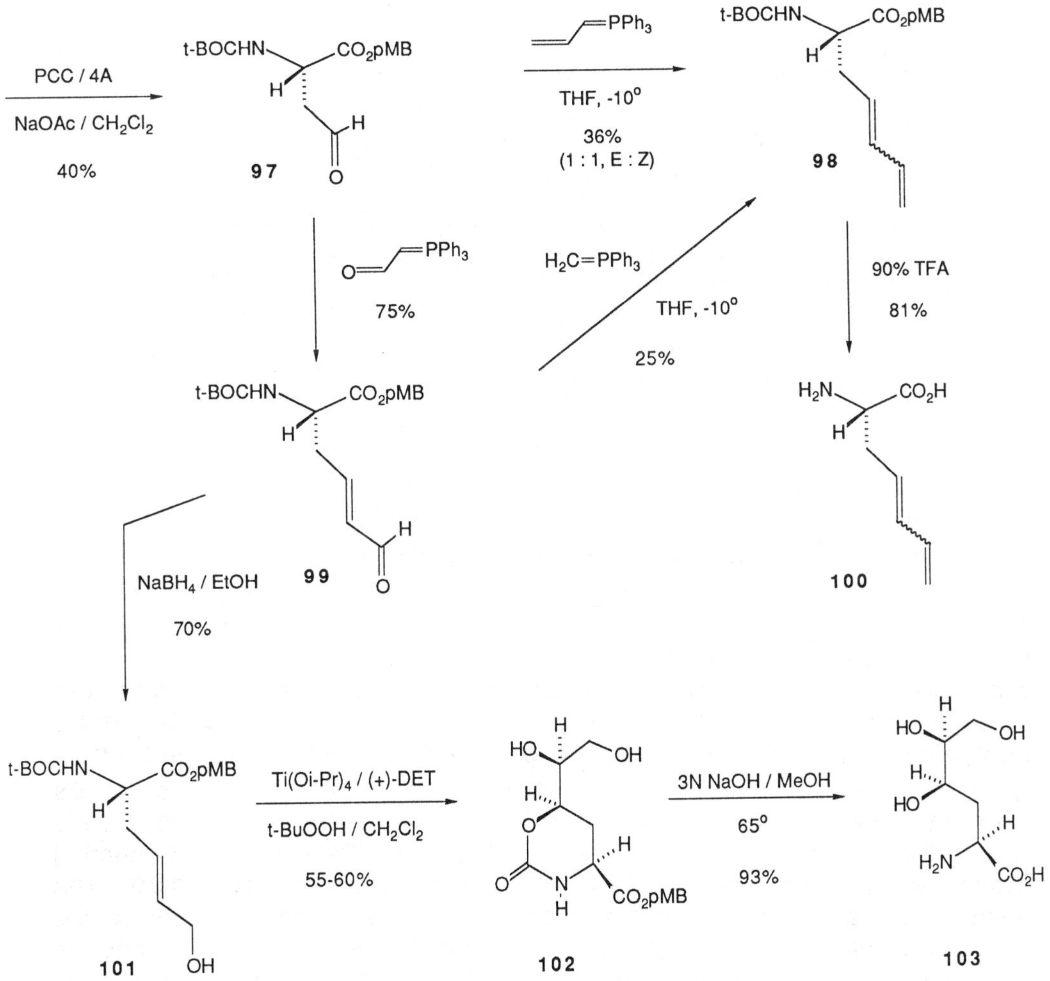

PCC / 4A
NaOAc / CH2Cl2
40%
t-BOCHN
CO2pMB
H
H
O
97
=PPh3
THF, -10°
36%
(1 : 1, E : Z)
t-BOCHN
CO2pMB
H
98
O
=PPh3
75%
H2C=PPh3
THF, -10°
25%
90% TFA
81%
t-BOCHN
CO2pMB
H
H
O
99
H2N
CO2H
H
100
NaBH4 / EtOH
70%
t-BOCHN
CO2pMB
H
OH
101
Ti(Oi-Pr)4 / (+)-DET
t-BuOOH / CH2Cl2
55-60%
H
HO
OH
H
O
O
N
H
H
CO2pMB
102
3N NaOH / MeOH
65°
93%
H
HO
OH
H
HO
H
H2N
CO2H
103

SCHEME 23

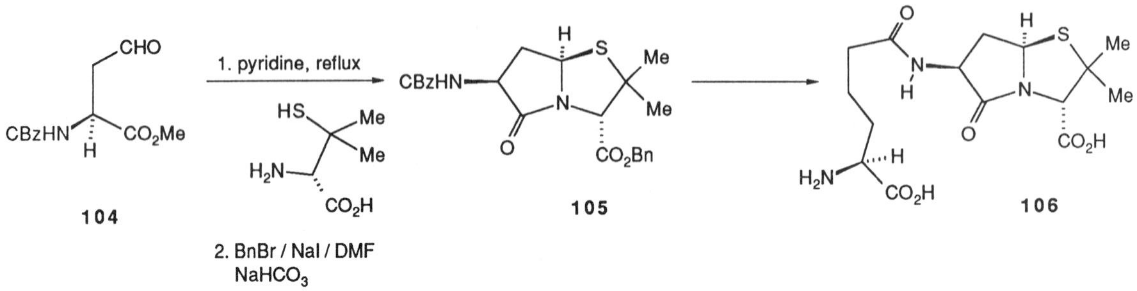

Baldwin, J.E.; Norris, W.J.; Freeman, R.T.; Bradley, M.; Adlington, R.M.; Long-Fox, S.; Schofield, C.J., *J.Chem.Soc.Chem.Comm.* (1988) 1128.

asymmetric epoxidation gave the cyclic urethane **102**; the stereochemistry of **102** was assigned based on the well-known enantiofacial selectivity of the Sharpless procedure. Saponification of **102** provided the triol **103** embracing the correct relative stereochemistry of bulgecine; efforts to convert **103** into the natural product have not yet appeared.

The second approach to the homoserine system from aspartic acid is detailed later in Scheme 29.

The useful aspartic acid β-semialdehyde equivalent (**97**) has also been utilized in the construction of interesting γ-lactam analogues of the β-lactam antibiotics. For example, Baldwin , et.al.,[22] have utilized the related aldehyde **104** to construct the bicyclic substance **105** by condensation with penicillamine in refluxing pyridine.

The homolgation at the δ-position has also been examined as shown in Scheme 24 by Olsen and associates [23]. γ-Benzyl-L-glutamate (**107**) is protected and hydrogenated to the alcohol **109**. Conversion to the key bromide proceeds in good yield. This substance was condensed with N-acetyl-O-benzylhydroxylamine to provide **111**. Subsequent deprotection afforded the acid **112** which can serve as a precursor to the ionophore antibiotic ferrichrome. Complete experimental details accompany this work.

Pyroglutamate derivatives have also enjoyed utility in the synthesis of non-proteinogenic amino acids and other chiral materials. Rapoport and associates [24] have utilized the Eschenmoser sulfide contraction protocol in a total synthesis of anatoxin as shown in Scheme 25. N-Benzyl pyroglutamate (**113**) is esterified and transformed into the thiolactam **114**. Alkylation of this material with methyl bromoacetate furnishes the

SCHEME 24

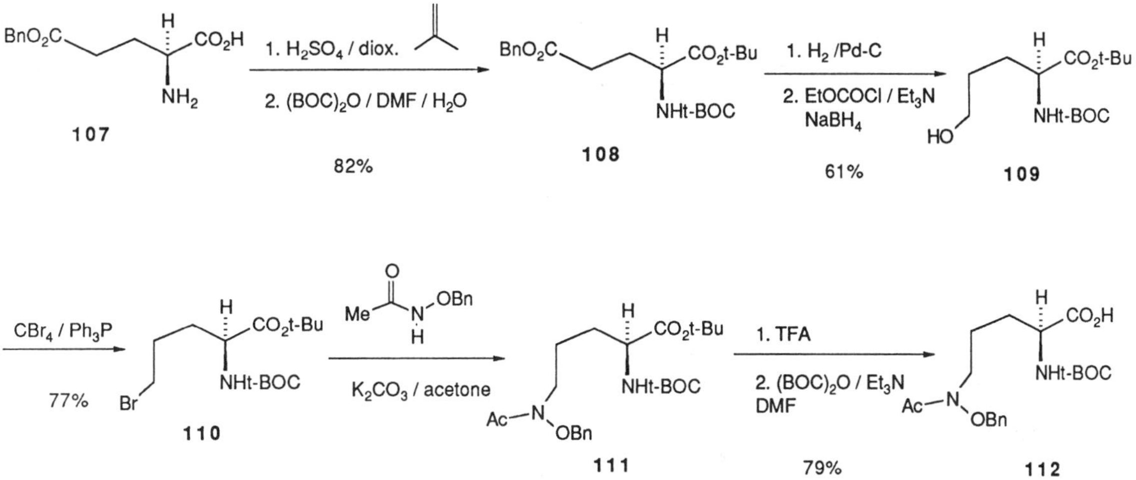

Olsen, R.K.; Ramasamy, K.; Emery, T., *J.Org.Chem.* (1984) **49**, 3527.

incipient adduct **115** that suffers cyclization to the episulfide and desulfurization to the unsaturated ester **116**. Stereoselective hydrogenation affords the *cis*-isomer **117** as the major product. This substance is subsequently converted into the natural neurotoxin anatoxin **(124)** as shown in the Scheme. This report, with full experimental details also delineates a synthetic route to the unnatural (-)-isomer of anatoxin.

Protected pyroglutamates can also directly serve as substrates for γ-homologations as recently demonstrated by Nozoe, and collaborators [25] (Schemes 26 and 27). Grignard additions to the protected pyroglutamates **125** occurs with excellent regiospecificity to furnish the ketones **126**. To illustrate the utility of this methodology, the authors have prepared the C-2 symmetric chiral amino acid (-)-pyrrolidine-2,5-dicarboxylic acid **(132)**; this substance is a natural product isolated from the red algae *Schizymenia dubyi*. The vinyl adduct **127** prepared as described in Scheme 26, is reduced to a diastereomeric mixture of carbinols **128**. Cyclization of the unstable mesylate derived from **128** provided a mixture of the *trans*- and *cis*-pyrrrolidines **129** and **130**. The major (*trans*) isomer was oxidized to the acid **131** and subsequently deprotected to **132**. This approach offers a potentially versatile and efficient method to prepare differentially protected amino acids via γ-homologation in a simple and straightforward manner.

SCHEME 25

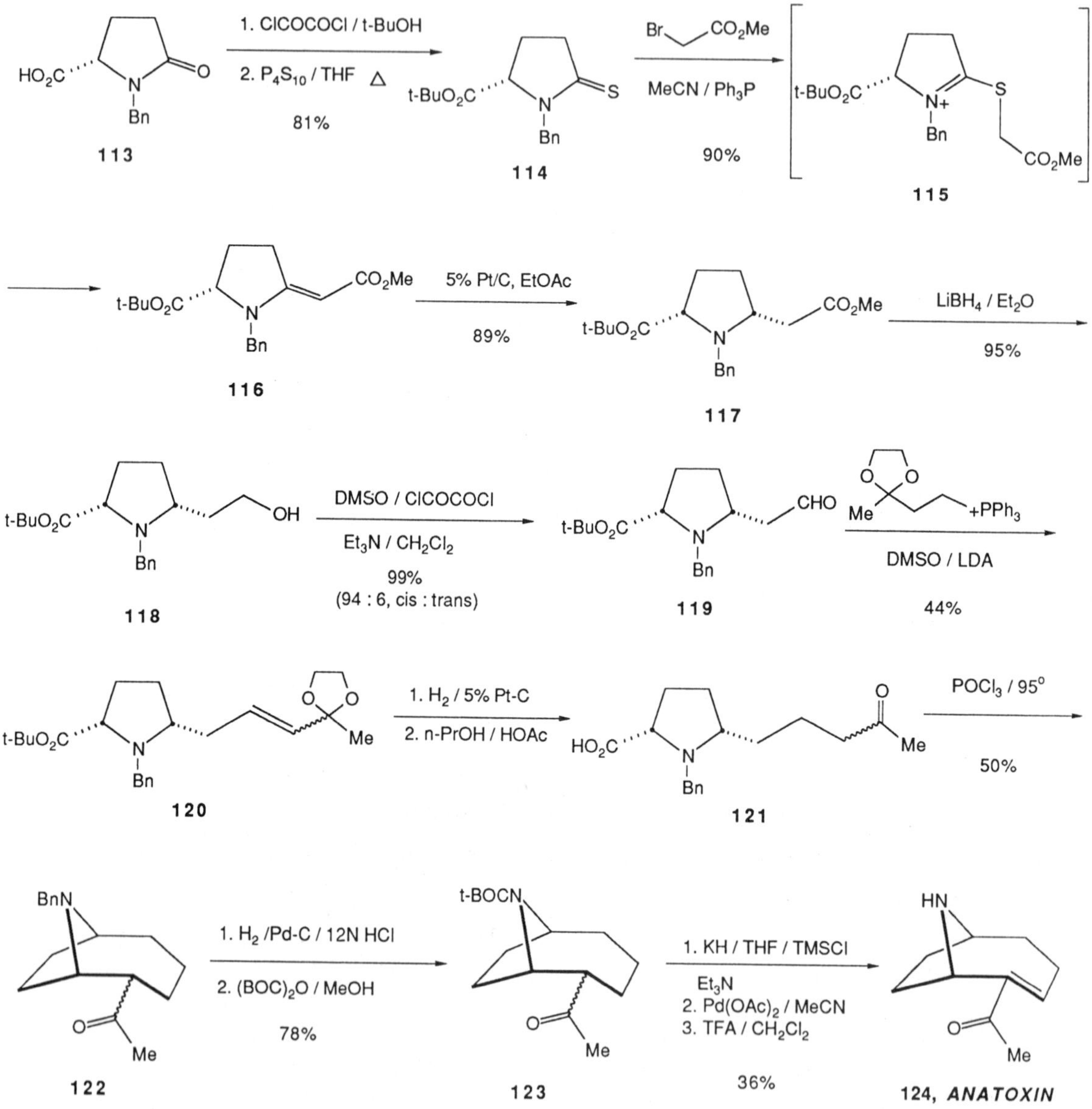

Petersen, J.S.; Fels, G.; Rapoport, H., *J.Am.Chem.Soc.* (1984) **106**, 4539.

SCHEME 26

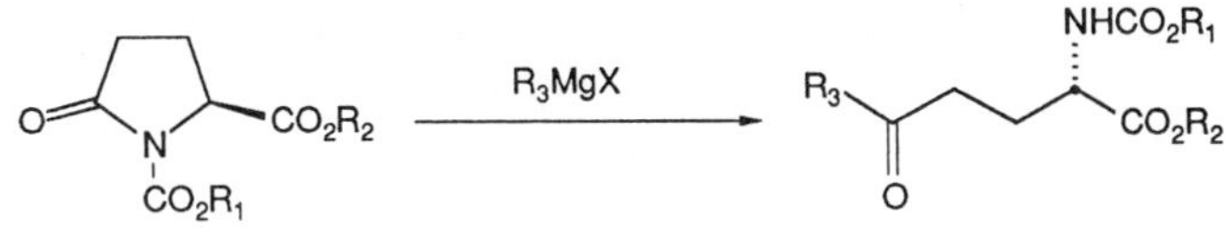

125 → **126**

Ohta, T.; Hosoi, A.; Kimura, T.; Nozoe, S., *Chemistry Lett.* (1987) 2091.

R$_1$	R$_2$	R$_3$MgX	YIELD (% 126)
t-Bu	Bn	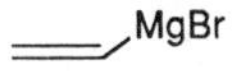 ══MgBr	82
Bn	t-Bu	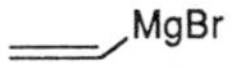 ══MgBr	55
Bn	Bn	 ══MgBr	52
t-Bu	p-NBn	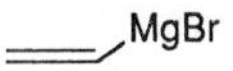 ══MgBr	78
t-Bu	Bn	MeMgI	78
t-Bu	Bn	 ══╱MgBr	70
t-Bu	Bn	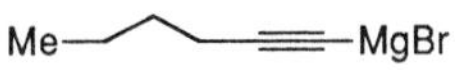 Me──══──MgBr	92

SCHEME 27

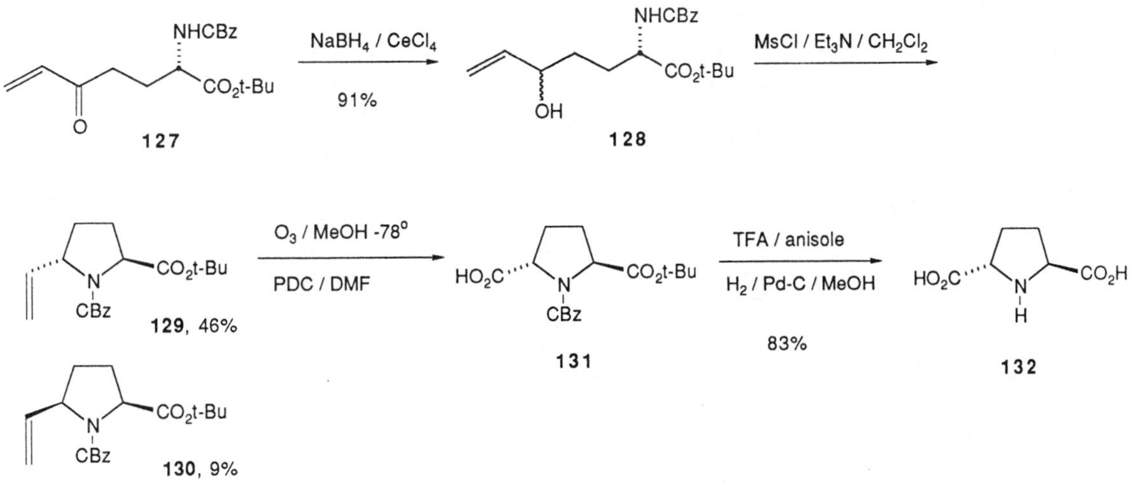

127 → NaBH$_4$ / CeCl$_4$, 91% → **128** → MsCl / Et$_3$N / CH$_2$Cl$_2$

129, 46%
130, 9%

O$_3$ / MeOH -78° ; PDC / DMF → **131** → TFA / anisole ; H$_2$ / Pd-C / MeOH, 83% → **132**

SCHEME 28

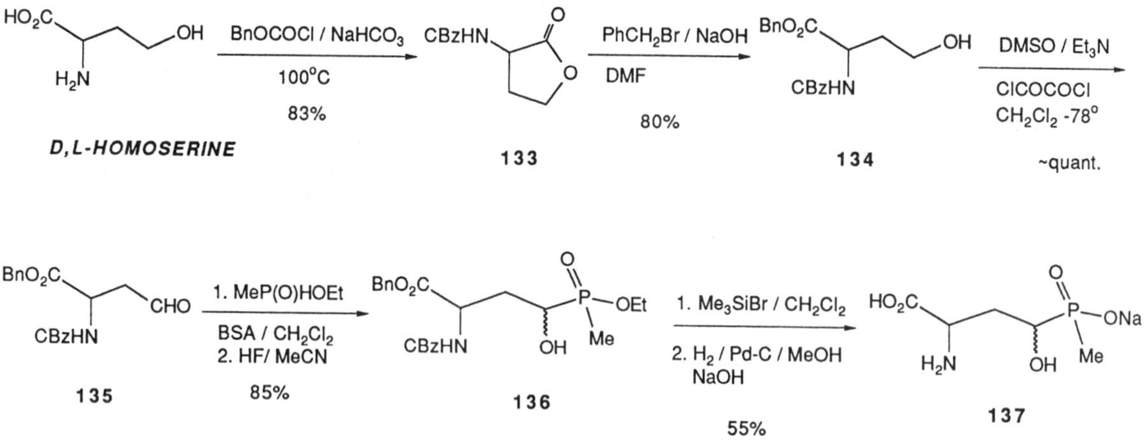

Walker, D.M.; McDonald, J.F.; Logusch, E.W., *J.Chem.Soc.Chem.Comm.* (1987) 1710.

Two recent reports [26,27] dealing with the introduction of a C-P bond at the γ-position have appeared as detailed in Schemes 28 and 29. The presence of phosphonate (or phosphinate) functionality at various positions has recently been effectively used to mimic the tetrahedral intermediate of amide hydrolysis during various metabolic processing of amino acids and peptides. These substances often display interesting enzyme inhibitory properties as transition state analogs. The first approach[26], although employing d,l-homoserine should be directly applicable to the optically active series, particularly with new syntheses of optically active homoserine recently becoming available. Logusch and associates[26] convert homoserine into the N-CBz lactone **133** and subsequently saponify and esterify to **134**. Swern oxidation of **134** proceeds in essentially quantitative yield affording the key aspartic acid β-semialdehyde equivalent **135**. Reaction of this substance with ethyl methylphosphinate in the presence of bis(trimethylsilyl)acetamide (BSA) provides the adduct **136** after removal of the silyl ether. This substance is obtained as a mixture of four chromatographically inseparable diastereomers. Deprotection of this mixture with trimethylsilyl bromide and hydrogenation afforded the sodium salt of d,l-γ-hydroxyphosphinothricin (**137**) as a 56 : 44 mixture of diastereomers. Compound **137** is reported to be a potent inhibitor of glutamine synthetase.

In a related study, Johns and co-workers [27] have prepared protected L-homoserine and the corresponding β-semialdehyde from L-aspartic acid. N-CBz-β-Methyl aspartic acid (**138**) is esterified and reduced via the mixed carbonic anhydride to provide the protected homoserine derivative **140** in 85% yield. Oxidation with chromium trioxide-pyridine affords the key aldehyde **141** in 60% yield after column chromatography. Reaction of this material with dimethyl- and diethyl trimethylsilyl phosphites afforded the phosphonates **142** as a 1 : 1 mixture of diastereomers. The hydroxyl group was subsequently removed with a Barton procedure to afford the phosphonate **145**. This substance can be converted into either the N-CBz free acid or the corresponding N-t-BOC acid **146**. Experimental details accompany this work.

The utilization of β- and γ- homologation templates for the preparation of a variety of natural and non-natural amino acids should find increased utility in the future. As several examples have elegantly demonstrated, the starting materials can be very inexpensive and easily manipulated. These systems, particularly those derived from glutamic acid, aspartic acid and methionine should be applicable to both large (process) scale industrial situations as well as the the more diverse and specialized needs of the basic research scientist.

SCHEME 29

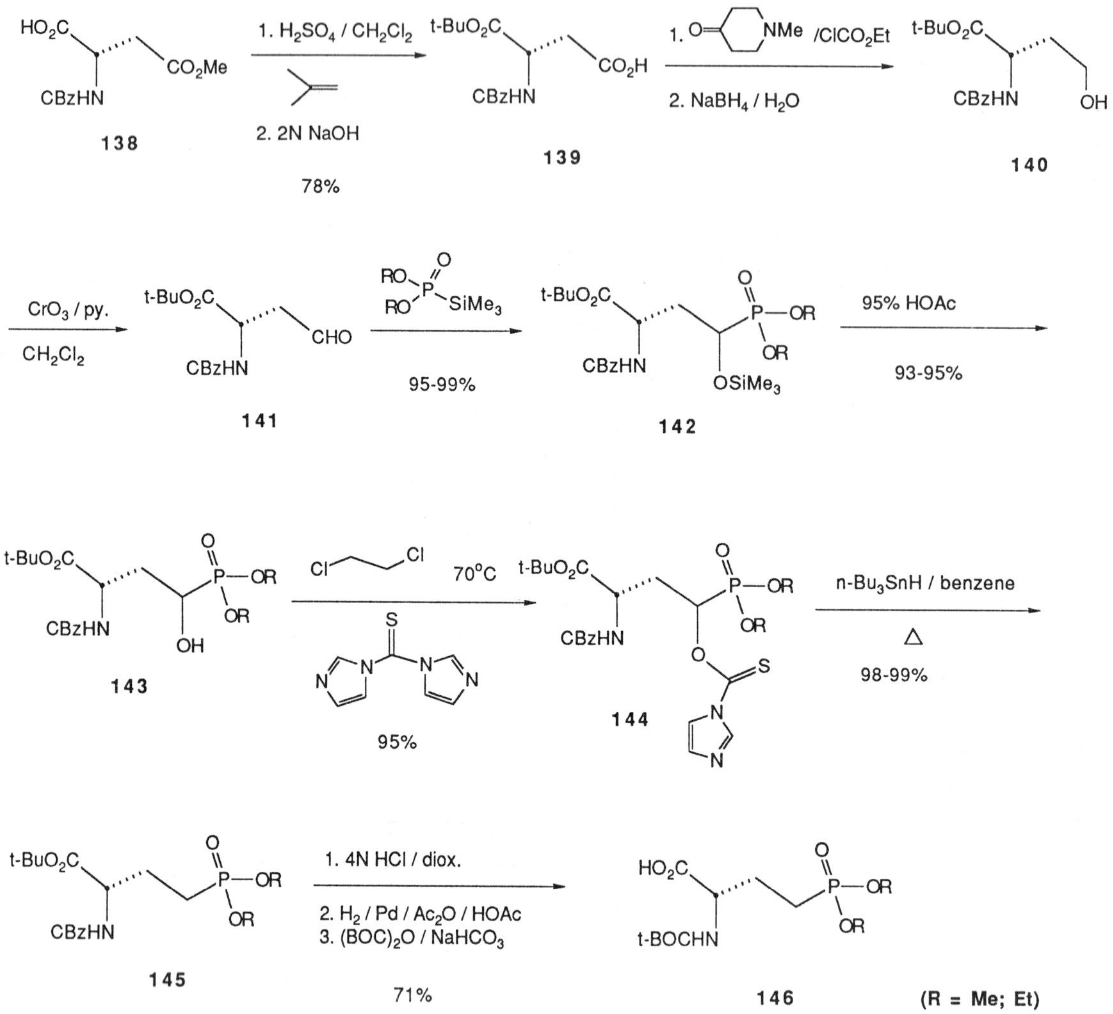

Valerio, R.M.; Alewood, P.F.; Johns, R.B., *Synthesis* (1988) 786.

References Chapter 2

1. Arnold, L.D.; Kalantar, T.H.; Vederas, J.C., *J.Am.Chem.Soc.* (1985) **107**, 7105.

2. Arnold, L.D.; May, R.G.; Vederas, J.C., *J.Am.Chem.Soc.* (1988) **110**, 2237.

3. Arnold, L.D.; Drover, J.C.G.; Vederas, J.C., *J.Am.Chem.Soc.* (1987) **109**, 4649.

4. Baldwin, J.E., *J.Chem.Soc.Chem.Comm.* (1976) 738.

5. Baldwin, J.E.; Adlington, R.M.; Robinson, N.G., *J.Chem.Soc.Chem.Comm.* (1987) 153.

6. Baldwin, J.E.; Adlington, R.M.; Birch, D.J.; Crawford, J.A.; Sweeney, J.B., *J.Chem.Soc.Chem.Comm.* (1986) 1339.

7. Mattingly, P.G.; Miller, M.J., *J.Org.Chem.* (1980) **45**, 410.

8. Baldwin, J.E.; Adlington, R.M.; Birch, D.J., *J.Chem.Soc.Chem.Comm.* (1985) 256.

9. Effenberger, F.; Weber, T., *Angew.Chem.Int.Ed.Engl.* (1987) **26**, 142.

10. Sasaki, N.A.; Hashimoto, C.; Potier, P., *Tetrahedron Lett.* (1987) **28**, 6069.

11. For leading references to the preparation and configurational stability of protected serinals, see: a) Garner, P.; Park, J.M., *J.Org.Chem.* (1987) **52**, 2361; b) Moriwake, T.; Hamano, S.; Saito, S.; Torii, S., *Chemistry Lett.* (1987) 2085.

12. Beaulieu, P.L.; Schiller, P.W., *Tetrahedron Lett.* (1988) **29**, 2019.

13. Garner, P.; Park, J.M., *J.Org.Chem.* (1988) **53**, 2979.

14. a) Maurer, P.J.; Takahata, H.; Rapoport, H., *J.Am.Chem.Soc.* (1984) **106**, 1095; b) for related use of threonine as a chiral starting material, see: Maurer, P.J.; Knudsen, C.G.; Palkowitz, A.D.; Rapoport, H., *J.Org.Chem.* (1985) **50**, 325.

15. Nakajima, K.; Takai, F.; Tanaka, T.; Okawa, K., *Bull.Chem.Soc.Jpn.* (1978) **51**, 1577.

16. Nakajima, K.; Tanaka, T.; Morita, K.; Okawa, K., *Bull.Chem.Soc.Jpn.* (1980) **53**, 283.

17. Nakajima, K.; Neya, M.; Yamada, S.; Okawa, K., *Bull.Chem.Soc.Jpn.* (1982) **55**, 3049.

18. a) Nakajima, K.; Oda, H.; Okawa, K., *Bull.Chem.Soc. Jpn.* (1983) **56**, 520; b) Nakajima, K.; Okawa, K., *Bull.Chem.Soc.Jpn.* (1983) **56**, 1565.

19. Parry, R.J.; Naidu, M.V., *Tetrahedron Lett.* (1983) **24**, 1133.

20. Baldwin, J.E.; North, M.; Flinn, A.; Moloney, M.G., *J.Chem.Soc.Chem.Comm.* (1988) 828.

21. Baldwin, J.E.; Flinn, A., *Tetrahedron Lett.* (1987) **28**, 3605.

22. Baldwin, J.E.; Norris, W.J.; Freeman, R.T.; Bradley, M.; Adlington, R.M.; Long-Fox, S.; Schofield, C.J., *J.Chem.Soc.Chem.Comm.* (1988) 1128.

23. Olsen, R.K.; Ramasamy, K.; Emery, T., *J.Org.Chem.* (1984) **49**, 3527.

24. a) Petersen, J.S.; Fels, G.; Rapoport, H., *J.Am.Chem.Soc.* (1984) **106**, 4539; b) see also: Shiosaki, K.; Rapoport, H., *J.Org.Chem.* (1985) **50**, 1229.

25. Ohta, T.; Hosoi, A.; Kimura, T.; Nozoe, S., *Chemistry Lett.* (1987) 2091.

26.   Walker, D.M.; McDonald, J.F.; Logusch, E.W., *J.Chem.Soc.Chem.Comm.* (1987) 1710.
27.   Valerio, R.M.; Alewood, P.F.; Johns, R.B., *Synthesis* (1988) 786.

CHAPTER 3

ELECTROPHILIC AMINATION OF ENOLATES

The electrophilic amination of chiral enolates is a very recent conceptual advance in the repertoire of synthetic methodology for the construction of α-amino acids. The reason for this is the general lack of *electrophilic* sources of nitrogen that readily participate in C-N bond-forming reactions with carbanions. The paucity of contributions in this area is somewhat surprising since, the electrophilic amination reaction of diethyl malonate with azodicarboxylates has been known since 1924 [1]. Subsequent, scattered reports of electrophilic amination of carbanions have since appeared [2] but the application to amino acid synthesis only very recently was recognized. This chapter outlines an extremely small collection of very recent papers that have appeared in the last few years. The vast arsenal of chiral enolate technology that is presently available signals that future chapters dealing with this approach will be considerably larger.

Boche and Schrott [3] examined the reaction of achiral benzylic carbanions with the optically active, electrophilic aminating reagent **3** (Scheme 1). This reagent is easily prepared from (-)-ephedrine (**1**), phosphorous oxychloride and N,N-dimethyl hydroxylamine. Reaction of this material with various lithio-carbanions effected enantiospecific amination resulting in the dimethylamino adducts **4**. The yields are moderate and the %ee's are very low indicating that the stereogenic centers of **3** are too far away from the nitrogen atom to allow for sufficiently high ΔΔG's of the diastereomeric transition states in the amination reaction. This method is unlikely to find widespread utility due to the low %ee's.

Four research groups simultaneously reported [4-9] a very effective solution to this general problem in 1986 involving the electrophilic amination of chiral enolates with azodicarboxylate esters.

As shown in Scheme 2, Gennari, and co-workers [4] condensed the ketenesilyl acetals of N-methyl ephedrine (**5**) with di-*tert*-butyl azodicarboxylate (**6**) in the presence of titanium tetrachloride in methylene chloride at low temperature to afford the α-hydrazido adducts

SCHEME 1

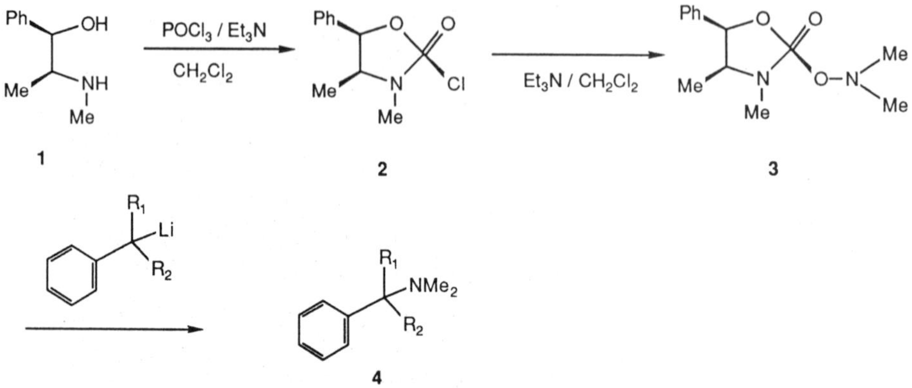

Boche, G.; Schrott,W., *Tetrahedron Lett.* (1982) **23**, 5403.

R$_1$	R$_2$	YIELD (%4)	%ee
H	CO$_2$Et	50	23
Me	CO$_2$Et	56	21
H	CN	62	8

7. The BOC groups were subsequently removed with trifluoro acetic acid and the chiral auxilliary hydrolyzed with lithium hydroxide; purification by ion-exchange chromatography and recrystallization affords the interesting α-hydrazino α-amino acids (**8**) in good overall yields and excellent %ee (>98%ee after recrystallization). The α-hydrazino amino acids have been found [7] to be inhibitors of certain amino acid-metabolizing enzymes such as the pyridoxal-dependent enzymes and ammonia lyases. Finally, the N-N bond can be reductively cleaved with hydrogen on a platinum catalyst to afford, after ion exchange purification, the α-amino acids **9**.

SCHEME 2

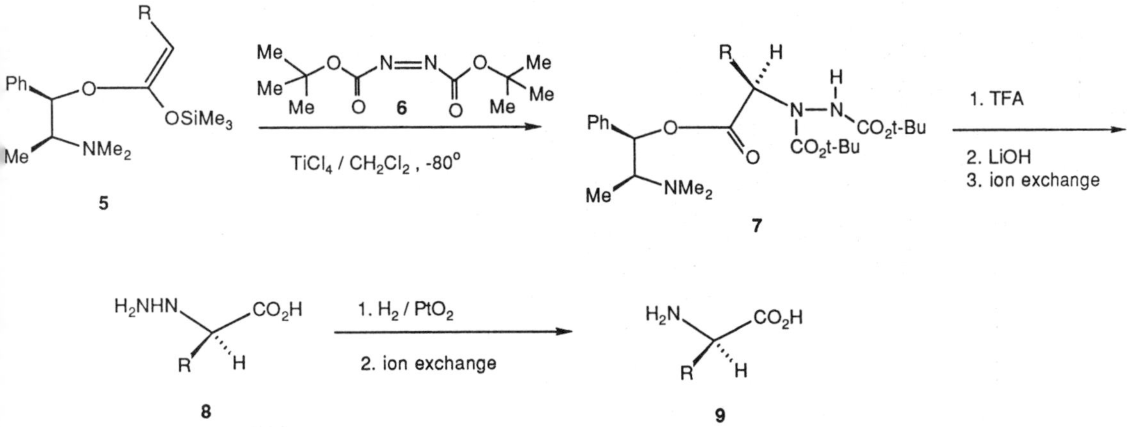

Ph–O, OSiMe₃, NMe₂, Me, R **5**

$$\xrightarrow[\text{TiCl}_4 \,/\, \text{CH}_2\text{Cl}_2 \,,\, -80°]{}$$

Me₃C–O–C(=O)–N=N–C(=O)–O–CMe₃ **6**

Ph–O, NMe₂, Me ... R, H, N(CO₂t-Bu)–NH–CO₂t-Bu **7**

1. TFA
2. LiOH
3. ion exchange

H₂NHN, CO₂H, R, H **8**

$$\xrightarrow[\text{2. ion exchange}]{\text{1. H}_2 \,/\, \text{PtO}_2}$$

H₂N, CO₂H, R, H **9**

Gennari, C.; Colombo, L.; Bertolini, G., *J.Am.Chem.Soc.* (1986) **108**, 6394.

R	YIELD (%7)	YIELD (%8)	YIELD (%9)	%ee(CRUDE)	%ee(recryst)	CONFIG.
Me	70	78	92	90.6	>98	R
Et	65	80	93	84	>98	R
n-Bu	45	78	90	78	>98	R
i-Bu	70	81	91	81.5	>98	R
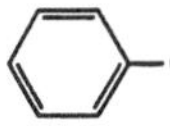 (Ph)–CH₂–	45	81	89	91	>98	R

Evans and associates [5,6] have applied the electrophilic amination of **6** to the versatile chiral carboximide enolates derived from **10** as shown in Scheme 3. Reaction of **10** with LDA in the usual way followed by the addition of **6**, afforded the α-hydrazido adducts **11** in excellent yields and excellent %de's (Table). Full experimental details have recently been published on this work [6]. The authors have examined several different reagents to effect the selective removal of the chiral auxilliary to provide the protected hydrazides **12**. Only in the racemization-prone phenylacetate cases (R=Ph) were difficulties encountered with transesterification; lithium hydroxide giving both high yields of the corresponding acids and little, if any, racemization. In very sterically

Electrophilic Amination of Enolates

SCHEME 3

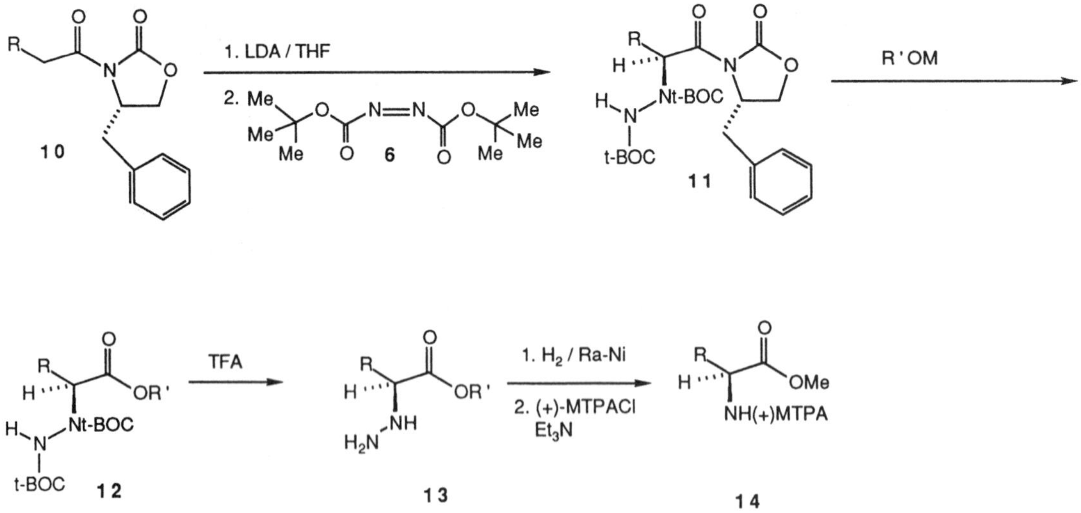

Evans, D.A.; Britton, T.C.; Dorow, R.L.; Dellaria, J.F., *J.Am.Chem.Soc.* (1986) **108**, 6395
Evans, D.A.; Britton, T.C.; Dorow, R.L.; Dellaria, J.F., *Tetrahedron* (1988) **44**, 5525.

R	KINETIC RATIO	YIELD (% 11)
Me	98 : 2	92
 (allyl)	98 : 2	94
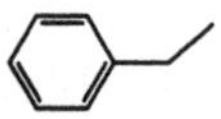 (benzyl)	97 : 3	91
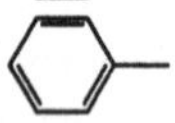 (phenyl)	97 : 3	96
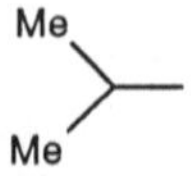 (isopropyl)	98 : 2	95
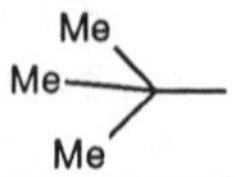 (tert-butyl)	>99 : 1	96

demanding cases, such as the isovalerate case (R=*t*-Bu) lithium peroxide proved to give outstanding results (91% yield, >99%ee). In less sensitive or sterically demanding substrates, hydrolysis to the acids with lithium hydroxide provides the acids and transesterification with bromomagnesium methoxide and lithium benzyloxide provide the corresponding methyl and benzyl esters, respectively.

The authors have rationalized the stereochemical outcome of these reactions by considering three different, chelated transition states T$_1$, T$_2$ and T$_3$. In the first example (T$_1$) , the lithium is chelated to the oxygen of the azodicarboxylate forming an 8-centered pericylic transition state. In the latter cases, chelation is to one of the nitrogen atoms forming a Zimmerman-Traxler-type 6-centered pericyclic transition state; the authors favor T$_2$ on steric and stereoelectronic grounds.

The optical purity of the final products can be consistently made to exceed 99%ee by chromatographic separation of the major and minor diastereomers produced in the amination reaction. The authors found that the major isomers **15**, are generally faster eluting and the minor isomers **16** are slower eluting. Thus, even in a case that displays modest %de's (rarely observed with the Evans enolates) the simple chromatographic purification following the amination step provides a reliable protocol to obtain optically pure products.

The BOC groups are easily removed with trifluoroacetic acid to afford the α-hydrazino derivatives **13** which are subsequently reduced to the α-amino acids with hydrogen on Raney-Nickel. The authors have combined the acylation of the crude amino acids obtained in the reduction by direct treatment with Mosher's acid chloride ((+)-MTPACl) to afford the amides **14** which are then examined for diastereomeric purity by capillary GLC.

15, Major Diastereomer
(Faster Eluting)

16, Minor Diastereomer
(Slower Eluting)

HYDROLYSIS AND TRANSESTERIFICATION OF HYDRAZIDE ADDUCTS (11 ⟶ 12)

R	REAGENT	R ' (12)	YIELD (% 12)	%ee
(benzyl/CH2Ph)	LiOH	H	82	>99
	BrMgOMe	Me	89	>99
	LiOBn	Bn	96	>99
(phenyl)	LiOH	H	84	98
	BrMgOMe	Me	71	93
	LiOBn	Bn	89	22
(isopropyl, Me Me)	BrMgOMe	Me	12	-
	LiOBn	Bn	82	>99
(tert-butyl, Me Me Me)	LiOH	H	16	>99
	LiOBn	Bn	51	>99
	LiOOH	H	91	>99

REDUCTION AND (+)-MTPA ACYLATION OF α-HYDRAZIDO ESTERS

R	R '	YIELD (% 14)	RATIO (2S : 2R)
(benzyl/CH2Ph)	Bn	94	>200 : 1
(phenyl)	Me	99	99 : 1
(isopropyl, Me Me)	Bn	83	>200 : 1
(tert-butyl, Me Me Me)	Bn	89	>200 : 1

Vederas and Trimble [7] have examined the Evans valinol-derived carboximide **17** for electrophilic amination with several azodicarboxylate esters as shown in Scheme 4. Treatment of **17** with LDA in the standard manner, forms the corresponding Z-lithium enolates; subsequent quenching with the azodicarboxylate provides the α-hydrazido adducts **18** in high yields. The authors found that the the kinetic diastereoselectivities improved with increasing steric bulk of the azodicarboxylate ester moiety (R '). Transesterification with lithium benzyloxide was found to cause some epimerization in several cases. An alternative procedure was found to obviate this problem involving cleavage with lithium hydrosulfide. The initial hydrolysis product is the thiol acid anion which is less prone to subsequent racemization than the esters (**19**, X=Bn). Conversion of these species to the free acids (**19**, X=H) is achieved with peracetic acid followed by medium-pressure chromatography, providing the carboxylic acids **19**. The benzyl esters (**19**, X=Bn) and hydrazides (**19**, R'=Bn) can be hydrogenated to the free α-hydrazino acids **20**. Conversion of the hydrazines **20** to the free α-amino acids **21** was accomplished by reduction on Raney-Nickel.

Oppolzer and Moretti [8,9] have employed the electrophilic amination reaction of ketenesilyl acetals derived from the Oppolzer chiral sulfamides **22** with **6** in the presence of titanium tetrachloride/titanium isopropoxide (Scheme 5). The chemical yields of the adducts **23** are consistently good and the crude (kinetic) diastereomer ratios are excellent in all but the very hindered adamantyl case. The BOC groups are removed with trifluoro acetic acid and the N-N bond of the incipient hydrazine is cleaved by reduction with hydrogen on a platinum catalyst to afford the α-amino esters **24**. The authors found that simple acidic hydrolysis of the esters directly afforded the corresponding amino acids, but destruction of the chiral auxilliary attended this procedure. Alternatively, transesterification with titanium ethoxide removed the chiral auxilliary that can be recovered in >95% yield. Hydrolysis of the resulting amino acid ethyl esters with hot 6N HCl afforded the amino acids **25** as their HCl salts. The chemical yields for these transformations are quite ggod and the optical purity of the final amino acids is uniformly excellent. Full experimental details have recently been published for this work [9].

Guanti, and associates [10] have recently studied the amination of (S)-β-hydroxybutyrates (**26**) with **6** as shown in Scheme 6. Formation of the dianions of **26** with LDA followed by addition of the aminating reagent **6** affords a mixture of the *anti*- and *syn*-hydrazides **27** and **28**. In every case, the *anti*-isomer **27** is the major product which is separated from

SCHEME 4

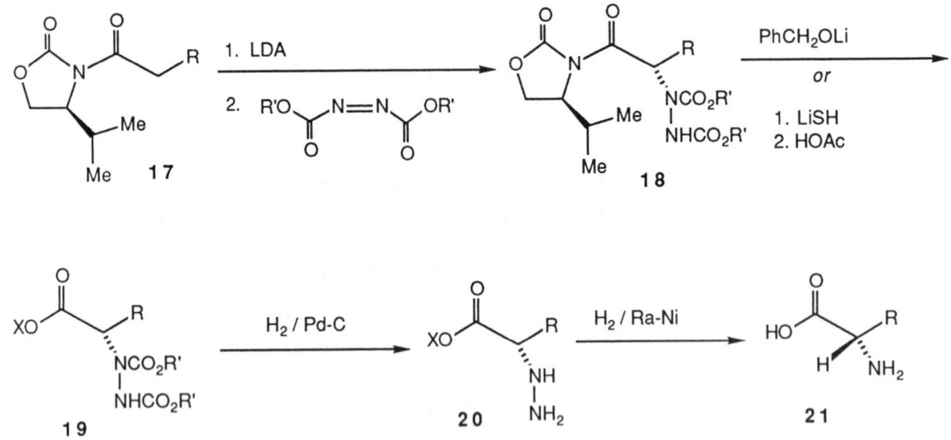

Trimble, L.A.; Vederas, J.C., *J.Am.Chem.Soc.* (1986) **108**, 6397.

R	R'	YIELD (%18)	DIASTEREOMER RATIO	YIELD (%19)	YIELD (%20)	ENANTIOMER RATIO
Me	⬡-CH₂	91	90 : 10	82	98	88 : 12
⬡-CH₂	Me	83	69 : 31	93	92	72 : 28
	Et	88	75 : 25	85	82	94 : 6
	⬡-CH₂	90	94 : 6	87	92	88 : 12
	Me₃C (tBu)	88	93 : 7	97	81	83 : 17
Me₃C (tBu)	⬡-CH₂	85	97 : 3	86	98	97 : 3
	Me₃C (tBu)	92		74	83	86 : 14

SCHEME 5

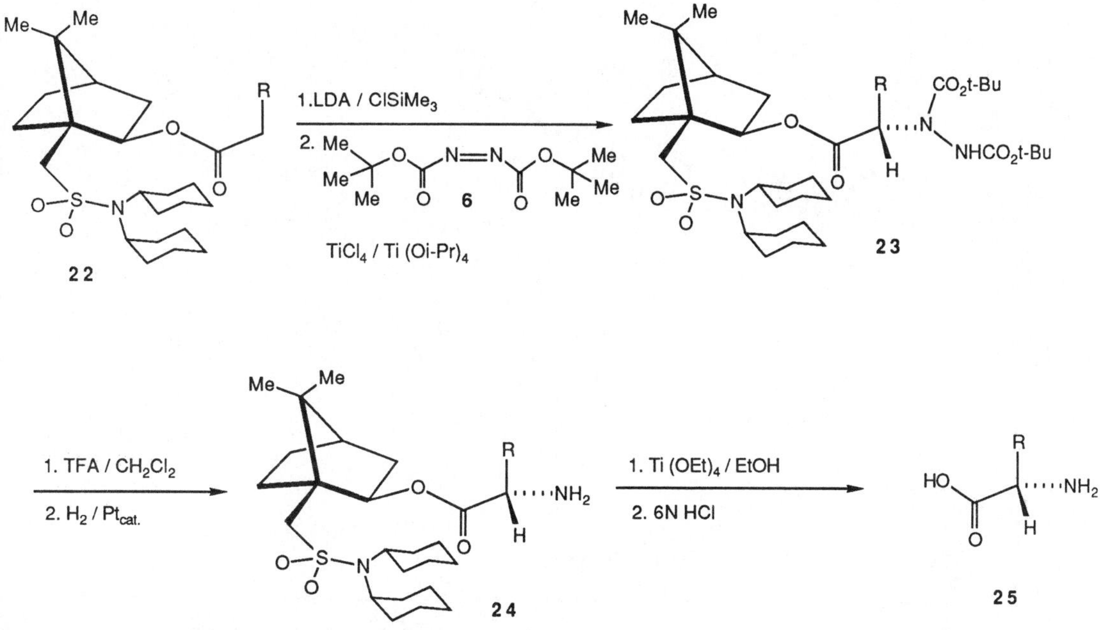

Oppolzer, W.; Moretti, R., *Helv.Chim.Acta.* (1986) **69**, 1923
Oppolzer, W.; Moretti, R., *Tetrahedron* (1988) **44**, 5541

R	DIASTEREOMER RATIO (Crude 23)	YIELD (%23)	%de (23) After Chrom.	YIELD (%24)	YIELD (%25)	%ee
Me	96.9 : 3.1	81	>99.5	83	83	95
Et	98.1 : 1.9	84	>99.5	77	91	99.7
n-Pr	98.2 : 1.8	72	>99.5	81	86	99.2
i-Pr	97.6 : 2.4	73	99	71	90	99.1
n-Bu	96.3 : 3.7	85	>99.5	80	95	97.6
i-Bu	96.6 : 3.4	71	>99.5	70	86	97.7
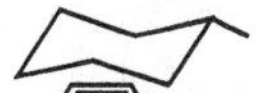 (cyclohexyl-CH₂)	98.0 : 2.0	69	>99.5	55	89	96.9
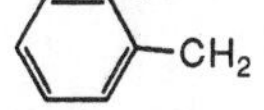 (benzyl, CH₂)	98.2 : 1.8	76	>99.5	64	86	98.4
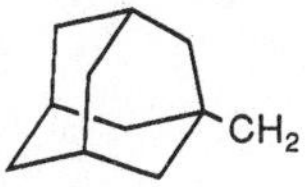 (adamantyl-CH₂)	82 : 18	65	>99.5	78	65	95.2

SOAA—G

SCHEME 6

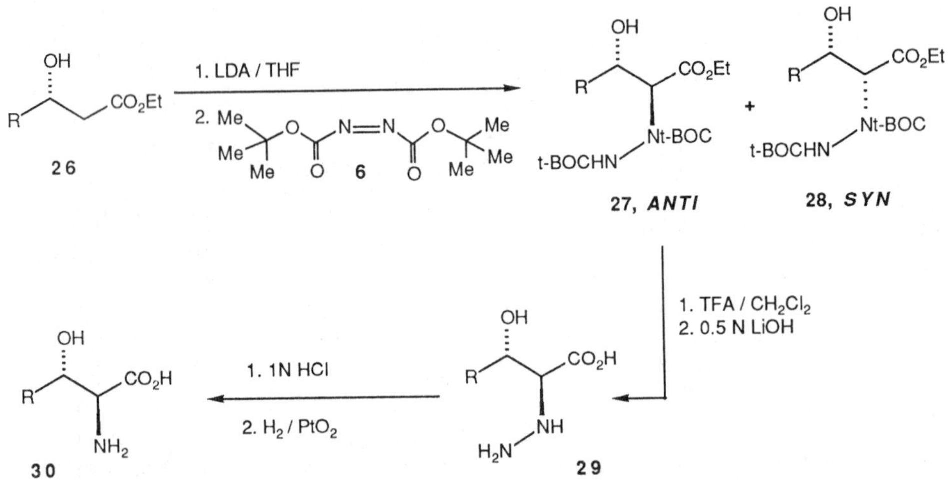

Guanti, G.; Banfi, L.; Narisano, E., *Tetrahedron* (1988) **44**, 5553.

R	Anti : Syn	YIELD (%27 + 28)	YIELD (%29)	AMINO ACID (%30)
Me	84 : 16	75	71	80
CF$_3$	87 : 13	62	66	78
n-Hexyl	90 : 10	74	54	76
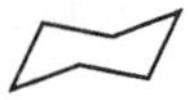 (cyclohexyl)	85 : 15	81	56	76

SCHEME 7

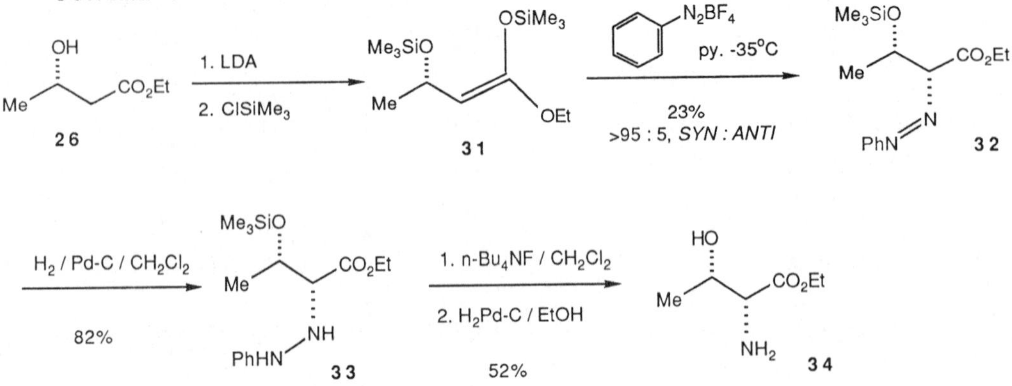

the minor isomer by chromatography. Removal of the t-BOC groups as above, is performed with trifluoro acetic acid. Saponification of the ethyl ester with lithium hydroxide affords the β-hydroxy-α-hydrazino acids **29** in reasonably good overall yields. Finally, these substances can be converted into the corresponding α-amino acids **30** by catalytic hydrogenation of the N-N bond with hydrogen on a platinum catalyst. The optical purity of the final amino acids is directly related to the optical purity of the starting esters **26** which can be obtained by various enantioselective reductions of β-ketoesters (ie., yeast reduction).

These authors have also examined the use of benzenediazonium tetrafluoroborate as an electrophilic aminating reagent as shown in Scheme 7. Conversion of **26** (R=Me) to the corresponding ketenesilyl acetal **31** followed by reaction with benzenediazonium tetrafluoroborate in pyridine at low temperature affords the adduct **32** in low yield. The diastereoselectivity of this transformation however is excellent, with a >95 : 5, *syn : anti* ratio. The authors comment that many attempts to improve the chemical yield of this stereoselective reaction were unsuccessful. Reduction of **32** with palladium on charcoal gave the N-phenyl hydrazine **33** in good yield; subsequent desilylation and reduction of the N-N bond provided enantiomerically pure *allo*-threonine ethyl ester **34** in 52% yield.

The methodologies reported on above give direct access to the interesting class of α-hydrazino amino acids that are emerging as important substances whose traditional syntheses from α-amino acids or other substrates often proves to be tedious and difficult. The major limitations of the above procedures for preparing α-amino acids reside in the reductive cleavage of the N-N linkage which will restrict the types of functionality that can be tolerated on accessible substrates . On the other hand, these approaches beautifully complement other amino acid syntheses and are not as restrictive of the nature of the α-R group as C-C bond forming approaches; the wide availability of substituted acetates which can be converted into the chiral enolate derivatives described above, provides a very useful, complementary approach to the synthesis of numerous types of amino acids.

In a recent variant on this general theme, Evans and Britton [11] have found an excellent protocol for the direct diastereoselective *azidation* of enolates as illustrated in Scheme 8. The optically active carboximide **10** is treated with potassium hexamethyldisilylazide followed by the addition of 2,4,6-triisopropylbenzenesulfonyl azide (**35**, 'trisyl azide'). The reaction is then quenched with glacial acetic acid to afford the α-azido carboximides **36**. The kinetic diastereoselectivities are excellent and the

Electrophilic Amination of Enolates

SCHEME 8

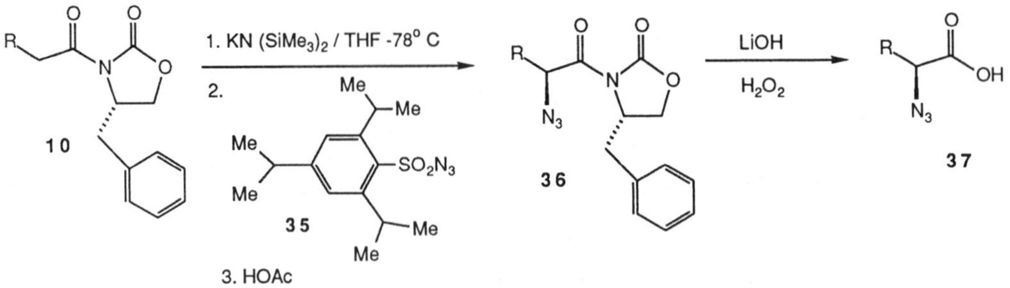

Evans, D.A.; Britton, T.C., *J.Am.Chem.Soc.* (1987) **109**, 6881.

R	KINETIC RATIO (36)	YIELD (%36)
Me	97 : 3	74
=CH₂ (allyl)	97 : 3	78
benzyl CH₂	97 : 3	91
phenyl	91 : 9	82
Me₂CH (isopropyl)	98 : 2	77
Me₃C (tert-butyl)	>99 :1	90

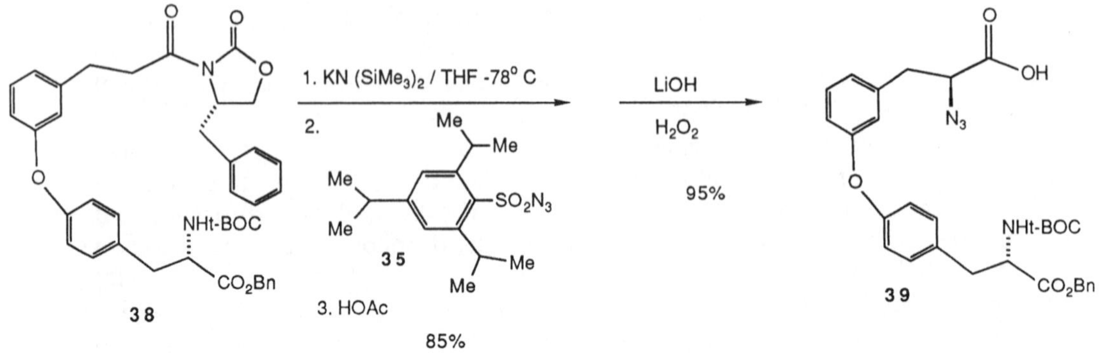

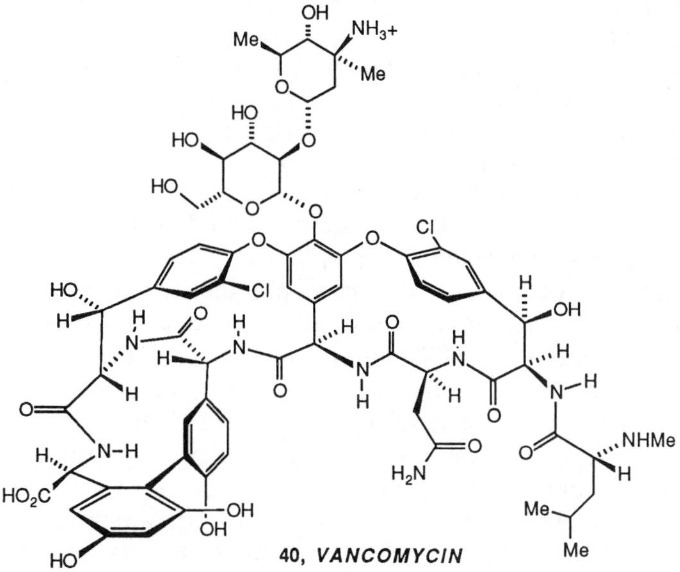

40, VANCOMYCIN

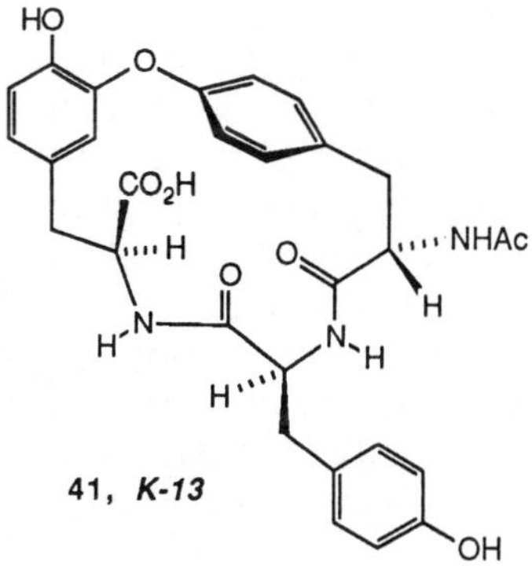

41, K-13

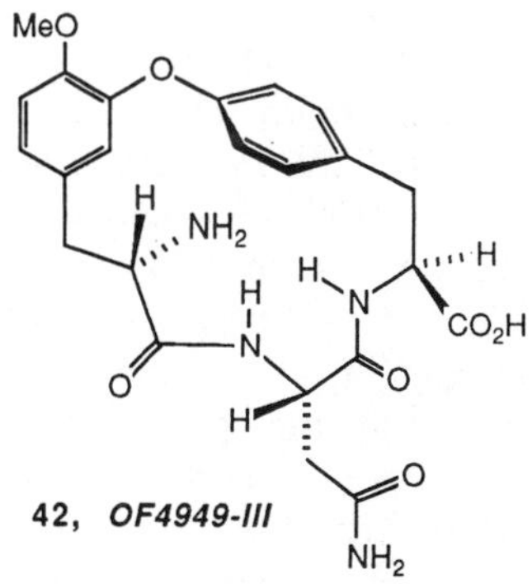

42, OF4949-III

chemical yields are good to excellent. The chiral auxilliary can be conveniently removed and recovered by hydrolysis with lithium peroxide to afford the α-azido acids **37**. The authors note that the yields for the azide transfer increases at the expense of the competing diazo transfer as the enolate counterion becomes more electropositive (Li<<Na<K) and as the azide transfer reagent becomes more sterically crowded (*p*-nitrobenzenesulfonyl azide< tosyl azide<trisyl azide). They also found that the quench reagent , glacial acetic acid , was an essential ingredient for a successful azidation reaction; the exact reasons for this are still obscure. This procedure combines the most desireable features of the azodicarboxylate amination discussed above and the nucleophilic azidation of α-substituted acids presented in Chapter 4.

A truly impressive and elegant demonstration of this new amino acid synthesis was very recently achieved by Evans and Ellman [12] in total syntheses of the complex cyclic tripeptides K-13 (**41**) and OF4949-III (**42**). These substances contain structural elements and synthetic challenges inherent in the fearsome glycopeptide antibiotics exemplified by vancomycin (**40**). The authors note at the outset, that the previously studied Ullman coupling of two derivatized tyrosine precursors proceeds in very low yield. These methods and syntheses of members of this class of natural products will be detailed elsewhere in this monograph. With these considerations in mind, the authors chose an alternate strategy of performing the biaryl ether coupling *prior* to introduction of the nitrogen atoms. As shown in Scheme 9, Ullman coupling of **43** and **44** proceeded in excellent yield to give the biaryl **45**. Hydrogenation of the olefins and benzyl ester followed by mixed anhydride formation and addition of lithio-oxazolidinone **46** provided the carboximide **47** in high overall yield. Subjecting **47** to the trisyl azide protocol described above proceeded in very good yield (diastereomer ratio= 97.5 : 5) to furnish the azide **48**. Transesterification with titanium benzyloxide cleanly afforded the benzyl ester **49.** The azide was then reduced by hydrogenation on Raney-Nickel followed by removal of the *t*-butyl ester . Protection of the amine as the corresponding t-BOC urethane gave **50** in 82-87% overall yield. It is significant that reduction of the benzyl ester does not attend the reduction of the azide. The acid was converted into the carboximide **51** through the mixed pivaloyl anhydride in excellent yield. Electrophilic azidation of **51** as above, produced the key intermediate **52** in astonishingly good yield (85%).

The t-BOC group of **52** was removed and coupled to N-t-BOC asparagine (**53**) with 1-(3-dimethylaminopropyl)-3-ethylcarbodiimide hydrochloride (EDC) and N-hydroxybenzotriazole (HOBt) to give the peptide **54** in 85-90% yield (Scheme 10). Lithium peroxide treatment of **54** effected removal of the chiral auxilliary; the corresponding acid was

SCHEME 9

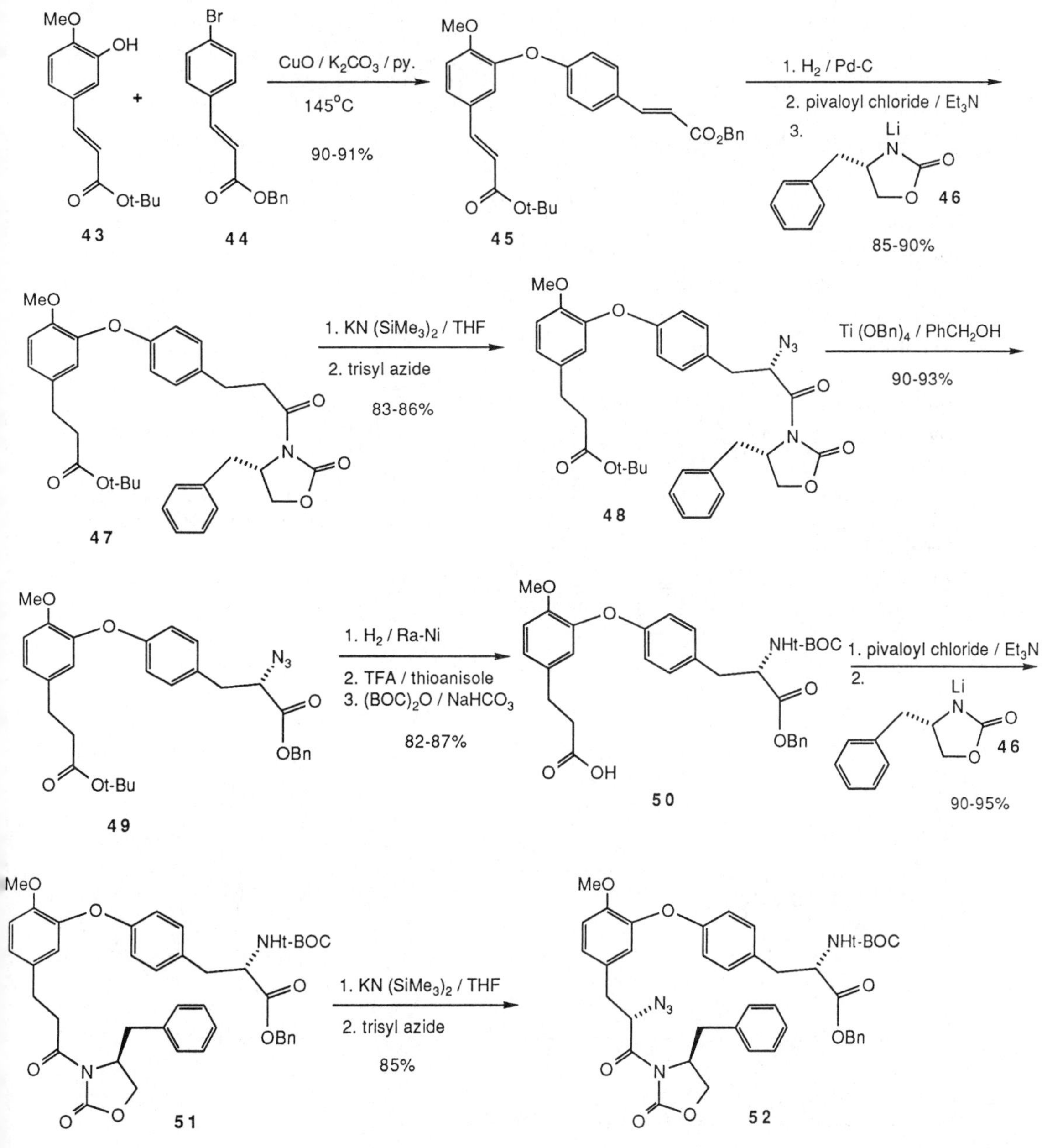

Evans, D.A.; Ellman, J.A., *J.Am.Chem.Soc.* (1989) **111**, 1063

Electrophilic Amination of Enolates

SCHEME 10

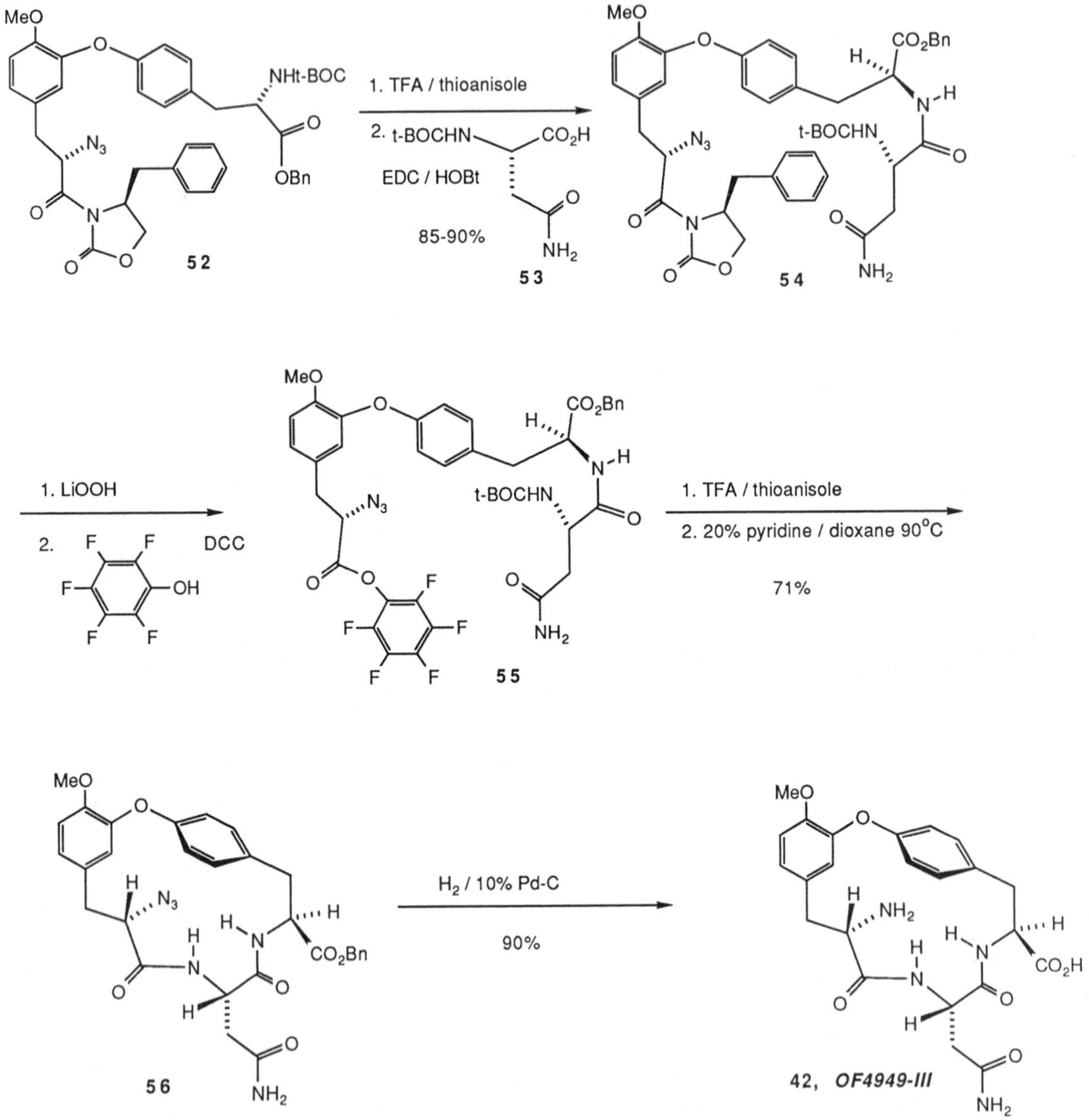

Evans, D.A.; Ellman, J.A., *J.Am.Chem.Soc.* (1989) **111**, 1063

converted into the active pentafluoro phenyl ester **55** by coupling pentafluoro phenol to the acid with DCC. The labile substrate **55** was immediately treated with trifluoro acetic acid and thioanisole to remove the t-BOC group and cyclized with 20% pyridine in dioxane at 90°C to afford the macrocycle **56** in 71% yield. Catalytic hydrogenation effected removal of the benzyl ester and reduction of the azide to afford OF4949-III (**42**) in 90% yield.

The key intermediate **52** was similarly transformed into K-13 as illustrated in Scheme 11. Removal of the chiral auxilliary from **52** was achieved in excellent yield with lithium peroxide and the acid esterified with diazomethane. The benzyl ester and azide were hydrogenated with palladium on charcoal and coupled with N-CBz-O-methyltyrosine pentafluoro phenyl ester (**58**) in 83% yield. Conversion of the remaining acid to the pentafluoro phenyl ester **60** was accomplished in high yield. Reduction of the N-CBz group and concomitant macrocyclization afforded the cyclic tripeptide **61** in 67% yield. Removal of the t-BOC protection, acylation and O-demethylation with aluminum tribromide/ethane thiol provided the natural product K-13 (**41**) in 89% yield. The present synthesis also elucidated the absolute stereochemistry of all stereogenic centers of **41** as (S).

These two examples brilliantly demonstrate the practical utility of the azidation of chiral enolates in highly functionalized systems. The α-azido carboximides are efficiently utilized as protected amino acids and can be easily converted into the corresponding amino acids in the presence of labile functional groups. Additional examples of the utilization of these excellent methodologies will undoubtedly be forthcoming.

Electrophilic Amination of Enolates

SCHEME 11

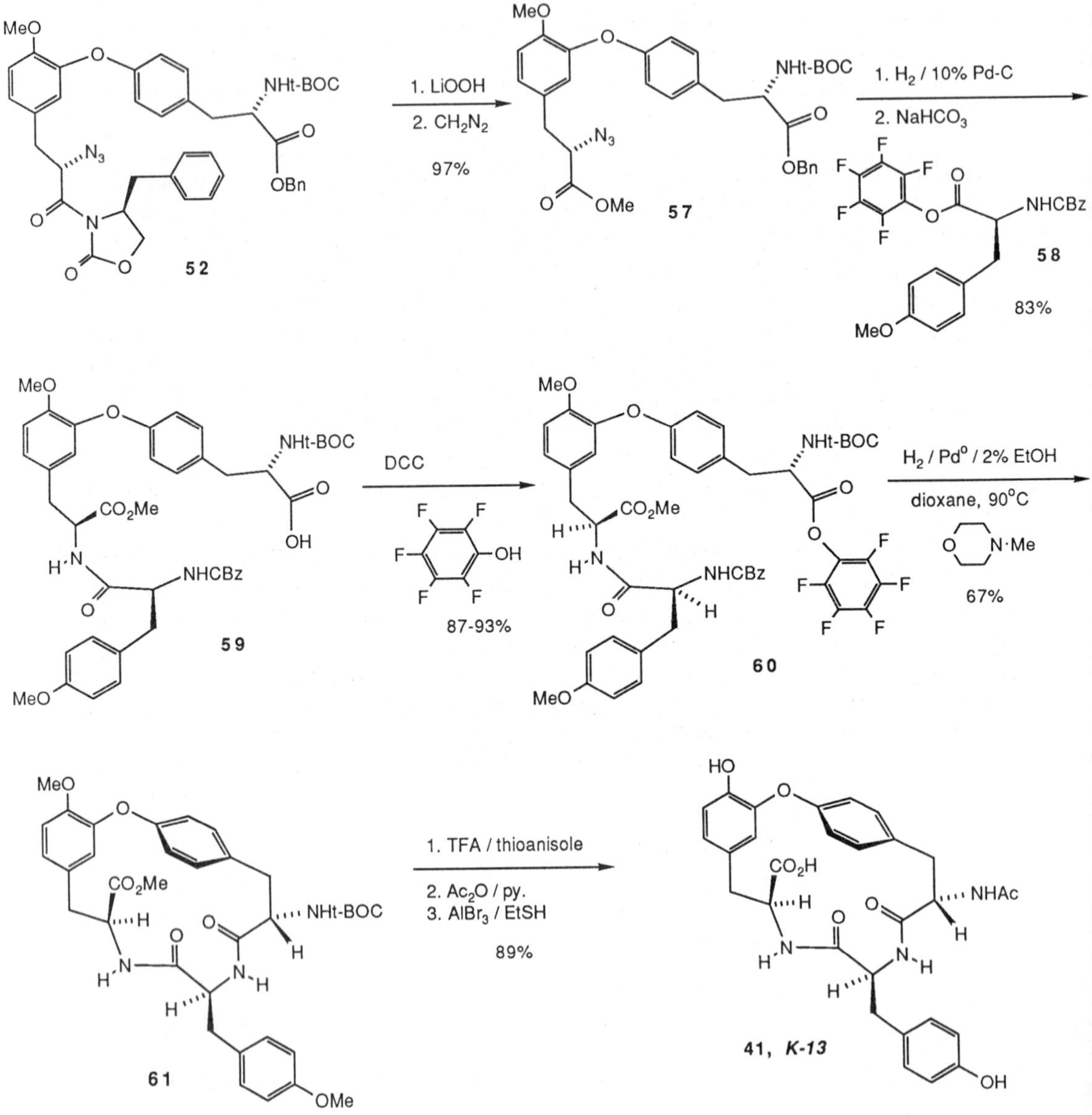

Evans, D.A.; Ellman, J.A., *J.Am.Chem.Soc.* (1989) **111**, 1063

References Chapter 3

1. Diels, O.; Behncke, H., *Chem.Ber.* (1924) **57**, 653.

2. a) Schroeter, S.H.; *J.Org. Chem.* (1969) **34**, 4012; b) Carlin, R.B.; Moores, M.S., *J.Am.Chem.Soc.* (1962) **84**, 4107; c) Huisgen, R.; Jacob, F.; Siegel, W.; Cadus, A., *Liebigs Ann.Chem* (1954) **590**, 1.

3. Boche, G.; Schrott, W., *Tetrahedron Lett.* (1982) **23**, 5403.

4. Gennari, C.; Colombo, L.; Bertolini, G., *J.Am.Chem.Soc.* (1986) **108**, 6394.

5. Evans, D.A.; Britton, T.C.; Dorow, R.L.; Dellaria, J.F., *J.Am.Chem.Soc.* (1986) **108**, 6395.

6. Evans, D.A.; Britton, T.C.; Dorow, R.L.; Dellaria, J.F., *Tetrahedron* (1988) **44**, 5525.

7. Trimble, L.A.; Vederas, J.C., *J.Am.Chem.Soc.* (1986) **108**, 6397.

8. Oppolzer, W.; Moretti, R., *Helv.Chim.Acta.* (1986) **69**, 1923.

9. Oppolzer, W.; Moretti, R., *Tetrahedron* (1988) **44**, 5541.

10. Guanti, G.; Banfi, L.; Narisano, E., *Tetrahedron* (1988) **44**, 5553.

11. Evans, D.A.; Britton, T.C., *J.Am.Chem.Soc.* (1987) **109**, 6681.

12. Evans, D.A.; Ellman, J.A., *J.Am.Chem.Soc.* (1989) **111**, 1063.

CHAPTER 4

NUCLEOPHILIC AMINATION OF α– SUBSTITUTED ACIDS

This chapter outlines a relatively small collection of papers dealing with the displacement of a leaving group α- to a carboxylic acid resulting in *optically active* α- amino acids and derivatives. The older literature dealing with the well-known Gabriel modification to access racemic amino acids will not be covered. The vast number of relatively new methods to prepare optically active α-hydroxy acids and related substances should undoubtedly result in forthcoming papers that employ this complementary strategy to access amino acids. The asymmetric versions of this approach are relatively new and few in number; an unfortunate manifestation of this is that, many of the papers are preliminary communications and complete experimental details are not available in every case.

Evans and associates [1] have devised an elegant and practical route to β-hydroxy-α-amino acids by carrying out highly diastereoselective aldol condensations on the chiral α-haloacetates **1** (Scheme 1). Formation of the dibutylboron enolates with di-n-butylboryl triflate followed by condensation with an aldehyde and oxidative removal of the boron, provides the α-halo aldol adducts **2**. The authors note that the reactions proceed to no more than 80% conversion with attendant recovery of *ca.* 20% of the carboximide **1**. NMR experiments revealed that the boron enolate is a 25 : 75, E : Z ratio. The sterically congested E-enolate reacts much more slowly than the corresponding Z-enolate and was also stereorandom in aldolizations. By using a slight excess of the enolates, high diastereoselectivities and high yields based on the aldehyde are achieved. To illustrate the utility of such adducts, the acetaldehyde adduct **3** was cleanly converted into the azide (**4**) of inverted stereochemistry at the α-position. The authors note that the bromo aldol adducts upon displacement with azide proceed without any loss of stereochemistry; whereas the chloro adducts exhibit some epimerization. This substance (**4**) could be hydrolyzed with lithium hydroxide in aqueous

SCHEME 1

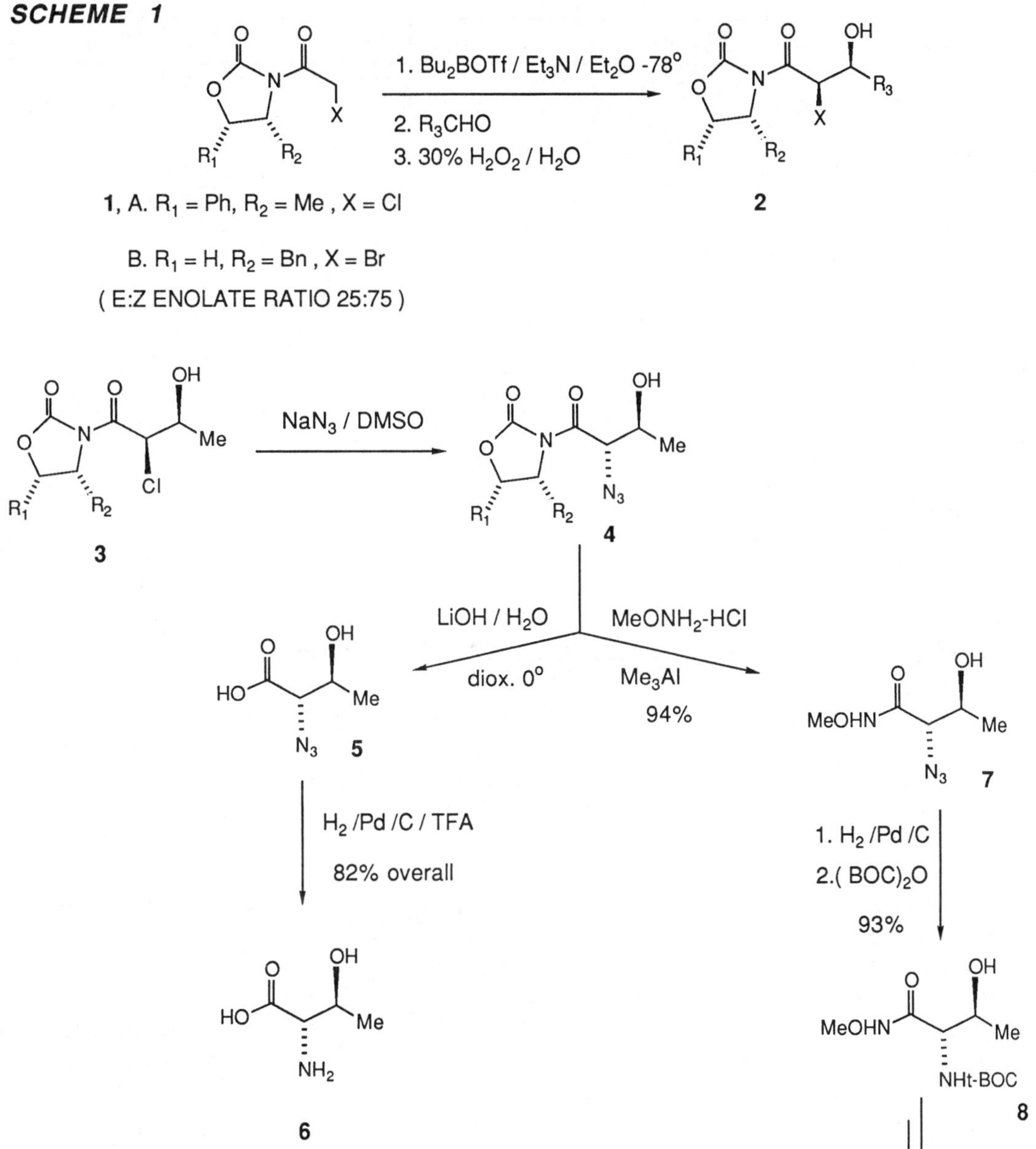

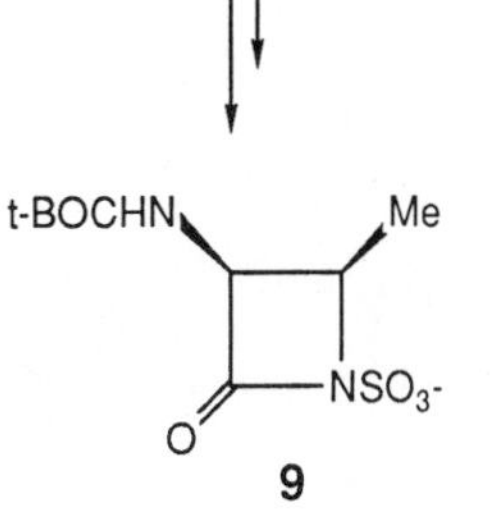

1, A. R_1 = Ph, R_2 = Me , X = Cl

B. R_1 = H, R_2 = Bn , X = Br

(E:Z ENOLATE RATIO 25:75)

Evans, D.A.; Sjogren, E.B.; Weber, A.E.; Conn, R.E.,
Tetrahedron Lett. (1987) **28**, 32.

IMIDE (1)	R₃CHO	RATIO	YIELD (% 2)
A	MeCHO	95:5	67
A	Me₂CHO	96:4	75
A	PhCHO	97:3 / 95:5	79 / 94
B	Me₂CHO	98:2	63
B	Me⌇CHO (Me-substituted)	94:6	63

SCHEME 2

dioxane (**5**) and hydrogenated to give L-*allo*-threonine (**6**) in 82% overall yield from **4**. Alternatively, the chiral auxilliary can be removed by transamination with methoxyamine hydrochloride and triethylaluminum to afford the N-methoxy amide **7** (94%). Reduction of the azide and *in situ* protection of the amine affords substance **8** in high yield. This material has been previously converted into the monobactam **9** by a Squibb group.

An additionally significant transformation of the halo aldol adducts is illustrated in Scheme 2 wherein N-alkyl amino acids may be prepared. Acylation of adduct **10** with methyl isocyanate affords the urethane **11** after removal of the chiral auxilliary with lithium hydroxide. Cyclization of the urethane is effected by treatment with base resulting in S_N2 displacement of the bromide providing the cyclic urethane **12**. Hydrolysis of the urethane in hot base affords the C-3-epimer of MeBmt (**13**).

Another recent contribution from the Cambridge group [2] that significantly expands the azide displacement protocol is shown in Schemes 3-5. Conversion of the acids **14** into the chiral carboximides **15** is performed in a previously described manner [2] . Formation of the di-n-butylboron enolate (**16**) is carried out with the corresponding triflate. Oxidation of these boron enolates with NBS proceeds in a highly stereocontrolled fashion yielding the α-bromo derivatives **17**. Azide displacement is carried out with tetramethylguanidinium azide and proceeds cleanly giving **18** after flash chromatography (>99 : 1 diastereomeric purity after chromatography). The authors note that under 'standard' azide displacement conditions (NaN_3, DMF or DMSO, 0ºC) 2-5% epimerization attends the reaction. The chiral auxilliary can then be removed and recovered by hydrolyzing the azides (**18**) with lithium hydroxide affording the α-azido acids **19**.

The azido carboximides **18** can then be processed in a variety of ways as shown in Scheme 4. Simple hydrolysis with base gives the corresponding acids **19**. Alternatively, transesterification with titanium(IV)benzyloxide provides the benzyl esters **21**. Only in the racemization-prone phenylglycine case (R=Ph), does some epimerization attend this reaction. The azide can be reduced to the corresponding amine and directly acylated with the Moshers' acid chloride (MTPACl) to give the amides **22**. These substances can be similarly saponified to the corresponding acids **23** or transesterified with the titanium alkoxide as above to yield **24**.

A beautiful extension of this methodology is further illustrated with the highly functionalized carboximide **25** (Scheme 5). Boron enolate formation, bromination, and azide displacement occurs in high overall yield and 96 : 4 diastereoselectivity affording the interesting substance **26**.

SCHEME 3

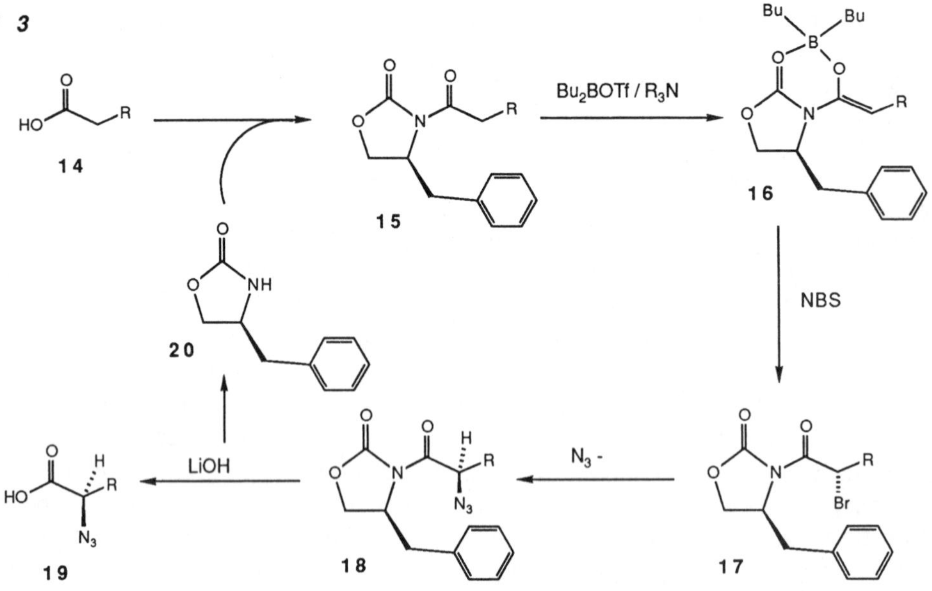

Evans, D.A.; Ellman, J.A.; Dorow, R.L., *Tetrahedron Lett.* (1987) **28**, 1123.

R (15)	DIASTEREOSELECTION (3S):(3R)	(4R):(4S)	YIELD (% 18)
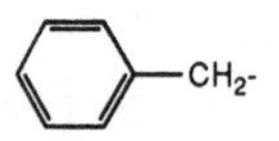 phenyl-CH₂-	95 : 5	94 : 6	83
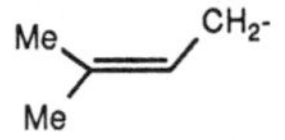 Me₂C=CH-CH₂-	95 : 5	95 : 5	86
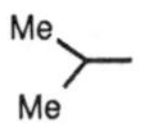 Me₂CH-	96 : 4	94 : 6	80
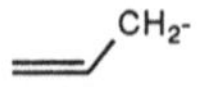 CH₂=CH-CH₂-	94 : 6	94 : 6	82
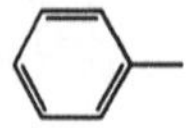 phenyl-	78 : 22	78 : 22	67

SCHEME 4

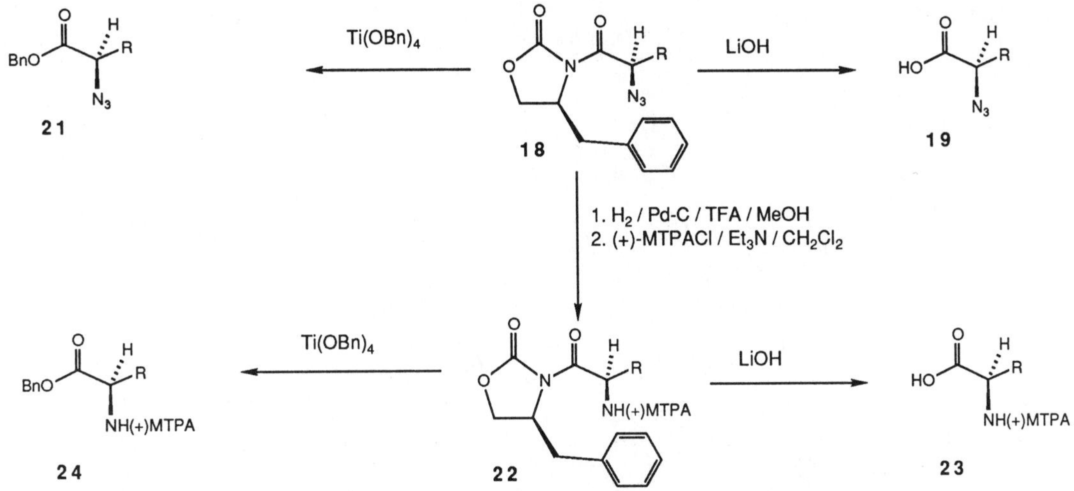

Evans, D.A.; Ellman, J.A.; Dorow, R.L., *Tetrahedron Lett.* (1987) **28**, 1123.

CARBOXIMIDE	REAGENT	YIELD,%	PRODUCT	RATIO (S):(R)
18 (R=Bn)	LiOH	97	**19**	>99 : 1
22 (R=Bn)	LiOH	95	**23**	>99 : 1
18 (R=Ph)	LiOH	97	**19**	99 : 1
22 (R=Ph)	LiOH	100	**23**	99 : 1
18 (R=Bn)	Ti(OBn)$_4$	93	**21**	>99 : 1
22 (R=Bn)	Ti(OBn)$_4$	89	**24**	>99 : 1
18 (R=Ph)	Ti(OBn)$_4$	94	**21**	82 : 18
22 (R=Ph)	Ti(OBn)$_4$	81	**24**	98 : 2

SCHEME 5

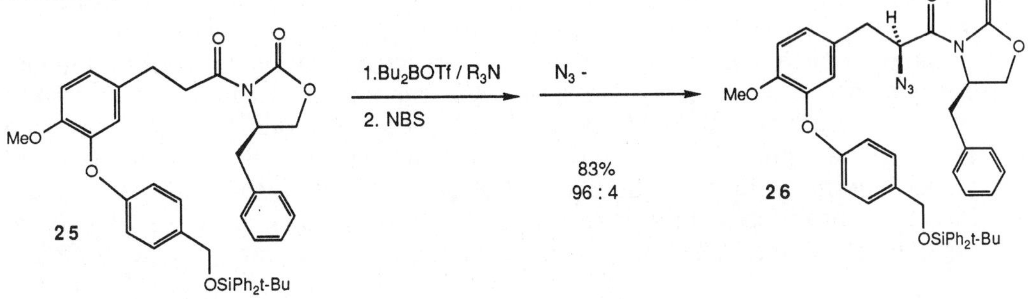

SCHEME 6

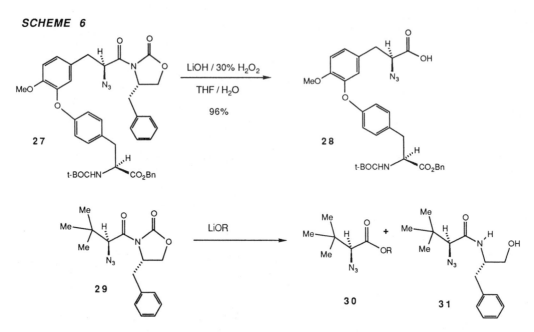

Evans, D.A.; Britton, T.C.; Ellman, J.A., *Tetrahedron Lett.* (1987) **28**, 6141.

LiOR	%30	%31
LiOOH	98	1
LiOBn	52	33
LiOH	42	52

The Evans group[3] has also recently found an improved method to remove the chiral auxilliaries from the azido adducts as shown in Scheme 6. In general , these workers have found that these substances (ie., **29**) are subject to hydrolytic cleavage at either the amide or urethane carbonyls resulting in **30** or **31**, respectively. When lithium peroxide is employed, virtually exclusive formation of the desired acid (**30**) is produced. Lithium benzyloxide and lithium hydroxide give significant amounts of urethane cleavage products (**31**).This is particularly compelling in the specific case of **29** where, the bulky *tert*-butyl group provides the maximum amount of steric hindrance to the desired hydrolysis. This is also highly significant for the practical implementation of the Evans chemistry since, regioselective hydrolysis at the carboxyl group rather than at the urethane group allow for a more nearly quantitative recovery of the chiral auxilliary; a procedure that is performed by simple extraction. The chiral auxilliaries (**20**) are now commercially available in both enantiomorphic forms making this technology applicable to preparing both D- and L-configured amino acids. The high yield for the very complex substrate **27** attests to this simple, yet very important experimental protocol.

Oppolzer and associates [4] have made a very useful contribution to this area by employing the stereocontrolled halogenation of 10-sulfonamido-isobornyl esters as shown in Scheme 7. Stereoselective halogenation of the ketene silyl acetals generated from the camphor-derived sulfamides **32** is performed with NBS or NCS providing the α-halo acetates **33** in high yields and excellent diastereoselectivities. The halides (**33**) can be conveniently recrystallized from hexane or ethyl acetate-ethanol. Displacement with sodium azide in DMF proceeds exceptionally well to provide the azides of inverted stereochemistry (**34**). Transesterification with titanium(IV) benzyloxide effects removal of the chiral auxilliary which can be easily recovered by extraction from the amino acid (**35**) produced after catalytic hydrogenation of the benzyl esters.

An elegant demonstration of the Oppolzer auxilliary is illustrated with an asymmetric synthesis of L-*allo*-isoleucine (**39**), an essential precursor for the synthesis of the psychotropic ergot peptide eticriptine. As shown in Scheme 8, the crotonate **36** is subjected to a highly stereoselective (97.8% ee) Michael-type addition of ethyl copper. The adduct is then converted to the corresponding ketene silyl acetal and brominated as above to give the bromide **37**. Azide displacement occurs in excellent yield to afford **38** which is subsequently transesterified and reduced to L-*allo*-isoleucine **39**. As noted previously in this monograph, there are very few good methods for controlling the relative stereochemistry at the β-position; the present example is an outstanding demonstration of a potentially general solution to this problem. These examples also demonstrate that the commercial availability of both enantiomorphs of the Oppolzer chiral auxilliary gives ready access to both D- and L-configured amino acids by simple selection of the appropriate chiral auxilliary.

Effenberger, et.al., [5] have examined the displacement reactions of α-substituted triflates with various amines as shown in Scheme 9. Optically active α-hydroxy acids (**40**) are converted into their corresponding triflate derivatives **41** with triflic anhydride in pyridine. These substances are reported as being fairly stable and can be stored for several months in the refrigerator. S_N2 displacement of the triflate with primary and secondary amines proceeds cleanly in high yields. The authors note that alternate leaving groups such as, bromide and mesylate give

SCHEME 7

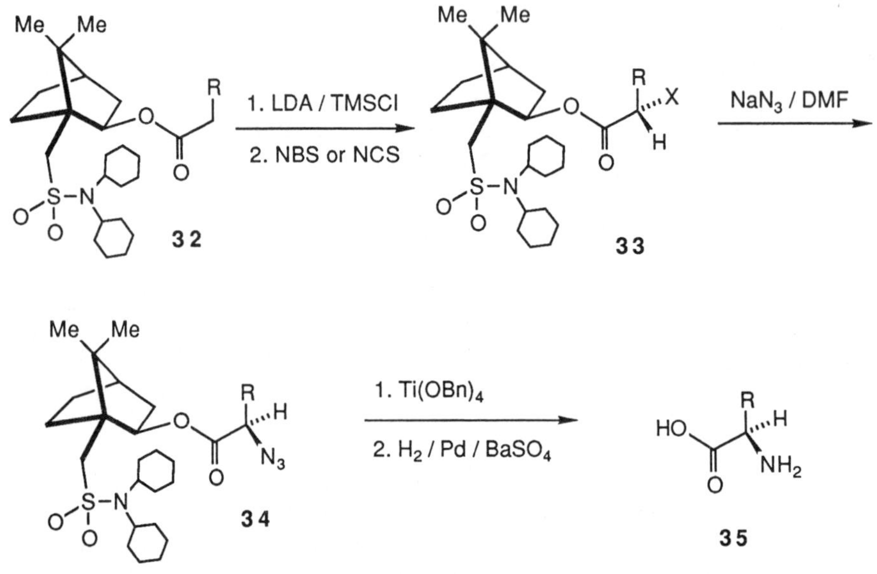

Oppolzer, W.; Pedrosa, R.; Moretti, R., *Tetrahedron Lett.* (1986) **27**, 831.

R	HALOGENATION YIELD	%de	AZIDE YIELD	%de	AMINO ACID	%ee	CONFIG
CH$_2$CH$_3$	75 (Cl)	>99	87	nd	72	94	R
n-C$_3$H$_7$	75 (Cl)	96.7	93	97	87	93.8	R
n-C$_4$H$_9$	77 (Br)	96	88	96	72	94	R
i-C$_4$H$_9$	72 (Cl)	>96	81	98	80	96	R
n-C$_6$H$_{13}$	82 (Cl)	>94	89	100	78	98	R
CH$_2$Ph	80 (Br)	nd	82	98	72	94.7	S
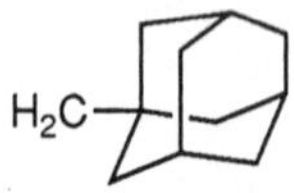 H$_2$C— (adamantyl)	54 (Cl)	nd	88	nd	73	96.4	S

SCHEME 8

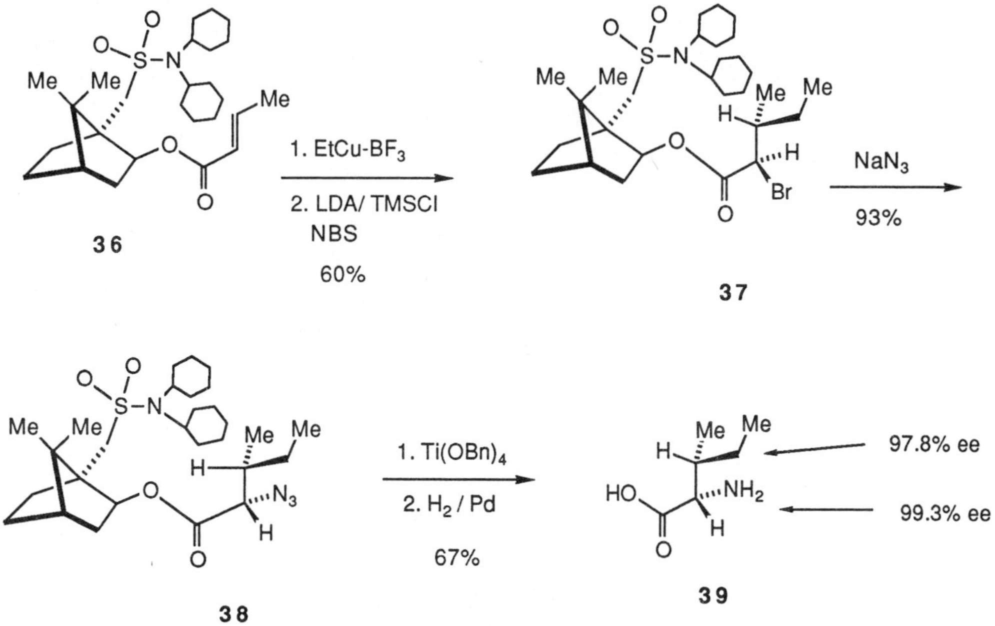

significant amounts of elimination in prone cases such as the phenylalanine and aspartic systems. The triflates obviate the problem of elimination making this a potentially very useful protocol. Final conversions to the free amino acids has not yet been described but seems straightforward from the N-benzylated amines which should yield to hydrolysis of the esters and reduction of the N-benzyl moiety.

Employing a complementary approach, Effenberger[6] and co-workers briefly described the asymmetric sulfenylation of acrylate esters of the Helmchen chiral bornyl auxilliary (Scheme 10). Reaction of **43** with phenyl sulfenyl chloride in methylene chloride at -95°C affords the α-chloro-β-sulfenyl propionate **44** in 65% yield (97% de). Reaction of this substance with azide followed by reduction and acylation provides the α-phthalimido-β-sulfenyl ester **45**.

Mukaiyama and Murakami[7] have reported a potentially very useful approach to β-hydroxy-α-amino acids by the regiospecific ring-opening of α-epoxy esters as shown in Scheme 11. Ethoxyacetylene (**46**) is converted into the α-chloro zinc enolate **48** by oxymercuration with pyridine N-oxide followed by zinc reduction. Aldolization of **48** with the chiral mannitol-derived aldehyde **49** furnishes a mixture of aldol adducts **50** in 68% overall yield. Cyclization with base provides a 1 : 4 mixture of the α-

Nucleophilic Amination of α-Substituted Acids

SCHEME 9

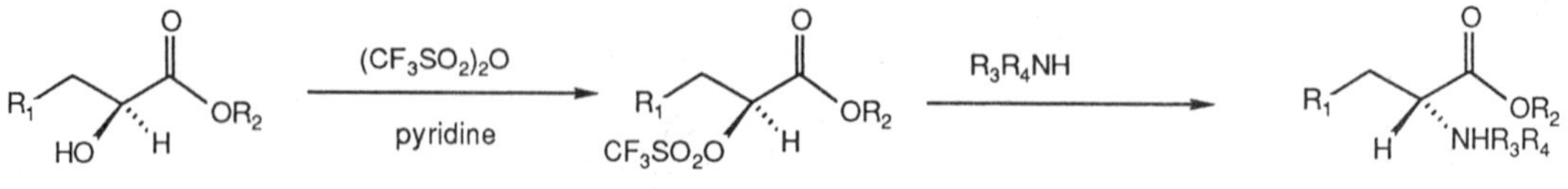

Effenberger, F.; Burkard, U.; Willfahrt, J., *Angew.Chem.Int.Ed.Engl.* (1983) **22**, 65.

R₁	R₂	R₃	R₄	YIELD (% 42)	CONFIG.
Me	Ph	H	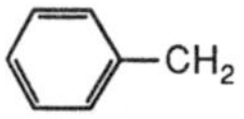 (benzyl) —CH₂	92	R
		H	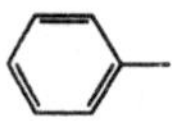 (phenyl)	93	R
		-(CH₂)₄- / Me	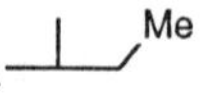 83	2R,2'R
Et	EtCO₂	H	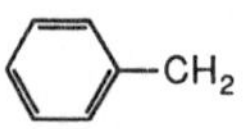 (benzyl) —CH₂	93	R
		H	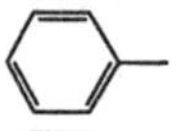 (phenyl)	94	R
Et	H	H	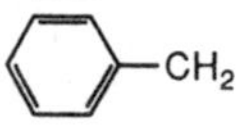 (benzyl) —CH₂	85	R
		-(CH₂)₄-	84	R
		Me	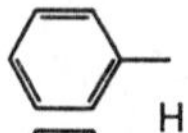 (phenyl)	75	R
		H	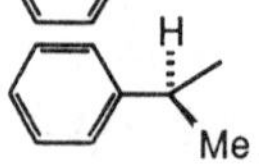 (phenyl)—CH(H)—Me	76	2R,1'S
		H	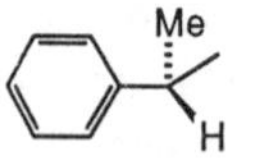 (phenyl)—CH(Me)—H	78	2R,1'R
		H	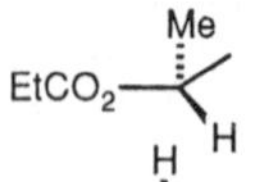 EtCO₂—CH(Me)—H	81	2R,2'S
		H	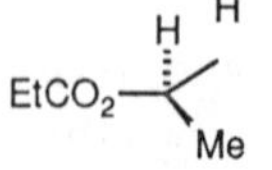 EtCO₂—CH(H)—Me	93	2S,2'S
		H	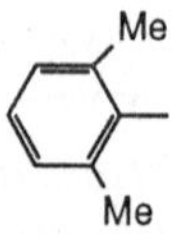 (dimethylphenyl) Me / Me	92	R

SCHEME 10

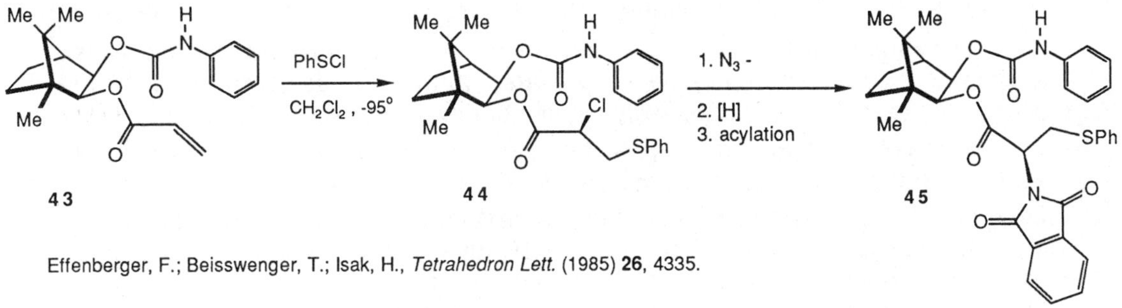

Effenberger, F.; Beisswenger, T.; Isak, H., *Tetrahedron Lett.* (1985) **26**, 4335.

SCHEME 11

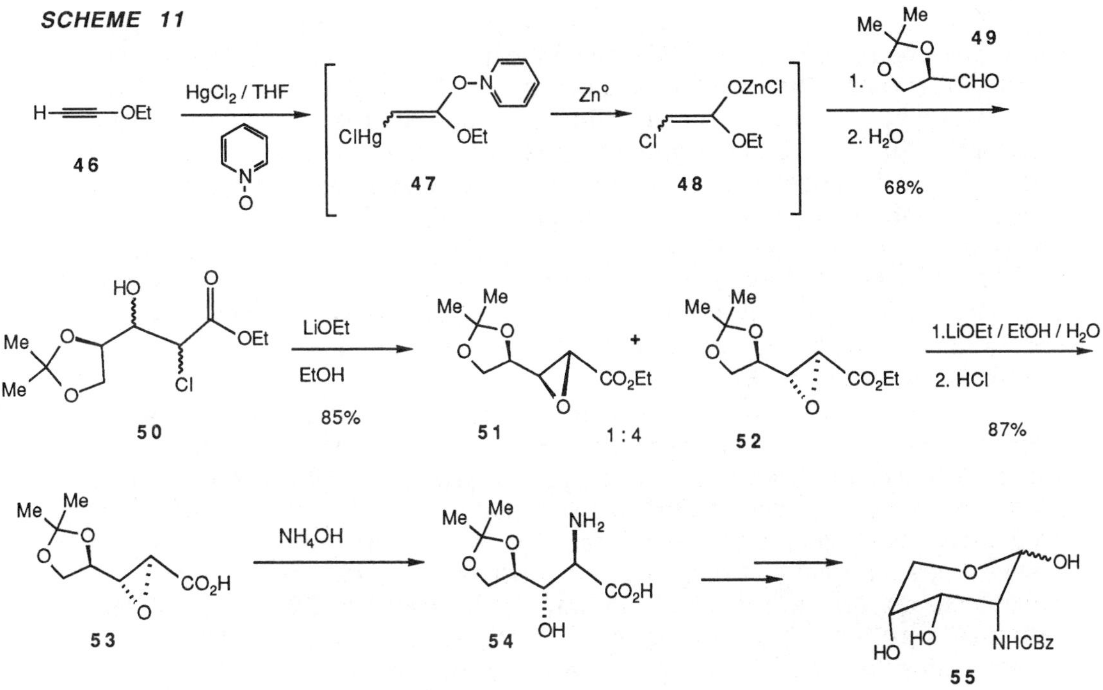

Murakami, M.; Mukaiyama, T. , *Chemistry Lett.* (1982) 1271.

epoxy acids **51** : **52** which were separated by flash chromatography. Hydrolysis of the major isomer (**52**) to the acid and aminolysis with aqueous ammonia proceeded in a regiospecific manner to furnish the amino acid **54** in high yield. This substance was subsequently converted into 2-amino-2-deoxy-D-ribose (**55**). In this context, it should be mentioned that the Evans aldol chemistry [1] described above in Scheme 1 provides a direct entry [7b] to optically active α-epoxy acids that should be readily expandable to a variety of β-hydroxy-α-amino acids via the epoxide-opening protocol[7a] described above. The employment of Sharpless asymmetric epoxidation chemistry provides another , powerful method to acces the same substances; this will be briefly discussed below.

Gennari, Scolastico and associates [8] have recently carried out a similar strategy to acces β-hydroxy-α-amino acids as illustrated in Scheme 12; full experimental details accompany this work. Compound **56** is prepared from norephedrine by a previously established procedure and diastereoselectively epoxidized with potassium hypochlorite in aqueous THF. Prolonged exposure to the oxidant effects a one-flask oxidation of the incipient epoxy aldehyde to the epoxy acid **57**. Treatment of **57** with aqueous ammonia at 70°C as above regioselectively opens the oxirane to furnish the amino acid **58** in quantitative yield. Protection of the nitrogen as a carbobenzoxy urethane and esterification with diazoethane provides **59** in 71% yield. The ephedrine auxilliary is removed with ethane dithiol and boron trifluoride to give the dithioacetal **60**. Silylation, and hydrolysis of the dithioacetal affords the key aldehyde **62**. This interesting substance should prove to ba a versatile γ-homologation reagent. The aldehyde was subsequently reduced, deprotected and converted into the acetonide **63**. By using the antipodal 1S,2R-ephedrine auxilliary, the enantiomorphic substance **65** was similarly prepared (Scheme 13). Solvolysis of **65** with methanolic ammonia, provided the known amide **66** which has previously been converted into the β-lactam antibiotic carumonam (**67**).

The corresponding *syn*-β-hydroxy-α-amino acid system was easily accessible via an intramolecular urethane inversion and is illustrated for **58** in Scheme 14. Treatment of **58** with phosgene provides **68** which is esterified to **69** with diazomethane. Treatment of this substance with hot methanolic KOH completely epimerizes C-2 to the *trans*-oxazolidinone and cleaves the urethane furnishing the *syn* isomer **70**. Processing this substance as above provides the key *syn*-aldehyde **72**. Thus all four optically active diastereoisomers of these potentially useful intermediates become available by these procedures.

A considerably more concise synthesis of the same *anti*-β-hydroxy-α-amino acids was reported by Torii, Moriwake and co-workers[9] as shown in Schemes 15 and 16. These workers have ingeniously exploited the symmetry of tartaric acid to rapidly construct the requisite amino acids. L- or D-tartaric acid is transformed into the optically pure *trans-*

SCHEME 12

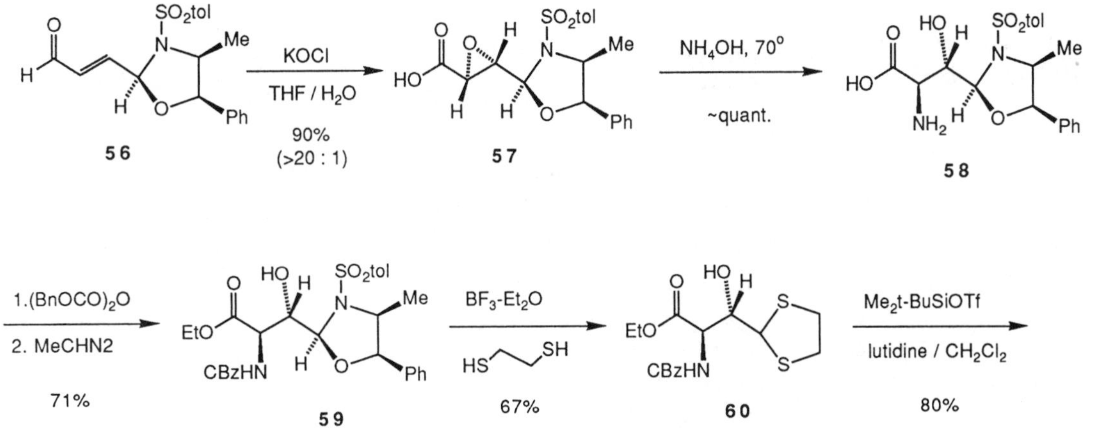

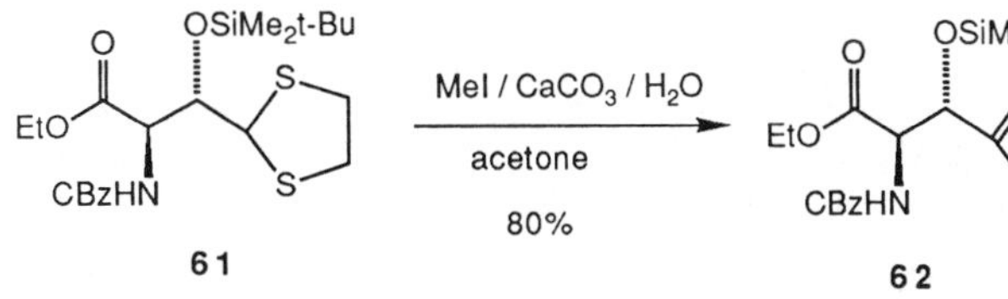

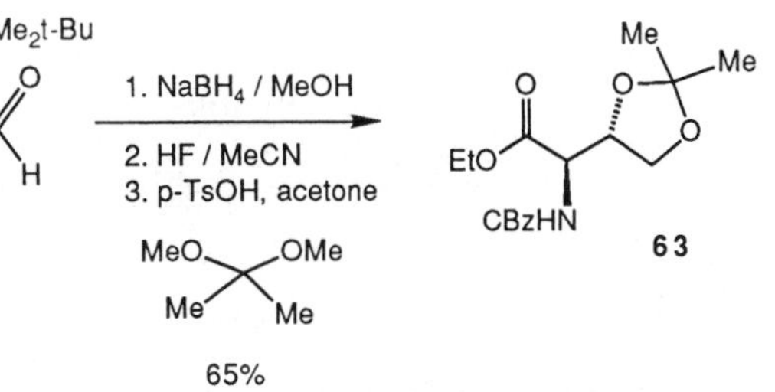

Cardani, S.; Bernardi, A.; Colombo, L.; Gennari, C.; Scolastico, C.;
Venturini, I., *Tetrahedron* (1988) **44**, 5563.

SCHEME 13

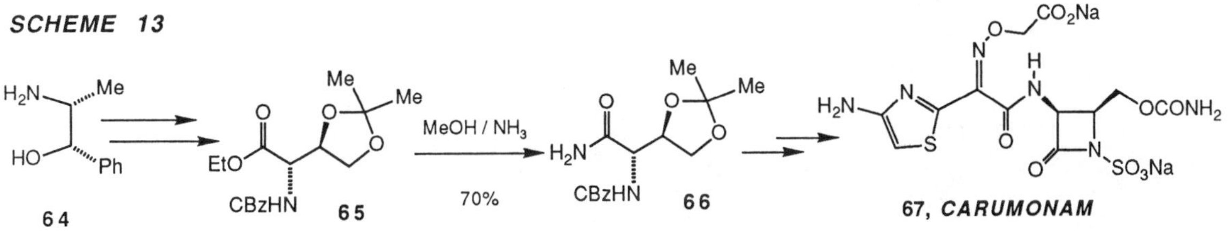

SCHEME 14

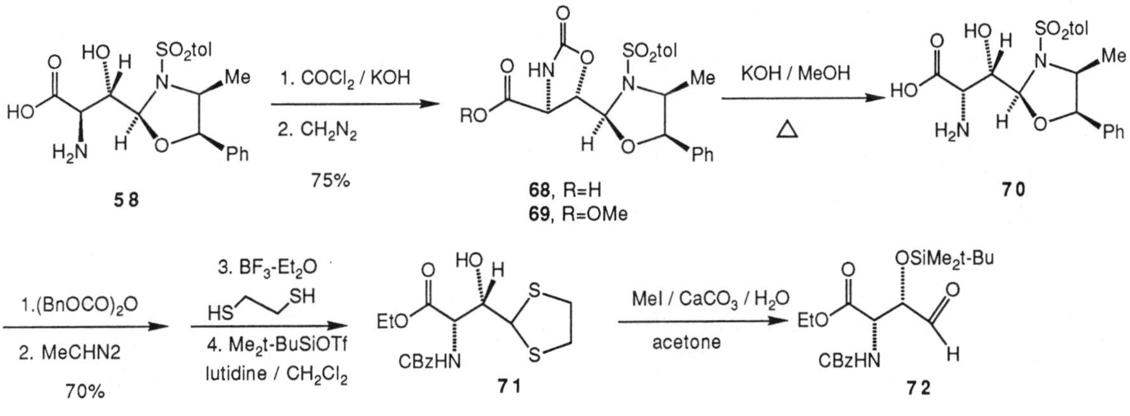

SCHEME 15

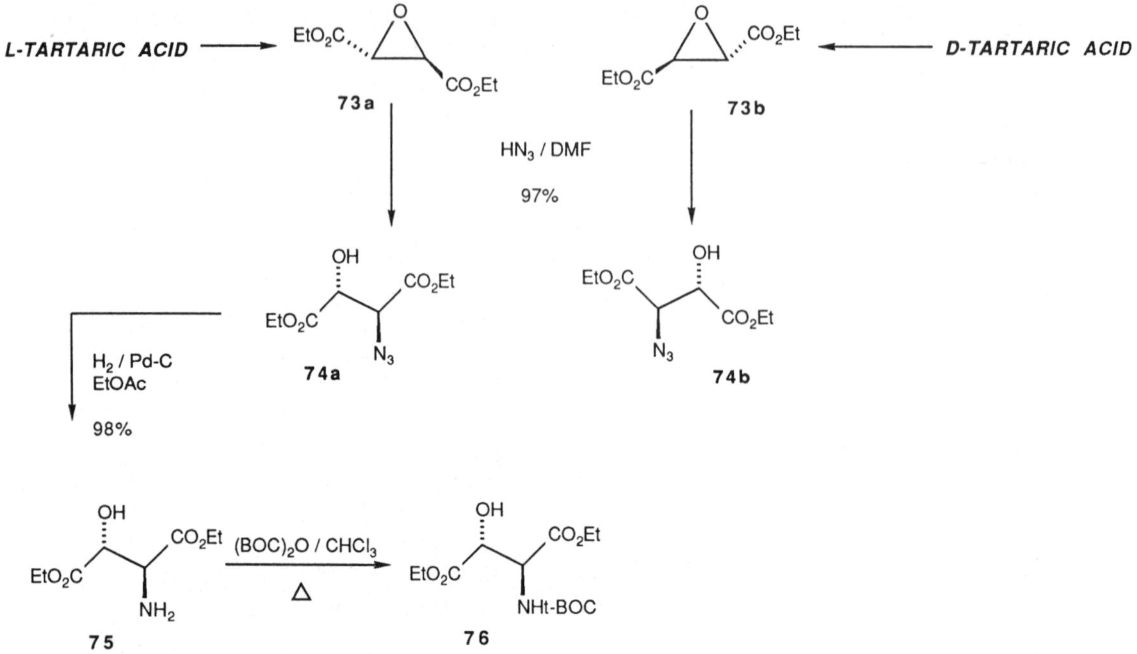

L-TARTARIC ACID ⟶ EtO₂C... [epoxide] CO₂Et   **73a**

[epoxide] EtO₂C, CO₂Et **73b** ⟵ D-TARTARIC ACID

HN₃ / DMF

97%

OH
EtO₂C CO₂Et
N₃
74a

OH
EtO₂C CO₂Et
N₃
74b

H₂ / Pd-C
EtOAc

98%

OH
EtO₂C CO₂Et
NH₂
75

(BOC)₂O / CHCl₃
Δ

OH
EtO₂C CO₂Et
NHt-BOC
76

Saito, S.; Bunya, N.; Inaba, M.; Moriwake, T.; Torii, S., *Tetrahedron Lett.* (1985) **26**, 5309.

SCHEME 16

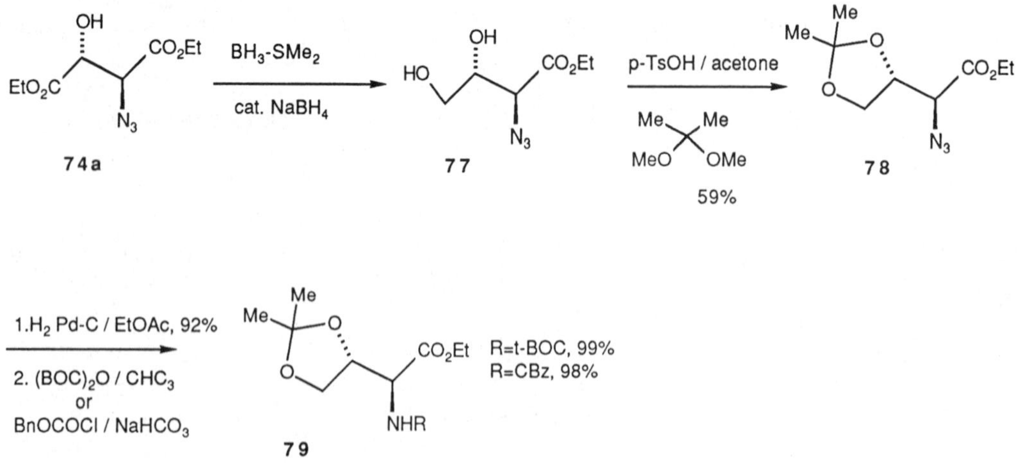

OH
EtO₂C CO₂Et
N₃
74a

BH₃-SMe₂

cat. NaBH₄

OH
HO CO₂Et
N₃
77

p-TsOH / acetone

Me   Me
MeO   OMe

59%

Me
Me O
O CO₂Et
N₃
78

1. H₂ Pd-C / EtOAc, 92%

2. (BOC)₂O / CHC₃
   or
   BnOCOCl / NaHCO₃

Me
Me O
O CO₂Et
NHR
79

R=t-BOC, 99%
R=CBz, 98%

2,3-epoxysuccinic esters **73a** and **73b**, respectively , following the procedure of Mori and Iwasawa. Nucleophilic ring-opening with hydrazoic acid in DMF proceeds in exceptional yield to furnish the azides **74**. Reduction of the azide by catalytic hydrogenation provides the amine **75** in 98% yield, which can then be protected as the corresponding urethane of β-hydroxy aspartic acid diethyl ester **76**. Alternatively, borane-methyl sulfide reduction of the β-ethyl ester of **74a** provides the labile azido diol **77** that is immediately ketalized to the acetonide **78** in 59% overall yield. Hydrogenation and acylation of **78** provides the requisite substance **79**. Thus, in a very concise manner, the protected *anti*-β-hydroxy-α-amino acid synthons **79** are readily available from inexpensive tartaric acid. The simplicity of the overall approach and ready availability of the starting materials makes this an attractive entry to these potentially versatile amino acids.

In an elegant demonstration of asymmetric synthesis of amino acids and total synthesis, Ohfune and Kurokawa [10] at the Suntory Institute in Japan applied the epoxy acid ring-opening strategy to construct one of the constituent amino acids of the oligopeptide antibiotic echinocandin D (**85**, Scheme 17). Epoxy alcohol **80** is prepared from the corresponding allylic alcohol by asymmetric Sharpless epoxidation [11]. Oxidation of this substance with pyridinium dichromate (PDC) affords the epoxy acid **81** in 80% yield. Aminolysis of this material with aqueous ammonia proceeds in good yield to afford the desired optically active amino acid **82**. This compound is then acylated, esterified, debenzylated and tosylated to afford the pre-cyclization intermediate **83**. Treatment of this substance with base cleanly affords the pyrrolidine **84** after acid/base removal of the methyl ester and urethane protection. The synthesis of the remaining constituent amino acids and the total synthesis of echinocandin D will be presented in the last chapter of this monograph.

The extremely practical application of the Sharpless asymmetric epoxidation protocol should render the general approach depicted in Scheme 17 as a versatile and predictable method to access numerous L- and D-configured amino acids. It should be noted that the epoxy alcohols, such as **80**, tend to exhibit regiospecific opening [12] at the 3-position; this selectivity can be markedly enhanced utilizing titanium alkoxides. The 2,3-epoxy acids, on the other hand, are more electrophilic at the 2-position and can be regioselectively opened with ammonia as the examples in Schemes 11 and 17 illustrate. Sharpless has also found [12] that the titanium-mediated opening of 2,3-epoxy acids with nitrogen nucleophiles provides a reliable method to ring-open at the 3-position. The titanium-mediated regioselective opening of 2,3-epoxy acids at the *3-position* has

SCHEME 17

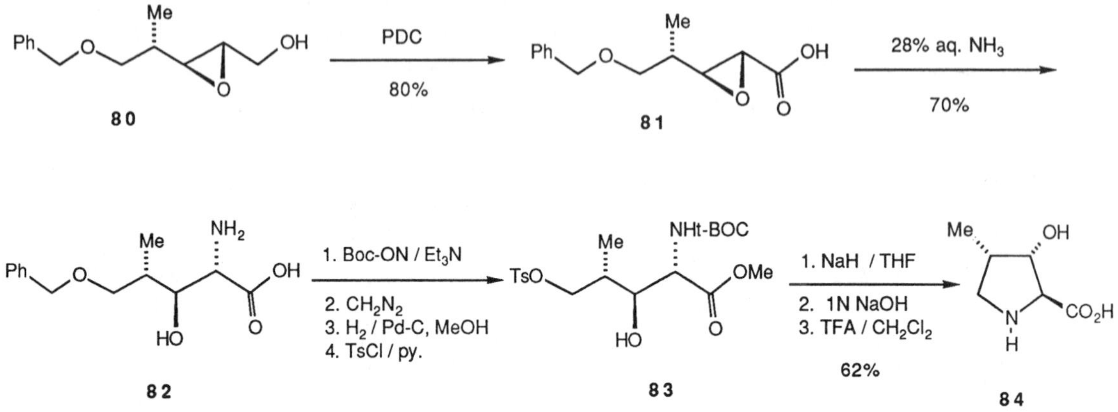

Kurokawa, N.; Ohfune, Y., *J.Am.Chem.Soc.* (1986) **108**, 6041.

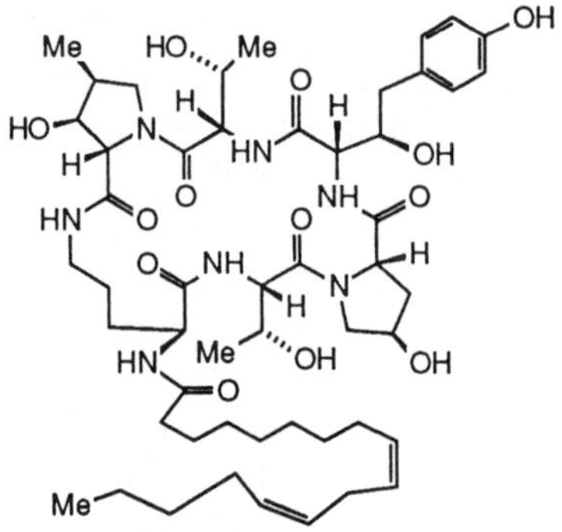

85, ECHINOCANDIN D

recently been exploited by Sharpless and associates[13] to access various amino acids as shown in Scheme 18.

These workers found that Ti[(Oi-Pr)$_2$(N$_3$)$_2$] affords very regioselective openings at the C-3 position of *trans*-2,3-epoxy alcohols to furnish the azido diols **87**. The regioselectivity was greatly improved over

SCHEME 18

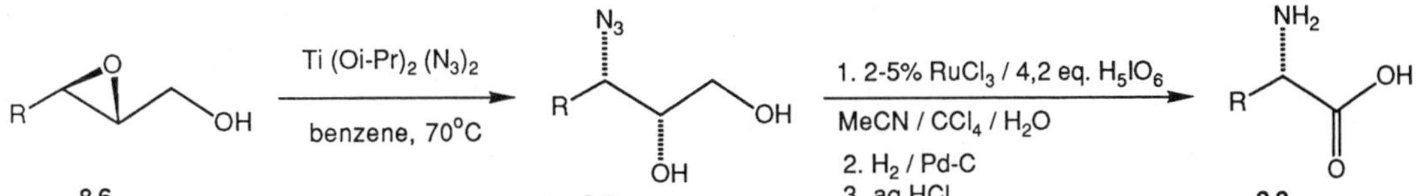

86 **87** **88**

Caron, M.; Carlier, P.R.; Sharpless, K.B., *J.Org. Chem.* (1988) **53**, 5185.

SUBSTRATE	REGIOSELECTIVITY C-3 : C-2	YIELD (%87)	AMINO ACID (% 88)	%ee
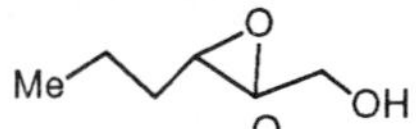	100 : 1 (36 : 1)	40 (88)		
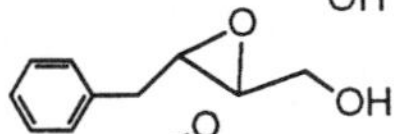	27 : 1	96	68	98
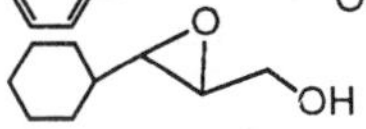	20 : 1	94		
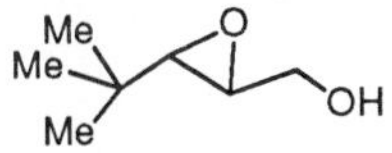	2 : 1	96		
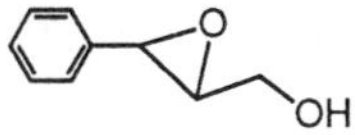	100 : 1	76	65	86
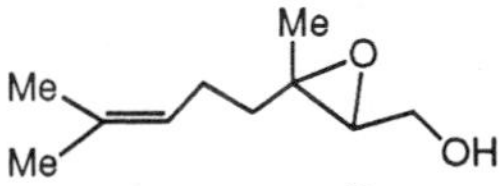	100 : 1	76		
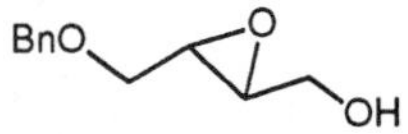	6 : 1	93		
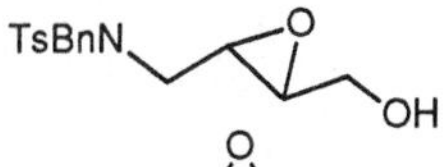	1.9 : 1	89		
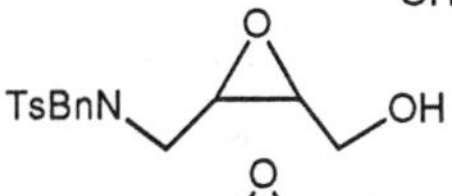	1 : 1	78		
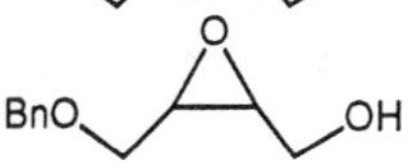	1 : 2	79		
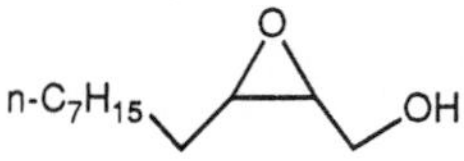	1 : 1.5	89		

'standard' reaction conditions employing sodium azide. For example, the *trans*-(cyclohexyl) 2,3-epoxy alcohol gave a 1.4 : 1 mixture of C-3 : C-2 azides with sodium azide but a 27 : 1 ratio with the titanium reagent. In contrast, the *cis*-2,3-epoxy alcohols gave very poor ratios of regioselective opening. The azido diols (**87**) can be converted into the corresponding α-amino acids (**88**) by a ruthenium catalyzed oxidative cleavage of the diol moiety followed by hydrogenolysis of the azide with palladium on charcoal. The slight amount of racemization that attends the sensitive, racemization-prone phenylglycine case (98% ee epoxy alcohol to 86% ee amino acid) is attributed to the expected lability of the incipient α-azido aldehyde produced in the ruthenium oxidation. The abscence of π-stabilizing 'R' groups, such as the phenylalanine case (98% ee) indicates that this methodology should be applicable to all but the most sensitive cases. The ruthenium oxidative sequence will also impose limitations on oxidizable substituents.

SCHEME 19

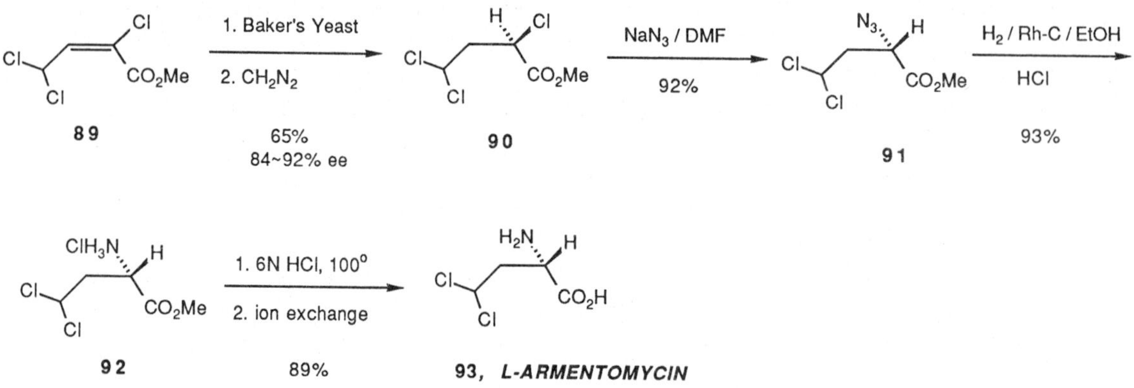

Utaka, M.; Konishi, S., Okubo, T.; Tsuboi, S.; Takeda, A.,
Tetrahedron Lett. (1987) **28**, 1447.

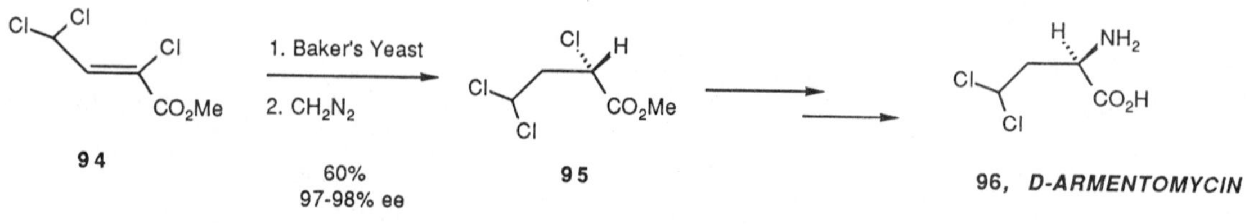

Takeda, Utaka and associates[14] have recently examined an interesting enantiospecific yeast reduction of 2,4,4-trichloro-2-butenoate (89) as shown in Scheme 19. The E-isomer (89) gave a 65% yield of 90 in 84~92%ee after silica gel chromatography. Stereospecific displacement of the α-chloro substituent is carried out under standard conditions with sodium azide in excellent yield [15] to afford 91. Reduction of the azide is carried out with rhodium on carbon to afford the amine salt 92 in high yield. Hydrolysis of the methyl ester and ion exchange purification afforded L-armentomycin (93) in high yield. The corresponding D-isomer (96) was similarly prepared by the yeast reduction of the Z-butenoate 94 in good yield and excellent %ee. The authors conclude that the relatively inferior results with the E-isomer (89) was a manifestation of the unique enantioselectivities of the reductase toward each geometrical isomer and not the result of partial isomerization of the 89 to 94 in the reaction culture.

Ottenheijm and co-workers [16,17] have examined the synthesis of N-hydroxy amino acids in optically active form by triflate displacement with O-benzyl hydroxylamine as shown in Scheme 20. N-Hydroxy amino acids are emerging as important substances as enzyme inhibitors, constituents of oxidized peptides, metal sequestering reagents and possible intermediates in the oxidative biosyntheses of numerous secondary metabolites. Optically active α-hydroxy acids (97) are converted into the corresponding triflates (98) with triflic anhydride and lutidine. Reaction of these substances with O-benzyl hydroxylamine proceeds in high yields with clean inversion furnishing the protected substances 99. The authors comment that the benzyl protecting group can be removed by hydrogenation after manipulation of the substrate in a peptide coupling sequence.

SCHEME 20

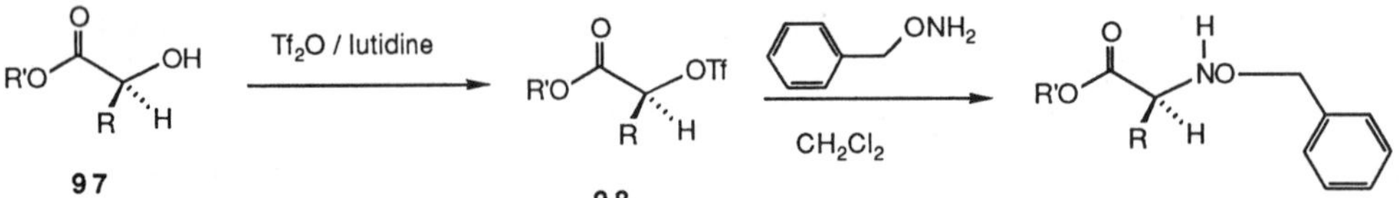

Feenstra, R.W.; Stokkingreef, E.H.M.; Nivard, R.J.F.; Ottenheijm, H.C.J., *Tetrahedron Lett.* (1987) **28**, 1215.
Feenstra, R.W.; Stokkingreef, E.H.M.; Nivard, R.J.F.; Ottenheijm, H.C.J., *Tetrahedron* (1988) **44**, 5583.

R	R'	YIELD (% 99)	OPTICAL PURITY
Me	Et	89	100
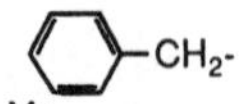 C6H5-CH2-	Me	84	100
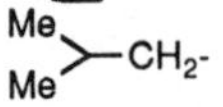 Me2CH-CH2-	Me	78	100
MeO2CCH2-	Me	88	95
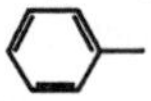 C6H5-	Me	88	76

References Chapter 4

1. Evans, D.A.; Sjogren, E.B.; Weber, A.E.; Conn, R.E.; *Tetrahedron Lett.* (1987) **28**, 32.

2. Evans, D.A.; Ellman, J.A.; Dorow, R.L., *Tetrahedron Lett.* (1987) **28**, 1123.

3. Evans, D.A.; Britton, T.C.; Ellman, J.A., *Tetrahedron Lett.* (1987) **28**, 6141.

4. Oppolzer, W.; Pedrosa, R.; Moretti, R., *Tetrahedron Lett.* (1986) **27**, 831.

5. Effenberger, F.; Burkard, U.; Willfahrt, J., *Angew.Chem.Int.Ed.Engl.* (1983) **22**, 65.

6. Effenberger, F.; Beisswenger, T.; Isak, H., *Tetrahedron Lett.* (1985) **26**, 4335.

7. a)Murakami, M.; Mukaiyama, T., *Chemistry Lett.* (1982) 1271; b) for a detailed study of α-epoxy acid syntheses utilizing the Evans protocol, see: Abdel-Magid, A.; Pridgen, L.N.; Eggleston, D.S.; Lantos, I., *J.Am.Chem.Soc.* (1986) 108, 4595.

8. Cardani, S.; Bernardi, A.; Colombo, L.; Gennari, C.; Scolastico, C.; Venturini, I., *Tetrahedron* (1988) **44**, 5563.

9. Saito, S.; Bunya, N.; Inaba, M.; Moriwake, T.; Torii, S., *Tetrahedron Lett.* (1985) **26**, 5309.

10. Kurokawa, N.; Ohfune, Y., *J.Am.Chem.Soc.* (1986) **108**, 6041.

11. Nagaoka, N.; Kishi, Y., *Tetrahedron* (1981) **37**, 3783.

12. a)Caron, M.; Sharpless, K.B., *J.Org.Chem.* (1985) **50**, 1557; b) Chong, J.M., Sharpless, K.B., *J.Org.Chem.* (1985) **50**, 1560; c) Ko. S.Y.; Masamune, H.; Sharpless, K.B., *J.Org.Chem.* (1987)**52**, 667, and references cited therein.

13. Caron, M.; Carlier, P.R.; Sharpless, K.B., *J.Org.Chem.* (1988) **53**, 5185.

14. Utaka, M.; Konishi, S.; Okubo, T.; Tsuboi, S.; Takeda, A., *Tetrahedron Lett.* (1987) **28**, 1447.

15. Heinzer, F.; Martin, P., *Helv.Chim.Acta.* (1981) **64**, 1379.

16. Feenstra, R.W.; Stokkingreef, E.H.M.; Nivard, R.J.F.; Ottenheijm, H.C.J., *Tetrahedron Lett.* (1987) **28**, 1215.

17. Feenstra, R.W.; Stokkingreef, E.H.M.; Nivard, R.J.F.; Ottenheijm, H.C.J., *Tetrahedron* (1988) **44**, 5583.

CHAPTER 5

ASYMMETRIC STRECKER SYNTHESES

The Strecker amino acid synthesis has been known since 1850 (Strecker, A., *Liebigs Ann.Chem.* (1850) **75**, 27) and has perhaps one of the most convenient experimental protocols for preparing amino acids on a preparative scale. It has also been invoked as the reaction responsible for the 'primordial' derivation of the amino acids. It is somewhat surprising that relatively few reports dealing with the development of asymmetric versions of the classical Strecker synthesis have appeared in the literature. Related enzyme-mediated Strecker preparations will be discussed in Chapter 7; this section will only detail the strictly 'chemical' methods.

The first asymmetric Strecker synthesis can be attributed to Harada [1a] in 1963. The basic concept that was developed involves the condensation of an optically active amine with an aldehyde forming the chiral Schiff base; subsequent addition of HCN forms an optically active α-amino nitrile that is then hydrolyzed to the α-amino acid. Harada employed the readily available D-(-)-α-methylbenzylamine as the chiral auxilliary. The very first experimental procedure described [1a] involved the condensation of D-(-)-α-methylbenzylamine with sodium cyanide in ice water followed by the addition of a methanolic solution of acetaldehyde. The system was stirred for five days and acidified to afford the crude N-(α-methylbenzyl) alanine hydrochloride. The chiral amine was destructively removed by catalytic hydrogenolysis of the N-(α-methylbenzyl) moiety providing L-alanine in 17% yield and 90% asymmetric yield (presumably %op).

This procedure was slightly modified [1b] to involve the reaction of the preformed racemic cyanohydrin with the same chiral amine; hydrolysis and reduction provided L-alanine in similar (low) yield and 86-99% optical purity after recrystallization. This basic strategy was later refined [2] as shown in Scheme 1. Various optically active benzylic amines (**2**) were examined for the extent of asymmetric induction as well as improvements in chemical yield. The experimental protocol involves formation of the

SCHEME 1

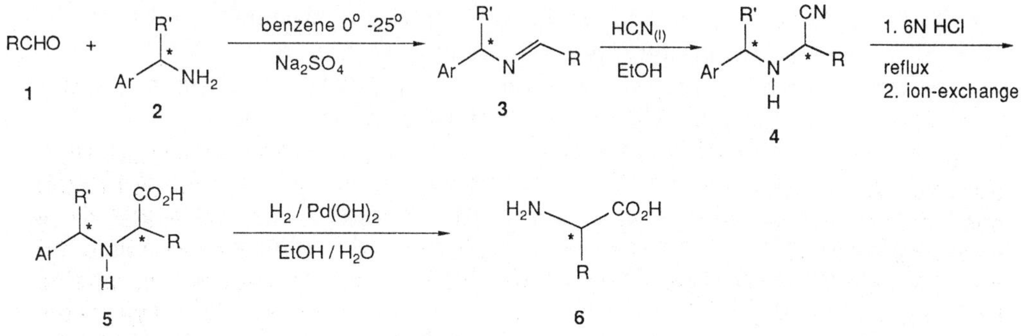

Harada, K.; Okawara, T., *J.Org.Chem.* (1973) **38**, 707
Harada, K.; Okawara, T.; Matsumoto, K., *Bull.Chem.Soc. Jpn.* (1973) **46**, 1865
Harada, K., *Nature* (1963) **200**, 1201
Harada, K.; Fox, S., *Naturwissenschaften* (1964) **51**, 106

BENZYLIC AMINE (2)	R (1)	YIELD (%6)	OPTICAL PURITY(6)	CONFIG.(6)
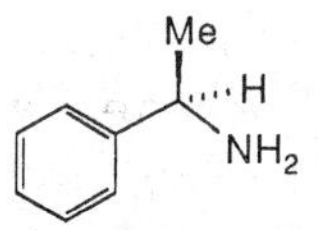	Me	17-54	40-95	R
	Et	14-49	48-100	R
	i-Pr	9-21	31-74	R
	i-Bu	14-40	24-78	R
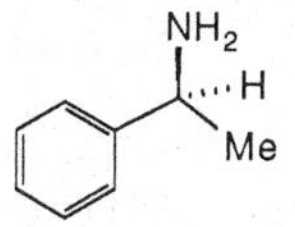	Me	58	41	S
	Et	43	49	S
	i-Pr	19	30	S
	i-Bu	37	23	S
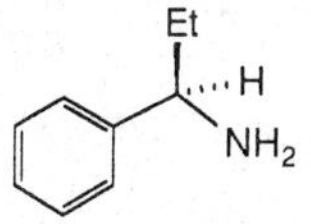	Me	52	53	R
	Et	41	55	R
	i-Pr	19	48	R
	i-Bu	35	30	R
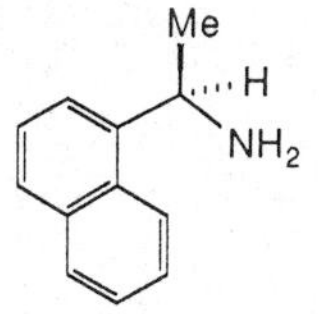	Me	54	37	R
	Et	45	53	R
	i-Pr	22	44	R
	i-Bu	34	22	R

Schiff base (**3**) by condensing the chiral amine with the aldehyde in benzene. The crude imine (**3**) is then treated with liquid HCN in absolute ethanol forming the adducts **4**. The crude, syrupy nitriles were then hydrolyzed in hot 6N HCl and purified by ion-exchange chromatography providing the N-benzylic amino acids **5**. Catalytic hydrogenolysis of the N-benzylic moiety afforded the optically active α-amino acids **6**. The Table tabulates the overall yields which ranged from 9-58% and optical purities ranging from 22-100% (op). The high optical purities in several cases are attributed to fractional crystallization of the diastereomeric salts **5** prior to hydrogenation. Thus, the asymmetric induction for the addition of HCN to the imines **3** is said to be rather modest. The authors note a straightforward correlation between the absolute configuration of the benzylic amine and the final amino acid. Full experimental details accompany this work.

The basic approach developed by Harada, has been re-examined by several groups with some notably significant improvements in yields and optical purities. Patel and Worsley [3] have examined virtually the same system with quite impressive improvements in optical yields a shown in Scheme 2. (R)- and (S)-α-methylbenzylamines (**7**) are condensed with three aliphatic aldehydes to form the corresponding imines **8**; the crude substances are obtained in very high yields and are directly utilized without further purification. Gaseous HCN is then introduced to an ethanolic solution of **8** providing the α-amino nitriles **9** , also in excellent yields. As above, the nitrile is hydrolyzed in hot, concentrated hydrochloric acid to furnish the N-(α-methylbenzyl) amino acids **10** in moderate yields. These substances are subsequently hydrogenated at high pressure to afford the optically active α-amino acids **11**. In every case the %op was in excess of 98%. From reading the experimental procedures, it is not at all clear why the system described by Patel and Worsley gives such consistently higher %ops' than the examples reported by Harada. The reaction conditions are virtually identical involving, HCN addition to the preformed imine at low temperature followed by reaction at ambient temperature for 20h. The authors[3] exclude the possibility of fractional crystallization of the diastereomeric substances **9** and **10** and report that virtually single diastereoisomers of **9** are produced. Full experimental details accompany this report.

Using the same general concept, Weinges and associates [4] examined the condensations of **7** with ketones (**12**) in the presence of methanolic sodium cyanide. The authors report that stereochemically homogeneous α-amino nitriles (**13**) are produced in these condensations. The authors demonstrate that the α-amino nitriles must be the result of kinetic

SCHEME 2

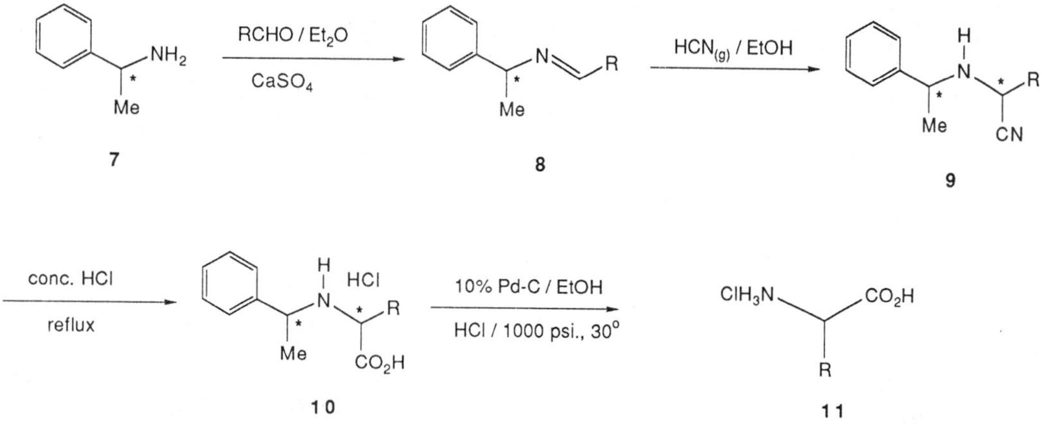

Patel, M.S.; Worsley, M., *Can. J.Chem.* (1970) **48**, 1881

R	CONFIG. (7)	YIELD (%8)	YIELD (%9)	YIELD (%10)	YIELD (%11)	%op (11)
n-Pr	R	96	94	46	41	99
n-Pr	S	99	98	48	43	99
n-Bu	R	96	94	47	43	99
n-Bu	S	95	94	49	43	98
i-Bu	R	97	97	64	58	99
i-Bu	S	95	94	56	49	99

control since, substantial equilibration to diastereomeric mixtures results when the adducts **13** are allowed to stand in solution for a day. Hydrolysis of the nitrile with sulfuric acid produces the corresponding amides **14**. The chiral benzylic moiety is destructively removed as above with catalytic hydrogenation affording the optically active amides **15**. The electron-rich substances (R$_1$, R$_2$ =OMe) suffer aromatic sulfonation during the nitrile hydrolysis. Only three examples cleanly suffer hydrolysis to homogeneous products; the other substances are reported to give complex reaction products. The authors note that hydrolysis does not proceed with hydrochloric acid and may be a manifestation of steric hindrance introduced by the α-methyl group. The amides **15** can be converted to the corresponding α-methyl-α-amino acids **16** by treatment

SCHEME 3

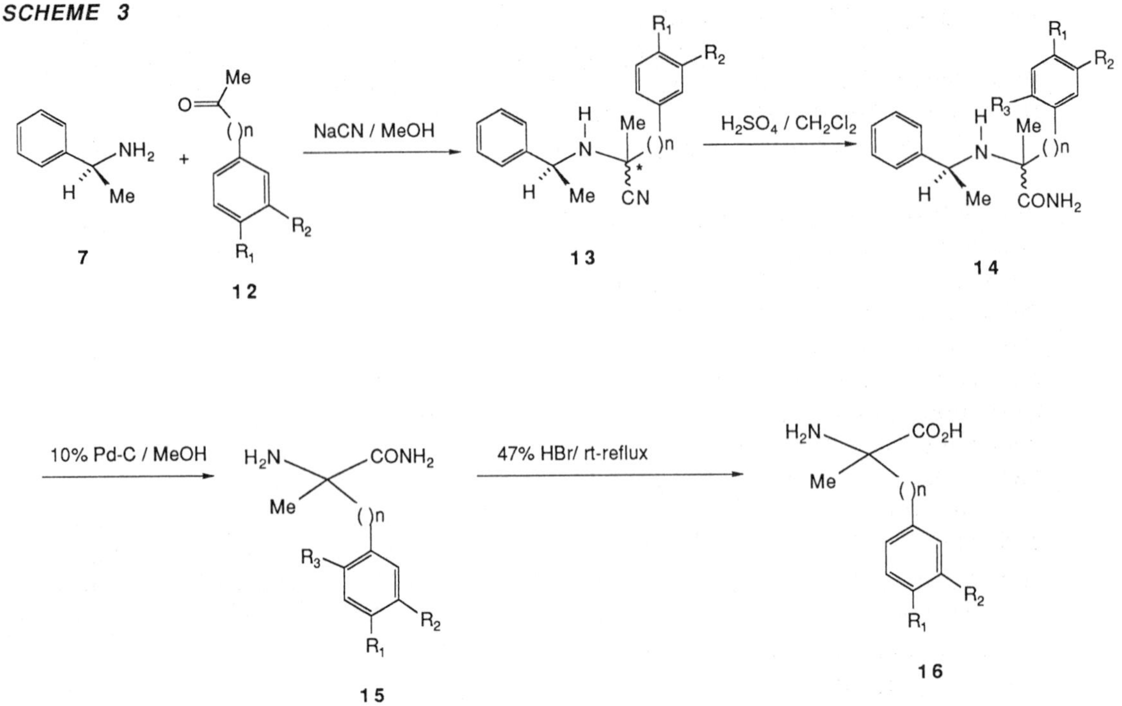

7 12 13 14 15 16

Weinges, K.; Gries, K.; Stemmle, B.; Schrank, W., *Chem.Ber.* (1977) **110**, 2098

R_1	R_2	n	YIELD (%13)	YIELD (%14)	R_3 (14 / 15)	YIELD (%15)	% (16)
H	H	1	75	82	H	95	84 (R)
OMe	H	1	84				
OMe	OMe	1	76	90	SO_3H	91	86 (S)
H	H	2	75	77	H (R_1 = SO_3H, R_2=H)	76	71 (R)
OMe	H	2	91				
OMe	OMe	2	88				

with hot, concentrated hydrobromic acid. This final hydrolysis also removes the sulfonic acid moiety introduced in the initial nitrile hydrolyses. While specific %ee's or optical purities are not reported for the final amino acids, they can be considered to be essentially optically pure since the nitriles are obtained as stereochemically homogeneous. Complete experimental details accompany this work.

More recently, Stout and associates [5] have examined the preparation of optically pure α-amino nitriles (**17**) with (R)- and (S)-α-methylbenzylamine (**7**) and several benzaldehyde derivatives. The conversion of the α-aryl-α-amino nitriles to the corresponding substituted phenylglycine derivatives in the present case is quite difficult since, the reductive protocol utilized above which removes the N-α-methylbenzyl moiety would also result in cleavage of the second benzylic residue present in **17** (where R=aryl). Under the conditions examined, the authors conclude that the α-amino nitriles (**17**) are formed under thermodynamic control and not kinetic control as in the case of the methyl ketones detailed above by Weinges.

A very recent synthetic procedure for the synthesis of almost optically pure α-methyl phenylalanine has been described by Subramanian and Woodard [6] as shown in Scheme 5. These workers found that when the hydrochloride salt of (-)-α-methylbenzylamine (**19**) is condensed with phenylacetone and a methanolic solution of sodium cyanide, exclusively one diastereomer (**20**) is formed. The authors note that an alternate procedure, employing sodium cyanide, the free base α-methylbenzylamine and phenylacetone in glacial acetic acid gave diastereomeric mixtures. Repeated recrystallizations of the major isomer only led to equilibrium mixtures. Hydrolysis of pure **20** obtained from **19** afforded the amide **21** which was subsequently hydrogenated to remove the chiral auxilliary. The amide was finally hydrolyzed to the free amino acid in hot, 48% HBr; ion-exchange purification afforded (R)-(+)-α-methyl phenylalanine **23** in very good overall yield. Application of the same procedure with the enantiomer of **19** afforded the enantiomer of **23**. A clear, complete and simple experimental procedure accompanies this work.

From the above examples that all employ α-methylbenzylamine as the chiral auxilliary, it can be concluded that methyl ketones consistently give the best results since their α-amino nitriles can be formed under kinetic control with a high degree of diastereoselectivity. The somewhat brutal conditions required for the hydrolytic manipulations and the reductive cleavage of the chiral benzylic moiety on the nitrogen will certainly limit the types of substrates that will be accomodated by these procedures. In the case of aldehyde substrates, fractional crystallization

Asymmetric Strecker Syntheses

SCHEME 4

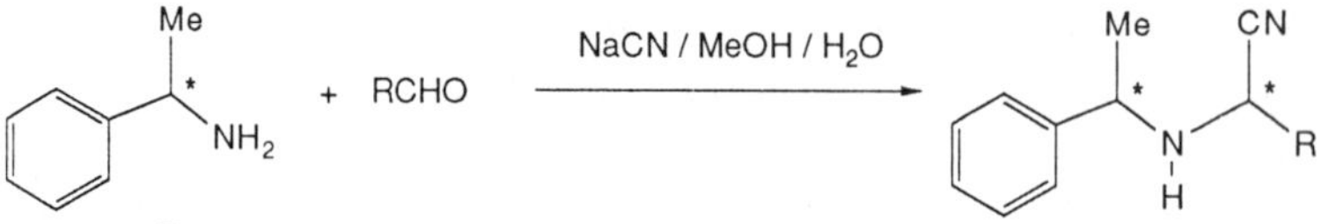

Stout, D.M.; Black, L.A.; Matier, W.L., *J.Org.Chem* (1983) **48**, 5369

RCHO	CONFIG.(7)	DIASTEREOMERIC RATIO	YIELD	YIELD OF PURE DIASTEREOMER	CONFIG. (17)
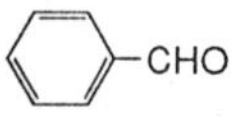 ⬡—CHO	R	3.2 : 1	93	33	R
	S	3.2: 1	97	26	S
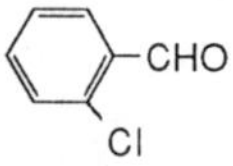 ⬡—CHO (Cl)	R	6.0 : 1	96	78	R
	S	4.5 : 1	96	19	S
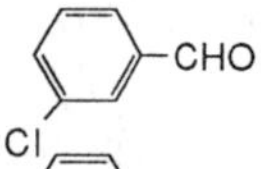 ⬡—CHO (Cl)	R	2.1 : 1	96	14	R
	S	2.7 : 1	89	29	S
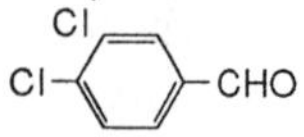 Cl—⬡—CHO	R	3.3 : 1	93	39	R
	S	3.2 : 1	85	21	S
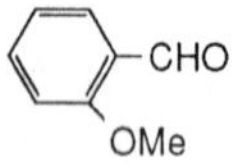 ⬡—CHO (OMe)	R	2.5 : 1	97	8	R
	S	2.3 : 1	97	5	S
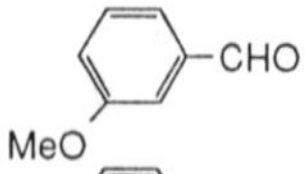 ⬡—CHO (MeO)	R	2.4 : 1	90	8	R
	S	2.5 : 1	94	3	S
 MeO—⬡—CHO	R	3.0 : 1	97	48	R
	S	3.0 : 1	97	46	S
 Me—⬡—CHO	R	3.0 : 1	87	45	R
	S	3.0 : 1	80	54	S
MeCHO	R	3.0 : 1	99	12	R
	S	3.3 : 1	92	25	S
EtCHO	R	3.6 : 1	99	13	R
	S	3.6 : 1	96	9	S

SCHEME 5

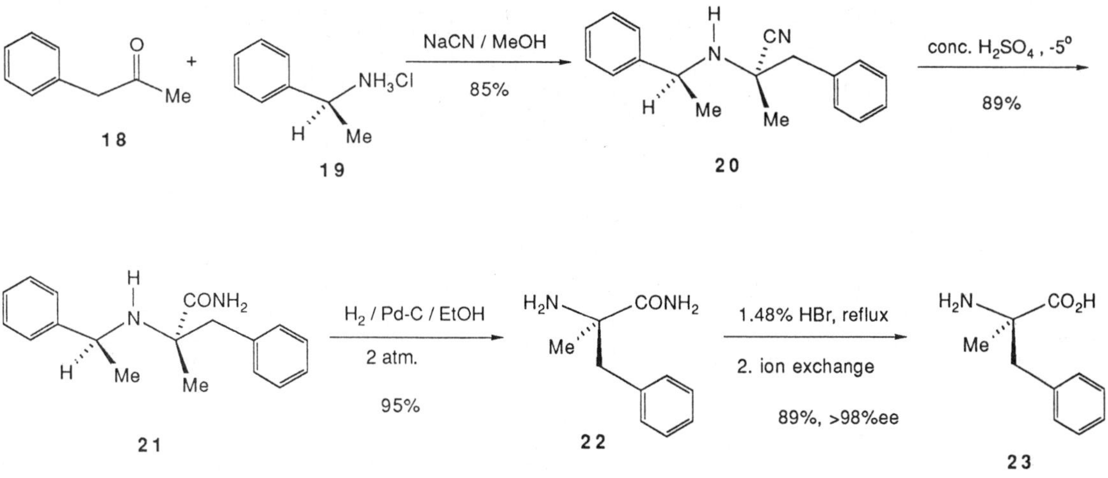

Subramanian, P.K.; Woodard, R.W., *Synth.Comm.* (1986) **16**, 337.

of partially diastereomeric mixtures of adducts prior to the hydrogenation step can lead to very practical preparations. It is surprising that none of these reports detail attempted dissolving metal reduction of the chiral auxilliary which would provide a complementary cleavage protocol that should be compatible with unsaturated 'R' groups of the final amino acids.

Weinges and co-workers [7-11] have extensively studied the use of an alternative chiral amine for the asymmetric Strecker synthesis. As shown in Scheme 6, [4S,5S]-(+)-5-amino-2,2-dimethyl-4-phenyl-1,3-dioxane (25) is condensed with methyl ketones (24) in the presence of methanolic sodium cyanide at 60°C to afford the α-amino nitriles 26. The chiral dioxane (25) is conveniently prepared on a preparative scale from S,S-*threo*-β-hydroxy phenylalinol (35); a substance available in large quantities as an intermediate in the industrial synthesis of chloramphenicol. The dioxane chiral auxilliary (25) offers the advantage over the α-methylbenzyl amines of exhibiting improved kinetic diastereoselectivity in the addition of cyanide to the imines. The authors demonstrate that the diastereomerically pure substances 26, when allowed to stand in solution for one day affords equilibrium mixtures. Treatment of the initially formed pure diastereomers (26) with concentrated hydrochloric acid effects hydrolysis of the nitrile, cleavage of the acetonide and cyclization to the *threo*-5-hydroxymethyl-6-phenyl-2,3,5,6-tetrahydro-4H-oxazin-2-ones (27) in good overall yields.

SOAA—H*

SCHEME 6

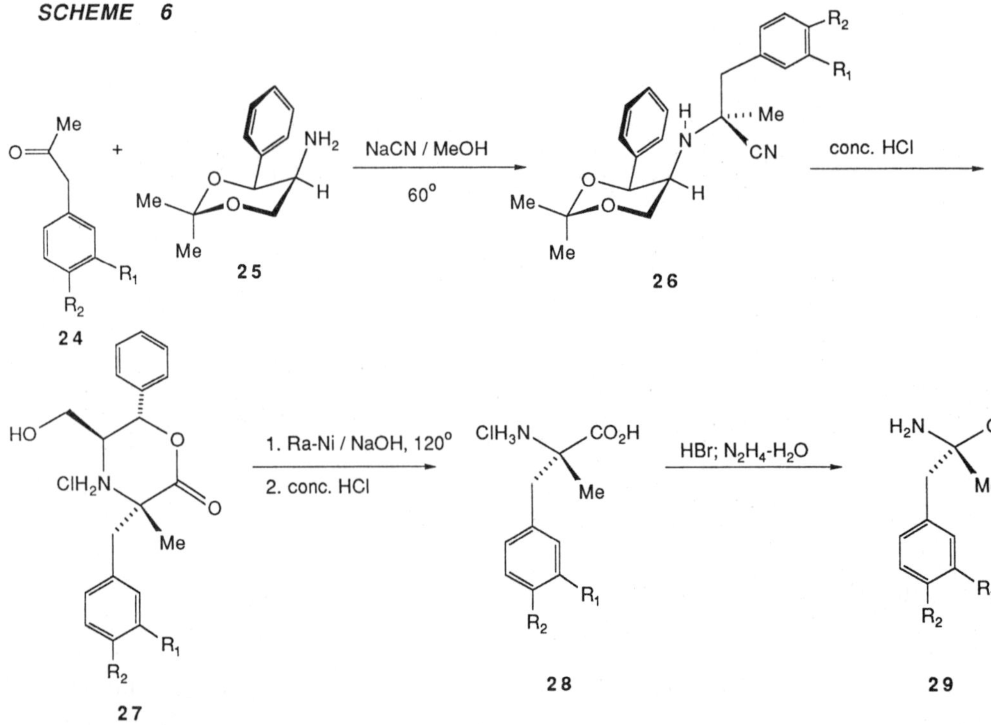

Weinges, K.; Graab, G.; Nagel, D.; Stemmle, B., *Chem.Ber.* (1971) **104**, 3594

R₁	R₂	YIELD(%26)	YIELD (%27)	YIELD (%28)	YIELD (%29)
OMe	OMe	82	83	98	81 (R₁=R₂=OH)
H	OMe	70	76	83	70 (R₁=H,R₂=OH)
H	H	80	75	80	100

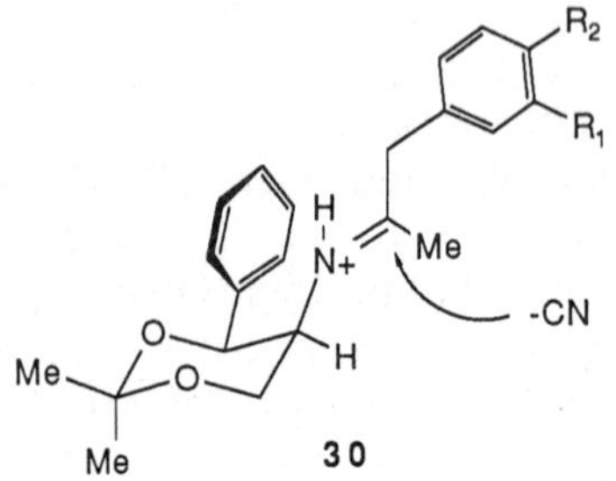

SCHEME 7

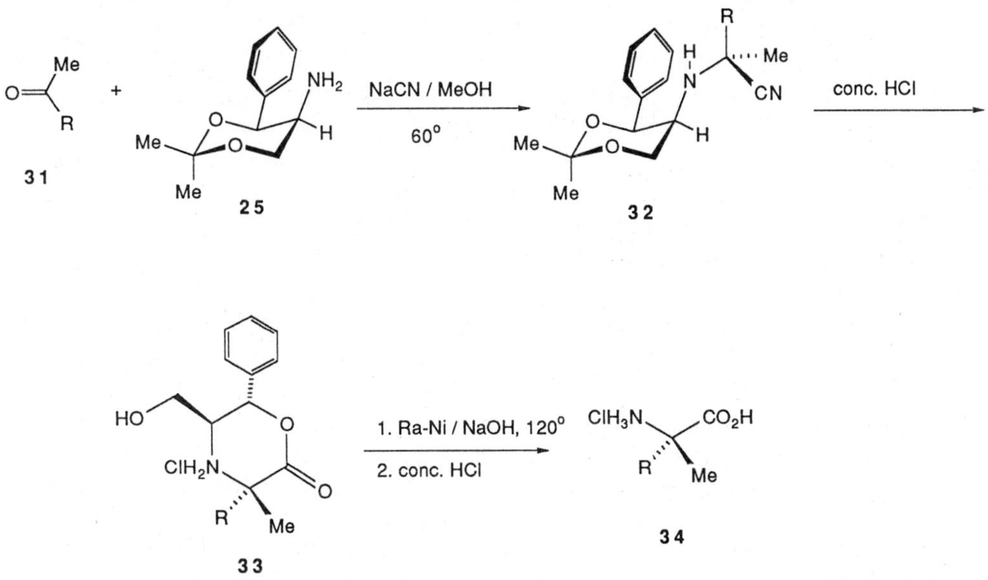

Weinges, K.; Stemmle, B., *Chem.Ber.* (1973) **106**, 2291

R	YIELD(%32)	YIELD (%33)	YIELD (%34)	CONFIG. (34)
Me	91	75	79	
Et	90	67	74	S
n-Pr	86	86	86	R
n-Bu	78	88	92	S
n-Pent.	90	63	75	R
i-Pr	93	-	-	S
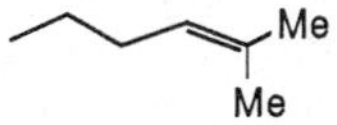 (alkenyl)	92	-	-	R

Treatment of these lactones with 2N sodium hydroxide solution and Raney-Nickel at 120°C results in the destructive removal of the chiral auxilliary furnishing the amino acids **28** in good to excellent yields. Substrates bearing methoxy groups are transformed into the corresponding phenols (**29**) by treatment with HBr and hydrazine; S-α-methyl-DOPA being one of the substances prepared. The authors speculate that the Schiff base assumes the conformation **30**; cyanide attack would therefore be reasonably expected to occur from the less hindered face opposite to the dioxane phenyl substituent. While specific %ee's or %op data is not reported, the compounds prepared are assumed to be of a high level of optical purity. Full experimental details accompany all of the papers reported from this group.

   In examining the reaction of simple aliphatic aldehydes (**31**) with **25**, these workers [8] found that the absolute configurations of the α-amino nitriles (**32**) and the subsequent amino acids (**34**) were directly related to the number of carbon atoms in the aliphatic chain. The authors report that the aldehydes bearing an odd number of carbon atoms result in products possessing the R-absolute stereochemistry; whereas substrates with an even number of carbon atoms result in products bearing the S-absolute stereochemistry. A rationale for this extremely curious behavior has not yet been suggested.

   A series of dioxanes **37** were prepared [9] from acetone; formaldehyde; acetophenone; t-butyl methyl ketone ; and t-butyl phenyl ketone and the amino diol **35** as shown in Scheme 8. The dioxanes show a marked tendency to dispose the bulkier group (R₁) in the equatorial position. Condensation of these materials with benzaldehyde furnishes the nicely crystalline imines **38** as diastereomerically pure substances. Addition of sodium cyanide to these substrates occurs with a high degree of diastereoselectivity providing the α-amino nitriles **39** in good to excellent yields. The experimental conditions to hydrolyze and cleave these adducts to optically active phenyl glycine is reported in another paper [11] to be detailed in Scheme 10. Spectroscopic investigation of the imines **38** reveals that these substances possess the E-geometry in the conformation depicted; the diastereoselective addition of cyanide to these substances under kinetic control is in accord with that depicted in structure **30**.

   A series of methyl ketones (**40**) are condensed [10] with **25** to afford the optically active α-amino nitriles **41**. The stereochemically pure adducts (**41**) were then allowed to stand in chloroform solution for one day at ambient temperature and the equilibrium ratios determined by NMR. This report is useful in that it uncovers a simple spectroscopic method to assign the absolute configuration of the newly formed stereogenic center in **41** which can be confidently extrapolated to the final amino acids

SCHEME 8

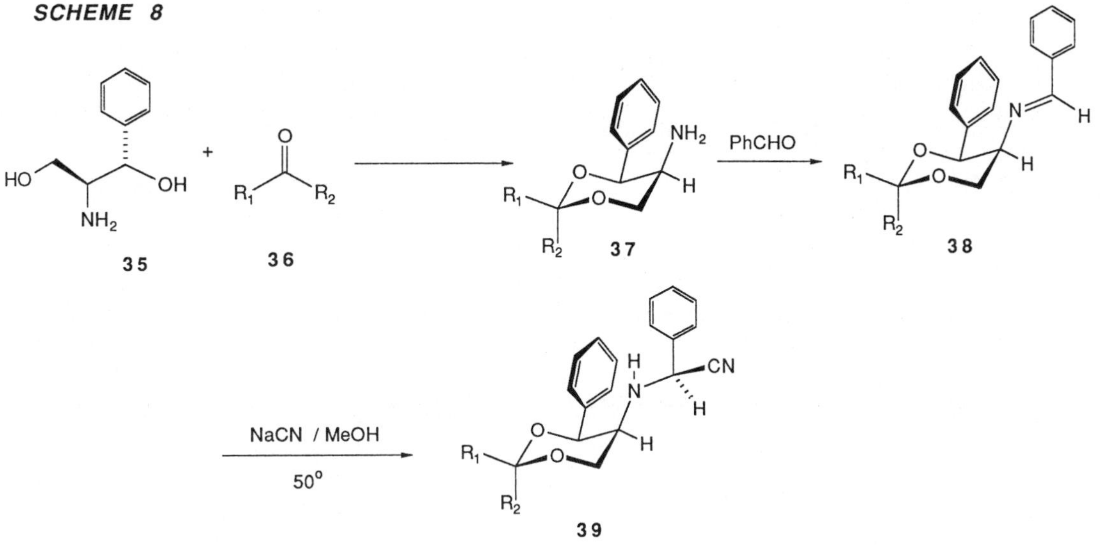

Weinges, K.; Klotz, K-P.; Droste, H., *Chem.Ber.* (1980) **113**, 710

R$_1$	R$_2$	YIELD (%37)	YIELD (%38)	YIELD (%39)
Me	Me		100	82
H	H	100	98	91
Ph	Me	87	87	92
t-Bu	Me	17	68	86
t-Bu	Ph	38	90	91

obtained after hydrolysis and removal of the chiral auxilliary. The chemical shift of the methyl group adjacent to the nitrile substituent appears at δ~0.6 ppm in the ^{1}H NMR spectrum for the L-configured isomers and at δ ~1.3 ppm for the D-configured substances. This large difference in chemical shift can be rationalized by consideration of the proximity of the

SCHEME 9

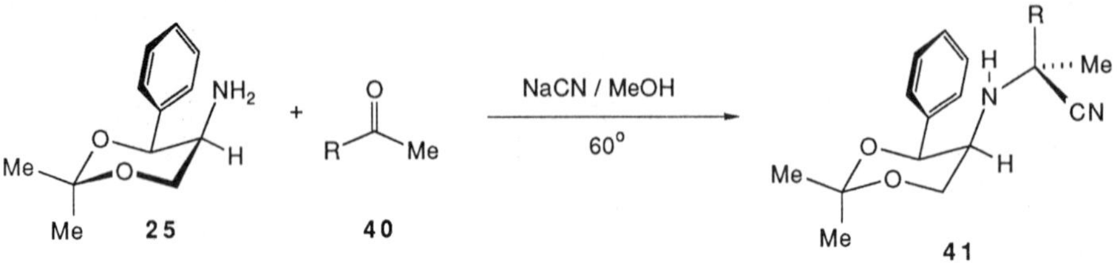

Weinges, K.; Blacholm, H., *Chem.Ber.* (1980) **113**, 3098

R	YIELD (%41)	CONFIG. (41)	L : D RATIO after 24 h (CHCl$_3$)
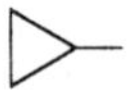	53	L (S)	61 : 39
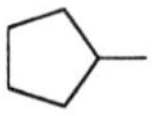	72	L (S)	82 : 18
	70	L (S)	80 : 20
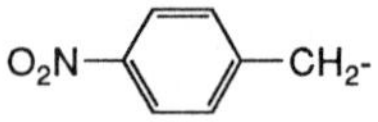 O$_2$N–⬡–CH$_2$-	61	L (S)	84 : 16
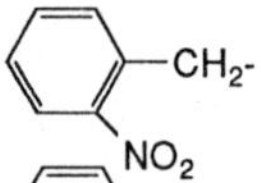 ⬡(NO$_2$)–CH$_2$-	63	L (S)	74 : 26
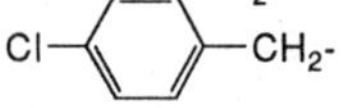 Cl–⬡(NO$_2$)–CH$_2$-	68	L (S)	73 : 27
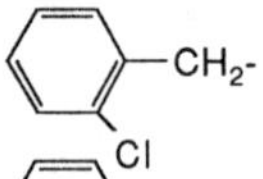 ⬡(Cl)–CH$_2$-	62	L (S)	76 : 24
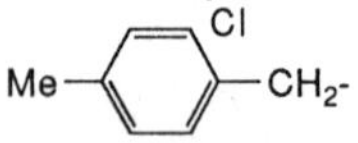 Me–⬡–CH$_2$-	82	L (S)	66 : 34
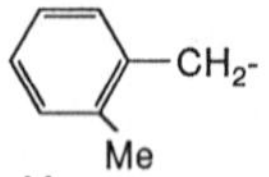 ⬡(Me)–CH$_2$-	83	L (S)	66 : 34
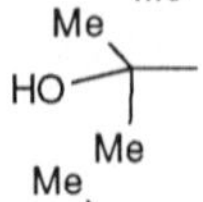 HO–C(Me)(Me)-	64	L (R)	63 : 37
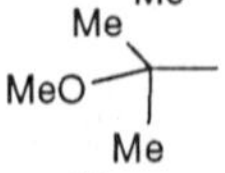 MeO–C(Me)(Me)-	60	L (R)	84 : 16
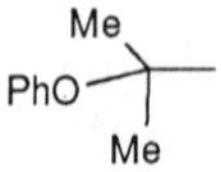 PhO–C(Me)(Me)-	69	L (R)	100 : 0
HOCH$_2$-	66	L (R)	63 : 37
MeOCH$_2$-	50	L (R)	85 : 15
PhOCH$_2$-	38* (D config. =S)	D (S)	0 : 100

SCHEME 10

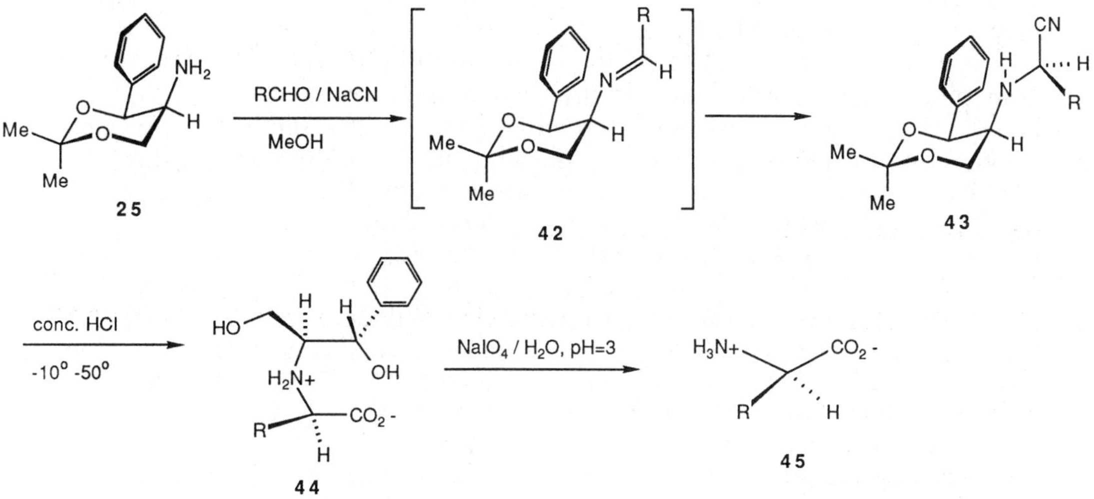

Weinges, K.; Brune, G.; Droste, H., *Liebigs Ann.Chem.* (1980) 212
Weinges, K.; Brachmann, H.; Stahnecker, P.; Rodewald, H.; Nixdorf, M.; Irngartinger, H., *Liebigs Ann.Chem.* (1985) 566

R	YIELD (%43)	YIELD (%44)	YIELD (%45)
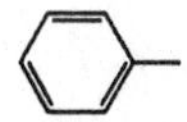	65	82	68
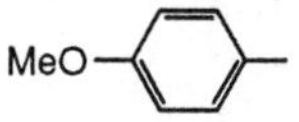 MeO—	60	73	49
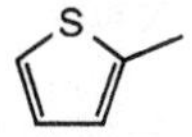	24	65	64
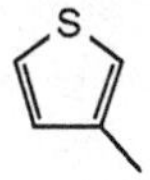	35	76	60

shielding cone of the dioxane phenyl ring to the methyl group in the L-configured isomers of **41**. This simple analytical method will be useful to workers preparing new α-methyl amino acids of unknown absolute stereochemistry as well as allowing for a convenient determination of diastereomeric ratios in the crude mixtures.

One of the most significant contributions of the Weineges asymmetric Strecker procedure is the discovery of an oxidative method [11] to cleave the chiral auxilliary allowing access to the difficult arylglycine derivatives. As shown in Scheme 10, condensation of aromatic aldehydes with **25** in the presence of sodium cyanide affords the adducts **43** which are obtained diastereomerically pure by crystallization. The nitrile and acetonide moieties are hydrolyzed to the corresponding oxazinone (see **33**, Scheme 8) which is opened with water to the hydroxy acids **44**. Treatment of these substances with sodium periodate in water at pH=3 results in the selective cleavage at the amino alcohol positions giving the enantiomerically pure aryl glycine derivatives **45**. This same oxidative procedure has been employed by Williams, et.al. in an aryl glycine synthesis described in Chapter 1 (see Scheme 117, Chapter 1). Arylglycines are amongst the most difficult amino acids to obtain in optically pure form due to their ease of racemization under basic conditions and their often poor solubility properties. The method described in Scheme 10 offers an attractive and simple method to obtain this class of compounds from aromatic aldehydes.

Very recently, Kunz and associates [12,13] have examined the asymmetric Strecker synthesis utilizing β-1-amino-*tetra*-O-pivaloyl galactose (**48**, Scheme 11) as the chiral matrix. The protocol is similar to those above, specifically involving imine formation to **49** and diastereoselective addition of cyanide to the imine forming the α-amino nitriles **50**. These workers have found that anomerization of the carbohydrate moiety (from β-to α-) can be minimized by using trimethylsilyl cyanide and a Lewis acid such as zinc chloride or tin tetrachloride giving high yields of the adducts **50** under mild conditions. The authors propose that the imines possess the E-geometry and adopt a conformation (**51**) that is stabilized by overlap of the C=N π orbital with the σ^* orbital of the glucose C_1-O bond. This conformation is further supported by a strong NOE in the ^{1}H NMR spectrum between the imine methine and the H_1 methine at the anomeric center of the sugar. The bulky pivaloyl group at C-2 would then hinder attack from this side forcing addition from the side of the ring oxygen; this hypothesis is in accord with the observed diastereoselectivity. A single example of hydrolyzing the nitrile with concomitant hydrolytic removal of the carbohydrate was

SCHEME 11

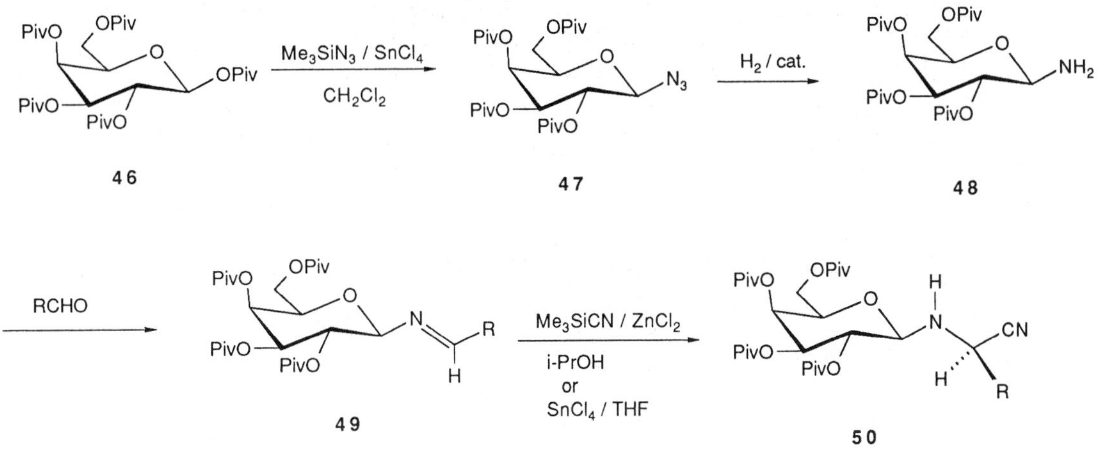

Kunz, H. ; Sager, W., *Angew.Chem.Int.Ed.Engl.* (1987) **26**, 557

RCHO	SOLVENT	CATALYST (mol%)	DIASTEREOMERIC RATIO (R : S)	YIELD (% PURE (R)-50)
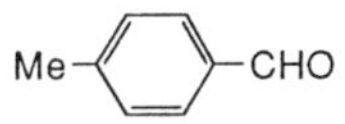 Me—⟨ ⟩—CHO	i-PrOH THF	ZnCl₂ (100) SnCl₄ (130)	6.5 : 1 12 : 1	78 87
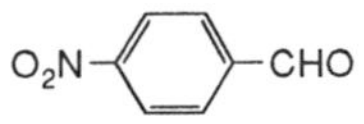 O₂N—⟨ ⟩—CHO	i-PrOH	ZnCl₂ (5)	7 : 1	80
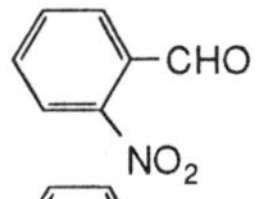 ⟨ ⟩—CHO NO₂	THF	SnCl₄ (130)	1 : 0	91
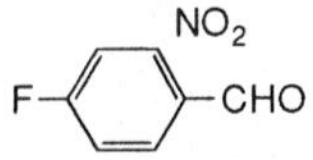 F—⟨ ⟩—CHO	i-PrOH THF	ZnCl₂ (5) SnCl₄ (130)	6.5 : 1 10 : 1	75 84
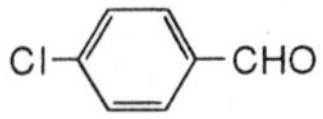 Cl—⟨ ⟩—CHO	THF	SnCl₄ (130)	11 : 1	84
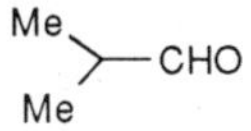 Me₂CH—CHO	THF	SnCl₄ (130)	8 : 1	74
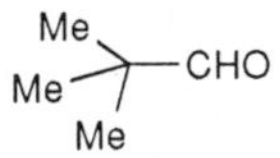 Me₃C—CHO	THF	SnCl₄ (130)	13 : 1	86

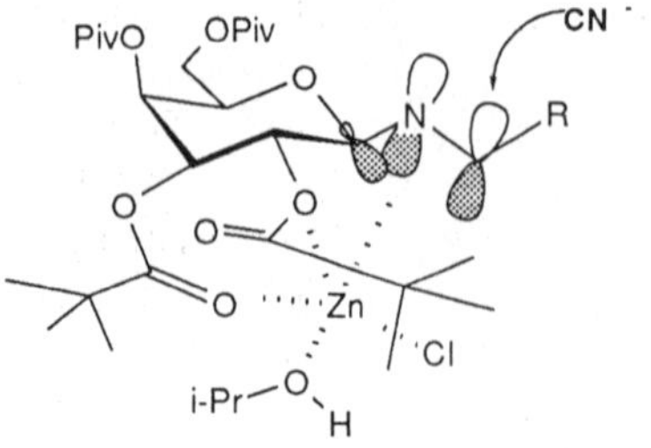

51

SCHEME 12

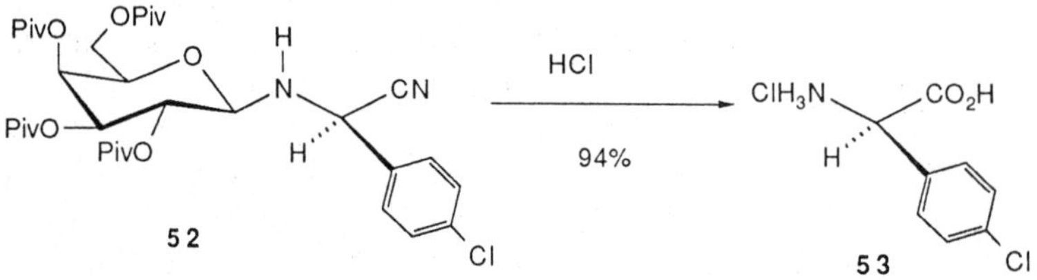

52   94%   **53**

SCHEME 13

49   75-90%   **54**

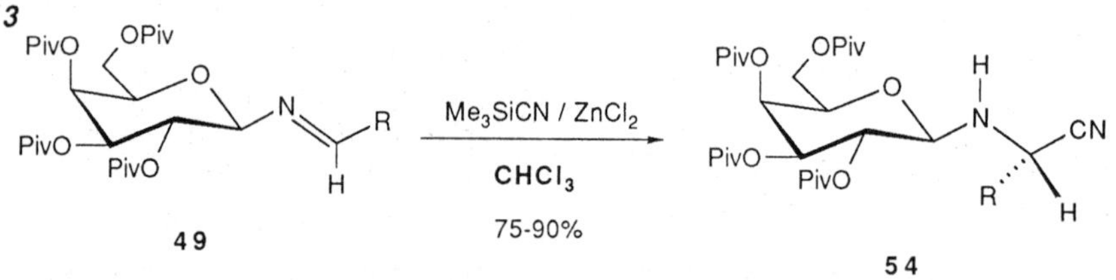

Kunz, H.; Sager, W.; Pfrengle, W.; Schanzenbach, D.; *Tetrahedron Lett.* (1988) **29**, 4397

R	ZnCl$_2$ (mol%)	DIASTEREOMERIC RATIO (54) (R : S)
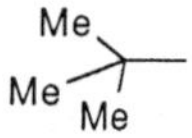 Me / Me / Me (tert-butyl)	5	1 : 9
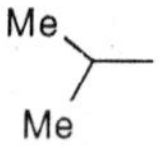 Me / Me (isopropyl)	100	1 : 5
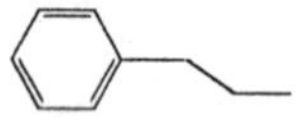 (phenyl ethyl)	100	1 : 3
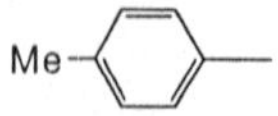 Me— (p-tolyl)	100	1 : 4.5
 F— (p-fluorophenyl)	100	1 : 3
 Cl— (p-chlorophenyl)	5 / 300	1 : 4 / 1 : 5
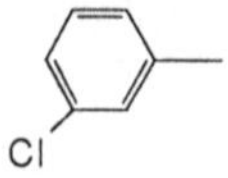 (m-chlorophenyl)	5 / 100	1 : 6 / 1 : 5

reported as shown in Scheme 12. Hydrolysis of the *para*-chlorophenyl adduct **52** with hydrochloric acid affords the enantiomerically pure phenylglycine derivative **53** in high yield. The carbohydrate moiety can be recovered as the corresponding hemi-acetal (**56**) in 70-90% yield by extraction with methylene chloride after hydrolysis. This method is nicely complementary to the systems detailed above since, the chiral auxilliary is removed under hydrolytic rather than reductive or oxidative conditions.

These workers [13] have also found an interesting solvent effect that *reverses* the diastereofacial selectivity of addition to the Schiff bases (**49**) as shown in Scheme 13. When the addition of trimethylsilyl cyanide to **49** is carried out in chloroform solution using zinc chloride as the Lewis acid, a preponderance of the S-isomer results. All of the examples performed on the *same* substrates in either tetrahydrofuran or isopropanol exhibited selectivity for the corresponding R-isomers. The authors demonstrate by ^{1}H NMR that the conformation of the imine (see **51**) is the same in THF as it is in chloroform. Thus, the reasons for this marked reversal in stereochemical outcome are not presently clear but may be a manifestation of aggregation and salt complexation differences in the different solvents. From a practical standpoint, this simple change in reaction solvent allows for convenient access to either the D- or L-configured amino acids from the same chiral imine. Full experimental details for this elegant work have not yet appeared.

In a related study employing the same chiral matrix, Kunz and associates [14] have examined the asymmetric Ugi condensation as shown in Scheme 14. The galactose derivative **48** is condensed with an aldehyde, an isonitrile , formic acid and zinc chloride to afford the Ugi products **55**. The chemical yields for the formation of **55** are high and the kinetic ratio of isomers produced is excellent. These substances (**55**) can be obtained diastereomerically pure by a simple recrystallization; the yields for the pure isomers is tabulated. Hydrolysis of the chiral auxilliary with methanolic HCl affords the amides **57** and the hemi-acetal **56** which can be recovered and recycled. Hydrolysis of the amides with hot 6N hydrochloric acid followed by ion-exchange chromatography furnishes the free amino acids **58**. It is very significant that two good examples of the racemization-prone phenyl glycine derivatives are reported. It can thus be inferred that numerous base-labile but acid-stable classes of amino acids should become accessible by this protocol. Full experimental details have been published for this work.

A very recent asymmetric synthesis of β-hydroxy glutamic acid has been reported by Hirobe and associates [15] as shown in Scheme 15. The approach differs considerably from those reported above, but maintains elements of the Strecker protocol. Coupling of the activated oxazolone **59**

SCHEME 14

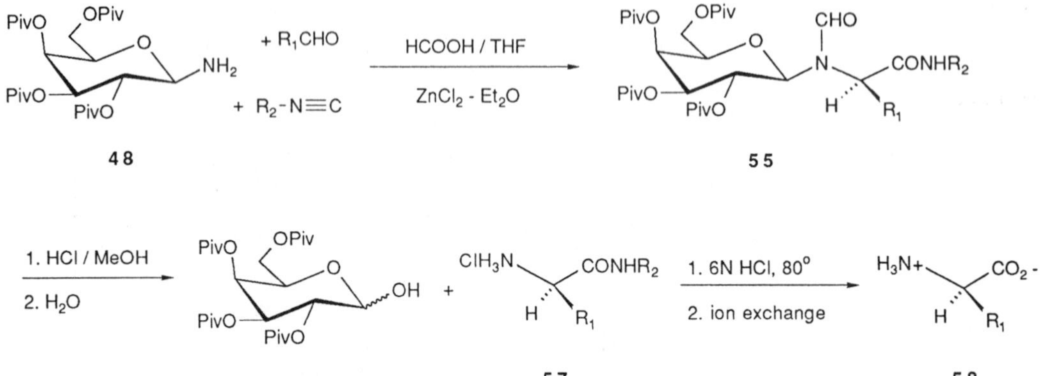

Kunz, H.; Pfrengle, W., *J.Am.Chem.Soc.* (1988) **110**, 651
Kunz, H.; Pfrengle, W., *Tetrahedron* (1988) **44**, 5487

R_1	R_2	KINETIC RATIO (55) (2R : 2S)	YIELD (%PURE 2R-55)	YIELD (%58)
n-Pr	t-Bu	94 : 6	80	
i-Pr	t-Bu	95 : 5	86	
t-Bu	t-Bu	96 : 4	80	90
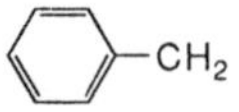 —CH$_2$ (benzyl)	t-Bu	95 : 5	80	82
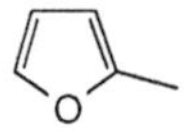 (furyl)methyl	t-Bu	95 : 5	90	
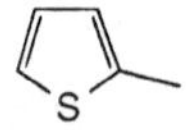 (thienyl)methyl	t-Bu	96 : 4	93	
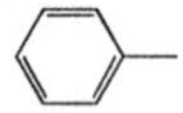 (phenyl)	t-Bu	91 : 9	81	85
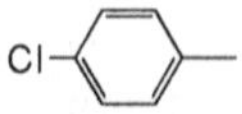 Cl—(phenyl)	t-Bu	97 : 3	92	90
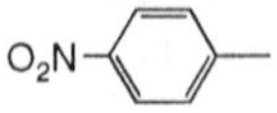 O$_2$N—(phenyl)	t-Bu	94 : 6	91	
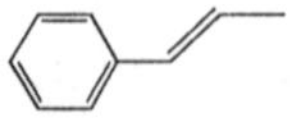 (cinnamyl)	t-Bu	95 : 5	75	
 NC—	Ph	93 : 7	80	87*

* R_1 of this amino acid = HOOC(CH$_2$)$_3$ -

to the ketopinic acid **60** provides the imide **61**. Bromination of this substance with bromine and trimethylsilyl triflate in trimethyl orthoacetate /methylene chloride at -100ºC afforded the diastereomers **62** and **63** in a 5 : 95 ratio. These could be separated by chromatography to obtain pure diastereomers. The major isomer was allylated with allyl tri-n-butyl tin under photolysis conditions to provide the oxazolone **64** after removal of the ketopinic acid moiety with dibutyl cuprate. The stereochemistry of the allyl addition proceeded with complete retention of configuration. The hemi-aminal was next coupled with trimethylsilyl cyanide in the presence of titanium tetrachloride to afford the trans carbomethoxy derivative **65** after hydrolysis of the incipient nitrile adduct. The nitrogen is protected as the t-BOC urethane and the cyclic urethane is hydrolyzed with methanolic cesium carbonate. Silylation of the resulting alcohol provided the protected amino acid **66** in good overall yield. Finally, the olefin is oxidatively processed to the acid and deprotected to the β-hydroxy glutamate **67**. Although the deacylation of the ketopinic acid in this preliminary protocol does not give back the acid **60** (the n-butyl ketone is produced), these substances are available in both enantiomorphic forms from inexpensive (-)- or (+)-camphor sulfonic acids. This work serves to illustrate that numerous chiral auxilliaries that can lead to asymmetric induction in Strecker-type applications are yet to be investigated.

SCHEME 15

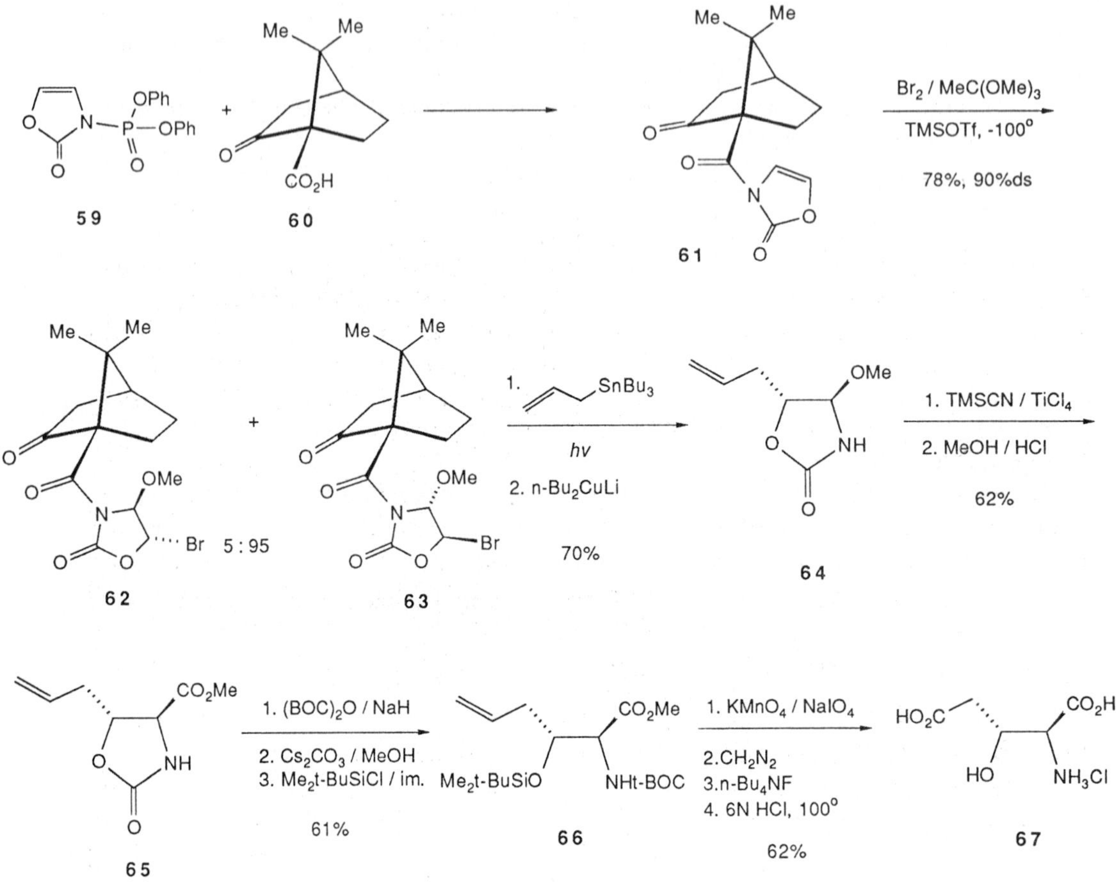

Kunieda, T.; Ishizuka, T.; Higuchi, T.; Hirobe, M., *J.Org.Chem.* (1988) **53**, 3381

References Chapter 5

1. a)Harada, K., *Nature* (1963) **200**, 1201; b) Harada, K.; Fox, S., *Naturwissenschaften* (1964) **51**, 106.

2. a)Harada, K.; Okawara, T., *J.Org.Chem.* (1973) **38**, 707; b)Harada, K.; Okawara, T.; Matsumoto, K., *Bull.Chem.Soc.Jpn.* (1973) **46**, 1865.

3. Patel, M.S.; Worsley, M., *Can.J.Chem.* (1970) **48**, 1881.

4. Weinges, K.; Gries, K.; Stemmle, B.; Schrank, W., *Chem.Ber.* (1977) **110**, 2098.

5. Stout, D.M.; Black, L.A.; Matier, W.L., *J.Org.Chem.* (1983) **48**, 5369.

6. Subramanian, P.K.; Woodard, R.W., *Synth.Comm.* (1986) **16**, 337.

7. Weinges, K.; Graab, G.; Nagel, D.; Stemmle, B., *Chem.Ber.* (1971) **104**, 3594.

8. Weinges, K.; Stemmle, B., *Chem.Ber.* (1973) **106**, 2291.

9. Weinges, K.; Klotz, K-P.; Droste, H., *Chem.Ber.* (1980) **113**, 710.

10. Weinges, K.; Blacholm, H., *Chem.Ber.* (1980) **113**, 3098.

11. a) Weinges, K.; Brune, G.; Droste, H., *Liebigs Ann.Chem.* (1980) 212; b) Weinges, K.; Brachmann, H.; Stahnecker, P.; Rodewald, H.; Nixdorf, M.; Irngartinger, H., *Liebigs Ann.Chem.* (1985) 566.

12. Kunz, H.; Sager, W.; *Angew.Chem.Int.Ed.Engl.* (1987) **26**, 557.

13. Kunz, H.; Sager, W.; Pfrengle, W.; Schanzenbach, D., *Tetrahedron Lett.* (1988) **29**, 4397.

14. a) Kunz, H.; Pfrengle, W., *J.Am.Chem.Soc.* (1988) **110**, 651; b) Kunz, H.; Pfrengle, W., *Tetrahedron* (1988) **44**, 5487.

15. Kunieda, T.; Ishizuka, T.; Higuchi, T.; Hirobe, M., *J.Org.Chem.* (1988) **53**, 3381.

CHAPTER 6

ASYMMETRIC HYDROGENATION OF DEHYDROAMINO ACIDS

The asymmetric hydrogenation of dehydroamino acids has provided an extremely useful, preparative approach to making a variety of optically active amino acids. Two basic conceptual approaches have evolved in this context:1) the heterogeneous hydrogenation of dehydroamino acids that contain an appended chiral auxilliary; and 2) the homogeneous hydrogenation of achiral dehydroamino acids using optically active, soluble hydrogenation catalysts. The latter topic has been extensively studied and reviewed; the present chapter will only highlight the most impressive results observed and will illustrate several notable applications of this technology.

A. HETEROGENEOUS HYDROGENATION OF DEHYDROAMINO ACIDS

Kagan and associates [1a] reported a brief, but impressive example of heterogeneous hydrogenation of a chiral dehydramino acid system in 1968 as shown in Scheme 1. *Erythro*-2-amino-1,2-diphenylethanol (1) can be obtained [1b] in >98%ee through a glutamate resolution reported by Tishler, et.al. [1c]. Condensation of this amino alcohol with dimethylacetylene dicarboxylate (2) afforded the dehydro lactone 3. Hydrogenation of this substance with Raney-Nickel and acidic work-up afforded D-β-methyl aspartate (4) in > 98% ee and in high yield. It is unfortunate that full experimental details and additional follow-up work developing this elegant asymmetric synthesis has not appeared in the literature.

Harada and Tamura [2] more recently reported on a similar system derived from phenylglycinol (5, Scheme 2). Condensation of dimethylacetylene dicarboxylate with 5 affords the dehydro lactone 6. The authors propose that 6 exists as the internally H-bonded Z-isomer (shown). Hydrogenation of 6 can either be done stepwise with Raney-Nickel or Aluminum amalgam through lactone 7, or in one step with a Pd⁰ catalyst to L-β-methyl aspartate 9; complete hydrolysis affords aspartic acid 8 in 12~17%ee. Comparing the Kagan system above with the

SCHEME 1

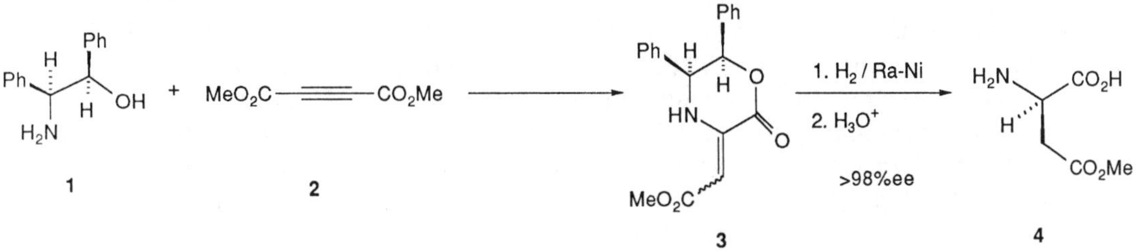

Vigneron, J.P.; Kagan, H.; Horeau, A., *Tetrahedron Lett.* (1968) 5681

SCHEME 2

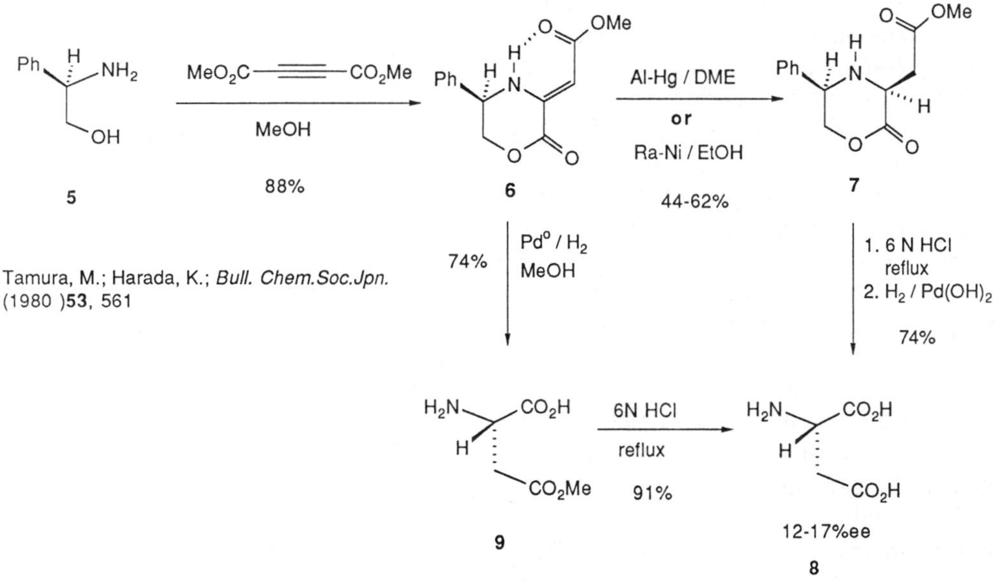

Tamura, M.; Harada, K.; *Bull. Chem.Soc.Jpn.*
(1980)**53**, 561

penylglycinol auxilliary clearly shows the advantage of the second phenyl ring which presumably provides considerably more steric shielding to the β-face of the lactone towards the catalyst surface. The primary disadvantage of both approaches is the reductive destruction of the chiral auxilliary in the last step. On the other hand, the experimental protocol for isolating the polar amino acid from the hydrocarbon residues is extremely easy and might somewhat offset the inherent disadvantage of sacrificing the chiral auxilliary.

In an effort to combine the advantages of the Kagan system with recovery of the chiral auxilliary, Corey and co-workers [3] employed the chiral N-aminoindolines **10** (Scheme 3). Condensation of α-keto esters (**11**) with **10** afforded the hydrazono lactones **12**. Aluminum amalgam reduction proceeded with a very high degree of asymmetric induction to furnish the hydrazino lactones **13**. Subsequent catalytic hydrogenation in acidic DME afforded the free amino acids **14** and the chiral indolines **15** that could be recovered and re-oxidized to the N-amino substrates **10**. The %ees' were uniformly high and the chemical yields quite good. This approach offers the versatility of employing α-ketoesters which can be prepared by a variety of straightforward techniques. Full experimental details accompany this work.

A series of papers dealing with the preparation and stereoselective reduction of dehydropiperazinediones have appeared in the literature.[4] It is well-known [4e] that N,N'-diacyl piperazinediones (**17**, Scheme 4) participate in aldolizations with aldehydes   to give the corresponding dehydropiperazinediones (**18**). The reaction is driven by   N/O acyl transfer of the N-acyl group to the proximal   incipient alkoxide; the resulting acetate generally eliminates directly to furnish   **18**. Izumiya and associates[4a-d] have extensively studied the catalytic hydrogenation of such substances with generally very high levels of asymmetric induction. The stereogenic center of the originally introduced amino acid directs hydrogenation to occur from the less-hindered opposite face providing the *syn*-products **20**. Subsequent hydrolysis of the piperazinedione in hot, 6N HCl provides the new amino acid (**21**) and the original chiral auxilliary amino acid which must be separated. The wide range in chemical yields likely reflects the experimental difficulties associated with separating the two amino acids. The method is generally confined to dehydropiperazinediones obtained from aldehydes since ketones are generally unreactive in forming the more hindered β,β-disubstituted dehydropiperazinediones.

A related, but more difficult approach in terms of achieving stereocontrol is illustrated in the hydrogenation of acyclic dehydropeptides reported by Harada [5]. The requisite dehydropeptides (**23**) are prepared by DBU-mediated elimination of HCl from β-chloroalanine residues. The results clearly show that the amino acid in the C-terminal position has a much larger contribution to determining the absolute sense and degree of asymmetric induction in producing the R-Ala moiety than the the N-terminal amino acid; regardless of the absolute configuration of the N-terminal residue, C-terminal (S)-proline derivatives gave (R)-Ala (**24**) in every case.

Didehydrotripeptides (**26**, Scheme 6) were also examined [5b] for asymmetric induction employing a C-terminal (S)-Pro residue. The

SCHEME 3

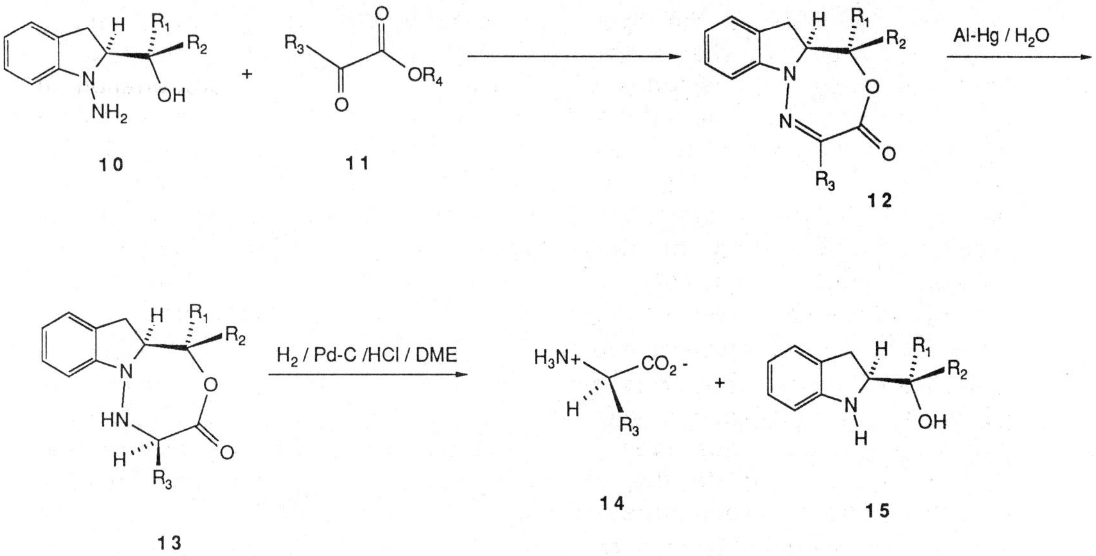

Corey, E.J.; McCaully, R.J.; Sachdev, H.S., *J.Am.Chem.Soc.* (1970) **92**, 2476
Corey, E.J.; Sachdev, H.S.; Gougoutas, J.Z.; Saenger, W., *J.Am.Chem.Soc.* (1972) **92**, 2488

R_1	R_2	R_3	R_4	YIELD (%12)	YIELD (%13)	YIELD (%14)	%ee
H	Me	Me	$-\!\!\bigcirc\!\!-NO_2$	70	95	78	96
H	Me	Et	$-\!\!\bigcirc\!\!-NO_2$	70.5	93	70	97
H	Me	i-Pr	Me	65	79	53.5	97
H	Me	i-Bu	Me	70	95	78	99
Me	H	Me	$-\!\!\bigcirc\!\!-NO_2$	62	97	90	92
Me	H	Et	$-\!\!\bigcirc\!\!-NO_2$	50	96	65	96

SCHEME 4

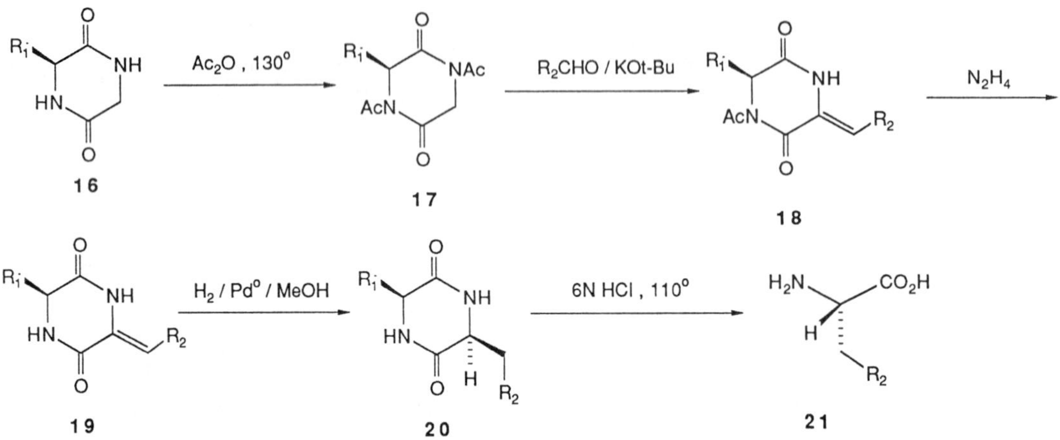

Kanmera, T.; Lee, S.; Aoyagi, H.; Izumiya, N., *Tetrahedron Lett.* (1979) 4483

	R_1	R_2	YIELD (%19)	%ee
	Me	Me	48	99
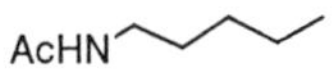 AcHN~~~~	Me	45	97
	Me	i-Pr	8	96
	Me	i-Bu	47	98
	i-Pr	i-Bu	61	>99
	i-Bu	i-Bu	69	98
AcHN~~~~	i-Bu	22	95
	Me	Ph	56	88
	i-Pr	Ph	63	94
	i-Bu	Ph	52	90
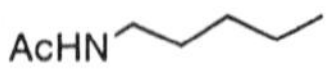 AcHN~~~~	Ph	26	77
	Me	PhCH$_2$CH$_2$	55	98
	i-Bu	PhCH$_2$CH$_2$	52	97
	Me	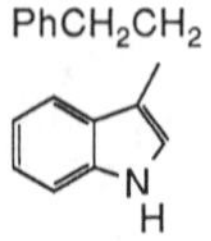 (indolyl)	49	71
	i-Bu	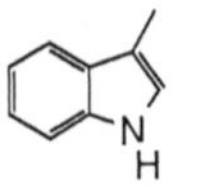 (indolyl)	18	66

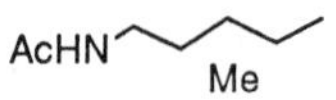

SCHEME 5

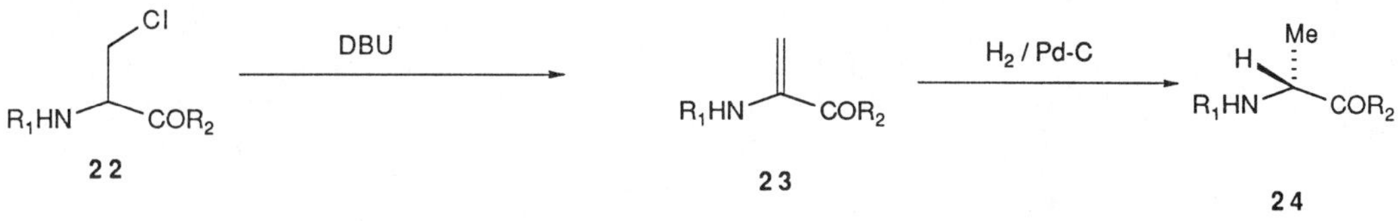

Takasaki, M.; Harada, K., *Chemistry Lett.* (1984) 1745.

R_1	R_2	YIELD (24)	%ee (24)
t-BOC-Gly-	-(S)-Pro-NH-t-Bu	96	84
t-BOC-(S)-Val-	-(S)-Pro-NH-t-Bu	97	87
t-BOC-(R)-Val-	-(S)-Pro-NH-t-Bu	97	81
t-BOC-(S)-Ile-	-(S)-Pro-NH-t-Bu	97	90
t-BOC-(R)-Phe-	-(S)-Pro-NH-t-Bu	92	89
t-BOC-(R)-Pro-	-(S)-Pro-NH-t-Bu	96	89
t-BOC-(R)-Ser-	-(S)-Pro-NH-t-Bu	86	82
t-BOC-(R)-Ser(t-Bu)-	-(S)-Pro-NH-t-Bu	96	93
t-BOC-Gly-	-(S)-Pro-N⟨⟩	92	43
t-BOC-(S)-Val-	-(S)-Pro-N⟨⟩	96	74

asymmetric induction at the internal Ala position was generally quite good with %ees' ranging from 89-95% ee. The butyrine residue, on the other hand, enjoyed a substantially lower degree of asymmetric induction with 54% ee being the best case. This is actually an impressive result if one considers that the N-terminal residue hydrogenates faster than the internal dehydroalanine moiety; this would then be a case of 1,7 asymmetric induction.

Ojima and co-workers[6a] have developed an elegant synthesis of various di- and tripeptides by the stereoselective formation and reduction of aryl-substituted β-lactams. As shown in Scheme 7, β-lactams of structure **30** are reduced with complete inversion of stereochemistry at N-benzylic residue with palladium on charcoal. This provides unique access to stereospecifically labelled phenylalnine-containing peptides.

Asymmetric Hydrogenation of Dehydroamino Acids

SCHEME 6

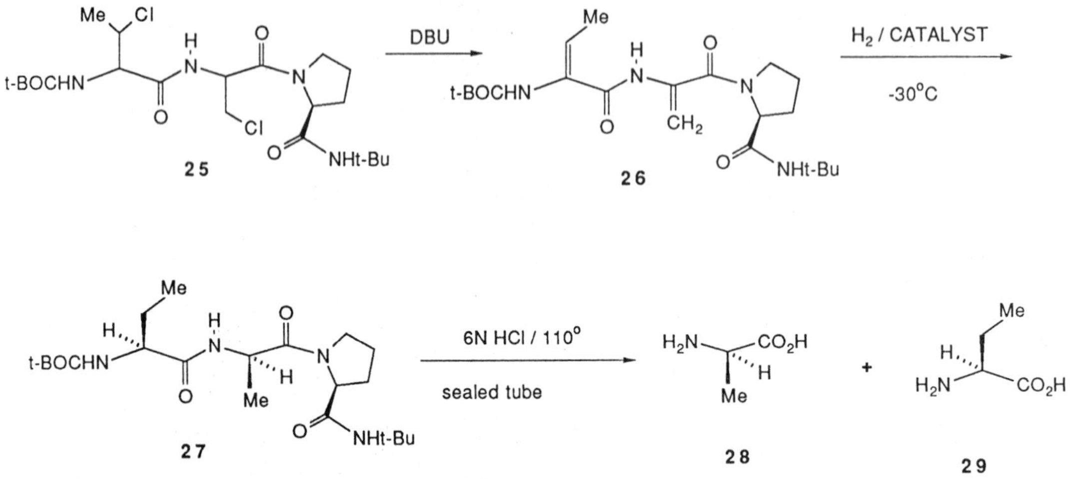

Takasaki, M.; Harada, K., *J.Chem.Soc.Chem.Comm.* (1987) 571.

CATALYST	YIELD (28)	%ee (28)	YIELD (29)	%ee (29)
W-1 Raney-Nickel	92	94	73	54
5% Pd-C	97	91	84	22
5% Pd(OH)$_2$/C	95	95	98	25
PtO$_2$	91	89	31	6

The absolute and relative stereochemistry of **31** was corroborated by homogeneous hydrogenation of (Z)-N-acetyl-dehydrophenylalanyl-3-d-(S)-alanine *tert*-butyl ester with PhCAPP-Rh+ and DIPAMP-Rh+ which provided authentic samples of both optically pure diastereomers of 31.

This strategy was further advanced [6b] with the asymmetric and diastereoselective β-lactam synthesis developed by Ikota[7a,b] and Evans[7c]. As shown in Scheme 8, the optically active oxazolidinone **32** provides a source of the chiral ketene **33** which, upon condensation with

SCHEME 7

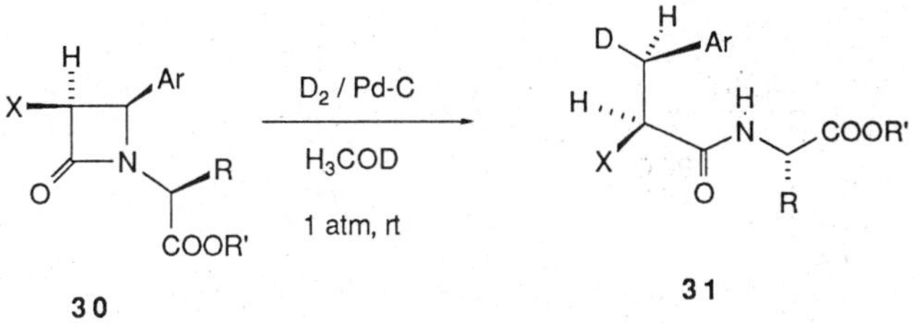

Ojima, I.; Shimizu, N., *J.Am.Chem.Soc.* (1986) **108**, 3100

SCHEME 8

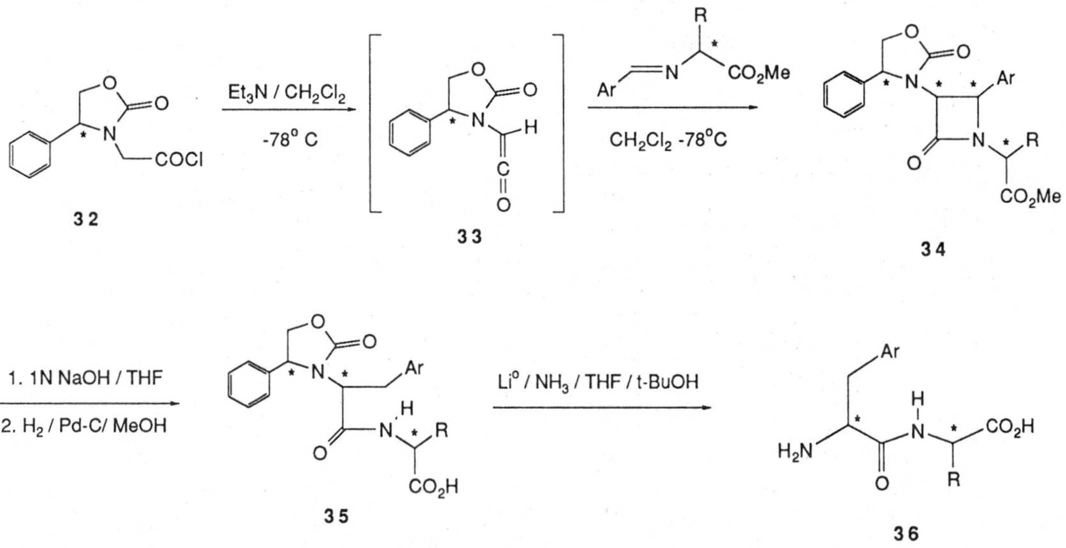

Ojima, I.; Chen, H-J, C. , *J.Chem.Soc. Chem.Comm.* (1987) 625

CONFIG (32)	R	YIELD (%34)
S	Me (S)	76
S	Me (R)	82
R	i-Pr (S)	92
R	i-Pr (R)	86
R	PhCH$_2$ (S)	91
R	MeS(CH$_2$)$_2$ (S)	79

SCHEME 9

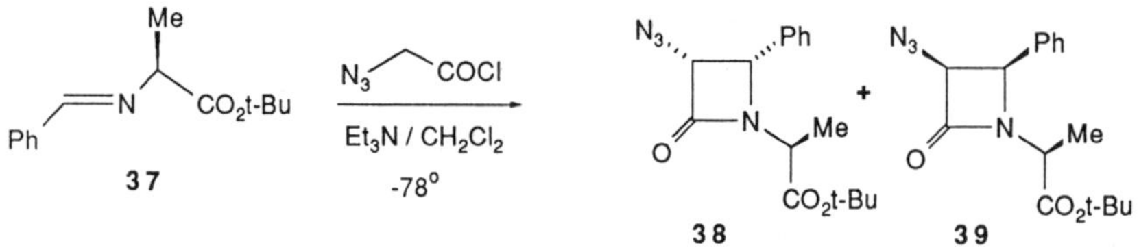

Ojima, I.; Nakahashi, K.; Brandstadter, S.M.; Hatanaka, N., *J.Am.Chem.Soc.* (1987) **109**, 1798

SCHEME 10

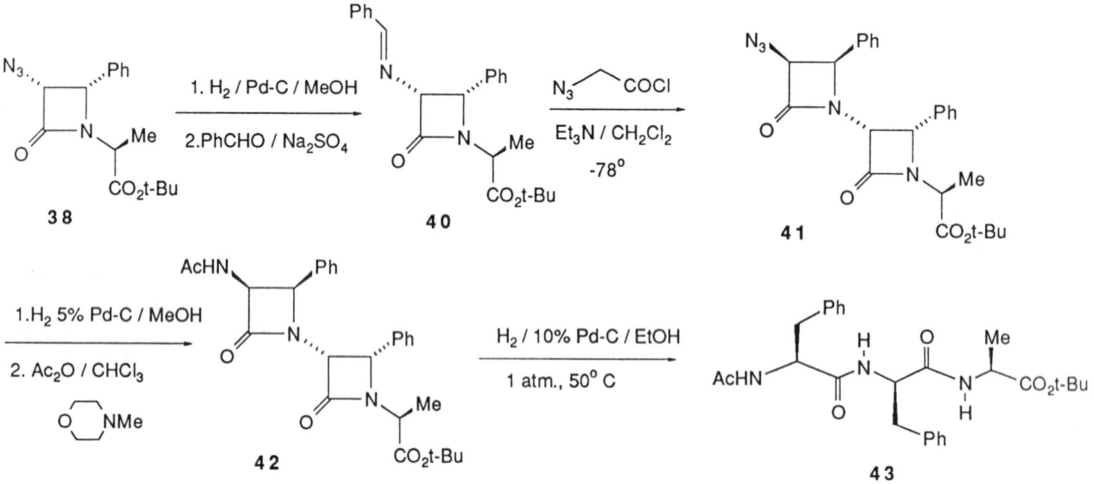

SCHEME 11

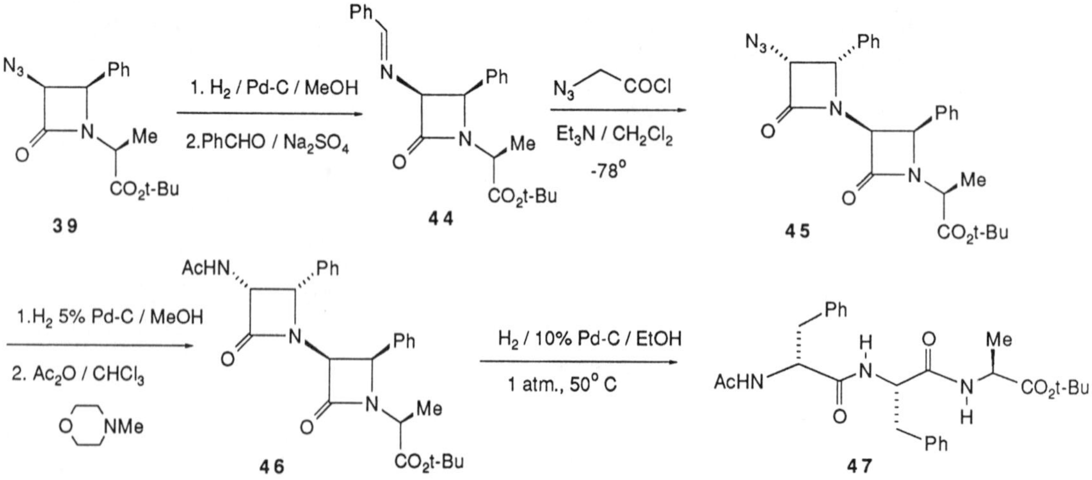

aryl aldehyde imines of amino acid methyl esters furnishes the β-lactams **34** in a highly stereocontrolled fashion. The authors note that the stereogenic center of the imine does not have any influence on the stereoselectivity of the cyclocondensation reactions; only the stereogenic center of the oxazolidinone auxilliary determines the absolute sense of asymmetric induction in the products **34**. These substances are reported to be produced in essentially diastereochemically pure form as other isomers could not be detected by HPLC and NMR. Heterogeneous hydrogenation of **34** furnishes the dipeptides **35** which can be completely reduced to the phenylalanyl dipeptides **36** through Birch reduction.

In a further permutation of this theme, these workers [8] explored the [2+2] cycloaddition reactions of chiral imines (**37**) with azidoketene as detailed in Schemes 9-11. Reaction of azidoacetyl chloride with **37** in the presence of triethylamine furnished a ~1:1 mixture of the *syn*-β-lactams **38** and **39** which were separated by silica gel chromatography. These were each in turn, reduced to the amines and condensed with benzaldehyde to afford the imines **40** (Scheme 10) and **44** (SCHEME 11), respectively. Reaction of these substances with azidoketene proceeded in a highly stereocontrolled manner (>99.5% stereoselectivity) to afford the bis-β-lactams **41** and **45**, respectively. In both cases, the newly formed *syn*-β-lactam has the opposite absolute configuration of the parent β-lactam. The authors propose that a non-bonded interaction of the β-lactam carbonyl with the oxygen atom of the incipient betaine controls the sense of asymmetric induction in the second cycloaddition. Stepwise reduction of the azide and reductive cleavage of the β-lactam benzylic residues affords the tripeptides **43** and **47**, respectively. While these systems are clearly limited to the production of phenylalnine (and derivatives) - containing peptides, these workers have ingeniously exploited the rich chemistry of modern asymmetric β-lactam syntheses to produce amino acids and derivatives in a stereocontrolled manner.

B. ASYMMETRIC HOMOGENEOUS HYDROGENATION OF DEHYDROAMINO ACIDS

The modification of Wilkinson's olefin hydrogenation catalyst with optically active phoshine ligands has constituted a field of study at the forefront of asymmetric synthesis for over a decade. This subject has been extensively studied from both a synthetic and mechanistic standpoint and numerous reviews have been published [9-11]. This section attempts to only highlight the state-of-the-art with some of the best cases reported.

The initial contribution by Knowles and associates at Monsanto culminated in the development of this reaction as a practical industrial synthesis of (S)-DOPA [17]. As shown in Scheme 12, the reaction generally

SCHEME 12

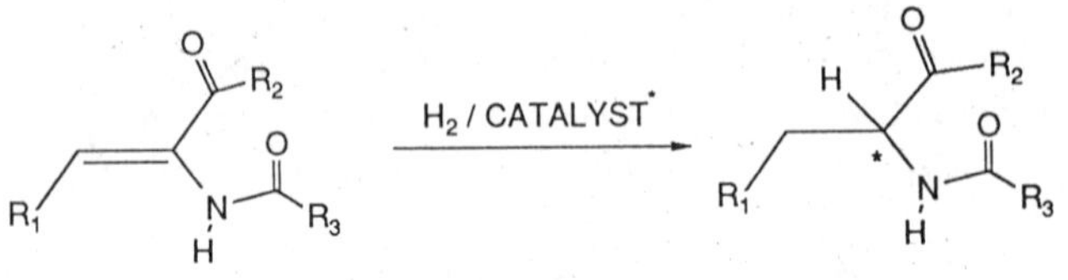

48 → 49

	% ENANTIOMERIC EXCESS		
SUBSTRATE	DIPAMP	DIOP	CHIRAPHOS
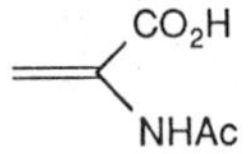 CO_2H / NHAc	94	73	91
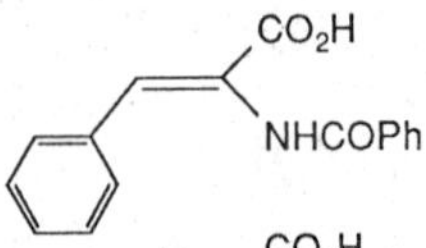 CO_2H / NHCOPh (Ph)	96	64	99
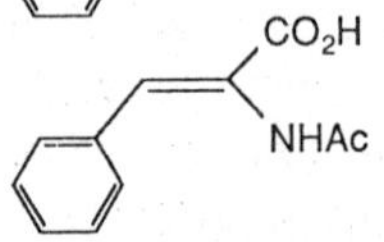 CO_2H / NHAc (Ph)	95	81	89
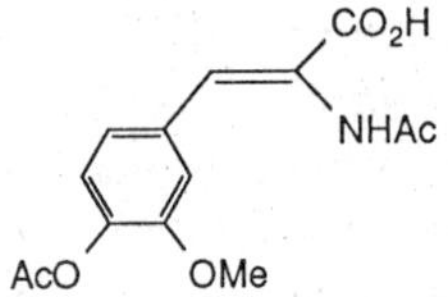 CO_2H / NHAc (AcO, OMe)	94	84	83

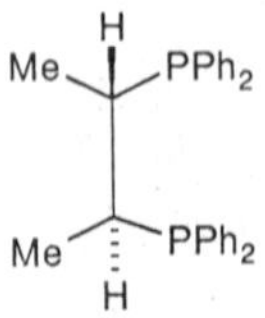

(S,S)-CHIRAPHOS

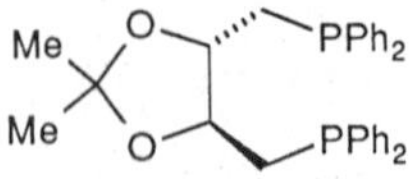

(+)-DIOP

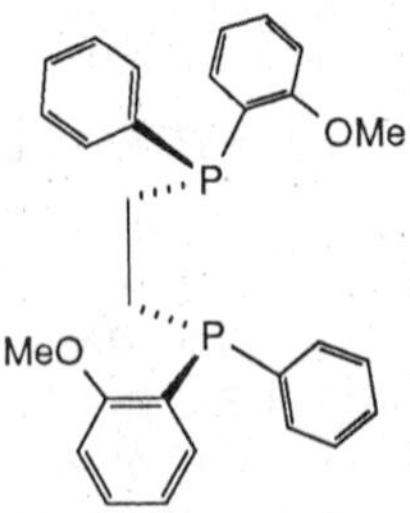

(R,R)-DIPAMP

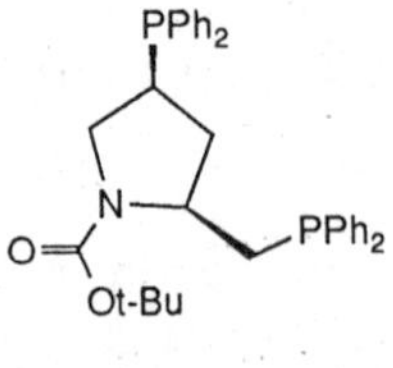

BPPM

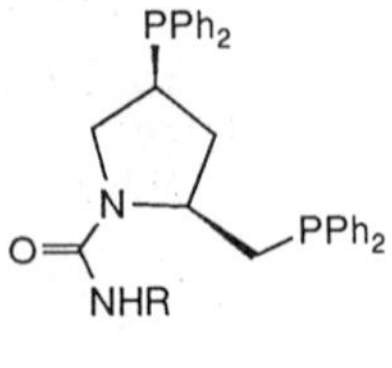

R-CAPP

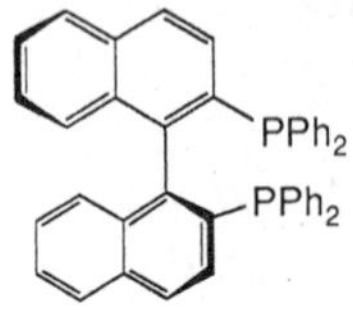

(S)-BINAP

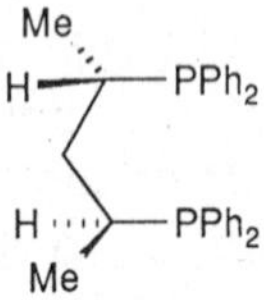

(S,S)-SKEWPHOS

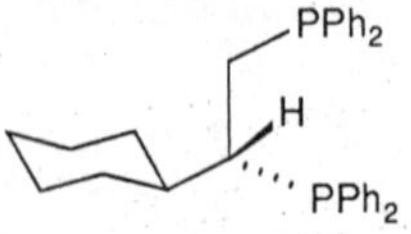

CYCPHOS

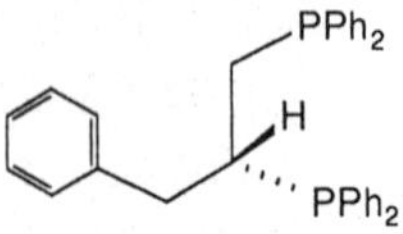

PHE-PHOS

48   **49**

ASYMMETRIC HYDROGENATION OF Z-DEHYDROALANINES

R₁	R₂	R₃	CATALYST LIGAND	%ee	ISOMER	Ref.
$(CH_3)_2CH$	OH	Me	(S,S)-CHIRAPHOS	100	R	12
Ph	OH	Me	(S,S)-CHIRAPHOS	99	R	12
H	OH	Me	(S,S,)-SKEWPHOS	99	R	13
			(S,S,)-R-CAPP	92-97	S	14
H	OH	Ph	(S)-BINAP	98	R	15
4-HO-C₆H₄	OH	Ph	(R)-CYCPHOS	98	S	16
Ph	OMe	Ph	(S)-BINAP	98	R	15
Ph	OMe	Me	(R,R)-DIPAMP	97	S	11
H	OH	CH₂Cl	(R,R)-DIPAMP	97	S	17
4-HO-C₆H₄	OH	Me	(R)-PHE-PHOS	96	R	18
n-Pr	OMe	Me	(R,R)-DIPAMP	96	S	19
$(CH_3)_2CHCH_2$	OH	Ph	(R)-CYCPHOS	96	S	16
6-Me-3-Indolyl	OEt	Me	(R,R)-DIPAMP	96	S	20
$(CH_3)_2CH$	NH₂	Ph	(R,R)-DIPAMP	95	S	17
Ph	OH	Ph	(S,S)-CHIRAPHOS	95	R	13
Ph	NH₂	Me	(R,R)-DIPAMP	94	S	17
3-YO-4-XOC₆H₃	OH	Me	(R,R)-DIPAMP	94	S	17
3-Indolyl	OH	Me	(R,R)-DIPAMP	93	S	11
MeOCH₂	OMe	Me	(S,S)-CHIRAPHOS	87	R	19
NC(CH₂)₂	OH	Ph	(R,R)-DIPAMP	85	S	17

SCHEME 13

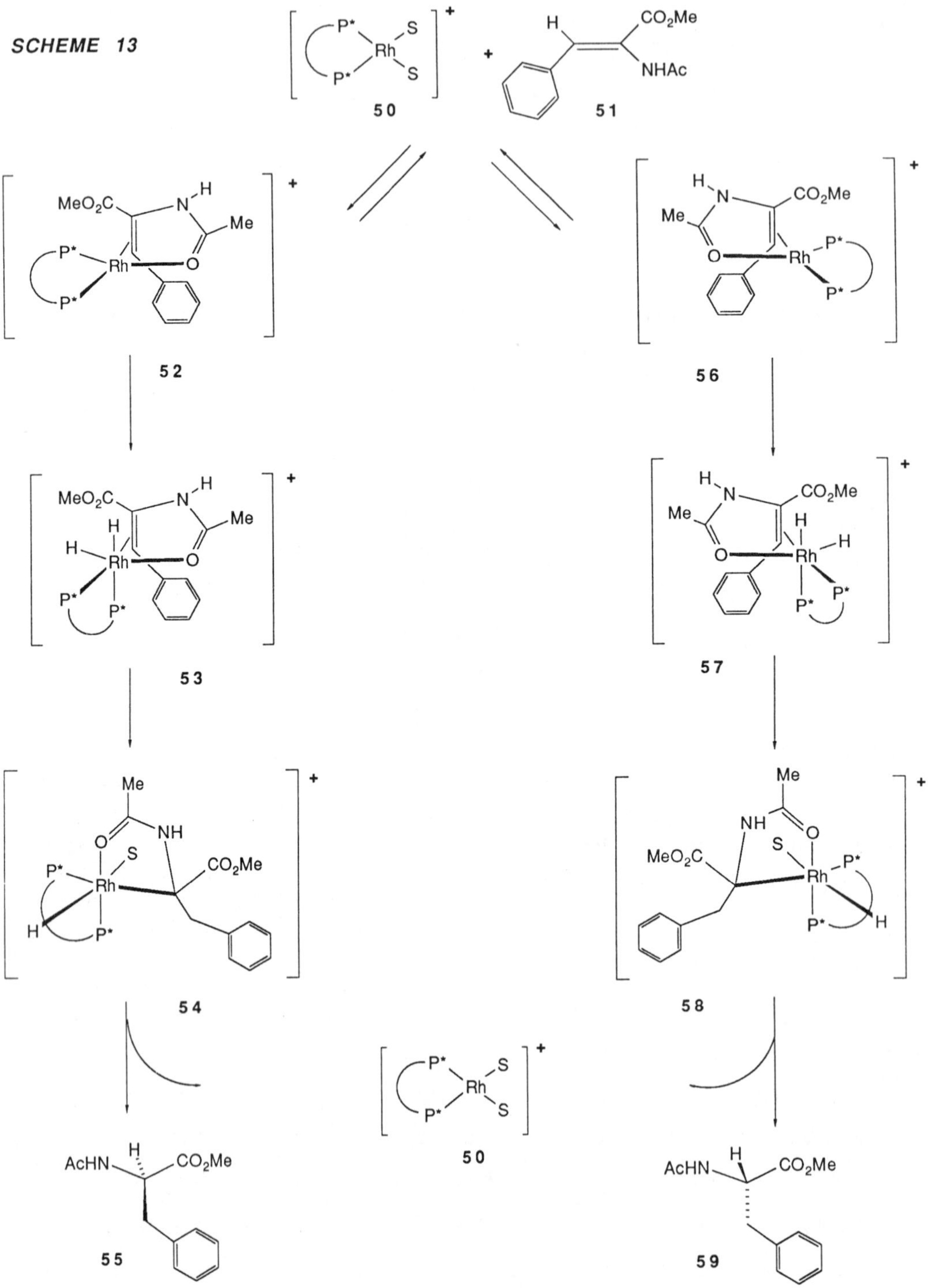

SCHEME 14

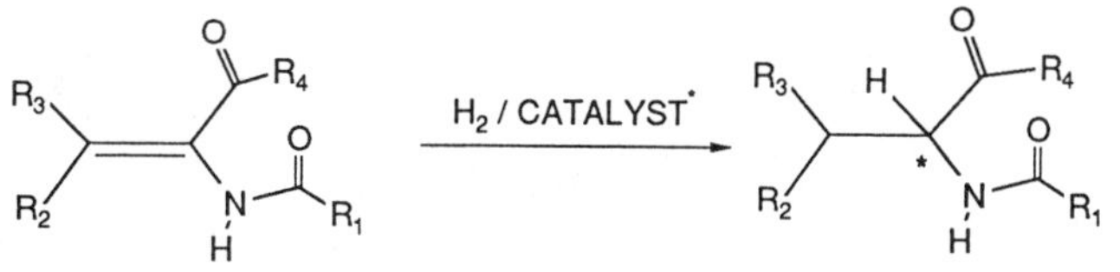

60   61

ASYMMETRIC HYDROGENATION OF E-SUBSTITUTED DEHYDROALANINES

R_1	R_2	R_3	R_4	CATALYST LIGAND	%ee	ISOMER	Ref.
Me	H	n-Pr	Me	(R,R)-DIPAMP	95	S	19
Me	H	MeOCH$_2$	Me	(R,R)-DIPAMP	94	S	19
Me	H	Ph	Me	(S)-BINAP	87	R	15
Me	H	(Me)$_2$CH	Me	(R,R)-DIPAMP	78	S	19
Me	CO$_2$Et	Me	Et	(S,S)-BPPM	58	R,R	21
Me	Me	Me	Me	(R,R)-DIPAMP	55	S	19

involves the asymmetric reduction of Z-N-acylated dehydroamino acids (**48**) with a cationic rhodium catalyst complexed to a chiral biphosphine ligand. The high %ee's have been generally restricted to the Z-isomers; elucidation of the underlying mechanistic factors for this have been the subject of a series of papers by Halpern, et.al.[10]. An extensive number of chiral biphosphines have been examined for efficacy in this process; the most commonly employed and successful systems are listed below Scheme 12. The chemical yields for these hydrogenations are often quantitative, and the turnover rates are often high. The reaction tends to be fairly insensitive to the nature of the R groups in the general structure **48** (see Tables) and various esters, free acids and amides have all been found to give excellent results. The mechanism of the reaction has been primarily worked out by Halpern [10] and is schematically detailed in Scheme 13. The first irreversible step in each of the diastereomeric manifolds is the oxidative addition of hydrogen to the coordinated olefin species (**52** to **53** and **56** to **57**, respectively); it is thus the relative rates of these competing parallel pathways that ultimately determines the enantioselectivity. E-Substituted and β,β-disubstituted dehydroamino acids generally give lower %ee's and lower yields of hydrogenation products (Table, Scheme 14).

SCHEME 15

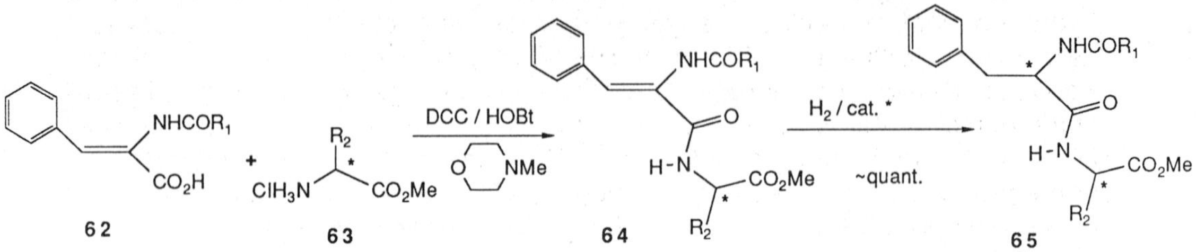

Ojima, I.; Suzuki, T., *Tetrahedron Lett.* (1980) **21**, 1239.

R$_1$	R$_2$	CATALYST	R,S : S,S
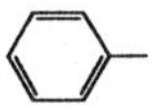 C$_6$H$_5$	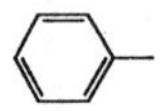 C$_6$H$_5$	p-Br-C$_6$H$_4$-CAPP- Rh$^+$	98.1 : 1.9
		BPPM-Rh^N	96.0 : 4.0
		(+)-DIOP-Rh$^+$	16.4 : 83.6
		(-)-DIOP-Rh$^+$	84.1 : 15.9
		diPAMP-Rh$^+$	2.2 : 97.8
		10% Pd-C	60 : 40
Me	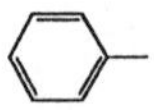 C$_6$H$_5$	BPPM-Rh$^+$	98.0 : 2.0
		BPPM-Rh^N	94.9 : 5.1
		(+)-DIOP-Rh$^+$	7.3 : 92.7
		10% Pd-C	60.2 : 39.8
Me	Me	p-Br-C$_6$H$_4$-CAPP- Rh$^+$	99.0 : 1.0
		p-Br-C$_6$H$_4$-CAPP- Rh^N	95.4 : 4.6
		(+)-DIOP-Rh$^+$	8.6 : 91.4
		10% Pd-C	61.2 : 38.8

Ojima and Suzuki [22] have examined the homogeneous asymmetric hydrogenation of several dehydrodipeptides with chiral rhodium catalysts as shown in Scheme 15. Each substrate (**64**) was hydrogenated with 10% Pd-C to derive the "intrinsic" asymmetric induction from the adjacent stereogenic center of the substrate; the diastereomer ratios were all near 60:40 with the heterogeneous catalyst. By employing the proper match of homegeneous hydrogenation catalyst with the substrate, excellent diastereoselectivities were obtainable in each case (Table) providing the dipeptides **65**. The cationic rhodium/p-Br-C_6H_4-CAPP system generally gave the superior result. The authors note that the chemical yields were nearly quantitative in each case. The substrates **64** satisfy the general requirements noted above (see structure **48**) for the dehydroamino acid substrates: 1) Z-olefin geometry; and 2) N-acyl group. The presence of the C-terminal amino acid moiety has a negligible effect on the asymmetric induction.

Kagan and associates[23] have carefully examined the asymmetric hydrogenation of numerous dehydrodi- , tri- and tetrapeptides in an effort to assess the subtle contributions of double asymmetric induction from interaction of the substrate and catalyst ligand. The approach used involves placing the the stereogenic center of a chiral amino acid on both the C-terminal and N-terminal sides of the dehydropeptide residue with varying numbers of "spacer" glycine residues (0-2) to assess the effects of remote asymmetric induction on the reaction. The substrates were prepared as illustrated in Schemes 16-19 and make extensive use of the azlactone method for coupling. By comparing the reduction of racemic dehydrodipeptides with an achiral catalyst, to that with a chiral catalyst, the intrinsic substrate contribution towards asymmetric induction was ascertained. The authors found that dehydroamino acid residues in an N-terminal position with a C-terminal chiral amino acid gives relatively modest substrate control with %de's in the 50% or less range. When the unsaturation is placed in the C-terminal position, the stereoselectivity is much higher. The outstanding example of double asymmetric induction is the reduction of Ac-(R,S)-Phe-Δ-Phe-OMe with Rh(R,R)-DIPAMP (91-99%ee). Much additional work needs to be done in this area to provide a clear and practical predictive model for matching the catalyst ligand with the intrinsic substrate stereofacial selectivity.

An elegant example of the potential of this approach was recently provided by Kagan and co-workers [24] in the asymmetric hydrogenation of a mono-dehydro Leu-enkephalin as illustrated in Scheme 21. The key dehydro tetrapeptide **90** was prepared by standard peptide coupling technology in reasonably good overall yield. The dehydro Phe moiety was introduced via the azlactone **87** prepared from phenylserine. Hydrogenation of **90** with the DIPAMP catalyst gave **91** with a diastereomeric excess of 93%. The authors note that hydrogenation with

SCHEME 16

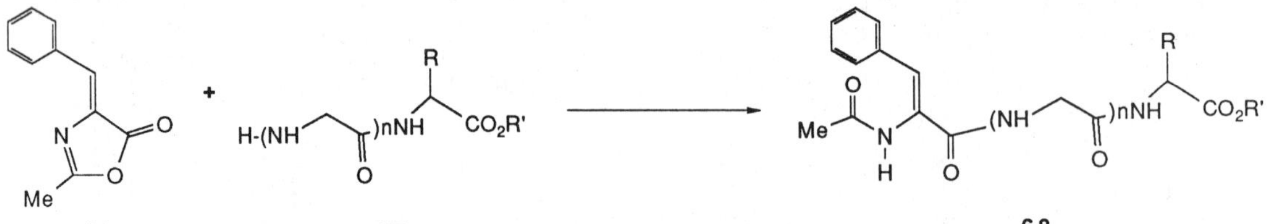

66 67 68

El-Baba, S.; Nuzillard, J.M.; Poulin, J.C.; Kagan, H.B., *Tetrahedron* (1986) **42**, 3851.

n	R	R'	YIELD (% 68)
0	CH_2 i-Pr	H	94
0	CH_2 i-Pr	Me	100
1	CH_2 i-Pr	H	63
1	CH_2 i-Pr	Me	68
2	CH_2 i-Pr	Me	75
1	H	Et	45
2	H	Et	67

SCHEME 17

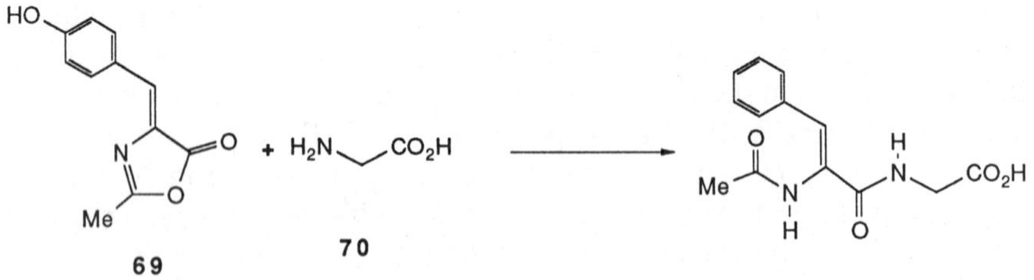

69 70 71

SCHEME 18

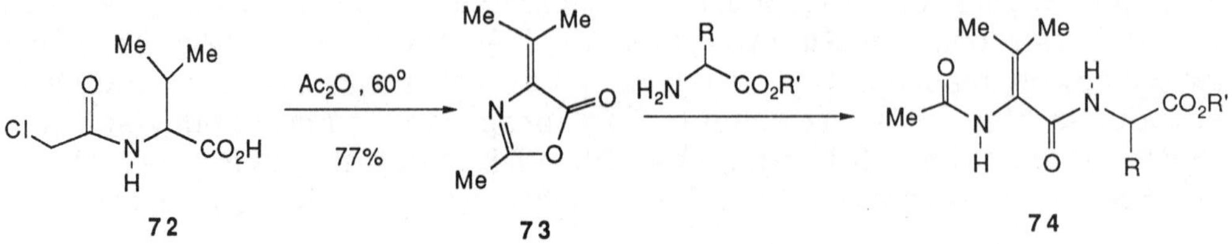

72 73 74

SCHEME 19

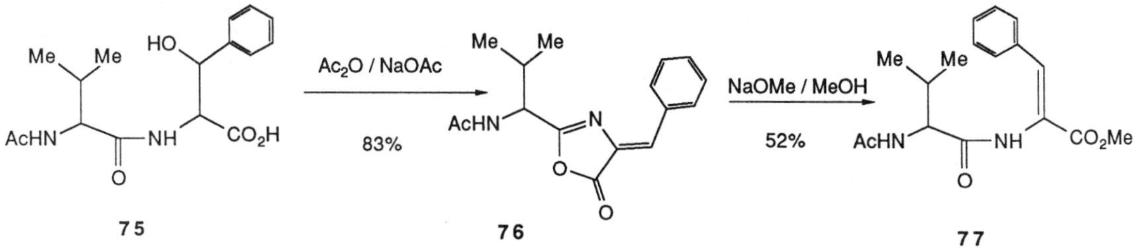

El-Baba, S.; Nuzillard, J.M.; Poulin, J.C.; Kagan, H.B., *Tetrahedron* (1986) **42**, 3851.

SUBSTRATE	CATALYST	(R,R /S,S) : (R,S / S,R)	%ee	R' / S'
Ac-(R)-Phe-Δ-Phe-OMe Ac-(S)-Phe-Δ-Phe-OMe	Rh(S,S)-DIOPCl	25.5 : 74.5	38.2 59.8	1 : 2.2 4.0 : 1
Ac-(R)-Phe-Δ-Phe-OMe Ac-(S)-Phe-Δ-Phe-OMe	Rh(S,S)-BPPMCl	48.9 : 51.1	45.8 50.0	2.7 : 1 3.0 : 1
Ac-(R)-Phe-Δ-Phe-OMe Ac-(S)-Phe-Δ-Phe-OMe	Rh(R,R)-DIPAMP[+]	51.8 : 48.2	91.8 98.8	1 : 22.8 1 : 166
Ac-(R)-Phe-Δ-Phe-OMe Ac-(S)-Phe-Δ-Phe-OMe	RhPh$_2$P(CH$_2$)$_4$PPh$_2$Cl[-]	29.0 : 71.0	42.0 42.0	1 : 2.4 2.4 : 1
Ac-(R)-Val-Δ-Phe-OMe Ac-(S)-Val-Δ-Phe-OMe	Rh(S,S)-DIOPCl	24.9 : 75.1	34.4 66.0	1 : 2.0 4.9 : 1
Ac-(R)-Val-Δ-Phe-OMe Ac-(S)-Val-Δ-Phe-OMe	Rh(R,R)-DIPAMP[+]	47.2 : 52.8	71.8 60.8	1 : 6.1 1 : 4.1
Ac-(R)-Val-Δ-Phe-OMe Ac-(S)-Val-Δ-Phe-OMe	Rh(S,S)-BPPMCl	46.7 : 53.3	20.9 34.1	1.5 : 1 2.1 : 1
Ac-Δ-Val-(R)-Phe-OMe Ac-Δ-Val-(S)-Phe-OMe	Rh(R,R)-DIPAMP[+]	51.1 : 48.9	46.2 50.6	1 : 2.7 1 : 3.0

SCHEME 20

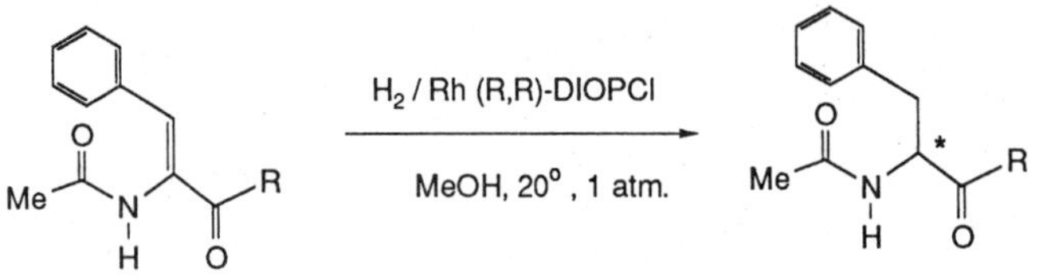

El-Baba, S.; Nuzillard, J.M.; Poulin, J.C.; Kagan, H.B., *Tetrahedron* (1986) **42**, 3851.

ASYMMETRIC HYDROGENATION OF DEHYDROPEPTIDES WITH Rh (R,R)-DIOPCI

SUBSTRATE	%ee	R'/S'
Ac-Δ-Phe-Gly-OH	64.0	4.6 : 1
Ac-Δ-Phe-(Gly)$_2$-OH	53.3	3.3 : 1
Ac-Δ-Phe-(S)-Leu-OH	57.7	3.7 : 1
Ac-Δ-Phe-Gly-(S)-Leu-OH	61.6	4.2 : 1
Ac-Δ-Phe-(Gly)$_2$-OEt	51.7	3.1 : 1
Ac-Δ-Phe-(Gly)$_3$-OEt	51.2	3.1 : 1
Ac-Δ-Phe-(R)-Leu-OMe	63.9	4.5 : 1
Ac-Δ-Phe-(S)-Leu-OMe	13.7	1.3 : 1
Ac-Δ-Phe-Gly-(R)-Leu-OMe	37.7	2.2 : 1
Ac-Δ-Phe-Gly-(S)-Leu-OMe	60.5	4.1 : 1
Ac-Δ-Phe-Gly$_2$-(R)-Leu-OMe	65.7	4.8 : 1
Ac-Δ-Phe-Gly$_2$-(S)-Leu-OMe	40.9	2.4 : 1

SCHEME 21

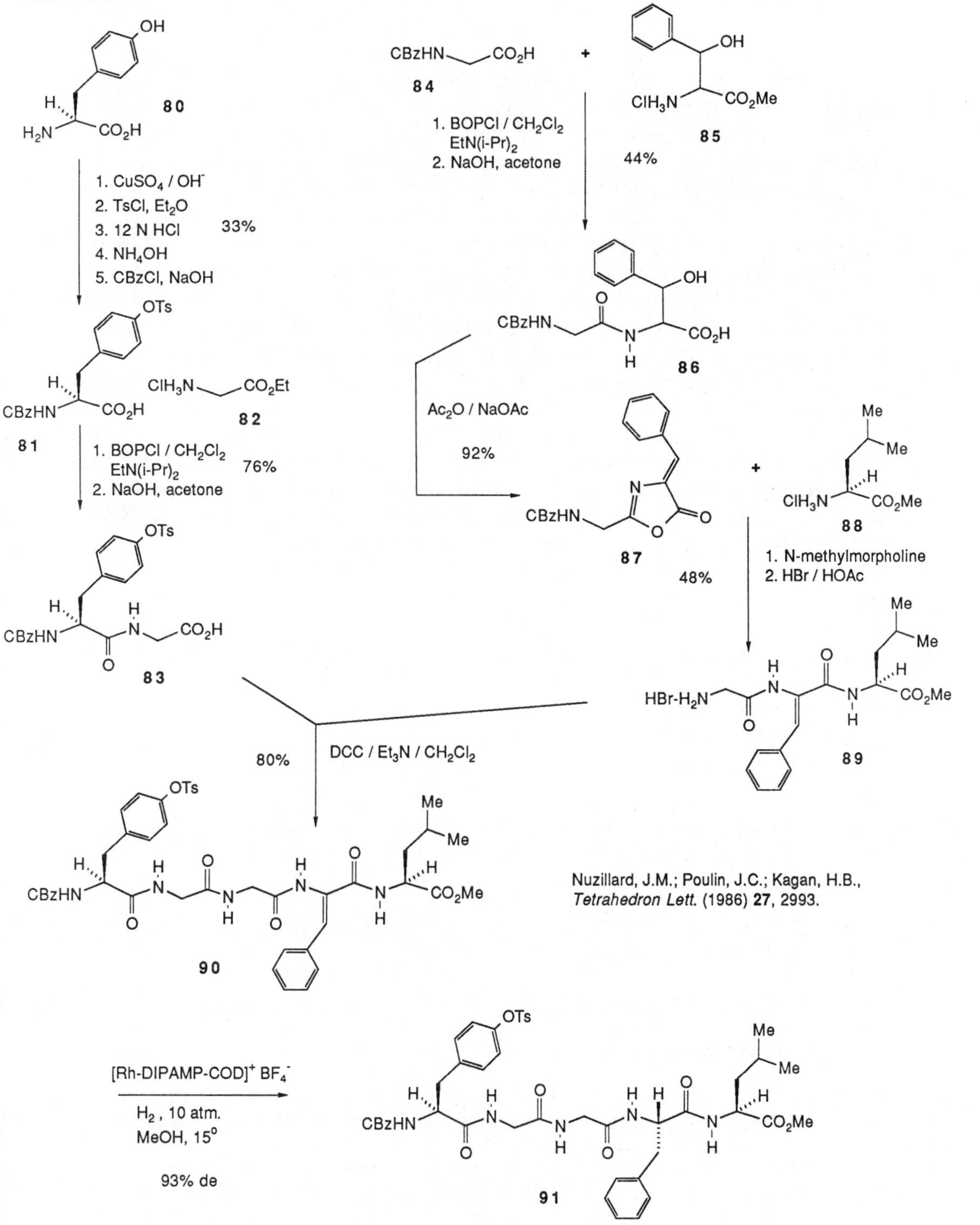

Nuzillard, J.M.; Poulin, J.C.; Kagan, H.B.,
Tetrahedron Lett. (1986) **27**, 2993.

RhCl(-)-BPPM gave, in quantitative yield, the opposite facial selectivity (R) but in only 63%de. These results are rationalized in terms of "catalyst control" of the reductions which overcomes the intrinsic asymmetric induction of the two chiral amino acids present in the substrate.

More elaborate applications of the homogeneous hydrogenations have appeared in the literature and in patents. Two recent examples are illustrated below in Schemes 22 and 23. Melillo and associates at Merck[25] have examined the synthesis of 4-O-methyl homotyrosine by homogeneous catalytic hydrogenation. As shown in Scheme 22, the requisite dehydroamino acid derivative (**94**) was prepared from the α-keto acid **93** in high yield. Hydrogenation of this substance with a Rh-CHIRAPHOS catalyst gave a selectivity of 9 : 1 providing the amino acid **95** (80%ee) . A single recrystallization of this substance from hexanes / ethyl acetate gave **95** in >98% ee in 80% yield from **94**. The overall yield of this sequence (58%) was slightly lower than an alternative sequence to the same substrate from aspartic acid (80% overall yield). Full experimental details accompany both routes to **95** (>100gm scale). This amino acid was required for the synthesis of the dopamine agonist **96**.

SCHEME 22

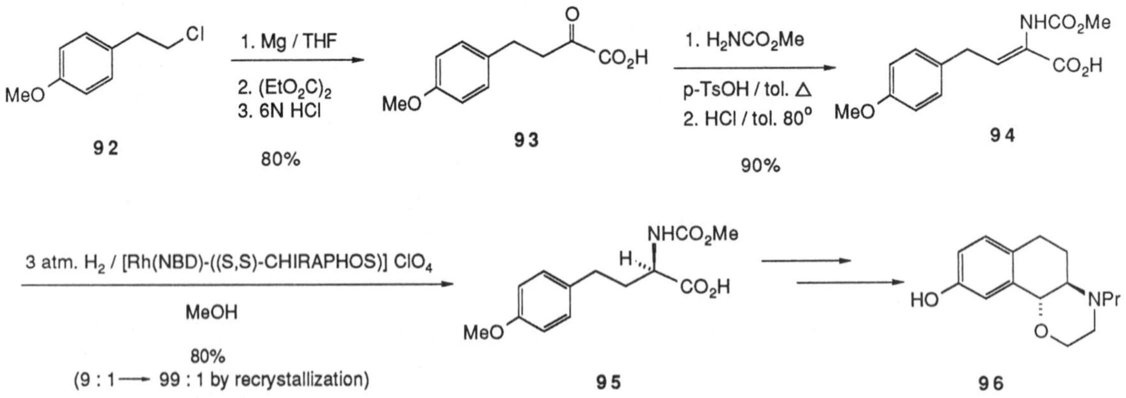

Melillo, D.G.; Larsen, R.D.; Mathre, D.J.; Shukis, W.F.; Wood, A.W.; Colleluori, J.R., *J.Org.Chem.* (1987) **52**, 5143.

An elegant, complex application of the asymmetric homogeneous hydrogenation reaction was reported by Schmidt and associates [26] in the total synthesis of the cyclopeptide aminopeptidase inhibitor OF4949-III (**110**) as shown in Scheme 23. The requisite Z- dehydroamino acid residues were prepared by aryl aldehyde condensations with phosphorylglycine [27] derivatives **98** and **102**. The authors report that the hydrogenations of **99**, **103** and **106** all proceed with >98%ee using the Rh-DIPAMP catalyst system in excellent yields. Ullmann couplings are employed to construct the biaryl ether **101** which is subsequently converted into the highly functionalized dehydroamino acid derivatives. This synthesis nicely demonstrates that the homogeneous asymmetric hydrogenation reaction is tolerant of various functionality including N-carbobenzoxy, N-*tert*-butyloxycarbonyl and benzyl esters; the N-CBz and benzyl esters are both reducible protecting groups.

Schmidt, et. al.[27], have developed a very useful protocol for preparing a wide variety of dehydroamino acids utilizing N-acylated-2-(dialkoxyphosphinyl)-glycine esters (**115**) as shown in Scheme 24. Since the dehydroamino acids (**116**) obtained through olefination of **115** with various aldehydes can all, in principle, be hydrogenated to optically active amino acids through asymmetric homogeneous hydrogenation, the wide variety of phosphonates (**115**) reported makes this a very useful complement to the azlactone method alluded to above. Condensation of **115** with a variety of aldehydes (R_4CHO) in the presence of a base (either NaH; LDA; or KOt-Bu) affords predominantly, the Z-substituted substances **116**. The method is directly applicable to incorporating other amino acid residues as the N-terminal "acyl" group and seems quite tolerant of several types of functionality.

It might be noted in closing, that the heterogeneous hydrogenation of chiral imines, oximes and hydrazones of α-keto acids has been extensively studied and reviewed [5c]. The enantiomeric excesses obtained by this approach have, in general been quite disappointing with ranges from 0~low 80% ee in the very best cases. This is primarily due to the conformational mobility of the chiral auxilliary comprising the imine moiety. The *cyclic* modification of this approach developed by Corey [3] was illustrated above in Scheme 3 and displayed significantly better results. It is surprising that additional extensions of *cyclic* versions of this general strategy that would undoubtedly enjoy improved stereocontrol have not emerged as complementary methodologies to prepare optically active amino acids. Due to the generally low selectivities reported for the acyclic systems, the reader is referred to the review [5c] on this subject by Harada; details will not be presented in this monograph.

SCHEME 23

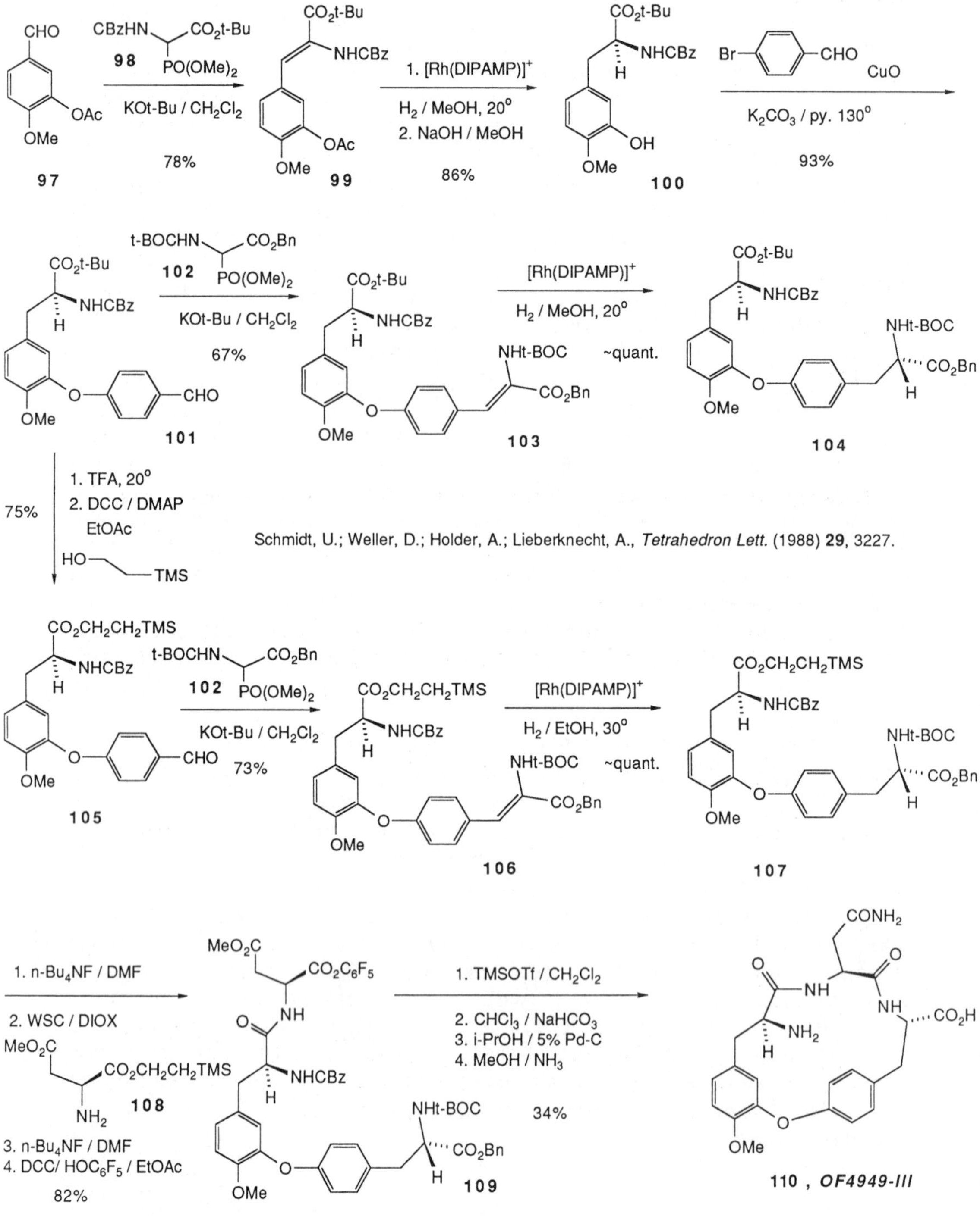

Schmidt, U.; Weller, D.; Holder, A.; Lieberknecht, A., *Tetrahedron Lett.* (1988) **29**, 3227.

SCHEME 24

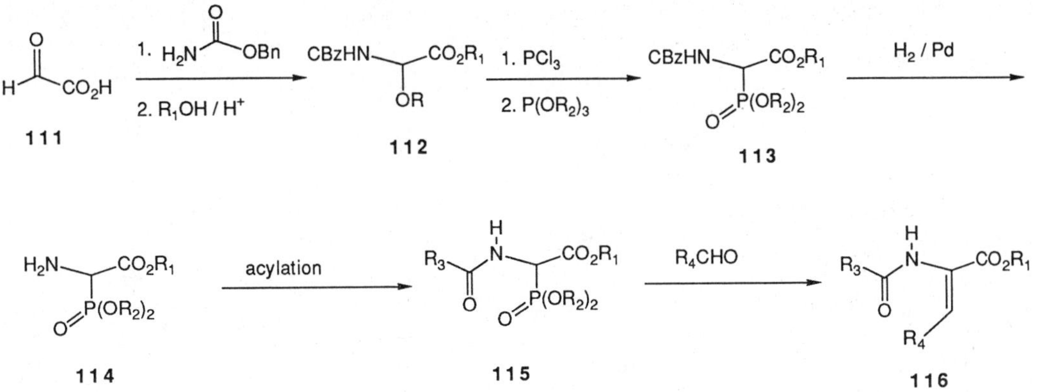

Schmidt, U.; Lieberknecht, A.; Wild, J., *Synthesis* (1984) 53.

R$_1$	R$_2$	R$_3$	YIELD (% 115)
Me	Me	BnO	80
Me	Me	t-BuO	80
Me	Me	Me	91
Me	Me	ClCH$_2$	82
Me	Me	H	81
t-Bu	Et	BnO	80
Me	Me	t-BOC-(S)-Leu	85
Me	Me	Ac-(S)-Leu	78
Et	Et	t-BOC-(S)-Ile	85
t-Bu	Et	t-BOC-(R,S)-Trp	80
H	Et	BnO	90
H	Me	BnO	84

R_1	R_3	R_4	BASE	Z : E RATIO	YIELD (%116)
Et	BnO	phenyl	NaH	1.1	81
Et	t-BuO	phenyl	KOt-Bu	2.5	88
Et	BnO	2-methoxyphenyl (OMe)	NaH	1.8	83
Me	H	MeO—phenyl (4-methoxyphenyl)	KOt-Bu	>20	79
Et	BnO	3-methyl-indole-Nt-BOC	LDA	1.5	84
Me	BnO	3-methyl-indole-Nt-BOC	KOt-Bu	>10	74
Me	Me	3-methyl-indole-Nt-BOC	KOt-Bu	>50	86
Me	Me	imidazole, t-BOCN	KOt-Bu	>20	55
Me	Me	furan	KOt-Bu	>50	82
Me	ClCH$_2$	furan	KOt-Bu	>50	93
Et	BnO	n-Pr	LDA	2.0	81
Me	BnO	Et	KOt-Bu	20	86
Et	t-BuO	n-Pr	LDA	1.3	77
Me	Me	n-Pr	KOt-Bu	>10	84
Me	Me	Cl—(CH$_2$)—dioxolane—pentyl	KOt-Bu	>50	78
Me	BnO	Cl—(CH$_2$)—dioxolane—pentyl	LDA	1.3	84
Me	Me	pentenyl	KOt-Bu	>20	84
Et	t-BOC-(S)-Leu	MeO—phenyl	LDA	0.35	60
t-Bu	t-BOC-(R,S)-Trp	(MeO$_2$CO)$_3$-phenyl	LDA	-	74
Me	Ac-(S)-Leu	(AcO)$_3$-phenyl	LDA	2.0	83

References Chapter 6

1. a) Vigneron, J.P.; Kagan, H.; Horeau, A., *Tetrahedron Lett.* (1968) 5681; b) both enantiomers of *erythro* 2-amino-1,2-diphenylethanol (1) are now commercially available; c) Weijlard, J.; Pfister, K.; Swanezy, E.F.; Robinson, C.A.; Tishler, M., *J.Am.Chem.Soc.* (1951) **73**, 1216.

2. Tamura, M.; Harada, K., *Bull.Chem.Soc. Jpn.* (1980) **53**, 561.

3. a) Corey, E.J.; McCaully, R.J.; Sachdev, H.S., *J.Am.Chem.Soc.* (1970) **92**, 2476; b) Corey, E.J.; Sachdev, H.S.; Gougoutas, J.Z.; Saenger, W., *J.Am.Chem.Soc.* (1970) **92**, 2488.

4. a) Kanmera, T.; Lee, S.; Aoyagi, H.; Izumiya, N., *Tetrahedron Lett.* (1979) 4483; b) Izumiya, N.; Lee, S.; Kanmera, T.; Aoyagi, H., *J.Am.Chem.Soc.* (1977) **99**, 8346; c) Lee, S.; Kanmera, T.; Aoyagi, H.; Izumiya, N., *Int.J.Peptide Protein Res.* (1979) **13**, 207; d) Kanmera, T.; Lee, S.; Aoyagi, H.; Izumiya, N. *Int.J.Peptide Protein Res.* (1980) **16**, 280; e) Gallina, C.; Liberatori, A., *Tetrahedron* (1974) **30**, 667; f) for a review on dehydroamino acids, see: Schmidt, U.; Hausler, J.; Ohler, E.; Poisel, H., in *Progress in the Chemistry of Organic Natural Products* , Herz, W.; Geisebach, H.; Kirby, G.W., Eds., Springer-Verlag (1979) **37**, 251.

5. a)Takasaki, M.; Harada, K., *Chemistry Lett.* (1984) 1745; b) Takasaki, M.; Harada, K., *J.Chem.Soc.Chem.Comm.* (1987) 571; c) for a review of this topic, see: Harada, K.,*Asymmetric Synthesis,* Vol. 5 (Chiral Catalysis), Academic Press , Orlando (1985) Chap.10.

6. a) Ojima, I.; Shimizu, N., *J.Am.Chem.Soc.* (1986) **108**, 3100; b) Ojima, I.; Chen, H-J.C., *J.Chem.Soc.Chem.Comm.* (1987) 625.

7. a)Ikota, N.; Hanaki, A., *Heterocycles* (1984) **22**, 2227; b) Ikota, N.; Hanaki, A., *Hetrocycles* (1984) **22**, 418; c) Evans, D.A.; Sjogren, E.B., *Tetrahedron Lett.* (1985) **26**, 3783.

8. Ojima, I.; Nakahashi, K.; Brandstadter, S.M.; Hatanaka, N., *J.Am.Chem.Soc.* (1987) **109**, 1798.

9. *Asymmetric Catalysis*, Bosnich, B, Ed., Martinus Nijhoff, Dordrecht (1986), Chap.2.

10. a) Halpern, J., *Inorganica Chim.Acta.* (1981) **50**, 11; b) Halpern, J., *Pure Appl.Chem.* (1983) **55**, 99; c) Halpern, J., *Asymmetric Synthesis,* Vol. 5 (Chiral Catalysis), Academic Press , Orlando (1985) Chap.2.

11. Koenig, K.E., *Asymmetric Synthesis,* Vol. 5 (Chiral Catalysis), Academic Press , Orlando (1985) Chap.3.

12. Fryzuk, M.D.; Bosnich, B., *J.Am.Chem.Soc.* (1977) **99**, 6262.

13. Bosnich, B.; Roberts, N.K., in *Catalytic Aspects of Metal Phosphine Complexes*, Chap. 21, p337, Am.Chem.Soc. (1982) Washington, D.C.

14. Ojima, I.; Yoda, N., *Tetrahedron Lett.* (1980) **21**, 1051.

15. Miyashita, A.; Yasuda, A.; Takaya, H.; Toriumi, K.; Ito, T.; Souchi, T.; Noyori, R., *J.Am.Chem.Soc.* (1980) **102**, 7932.

16. Oliver, J.D.; Riley, D.P., *Organometallics* (1983) **2**, 1032.

17. a) Vineyard, B.D.; Knowles, W.S.; Sabacky, M.J., *J.Mol.Catal.* (1983) **19**, 159; b) Vineyard, B.D.; Knowles, W.S.; Sabacky, M.J.; Bachman, G.L.; Weinkauff, D.J., *J.Am.Chem.Soc.* (1977) **99**, 5946; c) Knowles, W.S.; Sabacky, B.D.; Vineyard, B.D.; Weinkauff, D.J., *J.Am.Chem.Soc.* (1975) **97**, 2567.

18. Bergstein, W.; Kleemann, A.; Martens, J. *Synthesis* (1981) 76.

19. Scott, J.W.; Keith, D.D.; Nix, G.; Parrish, D.R.; Remington, S.; Roth, G.P.; Townsend, J.M.; Valentine, D.; Yang, R., *J.Org.Chem.* (1981) **46**, 5086.

20. Hengartner, U.; Valentine, D.; Johnson, K.K.; Larscheid, M.E.; Pigott, F.; Scheidl, F.; Scott, J.W.; Sun, R.C.; Townsend, J.M.; Williams, T.H., *J.Org.Chem.* (1979) **44**, 3741.

21. Achiwa, K., *Tetrahedron Lett.* (1978) 2583.

22. Ojima, I.; Suzuki, T., *Tetrahedron Lett.* (1980) **21**, 1239.

23. El-Baba, S.; Nuzillard, J.M.; Poulin, J.C.; Kagan, H.B., *Tetrahedron* (1986) **42**, 3851.

24. Nuzillard, J.M.; Poulin, J.C.; Kagan, H.B., *Tetrahedron Lett.* (1986) **27**, 2993.

25. Melillo, D.G.; Larsen, R.D.; Mathre, D.J.; Shukis, W.F.; Wood, A.W.; Colleluori, J.R., *J.Org.Chem.* (1987) **52**, 5143.

26. Schmidt, U.; Weller, D.; Holder, A.; Lieberknecht, A., *Tetrahedron Lett.* (1988) **29**, 3227.

27. a) Schmidt, U.; Lieberknecht,A.; Wild, J., *Synthesis* (1984) 53, and references cited therein; b) for a related study, see: Kober, R.; Steglich, W., *Liebigs Ann.Chem.* (1983) 599; c)for a recent review, see: Schmidt, U.; Lieberknecht, A.; Wild, J., *Synthesis* (1988) 159.

CHAPTER 7

ENZYMATIC SYNTHESES OF α– AMINO ACIDS

Syntheses of amino acids with purified, immobilized enzymes; cell-free enzyme preparations ; and whole cell systems have become an important commercial method to prepare amino acids on both large and intermediate scales. [1] Two major types of reactions can be used to classify the synthesis of optically active amino acids using biotransformation:

A. Enzymatic resolution of racemic amino acid derivatives.
B. Asymmetric bond-forming reactions on prochiral substrates catalyzed by an enzymatic system.

These can be further subdivided into the specific enzyme systems that perform the reactions to be discussed below:

A. Enzymatic resolution of racemic amino acid derivatives.
    a) amidases
    b) hydantoinases
    c) acylases
    d) esterases
    e) nitrilases and nitrile hydratases
B. Asymmetric bond-forming reactions on prochiral substrates catalyzed by an enzymatic system.
    a) transaminases
    b) ammonia lyases

Some of the requirements that a biocatalytic system should meet to be of serious utility include:
1. Good availability of the biocatalyst
2. Reasonable cost
3. Stability of the catalyst
4. High optical purity of the final product(s)

    5. Desireable substrate specificity
In addition, the substrate must be readily accessible, inexpensive and the process must meet certain standards for ease of operation, separation and purification of the products, as well as ease of recovery and re-use of the biocatalyst.

    This chapter briefly outlines some recent examples of enzyme-based syntheses of amino acids. Unfortunately, much of the detailed literature on the experimental procedures to perform these syntheses are obscured in the patent literature. Proprietary cell lines have been developed specifically for certain commercial applications; information and availability of the relevant technologies are not always accessible in the public domain. This chapter is necessarily brief for this reason; additionally, a forthcoming monograph in the *Organic Chemistry Series* , "Enzymes in Synthetic Organic Chemistry", will cover this topic in more depth.

    The accompanying Chart details some recent, known amino acids that are produced industrially using biocatalytic systems. Some of these will be touched on below.

A. ENZYMATIC RESOLUTION OF RACEMIC AMINO ACID DERIVATIVES

    The ease of preparing racemic α-amino amides by the Strecker synthesis has made the use of selective amidases in effecting a kinetic resolution of considerable importance. A Dutch group[2] has extensively developed this approach in the synthesis of numerous optically active amino acids as illustrated in Scheme 1. Racemization-prone phenylglycine has been prepared via the efficient hydrolysis of d,l-phenylglycine amide (3) with an L-specific aminopeptidase from *Pseudomonas putida*. This process produces in quantitative yield, L-phenylglycine (4) and D-phenylglycine amide (5). The authors report that the crude enzyme preparation displays nearly 100% stereoselectivity in hydrolyzing the L-amide to the L-acid. The two components (4 and 5) are separated by Schiff base formation with benzaldehyde; the Schiff base of the amide (5) was fortuitously found to be completely insoluble in water and is simply filtered off. After acidic hydrolysis at 100°C, the D-amino acid (7) is obtained without racemization. It is noted that the amino acid amide must contain an α-H atom and have an unsubstituted amino group to be a substrate for the aminopeptidase of *Pseudomonas putida*. Thus, both N-alkyl amino acid derivatives and the interesting α-substituted amino acids can not be prepared by this method.   D-Phenylglycine (7) is a side chain precursor to ampicillin and enjoys a significant world-wide demand. The unwanted L-isomer (4) does not have such a large market demand and

Some current, commercially operated biocatalytic syntheses of α-amino acids

SUBSTRATE	PRODUCT	BIOCATALYST
		IMMOBILIZED *E.Coli* CELLS (Aspartase)
		IMMOBILIZED AMINOACYLASE (*Aspergillus oryzae*)
		Candida Humicola CELLS (L-Aminocaprolactam Hydrolase) *Alcaligenes feacalis* CELLS (D-Aminocaprolactam racemase)
		Bacillus brevis CELLS (D-Hydantoinase)
		IMMOBILIZED *Ps. dacunhae* CELLS (L-Aspartate β-decarboxylase)
		IMMOBILIZED *Rhodotorula rubra* CELLS (Phenylalanine Ammmonialyase)
		IMMOBILIZED CELLS (Transaminase)
		Tryptophan Synthetase
		Ps. thiazolinophilum CELLS

SCHEME 1

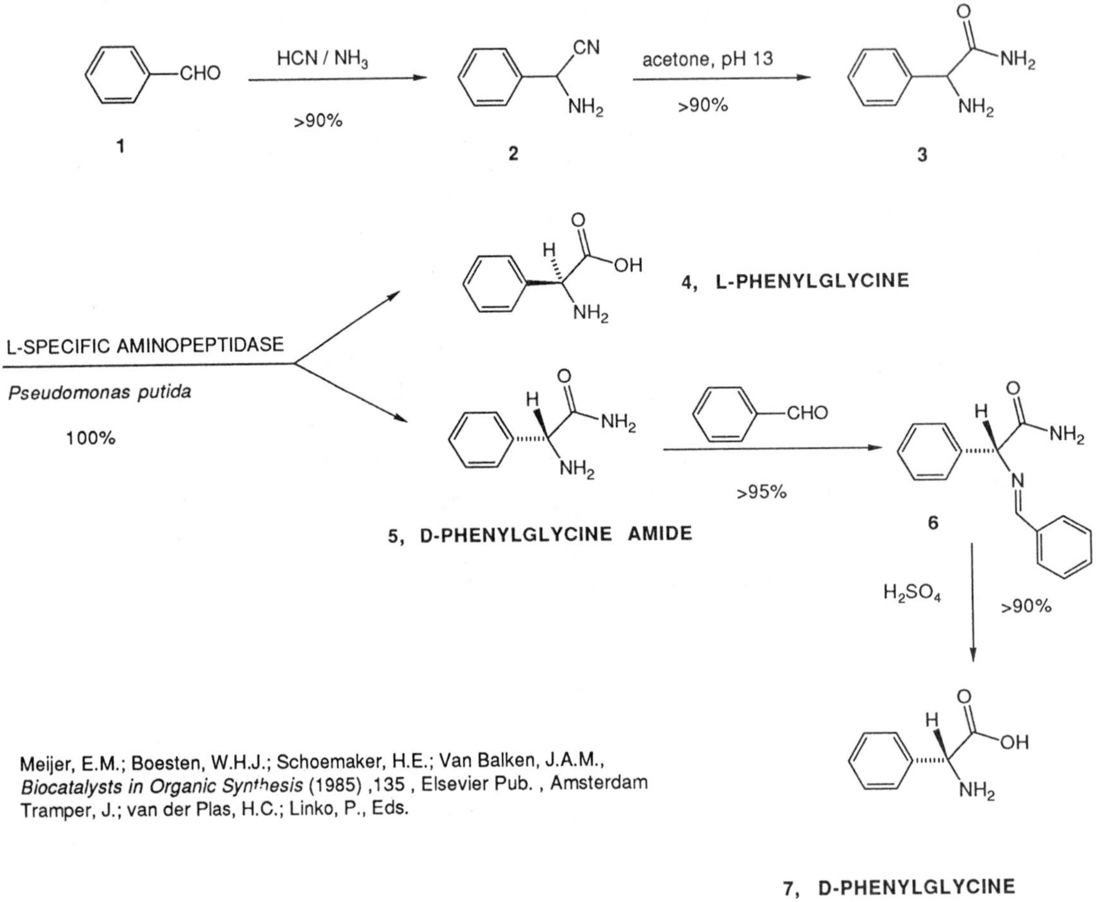

Meijer, E.M.; Boesten, W.H.J.; Schoemaker, H.E.; Van Balken, J.A.M.,
Biocatalysts in Organic Synthesis (1985) ,135 , Elsevier Pub. , Amsterdam
Tramper, J.; van der Plas, H.C.; Linko, P., Eds.

is therefore racemized , esterified to the methyl ester, subjected to amminolysis forming d,l-**3** and is reused in the above resolution.

It has been recently reported [3] that a new aminopeptidase from *Mycobacterium neoaurum* ATCC 25795 is capable of hydrolyzing a range of α-substituted L-amino acid amides stereoselectively. A shown in Scheme 2, the α-substituted d,l-amino acid amides (**8**) are treated with whole cells of *Mycobacterium neoaurum* in water at 37°C. The two components produced, the L-acid (**9**) and the D-amide (**10**) are separated by differences in solubility in an organic solvent such as chloroform (the acid is insoluble in chloroform). These workers also report a fast and simple ^1H NMR method to determine the %ees' on the product amino acids

SCHEME 2

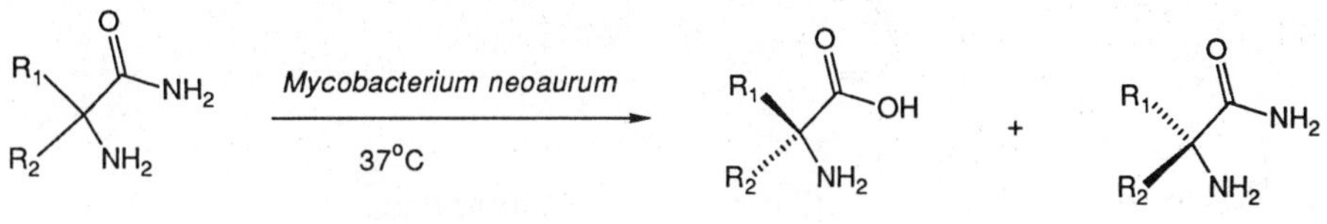

8 → 9, L-ACID + 10, D-AMIDE

Mycobacterium neoaurum

37°C

Kruizinga, W.H.; Bolster, J.; Kellogg, R.M.; Kamphius, J.; Boesten, W.H.J.; Meijer, E.M.; Schoemaker, H., *J.Org.Chem.* (1988) **53**, 1826.

R_1	R_2
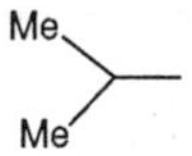 Me₂CH— (isopropyl)	Me
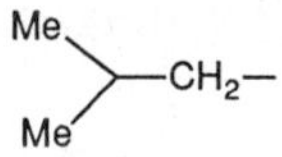 Me₂CH—CH₂—	Me
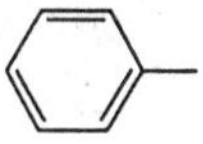 C₆H₅—	Me
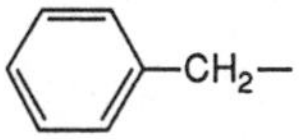 C₆H₅—CH₂—	Me
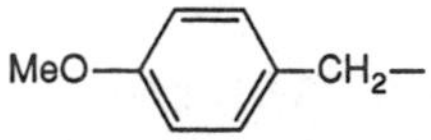 MeO—C₆H₄—CH₂—	Me
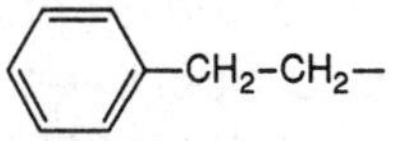 C₆H₅—CH₂—CH₂—	Me
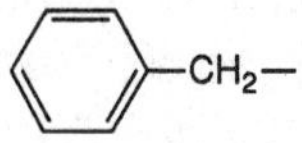 C₆H₅—CH₂—	Et

by acylation with (S)-2-chloropropionyl chloride. Resolvable doublets for the *CH₃*CHCl residue can be accurately integrated to within ± 2%. The microfilm edition of this *J.Org.Chem.* paper contains complete experimental details.

SCHEME 3

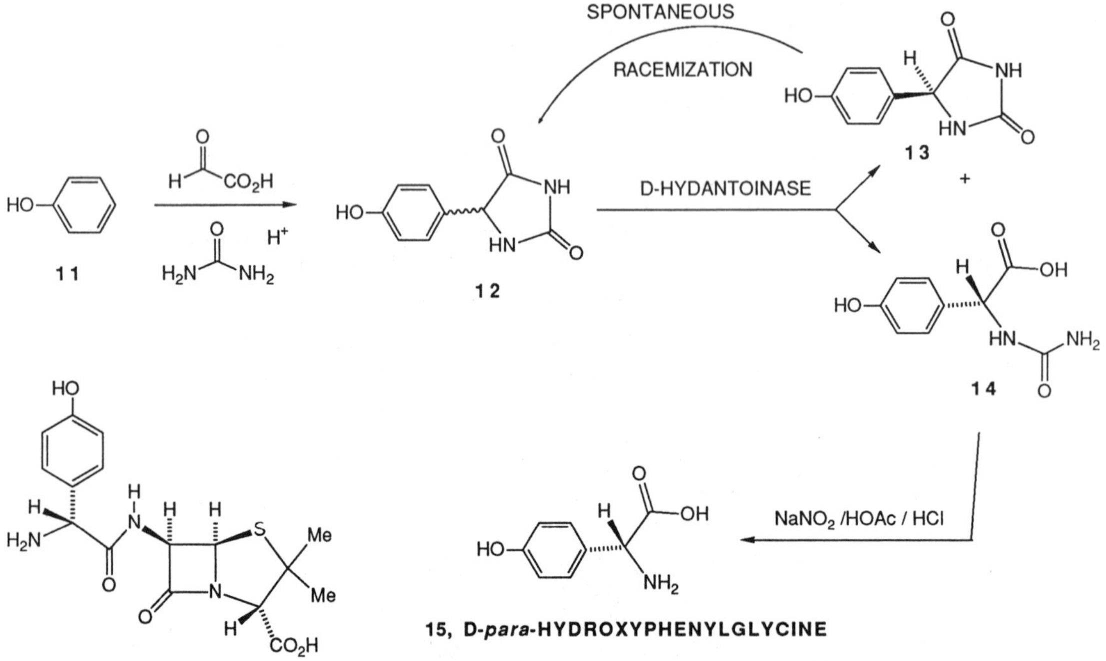

16, AMOXICILLIN

15, D-para-HYDROXYPHENYLGLYCINE

Meijer, E.M.; Boesten, W.H.J.; Schoemaker, H.E.; Van Balken, J.A.M., *Biocatalysts in Organic Synthesis* (1985) ,135 , Elsevier Pub. , Amsterdam Tramper, J.; van der Plas, H.C.; Linko, P., Eds.

The β-lactam antibiotic amoxicillin (**16**) contains the amino acid D-*para*-hydroxyphenylglycine (**15**) as a side chain residue. An interesting enzymatic synthesis of this amino acid has been developed [2] as shown in Scheme 3. Racemic (*para*-hydroxyphenyl) hydantoin (**12**) is prepared from phenol, glyoxylic acid and urea under acidic conditions. This substrate is selectively hydrolyzed to D-N-carbamoyl-*para*-hydroxyphenylglycine (**14**) by a D-specific hydantoinase from *Bacillus brevis*. The unreacted L-isomer **13** spontaneously racemizes under the conditions of the enzymatic reaction. Thus, a theoretical 100% turnover of the d,l-substrate **12** can be achieved by this protocol. The carbamoyl product **14** can be chemically converted into D-*para*-hydroxyphenylglycine (**15**) with sodium nitrite in a mixture of hydrochloric and acetic acids. Alternatively, a two-enzyme, one-reactor procedure [4,5] with cells of *Agrobacterium radiobacter* directly converts **12** into **15**. This microbe produces both the D-hydantoinase and N-carbamoyl-D-amino acid hydrolase.

It has also been reported [4] that *Agrobacterium rhizogenes* IFO 13259, *Coryne bacterium sepedonicum* IFO 3306 and *Mycobacterium smegmatis* ATCC 607 will successfully hydrolyze d,l-N-carbamoyl-methionine producing L-methionine and D-N-carbamoyl-methionine. Guivarch and collaborators [6] have also found hydantoinase activity in a *Pseudomonas sp.* and *Arthrobacter globiformis*. These workers provide evidence for a racemase that causes complete consumption of the d,l-hydantoin substrate. Tryptophan, phenylalanine, methionine, 2-amino-6-hydroxy-hexanoic acid and 2-amino-5-cyano-pentanoic acid were the amino acids studied.

SCHEME 4

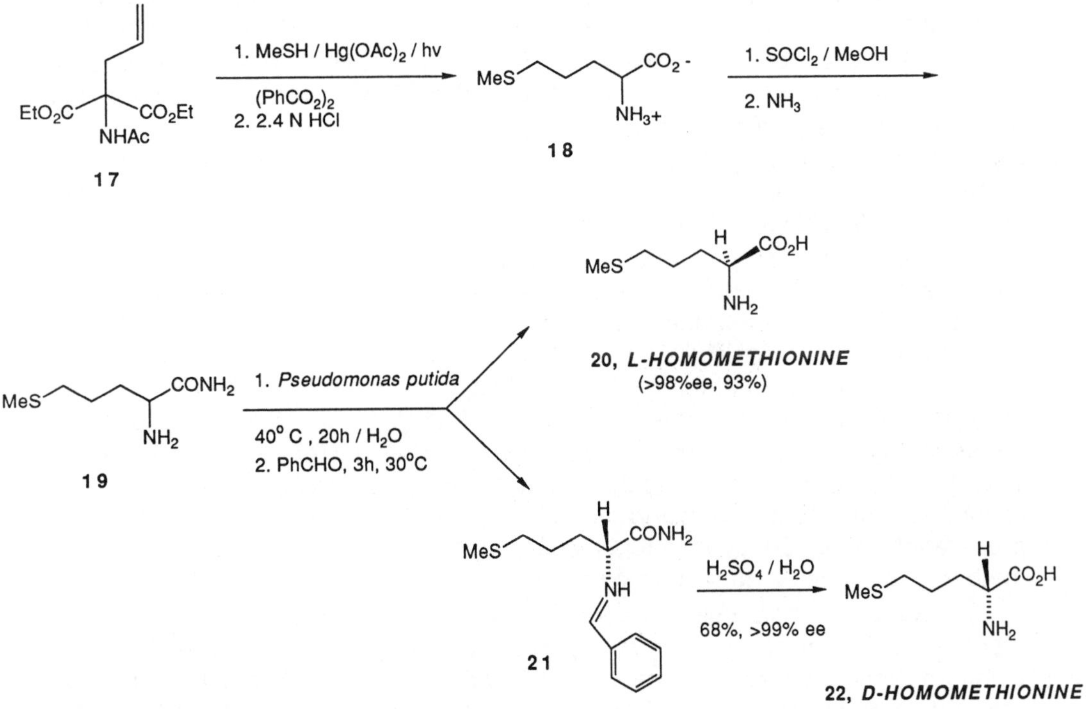

20, *L-HOMOMETHIONINE*
(>98%ee, 93%)

22, *D-HOMOMETHIONINE*

Vriesema, B.K.; ten Hoeve, W.; Wynberg, H.; Kellog, R.M.; Boesten, W.H.J.; Meijer, E.M.; Schoemaker, H.E., *Tetrahedron Lett.* (1986) **26**, 2045.

In addition to the method outlined in Scheme 3 for the synthesis of the hydantoin, the classical Bucherer-Bergs reaction is most commonly employed for the preparation of the substrate hydantoins. The hydantoinases are also remarkable in that, they are one of the rare classes of proteases that will hydrolyze a cis-amide-type linkage.

L-Homomethionine (**20**) is a naturally occurring amino acid that is a constituent of black mustard seed , horseradish root and serves as a biosynthetic precursor of sinigrin. Kellogg and co-workers [7] have prepared this substance using the L-aminopeptidase produced by *Pseudomonas putida*. As shown in Scheme 4, free radical addition of methane thiol to **17** and hydrolysis afforded racemic homomethionine (**18**). Formation of the corresponding amide (**19**) and treatment with *Pseudomonas putida* cells at 40°C for twenty hours gave a mixture of L-homomethionine (**20**) in >98% ee and the unreacted D-amide. These were separated by addition of benzaldehyde which formed the insoluble Schiff base (**21**) that was removed by simple filtration. Hydrolysis of the latter furnished D-homomethionine (**22**) in >99% ee and 68% chemical yield.

Mori and Iwasawa [8] have prepared both enantiomers of *threo*-2-amino-3-methylhexanoic acid (**26** and **28**) via the acylase-catalyzed deacylation of the d,l-N-acetyl derivative **25** as shown in Scheme 5. The reaction is performed at pH 6.7 at 37°C for four days in the presence of a trace of $CoCl_2$. The enzyme selectively deacylates the L-isomer leaving the partially resolved D-isomer **27**. This substance was fractionally crystallized and resubjected to the enzymatic reaction. In this way, both isomers of high optical purity (98%) can be obtained on a preparative scale. Complete experimental details are reported.

Whitesides and co-workers [9] have recently introduced a convenient and practical method to carry out enzymatic reactions. The procedure, named MEEC, or Membrane-Enclosed Enzymatic Catalysis, calls for placing the enzyme in commercially available dialysis membranes. This method combines the advantages of soluble enzymes with those of immobilized enzymes. A single example of the resolution of d,l-N-(chloroacetyl)-α-methylmethionine (**29**) with Acylase I was carried out on a 10 mMol scale. This technique will undoubtedly result in increased use of enzymes for amino acid synthesis on laboratory scales, since crude or purified enzyme preparations may be used and is operationally much simpler than classical methods of enzyme immobilization.

Miyazawa and collaborators [10] have recently reported a method to resolve d,l-2-chloroethyl esters of N-CBz amino acids (**32**) with lipase obtained from *Aspergillus niger*. As shown in Scheme 7, the L-acids (**33**) are obtained in 85-96%ee. These workers also examined acylases from *Pseudomonas fluorescens* and *Candida cylindracea*; the results clearly

SCHEME 5

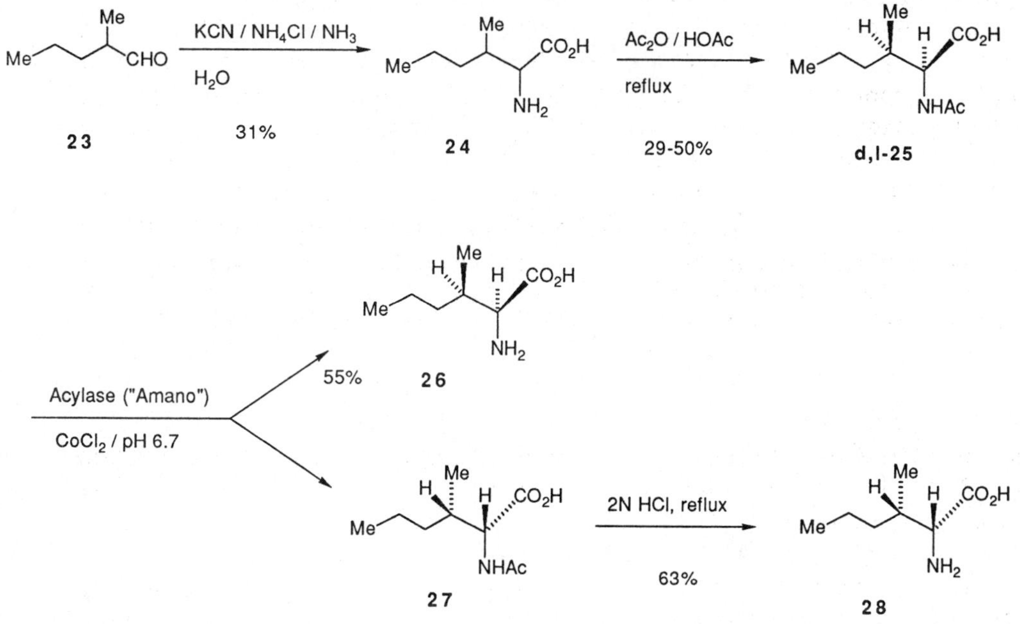

Mori, K.; Iwasawa, H., *Tetrahedron* (1980) **36**, 2209.

SCHEME 6

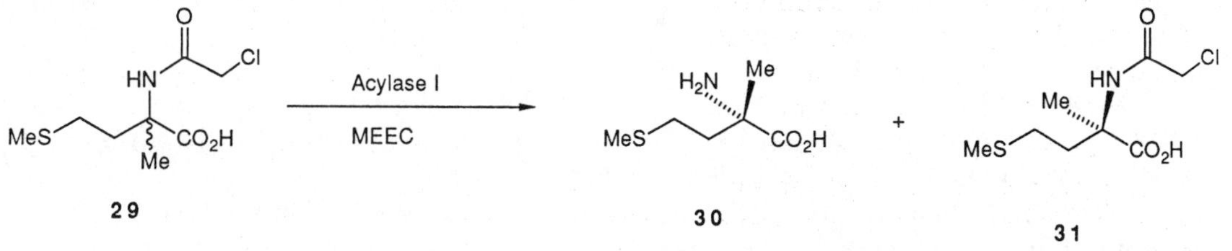

Bednarski, M.D.; Chenault, H.K.; Simon, E.S.; Whitesides, G.M., *J.Am.Chem.Soc.* (1987) **109**, 1283.

SCHEME 7

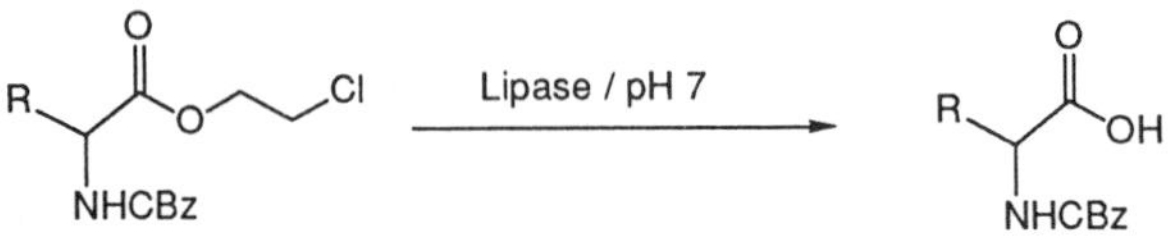

32   33

Miyazawa, T.; Takitani, T.; Ueji, S.; Yamada, T.; Kuwata, S., *J.Chem.Soc.Chem.Comm.* (1988) 1214.

R	*Aspergillus niger*		*Pseudomonas fluorescens*		*Candida cylindracea*	
	%conversion	%ee	%conversion	%ee	%conversion	%ee
Me	36	89	46	16	38	7
Et	40	96	49	52	36	7
n-Pr	27	86	43	58		
n-Bu	32	85	52	60	42	30
n-Pent	44	95	25	80	27	66
n-Hex	31	94	47	87		
i-Pr	14	92	8	53		
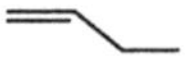 ⌇ (allyl)	34	86	40	52	31	27
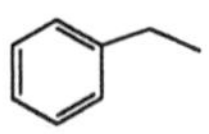 ⌇ (benzyl)	32	94	12	57	35	63
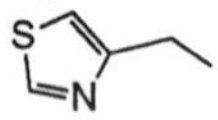 ⌇ (thiazolyl-ethyl)	37	94	40	70	35	43

show the overall superior broad substrate specificity and enantioselectivity of the *Aspergillus niger* system.

A very practical resolution of racemic amino acid esters (**34**) has recently been disclosed [11] that employs alkaline protease from *Bacillus lichenformis*. As shown in Scheme 8, methyl and benzyl L-esters are enantioselectively hydrolyzed in pH 8 buffer at 30° C for only ca. 20 minutes. The unreacted D-esters (**36**) are obtained in very high optical purity. It is important to monitor the reaction and stop it at as close to

SCHEME 8

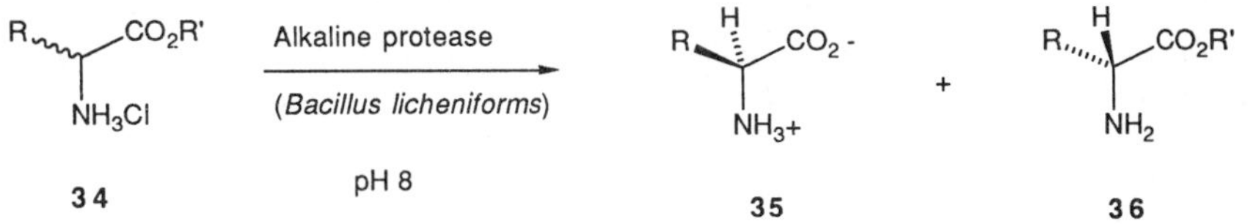

Chen, S-T.; Wang, K-T.; Wong, C-H., *J.Chem.Soc.Chem.Comm.* (1986) 1514.

R	R'	YIELD (% 35)	%ee (35)	YIELD (%36)	%ee (36)
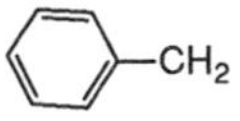 [benzyl] —CH$_2$	Me	96	90	85	100
 HO—[phenyl]—CH$_2$	Me	95	91	86	100
Me	CH$_2$Ph	98	86	75	93

50% conversion as possible. The authors note that benzyl esters are hydrolyzed ca. three times faster than the corresponding methyl esters of the same amino acid. Organic co-solvents such as DMF, dioxane and acetonitrile up to 30% v/v do not significantly lower the enzyme activity. In addition to the three examples listed, the methyl esters of Ala, Phe, Tyr, and Trp are hydrolyzed faster than Asp, Leu, Glu and Pro. Met, Thr, and Ser esters are hydrolyzed much slower; Val, Arg, Ile, Lys, Cys and His are not hydrolyzed at all. The authors list high turnover rates (100 gm/h/AU); low enzyme cost; ease of substrate preparation; enzyme stability and ease of product isolation as significant advantages of this method over currently existing processes.

Wong and associates [12] have also briefly reported on the utility of papain in 20% DMF to be effective in resolving d,l-**37** into the corresponding L-acid (**38**) and the D-ester **39**. The authors report that N-benzoyl, N-ethoxycarbonyl and N-benzyloxycarbonyl substrates **37** all give equally good results. The methyl esters **39** were obtained in >98% ee and 31-44% yields; the optical purity of this material was ascertained by NMR analysis using a chiral shift reagent.

SCHEME 9

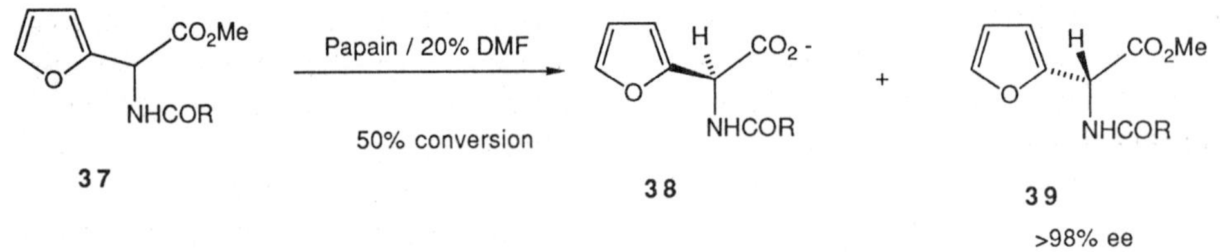

furan–CH(CO_2Me)(NHCOR) $\xrightarrow[\text{50\% conversion}]{\text{Papain / 20\% DMF}}$ furan–CH(CO_2^-)(NHCOR)   +   furan–CH(CO_2Me)(NHCOR)

37   **38**   **39**

                                                           >98% ee

Drueckhammer, D.G.; Barbas, C.F.; Nozaki, K.; Wong, C-H., *J.Org.Chem.* (1988) **53**, 1607.

SCHEME 10

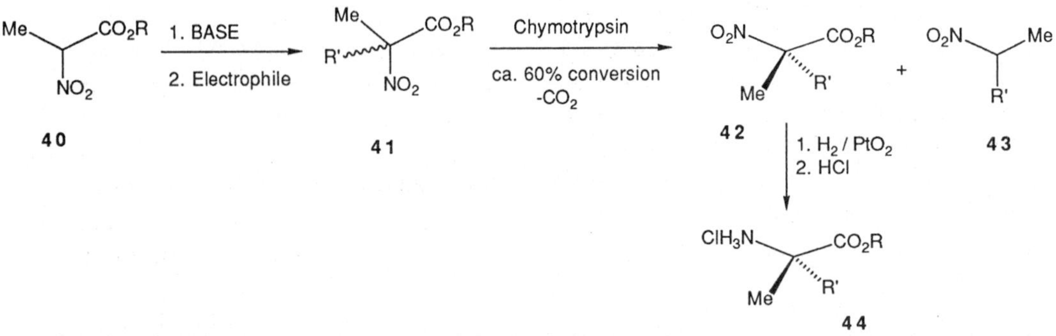

Me–C(CO_2R)(H)(NO_2) $\xrightarrow[\text{2. Electrophile}]{\text{1. BASE}}$ Me–C(CO_2R)(R')(NO_2) $\xrightarrow[\substack{\text{ca. 60\% conversion}\\-CO_2}]{\text{Chymotrypsin}}$ O_2N–C(Me)(CO_2R)(R')   + O_2N–CH(Me)(R')

40   **41**   **42**   **43**

$\downarrow$ 1. H_2 / PtO_2 2. HCl

ClH_3N–C(Me)(CO_2R)(R')

44

Lalonde, J.J.; Bergbreiter, D.E.; Wong, C-H., *J.Org.Chem.* (1988) **53**, 2323.

ELECTROPHILE	R'	YIELD (%41)	R	%ee (44)	AMINO ACID
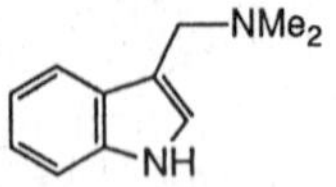 (indol-3-yl)-CH₂–NMe₂	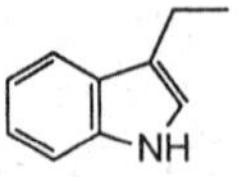 (indol-3-yl)-CH₂–	93	Bu	>95	α-methyltryptophan
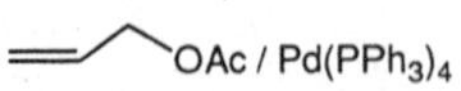 CH₂=CH–CH₂–OAc / Pd(PPh₃)₄	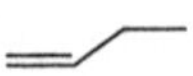 CH₂=CH–CH₂–	62	Bu	>95	α-methylallylglycine
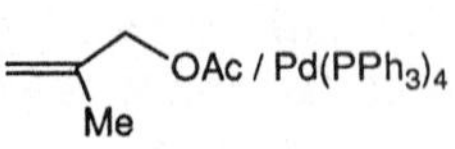 CH₂=C(Me)–CH₂–OAc / Pd(PPh₃)₄	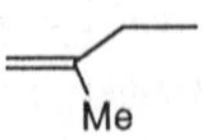 CH₂=C(Me)–CH₂–	94	Bu	85	α-methylleucine
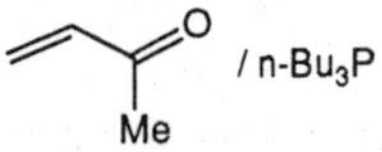 CH₂=CH–C(O)Me / n-Bu₃P	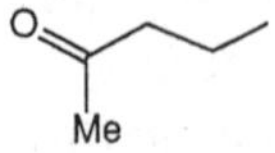 CH₃–C(O)–CH₂–CH₂–	94	Bu	75	δ-keto-α-amino-α-methyl-hexanoic acid
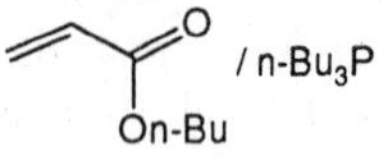 CH₂=CH–C(O)–On-Bu / n-Bu₃P	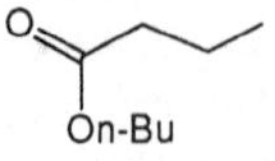 n-BuO–C(O)–CH₂–CH₂–	92	Me	39	α-methylglutamic acid
(C_6H_5)₃BiCl₂ / tetramethylguanidine	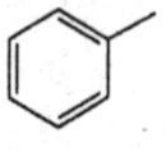 C₆H₅–	81	Bu	>95	α-methyl-α-phenylglycine

A very interesting enzymatic resolution of α-nitro-α-methyl carboxylic acid esters was recently reported by Lalonde, Bergbreiter and Wong [13]. As shown in Scheme 10, the 2-nitro propionates (**40**) are alkylated to afford the racemic substrates **41**. Treatment of these substances with α-chymotrypsin effects hydrolysis of the D-esters to the corresponding free acids. Under the reaction conditions (0.25 M pH 7 phosphate buffer in 2:1 buffer : DMSO), these compounds spontaneously decarboxylate leaving a mixture of **42** and **43**; these substances were then separated by radial chromatography. The authors note that lipases and esterases were found to be ineffective. The optically active nitro esters **42** can then be reduced to the corresponding α-methyl-α-amino acid esters (**44**) using Adams catalyst. The authors note that substrates containing unsaturated R' groups give the best results; this observation is consistent with the well-known substrate specificity of chymotrypsin. This procedure provides a potentially very powerful and economical method to synthesize the important α-methyl-α-amino acids; α-methyl-α-amino acid derivatives are commonly poor substrates for hydrolytic enzymes compared to their unsubstituted counterparts. The enantiomeric excess of the product amino acids (**44**) was high, usually >95% ee. This approach is limited to α-methylated amino acids since the mono-α-substituted substrates have a very acidic α-proton that readily undergoes exchange resulting in racemic products. Experimental details accompany this report.

An emerging area of great potential promise are the nitrilases and nitrile hydratases which effect the conversion of nitriles to carboxylic acids (**46**) and amides (**47**), respectively. The substrate α-amino nitriles (**45**) are easily prepared from aldehydes, ammonia and cyanide. The problems associated with rendering these hydrolases practical for the preparation of amino acids have to do with controlling the equilibrium that is established in solution with the α-amino nitrile and the starting aldehyde component. Polymerization of HCN can drive the equilibrium in the undesired direction resulting in loss of the substrate. It is not yet clear if nitrilases and /or nitrile hydratases have been identified that display enantioselectivity toward the substrate α-amino nitriles. For example, Jallageas and associates [14] reported that L-amino acids could be obtained from α-amino nitriles using a "nitrilase" from *Pseudomonas putida* and *Brevibacterium sp.* with 50% conversion. This result has been reinterpreted[4] as involving non-stereospecific hydrolysis of the nitrile to the d,l-amide (**47**) followed by enantioselective hydrolysis of the L-amide to the L-acid by an L-specific amidase; both of these species are known to produce L-specific amidases. More recently, Macadam and Knowles [15] have reported a synthesis of L-alanine (**49**, Scheme 11) from α-aminopropionitrile using an *Acinetobacter sp.* that displayed nitrilase

Enzymatic Syntheses of α-Amino Acids

SCHEME 11

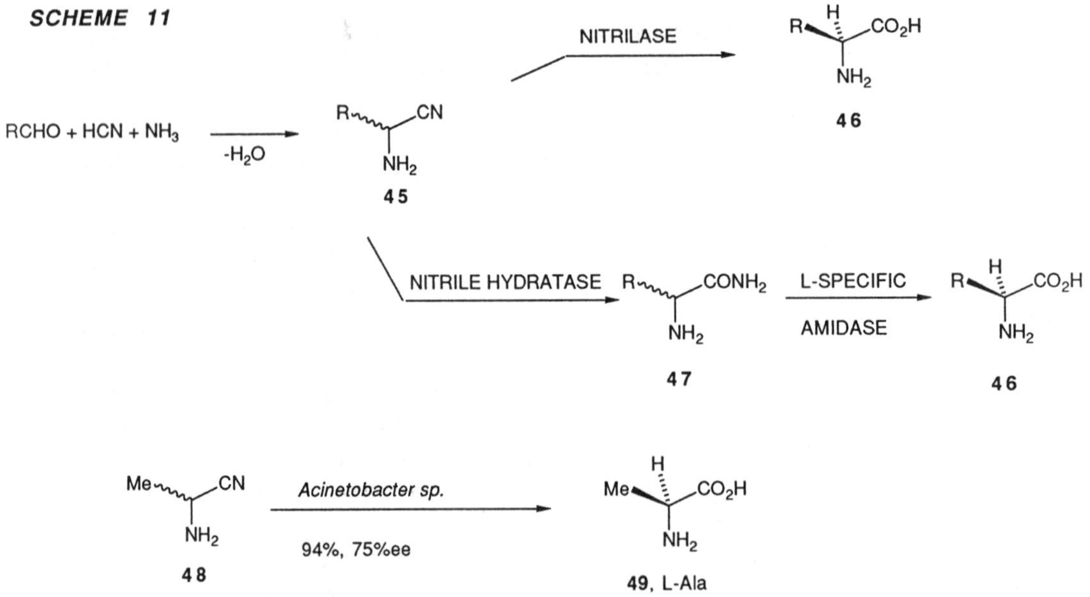

$$RCHO + HCN + NH_3 \xrightarrow{-H_2O}$$

45

NITRILASE → 46

NITRILE HYDRATASE → 47   L-SPECIFIC AMIDASE → 46

48   *Acinetobacter sp.*   94%, 75%ee   49, L-Ala

Macadam, A.M.; Knowles, C.J., *Biotechnology Lett.* (1985) **7**, 865.

activity. The chemical conversion was very high (94%) and the %ee of the alanine obtained was 75%ee. The high conversion and lack of unreacted D-amide in this system provides some evidence that a true enantiospecific nitrilase is at work and not the combination of a nitrile hydratase/ L-specific amidase system. However, it is also possible that a racemase is present that, coupled with the nitrile hydratase/amidase, would give the same result. This relatively new area needs considerably more investigation; the low cost of the raw materials for such a process should inspire more detailed examination.

B. ASYMMETRIC BOND-FORMING REACTIONS ON PROCHIRAL SUBSTRATES CATALYZED BY AN ENZYMATIC SYSTEM

Several important processes for the bulk production of amino acids have emerged that require nicotinamide and pyridoxal phosphate co-factors. Both whole cell[16] and immobilized enzyme systems have been utilized. L-Phenylalanine has been the subject of numerous approaches due to the great recent demand for low cost production of this amino acid for the synthesis of Aspartame. For example, phenylalanine can be produced[17] by the deacylation of acetamidocinnamic acid to α-amino cinnamic acid

SCHEME 12

+ Me OH + NH_3
IMMOBILIZED *E.coli*
NH
H
CO_2H
N
H
NH_2

50 **51** **52**, L-Trp

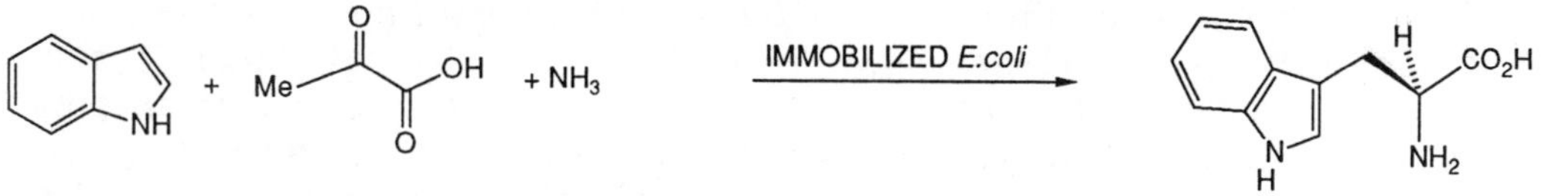

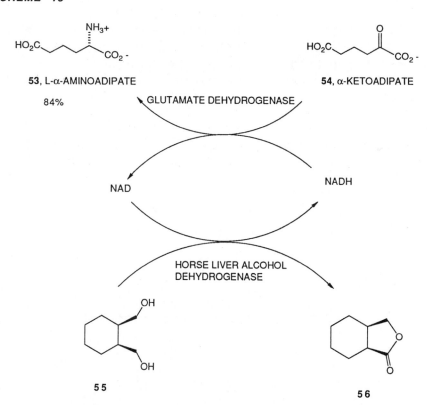

Azerad, R.; Calderon-Seguin, R.; Decottignies-Le Marechal, P., Bull.Chim.Soc.Fr. (1980) II-83

SCHEME 13

NH_3+
HO_2C
CO_2 -
53, L-α-AMINOADIPATE
84%
HO_2C
CO_2 -
54, α-KETOADIPATE
GLUTAMATE DEHYDROGENASE
NAD
NADH
HORSE LIVER ALCOHOL
DEHYDROGENASE
OH
OH
55
O
O
56

Matos, J.R.; Wong, C-H., *J.Org.Chem.* (1986) **51**, 2388

which spontaneously hydrolyzes to phenylpyruvic acid. A single microbial species (*Alcaligenus faecalis* S-7 and *Bacillus sphaericus* N-7) converts the phenylpyruvic acid to L-phenylalanine with a transaminase enzyme. A similar preparation of L-tryptophan (**52**) has been developed that employs immobilized *E.coli* cells as shown in Scheme 12. Indole (**50**), pyruvic acid (**51**) and ammonia were fed into a continuous reactor system that absorbs the L-Trp produced and recycles the excess pyruvic acid and ammonia.

In purified enzyme systems, it is necessary to devise means to drive the inherent equilibrium in the desired direction and regenerate the co-factor that is consumed in the reductive amination of the α-keto acid. Wong and Matos[18] have recently reported a useful example illustrated in Scheme 13 for the synthesis of L-α-aminoadipate (**53**). α-Ketoadipate (**54**) plus ammonia and glutamate dehydrogenase produce α-aminoadipate in optically pure form in 84% yield. The consumed NADH is regenerated by coupling the reductive transamination half-reaction with the oxidative cyclization of the meso-diol **55** to the optically active lactone **56** with horse liver alcohol dehydrogenase (HLADH) which requires NAD+. This reaction is carried out in a biphasic system that allows for easy separation of the product components and obviates the normal problems of product inhibition. Complete experimental details for a 50 mMol scale preparation are included.

A new industrial scale (600 ton) synthesis of L-phenylalanine has been developed[2] that employs a pyridoxal phosphate-dependent transaminase system as shown in Scheme 14. Benzylidene hydantoin (**57**) is hydrolyzed to phenylpyruvic acid and then transaminated with aspartic acid as the amino donor. The oxaloacetic acid (**61**) must then be chemically or enzymatically decarboxylated to pyruvic acid.

SCHEME 14

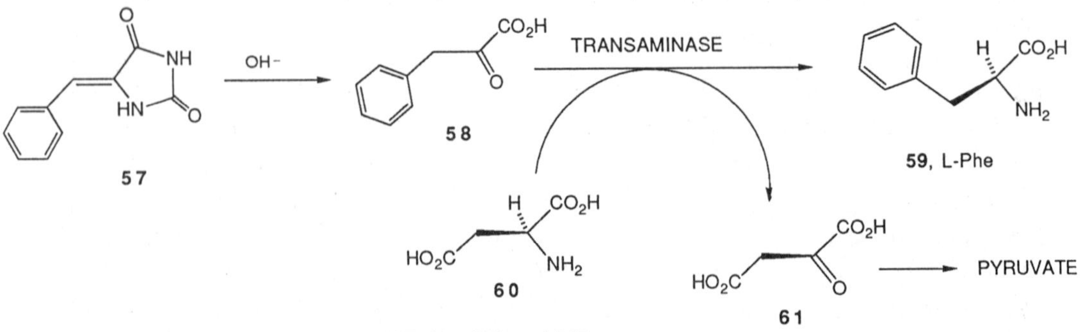

Meijer, E.M.; Boesten, W.H.J.; Schoemaker, H.E.; Van Balken, J.A.M.,
Biocatalysts in Organic Synthesis (1985) ,135 , Elsevier Pub. , Amsterdam
Tramper, J.; van der Plas, H.C.; Linko, P., Eds.

SCHEME 15

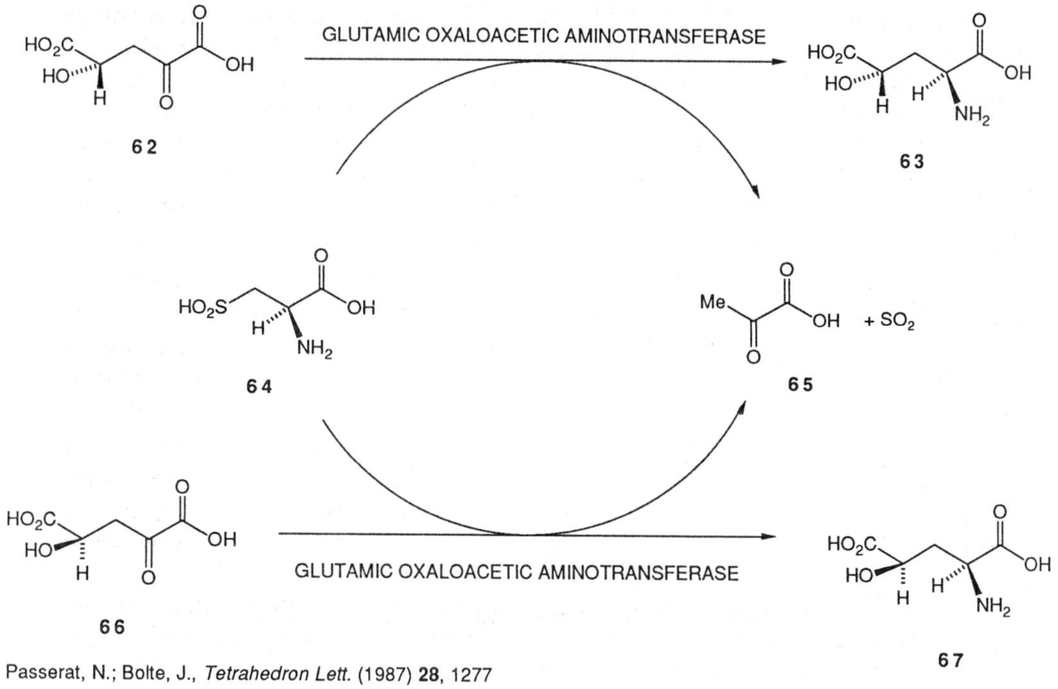

Passerat, N.; Bolte, J., *Tetrahedron Lett.* (1987) **28**, 1277

In a related approach, a NAD+-dependent phenylalanine dehydrogenase was detected [19] in a *Brevibacterium sp.* Phenylpyruvic acid plus ammonia as amino donor was converted into L-Phe at pH 8.5. The authors note that ammonia could not be replaced L-Glu or L-Asp as amino donor. It has been suggested [2] that the enzyme from this *Brevibacterium sp.* could be coupled to formate dehydrogenase to regenerate the NADH in a membrane-type reactor for the efficient production of L-Phe.

Passerat and Bolte [20] have recently described an interesting application of immobilized glutamic oxaloacetic aminotransferase to the synthesis of γ-hydroxy L-glutamic acids. As shown in Scheme 15, γ-hydroxy-α-ketoglutaric acids (**62** and **66**) are converted into the corresponding γ-hydroxy L-glutamic acids (**63** and **67**, respectively) using glutamic oxaloacetic aminotransferase obtained from rat liver and immobilized in a polyacrylamide (PAN) gel according to Whitesides[21]. The equilibrium is driven by using cysteine sulfinic acid (**64**), an analog of Asp , as amino donor; upon conversion to the corresponding α-keto acid, this material spontaneously loses sulfur dioxide giving pyruvic acid. The

SCHEME 16

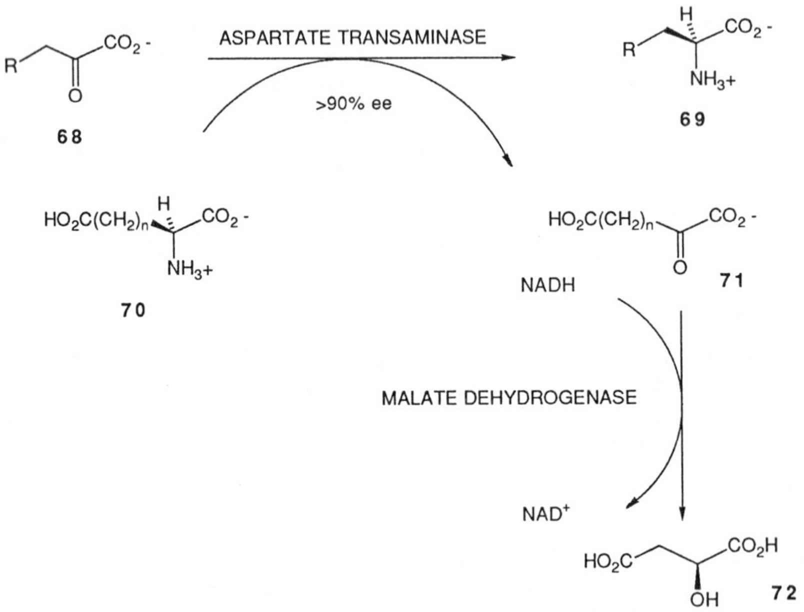

Baldwin, J.E.; Dyer, R.L.; Ng, S.C.; Pratt, A.J.; Russell, M.A., *Tetrahedron Lett.* (1987) **28**, 3745

R	YIELD (% 72)
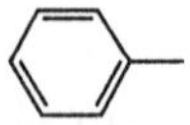 (phenyl)	84
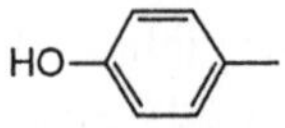 HO-(phenyl)	80
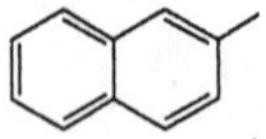 (naphthyl)	-
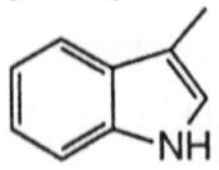 (indolyl)	72
HO_2C-	-
HO_2CCH_2-	-
H	35
Me	37
Et	32
i-Pr	40
$MeSCH_2-$	58

SCHEME 17

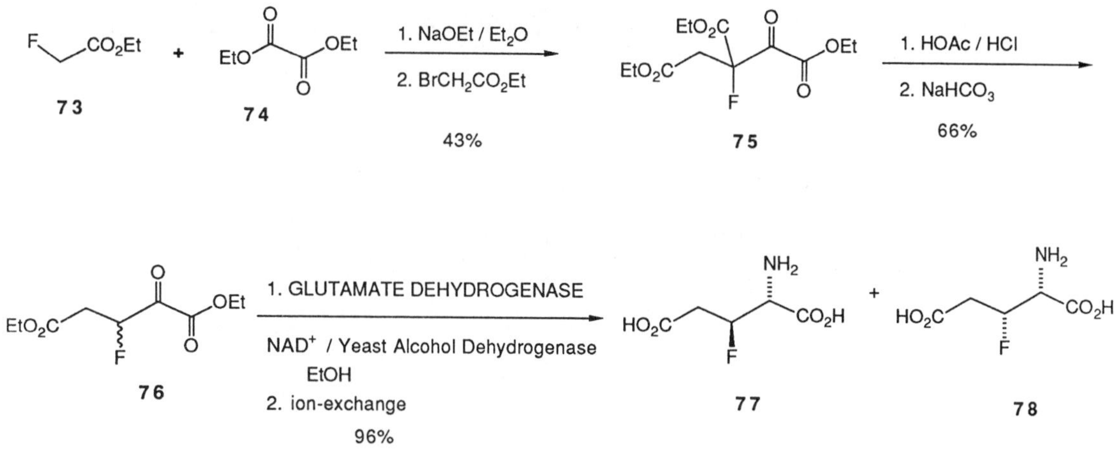

Vidal-Cros, A.; Gaudry, M.; Marquet, A., *J.Org.Chem.* (1989) **54**, 498

diastereomeric products (**63** and **67**) must be separated by chromatography or by chemical means (lactonization of **63**). A limitation of this approach is the expense of cysteine sulfinic acid.

Baldwin and associates [22] have recently reported the use of cloned aspartate transaminase from *E.coli* for the synthesis of several α-amino acids from the corresponding α-keto acids as shown in Scheme 16. L-Glu or L-Asp are employed as amino donors; the enzyme displays broad substrate specificity and produces amino acids in >90% ee. The aliphatic substrates which were previously thought not to be suitable substrates for the enzyme are converted by increasing the enzyme concentration 10-15 fold. The large scale availability of the cloned enzyme makes this economically feasible.

Both diastereomers of β-fluoroglutamic acid (**77** and **78**) have been prepared [23] by the reductive transamination of racemic 2-oxo-3-fluoroglutarate (**76**) with glutamate dehydrogenase. The consumed NADH was regenerated *in situ* by coupling a yeast alcohol dehydrogenase/ethanol system. The authors note that undesirable fluoride elimination in the NADH system is unlikely which contrasts with a PLP-dependent transaminase. The final products (**77** and **78**) were separated by ion-exchange chromatography and were each shown to optically pure. Complete experimental details are reported (5 mMol scale).

SCHEME 18

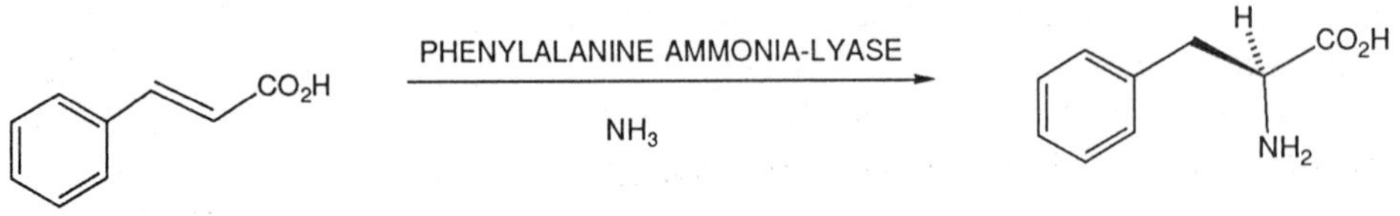
CO2H
PHENYLALANINE AMMONIA-LYASE
NH3
H
CO2H
NH2

79 **80**, L-Phe

Yamada, S.; Nabe, K.; Izuo, N.; Nakamichi, K.; Chibata, I., *Appl.Environ. Microbiol.* (1981) **42**, 773
Hamilton, B.K.; Hsiao, H.; Swann, W.E.; Anderson, D.M.; Delente, J.J., *Trends in Biotechnol.* (1985) **3**, 64

SCHEME 19

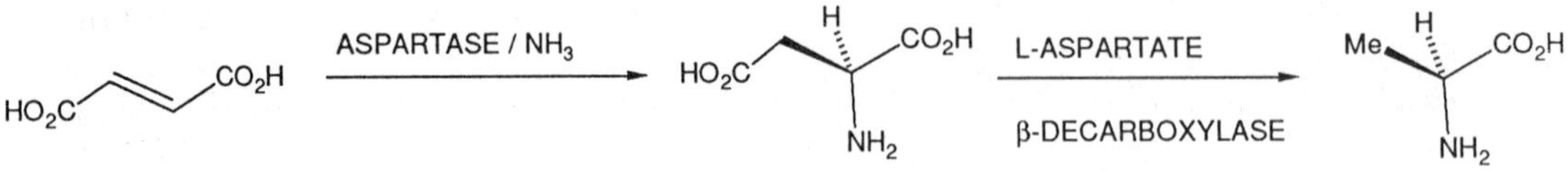
HO2C
CO2H
ASPARTASE / NH3
H
HO2C
CO2H
NH2
L-ASPARTATE
β-DECARBOXYLASE
Me
H
CO2H
NH2

81, FUMARIC ACID **82**, L-Asp **83**, L-Ala

Kitahara, K.; Fukui, S.; Misawa, M., *J.Gen.Appl. Microbiol.* (1959) **5**, 74
Yokote, Y.; Maeda, S.; Yabushita, H.; Noguchi, S.; Kimura, K.; Samejima, H., *J.SolidPhaseBiochem.* (1978) **3**, 247
Watanabe, S.; Isshiki, S.; Osawa, T.; Yamamoto, S., *Hakko Kogakuzahi* (1965) **43**, 697

SCHEME 20

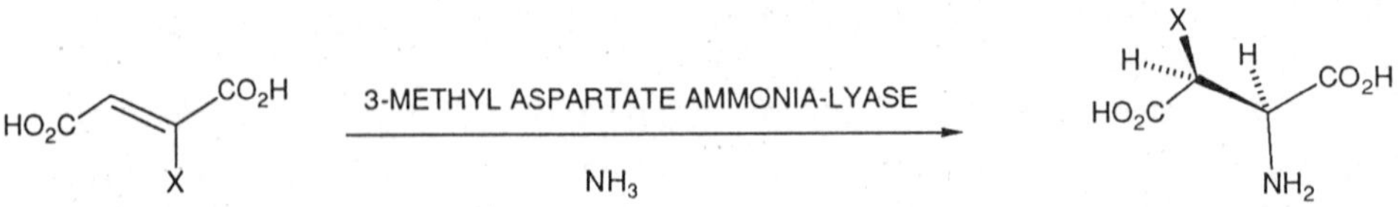
HO2C
CO2H
X
3-METHYL ASPARTATE AMMONIA-LYASE
NH3
X
H
H
HO2C
CO2H
NH2

84, X=Cl; Br **85**

Akhtar, M.; Cohen, M.A.; Gani, D., *Tetrahedron Lett.* (1987) **28**, 2413

Several enzymatic systems are available for converting unsaturated carboxylic acids into amino acids via stereospecific addition of ammonia across the olefin. For example, L-phenylalanine can be synthesized [24] from *trans*-cinnamic acid with L-phenylalanine ammonia-lyase obtained from *Rhodococcus glutinis*. The commercialization of this process has been hampered by the low conversion (70%), the poor stability of the enzyme and the inhibition caused by **79**. An improved version [25] of the same synthesis that proved suitable for commercialization has been achieved with a cell-free preparation of L-phenylalanine ammonia-lyase from *Rhodotorula rubra* (Scheme 18).

An industrially important synthesis [26] of L-aspartic acid is shown in Scheme 19. Aspartase from *E.coli* effects the efficient enantiospecific amminolysis of fumarate (**81**) giving L-Asp (82). Very high conversions (99%) on a production scale have been reported. From the same type of system, L-alanine can be prepared commercially utilizing L-aspartate-β-decarboxylase from *Xanthomonas oryzae* No.531; *Pseudomonas dacunhae*; and *Achromobacter pestifer* [27].

Gani and associates[28] have deployed 3-methylaspartate ammonia-lyase in an interesting synthesis of β-chloro and bromo aspartic acids (**85**) as shown in Scheme 20. This enzyme catalyzes the reversible elimination of ammonia from β-methylaspartic acid to give mesaconic acid. As expected, bromo- and chloro-fumarates add ammonia in a retro-physiological reaction to form the 2-(R),3-(S)-β-halo aspartates (**85**).

It is clear that enzymatic syntheses of natural as well as some unnatural amino acids will continue to be exploited on both a laboratory and industrial scale. The emergence of improved technologies to select for a given type of enzymatic system and the ability to clone and produce these valuable catalysts cheaply will play an increasingly important role in the production and utilization of amino acids. Improved methods to broaden the substrate specificity of these systems and simplifying the experimental protocols for manipulating the biocatalyst will be important for broadening the appeal of these approaches for the synthetic laboratory chemist.

References   Chapter   7

1. For a review on the industrial production and utilization of amino acids, see; Izumi, Y.; Chibata, I.; Itoh, T., *Angew.Chem.Int.Ed.Engl.* (1978) **17**, 176.

2. a) Meijer, E.M.; Boesten, W.H.J.; Schoemaker, H.E.; VanBalken, J.A.M.; *Biocatalysts in Organic Synthesis* (1985)135, Elsevier Pub., Amsterdam; Tramper, J.; van der Plas, H.C.; Linko, P., Eds.; b) Sheldon, R.A.; Porskamp, P.A.; ten Hoeve, W.,*Biocatalysts in Organic Synthesis* (1985)59, Elsevier Pub., Amsterdam; Tramper, J.; van der Plas, H.C.; Linko, P., Eds.

3. Kruizinga, W.H.; Bolster, J.; Kellog, R.M.; Kamphius, J.; Boesten, W.H.J.; Meijer, E.M.; Schoemaker, H.E., *J.Org.Chem.* (1988) **53**, 1826.

4. Sheldon, R.A.; Schoemaker, H.E.; Kamphius, J.; Boesten, W.H.J.; Meijer, E.M., *Stereoselectivity of Pesticides; Biological and Chemical Problems* (1988), Chap. 14 (pp 409-451),Elsevier Pub., Amsterdam; Ariens, E.J.; van Rensen, J.J.S.; Welling, W., Eds.

5. Olivieri, R.; Fascetti, E.; Angelini, L.; Degen, L., *Biotechnol.Bioengin.* (1981) **23**, 2173.

6. Guivarch, M.; Gillonnier, C.; Brunie, J-C., *Bull.Soc.Chim.Fr.* (1980)II-91.

7. Vriesema, B.K.; ten Hoeve, W.; Wynberg, H.; Kellogg, R.M.; Boesten, W.H.J.; Meijer, E.M.; Schoemaker, H.E., *Tetrahedron Lett.* (1986) **26**, 2045.

8. Mori, K.; Iwasawa, H., *Tetrahedron* (1980) **36**, 2213.

9. Berdnarski, M.D.; Chenault, H.K.; Simon, E.S.; Whitesides, G.M., *J.Am.Chem.Soc.* (1987) **109**, 1283.

10. Miyazawa, T.; Takitani, T.; Ueji, S.; Yamada, T.; Kuwata, S., *J.Chem.Soc.Chem.Comm.* (1988) 1214.

11. Chen, S-T.; Wang, K-T.; Wong, C-H., *J.Chem.Soc.Chem.Comm.* (1986) 1514.

12. Drueckhammer, D.G.; Barbas, C.F.; Nozaki, K.; Wong, C-H., *J.Org.Chem.* (1988) **53**, 1607.

13. Lalonde, J.J.; Bergbreiter, D.E.; Wong, C-H., *J.Org.Chem.* (1988) **53**, 2323.

14. a) Arnaud, A.; Galzy, P.; Jallageas, J-C., *Bull.Soc.Chim.Fr.* (1980) II-87; b) Jallageas, J-C.; Arnaud, A.; Galzy, P., *Adv.Biochem Engin.* (1980) **14**, 1.

15. Macadam, A.M.; Knowles, C.J., *Biotechnology Lett.* (1985) **7**, 865.

16. Chibata, I.; Tosa, T., *Ann.Rev.Biophys.Bioeng.* (1981) **10**, 197.

17. Nakamichi, K.; Nabe, K.; Yamada, S.; Tosa, T.; Chibata, I., *Appl.Microbiol.Biotechnol.* (1984) **19**, 100.

18. Matos, J.R.; Wong, C-H., *J.Org.Chem.* (1986) **51**, 2388.

19. Hummel, W.; Weiss, N.; Kula, M-R., *Arch. Microbiol.* (1984) **137**, 47.

20. Passerat, N.; Bolte, J., *Tetrahedron Lett.* (1987) **28**, 1277.

21. Pollak, A.; Blummenfeld, H.; Wax, M.; Baughn, R.L.; Whitesides, G.M., *J.Am.Chem.Soc.* (1980) **102**, 6324.

22. Baldwin, J.E.; Dyer, R.L.; Ng, S.C.; Pratt, A.J.; Russell, M.A., *Tetrahedron Lett.* (1987) **28**, 3745.

23. Vidal-Cros, A.; Gaudry, M.; Marquet, A., *J.Org.Chem.* (1989) **54**, 498.

24. Yamada, S.; Nabe, K.; Izuo, N.; Nakamichi, K.; Chibata, I., *Appl. Environ.Microbiol.* (1981) **42**, 773 .

25. Hamilton, B.K.; Hsiao, H,; Swann, W.E.; Anderson, D.M.; Delente, J.J., *Trends in Biotechnol.* (1985) **3**, 64.

26. a) Kitahara, K.; Fukui, S.; Misawa, M., *J.Gen.Appl.Microbiol.* (1959) **5**, 74; b) Yokote, Y.; Maeda, S.; Yabushita, H.; Noguchi, S.; Kimura, K.; Samejima, H., *J.Solid Phase Biochem.* (1965) **43**, 697.

27. a) Watanabe, S.; Isshiki, S.; Osawa, T.; Yamamoto, S., *Hakko Kogakuzahi* (1965) **43**, 697; b) Chibata, I.; Kakimoto, T.; Kato, J., *Appl.Microbiol.* (1965) **13**, 638; c) Jandel, A.S.; Hustedt, H.; Wandrey, C., *Eur.J.Appl.Microbiol.Biotech.* (1982) **15**, 59.

28. Akhtar, M.; Cohen, M.A.; Gani, D., *Tetrahedron Lett.* (1987) **28**, 2413.

CHAPTER 8

MISCELLANEOUS METHODS

A variety of innovative syntheses of α-amino acids have appeared that do not readily fit any of the other major conceptual groupings utilized in this monograph. This chapter briefly outlines some of the most notable recent contributions that show promise for future developments. In addition, several miscellaneous syntheses of very difficult types of amino acids in either optically active or racemic form are detailed.

Matteson and Beedle [1] have recently reported a very interesting asymmetric synthesis employing chiral boronic esters. As shown in Scheme 1, (S)-pinanediol is deployed as the chiral auxilliary. Treatment of **1** with (dichloromethyl)lithium proceeds with a high degree of stereoselectivity affording the chain-extended boronic esters **2**. Stereospecific azide displacement affords **3** which is subjected to a second (dichloromethyl)lithium chain extension to **4**. Oxidative removal of the 1-chloroboronic ester moiety with sodium chlorite provides the optically active α-azido acids **5** which can be easily reduced to the corresponding α-amino acids. The overall chemical yields to the amino acids (**6**) are very good and the %ees' of the amino acids are 92-96%ee.

Georg and co-workers [2] have recently deployed a Schmidt rearrangement of optically active α,α-disubstituted β-keto esters (**8**) to prepare α,α-disubstituted α-amino acids (**10**) as shown in Scheme 2. The requisite optically active β-keto ester substrates **8** are prepared utilizing the asymmetric alkylation of the chiral enamines **7** developed by Koga, et.al. [3]. The Schmidt rearrangement proceeds with retention of configuration providing the amides **9**; subsequent hydrolysis furnishes the α-amino acids **10** in very good yields and high %ees' for benzylic substrates. The one aliphatic case leading to aspartic acid proceeds with only 70% ee.

O'Donnell and associates [4] have recently examined the use of chiral phase-transfer catalysts in the enantiospecific alkylation of the glycine Schiff base **11**. The method (Scheme 3) involves the use of the cinchona-derived phase transfer catalysts **13** and **14** developed at Merck [5] for the enantioselective alkylation of an indanone enolate. The experimental

SCHEME 1

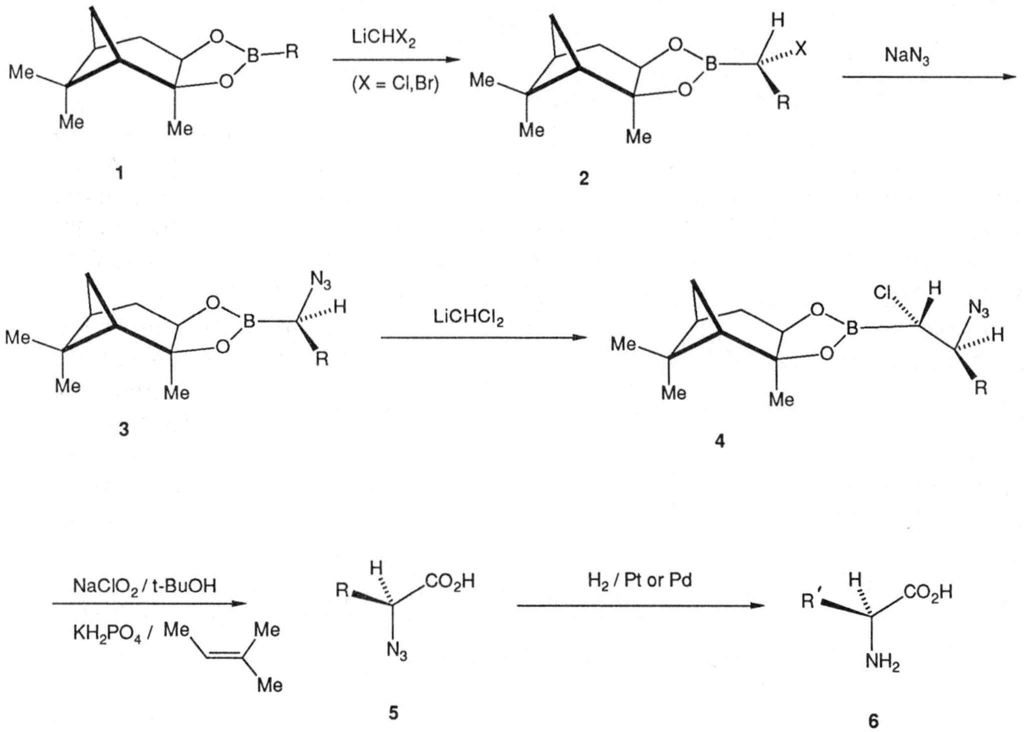

Matteson, D.S.; Beedle, E.C., *Tetrahedron Lett.* (1987) **28**, 4499.

R	R '	X	YIELD (%6)	ISOMER RATIOS (3)	(6)
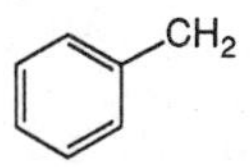 PhCH₂ (benzyl)	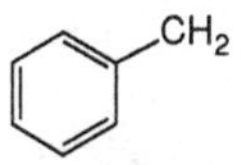 PhCH₂ (benzyl)	Cl	63	25 : 1	25 : 1
(CH₃)₂CH	(CH₃)₂CH	Br	57	>50 : 1	50 : 1
PhCH₂OCH₂	HOCH₂	Br	39	30 : 1	25 : 1
t-BuO₂CCH₂CH₂	HO₂CCH₂CH₂	Br	32	-	30 : 1

SCHEME 2

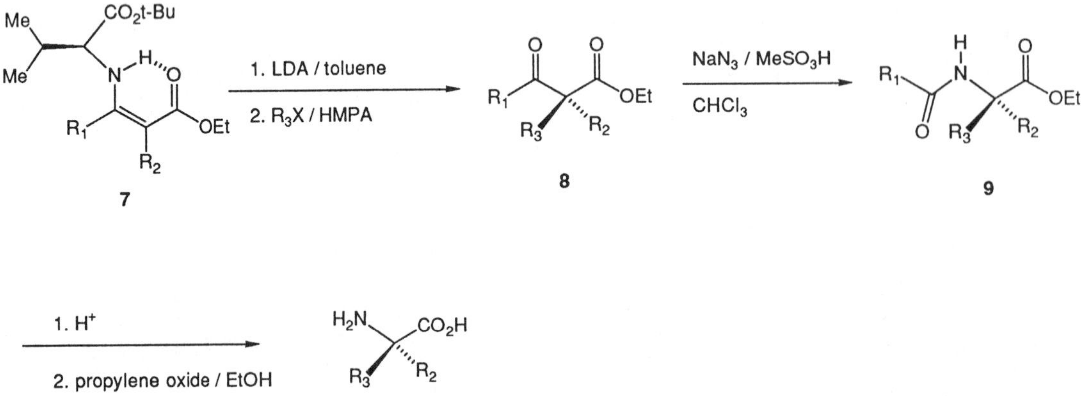

Georg, G.I.; Guan, X.; Kant, J., *Tetrahedron Lett.* (1988) **29**, 403
Tomioka, K.; Ando, K.; Takemasa, Y.; Koga, K., *J.Am.Chem.Soc.* (1984) **106**, 2718.

R$_1$	R$_2$	R$_3$	YIELD (%9)	%ee (9)	YIELD (%10)	%ee (10)
Me	Me	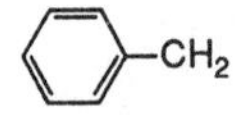 (benzyl)–CH$_2$	95	>95	95	97.8
Me	Me	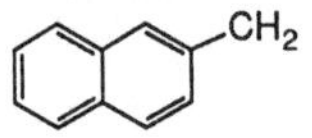 (naphthyl)–CH$_2$	89	>95	89	98.6
(CH$_2$)$_4$		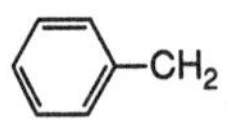 (benzyl)–CH$_2$	97	>95	99	>95
(CH$_2$)$_4$		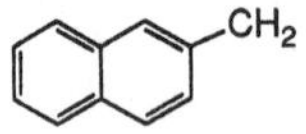 (naphthyl)–CH$_2$	71	>95	88	>95
Me	Me	MeO$_2$CCH$_2$	88	70	80	69

protocol is straightforward involving reaction in 50% aqueous sodium hydroxide and methylene chloride at ambient temperature. The chemical yields with a range of alkyl bromides are good and the %ees' for the crude products range from 42-66%ee. The authors note that catalyst **13** gives a preponderance of the (R)-enantiomer and catalyst **14** gives mainly the (S)-enantiomer. Although the direct %ees' for these alkylations are modest, the authors have illustrated an experimental protocol for obtaining amino acids in >99% ee. As shown in Scheme 4, *para*-chloro benzyl bromide reacts with **11** in the presence of **13** to afford a mixture

SCHEME 3

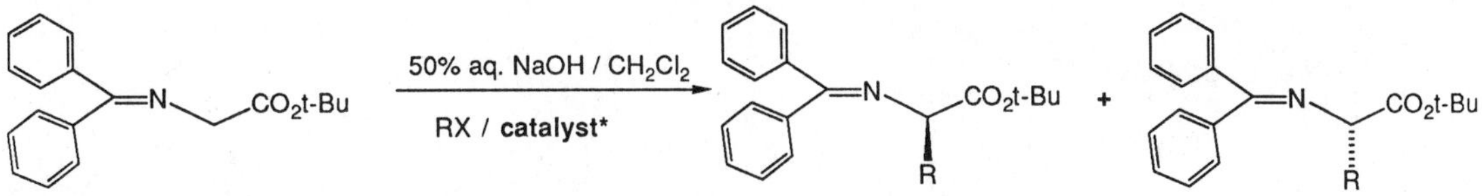

O'Donnell, M.J.; Bennett, W.D.; Wu, S., *J.Am.Chem.Soc.* (1989) **111**, 2353

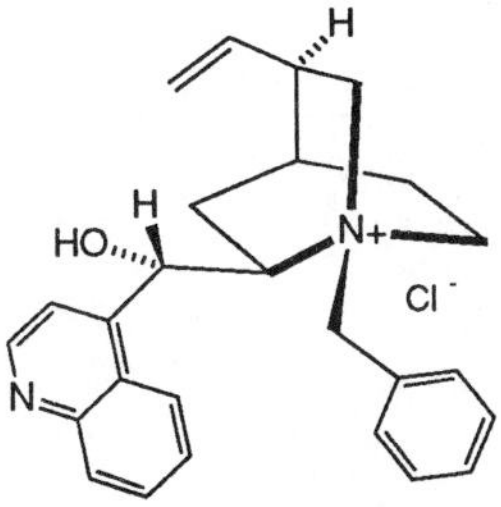

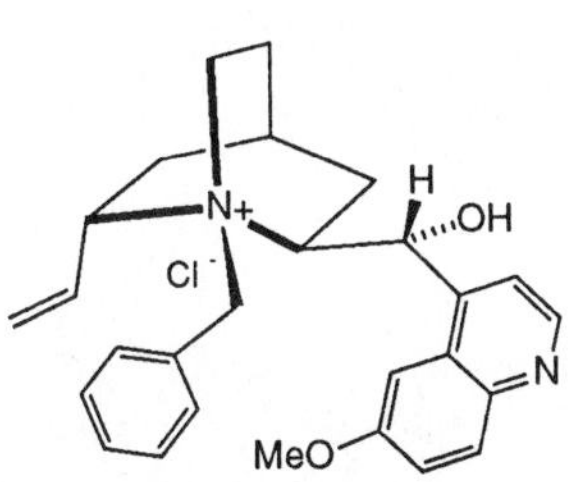

RX	CATALYST	%ee	YIELD (%12)
 ⟋⟍Br	**13**	66	75
 ⟋⟍Br	**14**	62	78
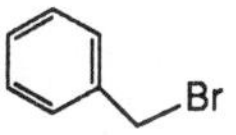 Ph⟍Br	**13**	66	75
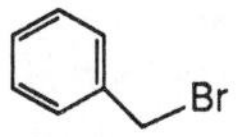 Ph⟍Br	**14**	64	85
MeBr	**13**	42	60
n-BuBr	**13**	52	61
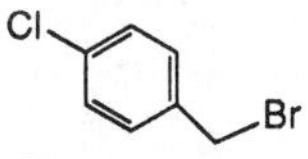 Cl–C6H4–CH2Br	**13**	66	81
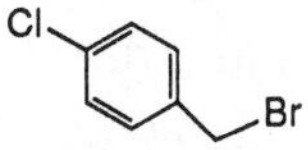 Cl–C6H4–CH2Br	**14**	62	82
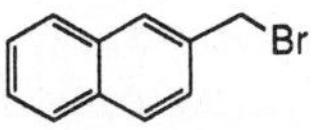 naphthyl-CH2Br	**13**	54	82
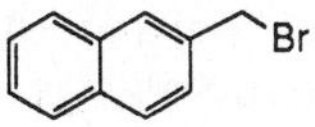 naphthyl-CH2Br	**14**	48	81

SCHEME 4

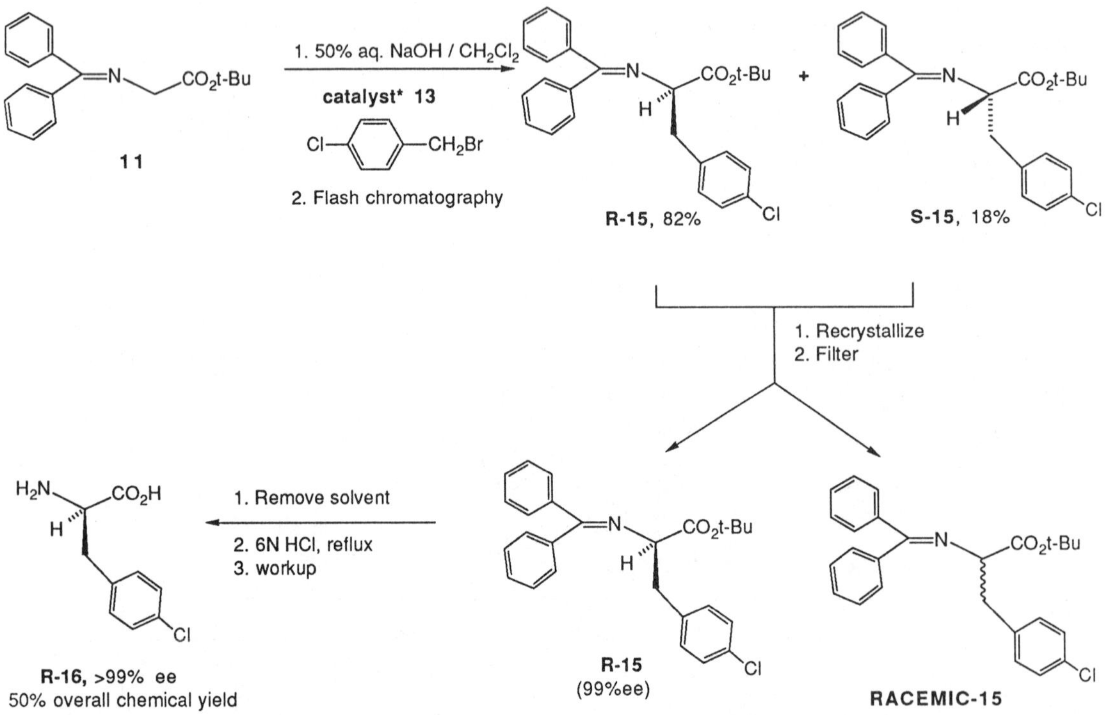

O'Donnell, M.J.; Bennett, W.D.; Wu, S., *J.Am.Chem.Soc.* (1989) **111**, 2353

of (R)-**15** (82%) and (S)-**15** (18%). Recrystallization of the product following a flash chromatographic separation gives (R)-**15** in >99%ee and racemic **15** remains in the mother liquors. Removal of the solvent and hydrolysis affords *para*-chlorophenylalanine (**16**) in 50% overall chemical yield and >99% ee. Presumably, the benzhydryl Schiff base moiety provides a consistent series of crystalline derivatives (**12**) that can be optically enriched through recrystallization. The microfilm edition of this paper provides experimental details for the synthesis of **16**.

Woodard, et.al.[6] have recently reported a multi-gram synthesis of (S)- and (R)-[2-²H] glycines (**21** and **26**) as shown in Schemes 5 and 6. The authors cite over 11 reported syntheses of chiral glycine; the advantage of the present approach being the high isotopic and good optical purity of the final amino acids. The authors deploy (S)-(-)-and (R)-(+)--Alpine Boranes to reduce the deuteriated aldehydes **18** and **23**, respectively. The %ees' for the obtained alcohols are 76% ee and 82% ee, respectively. These numbers

SCHEME 5

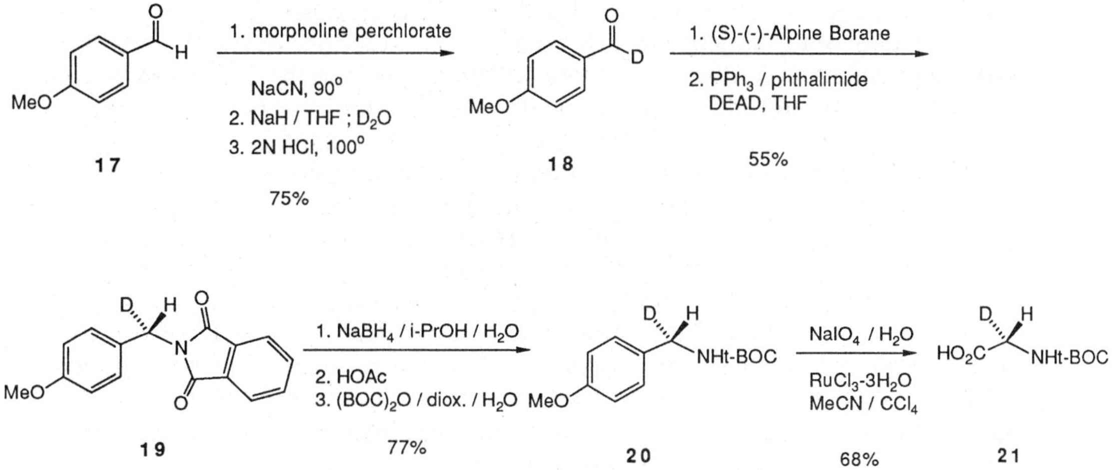
MeO
17
1. morpholine perchlorate
NaCN, 90°
2. NaH / THF ; D2O
3. 2N HCl, 100°
75%
MeO
18
1. (S)-(-)-Alpine Borane
2. PPh3 / phthalimide
DEAD, THF
55%
MeO
19
1. NaBH4 / i-PrOH / H2O
2. HOAc
3. (BOC)2O / diox. / H2O
77%
NHt-BOC
MeO
20
NaIO4 / H2O
RuCl3-3H2O
MeCN / CCl4
68%
HO2C
NHt-BOC
21

SCHEME 6

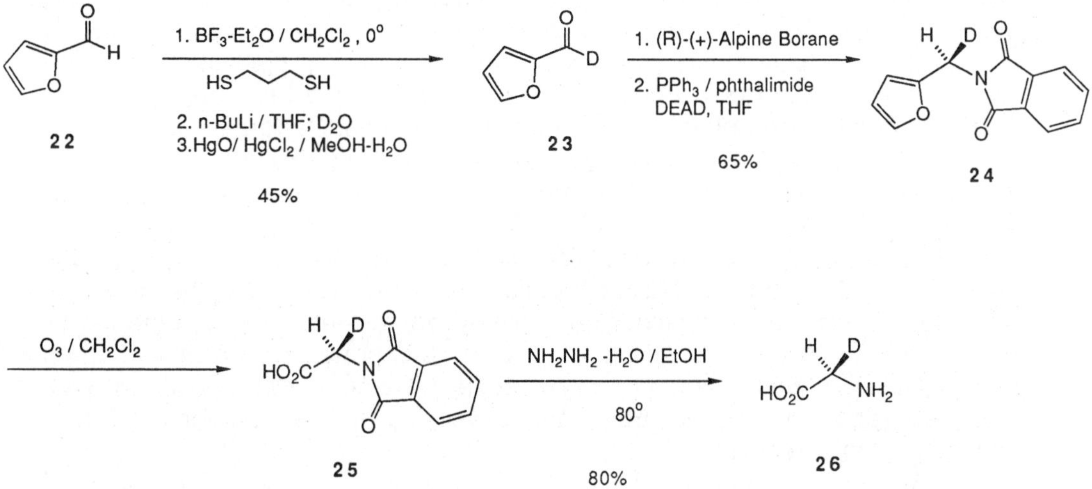
22
1. BF3-Et2O / CH2Cl2 , 0°
HS SH
2. n-BuLi / THF; D2O
3.HgO/ HgCl2 / MeOH-H2O
45%
23
1. (R)-(+)-Alpine Borane
2. PPh3 / phthalimide
DEAD, THF
65%
24
O3 / CH2Cl2
HO2C
25
NH2NH2 -H2O / EtOH
80°
80%
HO2C
NH2
26

Ramalingam, K.; Nanjappan, P.; Kalvin, D.M.; Woodard, R.W., *Tetrahedron* (1988) **44**, 5597 also represent the maximum %ees' of the final chiral glycines. The disadvantages of the two routes shown are the high costs of the deuteriated aldehydes and the moderate asymmetric induction of the borane reductions. On the other hand, the final glycine products are not contaminated with significant amounts of di-deuterio glycine which is a major contaminant in the alternative enzymatic exchange syntheses. This method is not readily applicable to the synthesis of chiral tritiated glycine, however; an alternate and more direct synthesis reported by Williams [7] (see Chapter 1, Scheme 108) should be considered for the latter application. Complete experimental details accompany this work.

The synthesis of optically pure vinylglycine (37) has been a challenging and important synthetic problem. In addition to the enzyme-inhibitory and antibiotic properties that this simple amino acid displays, it has become an important chiral starting material for a variety of other amino acid syntheses and optically active products. Three practical syntheses of L-vinylglycine have been recorded as shown in Schemes 7-11. Barton and associates [8] have employed an interesting oxidative decarboxylation reaction on two different L-glutamic acid derivatives (Schemes 7-9). Treatment of the protected L-glutamate 27 with N-hydroxy-2-selenopyridine (28) and isobutylchlorformate followed by photolysis afforded the selenide 29 in good yield. Ozonolysis of this substance afforded N-t-BOC vinylglycine benzyl ester (30) in 81% yield. In a related sequence, the oxazolidinone 31 was transformed into the corresponding vinyl substrate 33; this substance was previously reported by Hanessian [10] (Scheme 11) and shown to be convertible into L-vinylglycine via acid hydrolysis. In a third iteration, a modification of the Hunsdiecker reaction developed by Barton, afforded the bromide 35 in good yield. Selenide formation and oxidative elimination afforded 30 in high yield. Final hydrolysis of this substance in refluxing 6N HCl afforded L-vinylglycine-hydrochloride in 70% yield (1mMol scale reported). Complete experimental details are provided.

Rapoport [9] realized the first practical synthesis of L-vinylglycine as shown in Scheme 10. N-CBz methionine methyl ester (38) is converted into the corresponding sulfoxide and pyrolyzed to yield the protected vinyl glycine 40 in 80% yield. Acidic removal of the protecting groups afforded L-vinylglycine in 88% yield (10 mMol scale). These workers also reported the preparation of the N-t-BOC derivative 41; full experimental details accompany this report.

Hanessians' approach [10] is similar to that subsequently deployed by Barton and begins with L-glutamic acid (Scheme 11). N-CBz-L-glutamic acid methyl ester (43) is oxidatively decarboxylated with lead tetraacetate in hot benzene to directly furnish the protected vinylglycine derivative 44. Acidic hydrolysis of this substance afforded L-vinylglycine (10 mMol scale reported). Of the three approaches, the Hanessian

SCHEME 7

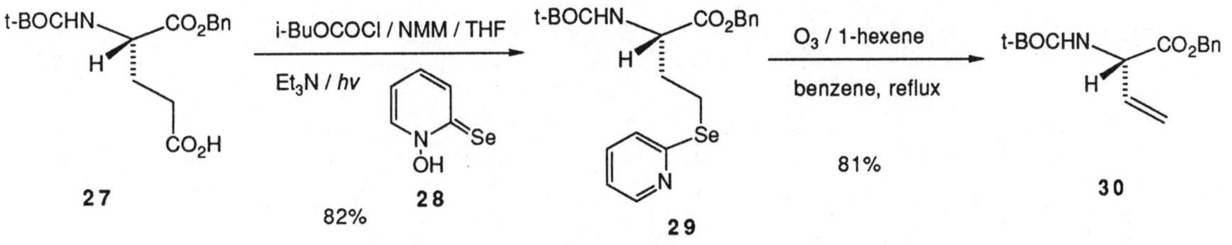

| 27 | 82% 28 | 29 | 81% | 30 |

SCHEME 8

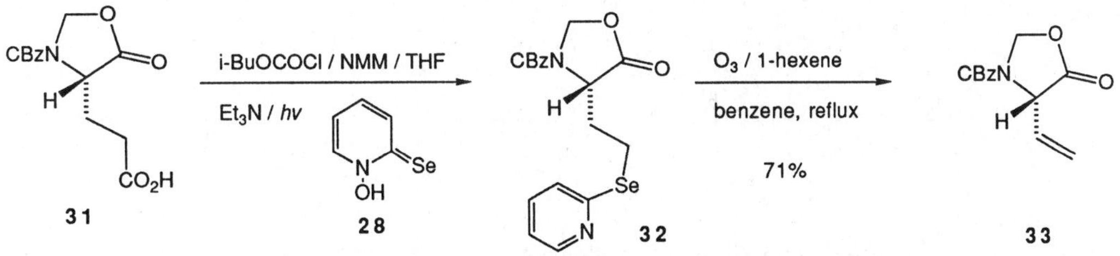

| 31 | 28 | 32 | 71% | 33 |

SCHEME 9

| 27 | 82% | 35 | ~quant. | 36 |

| 30 | 70% | 37, *L-VINYLGLYCINE* |

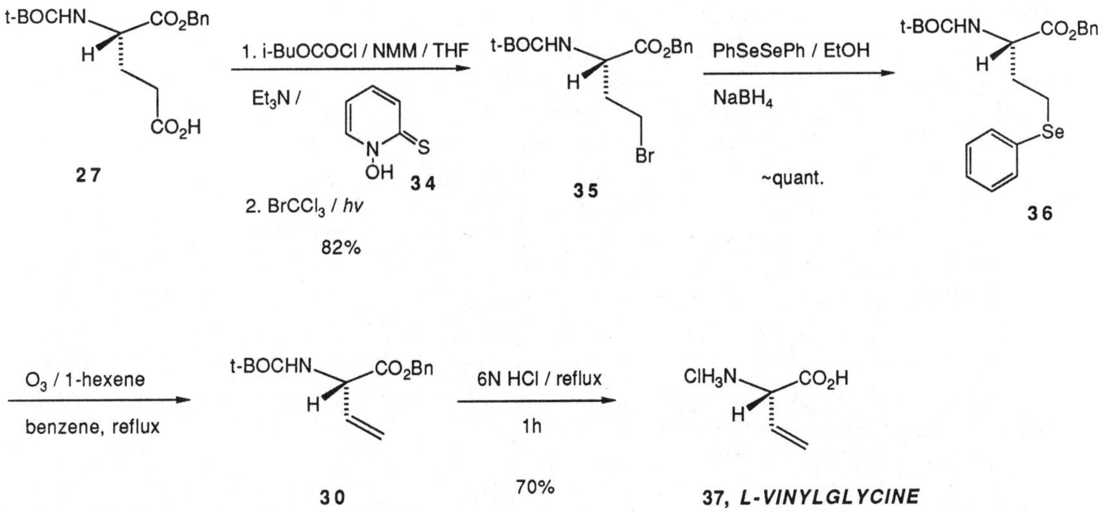

Barton, D.H.R.; Crich, D.; Herve, Y.; Potier, P.; Thierry, J., *Tetrahedron* (1985) **41**, 4347

Miscellaneous Methods

SCHEME 10

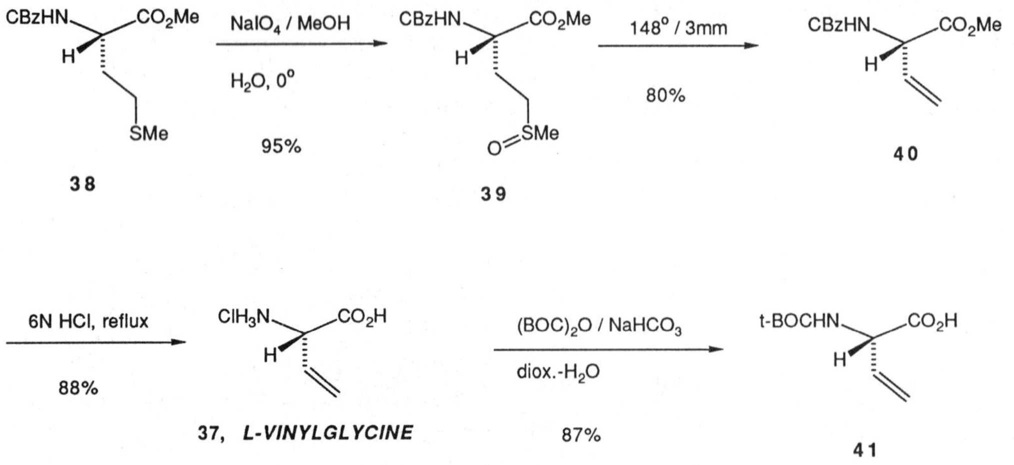

Afzali-Ardakani, A.; Rapoport, H., *J.Org.Chem.* (1980) **45**, 4817

SCHEME 11

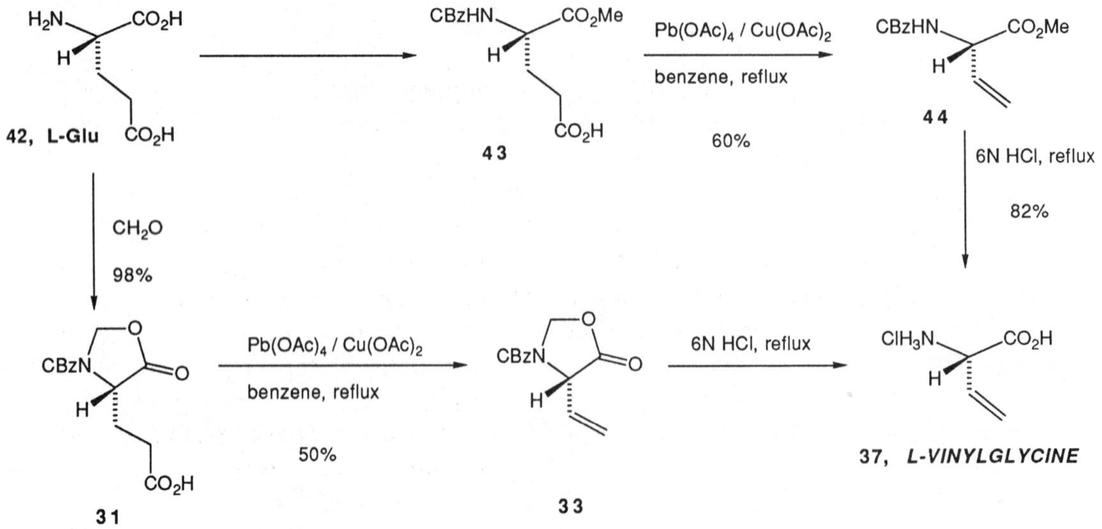

Hanessian, S.; Sahoo, S.P., *Tetrahedron Lett.* (1984) **25**, 1425

synthesis appears to be the most convenient and economical. This report provides the experimental procedure.

Harding, et.al.[11], have examined the intramolecular oxymercuration of homoallyic alcohols to access *erythro-* and *threo-*γ-hydroxynorvaline derivatives (**50** and **52**, Schemes 12 and 13). Although racemic substrates were prepared, the wide range of technologies available for preparing optically active homoallylic alcohols renders this approach of potential utility to prepare optically active substances. The authors note an interesting change in diastereoselectivity during the intramolecular mercuration by the addition or deletion of HBr (kinetic and thermodynamic control was invoked).

Hirama and associates [12] have developed an interesting intramolecular Michael-type addition of trichloromethyl carbamoyl sulfones (**54**) to afford the cyclic urethanes **55**. The key substrates are prepared from optically active α-hydroxy aldehydes and methylthiomethyl tolylsulfone. Pummerrer oxidation of the cyclization products provides the thiol esters (**56**) which can be hydrolyzed to the corresponding *syn-*β-hydroxy-α-amino acid derivatives. Using the same basic protocol, the dihydroxy substance **59** was prepared from **58** and the trihydroxy norvaline derivative (**61**) was prepared from **60**.

Ohfune and Shimamoto [13] have developed a method to oxidize tyrosine derivatives at the β-position through the agency of potassium peroxodisulfate (Scheme 15). The authors note that this reaction is restricted to tyrosine derivatives since, the corresponding phenylalanine derivatives were unreactive under the same conditions. The oxidation displays high *threo-*selectivity which can be rationalized by considering the least hindered conformer of the incipient benzylic cation which is trapped intramolecularly by the urethane moiety.

An extremely interesting and challenging class of amino acids for which there are no suitable asymmetric synthetic solutions for are the α-allenic amino acids and the α-ethynyl amino acids. Both systems have been shown to be potent irreversible inhibitors of pyridoxal-dependent enzymes. Chemically these substances are quite labile to racemization, isomerization to α,β-dehydroamino acids and other decomposition pathways. A Syntex group[13] has developed a useful racemic synthesis of α-substituted-α-allenic amino acids (**70**, Scheme 16) by a Claisen rearrangement of α-benzamido propargylic esters. The authors state that the yields for the Claisen rearrangement are 50-100%. Deprotection of the sensitive allenic products (**69**) to the free amino acids was accomplished in 50-80% yields via treatment of **69** with Meerweins' reagent and gentle hydrolysis.

Cazes and co-workers [14] have deployed a palladium-catalyzed coupling of the glycinate anion **72** with the allenic phosphates **71** in good

Miscellaneous Methods

SCHEME 12

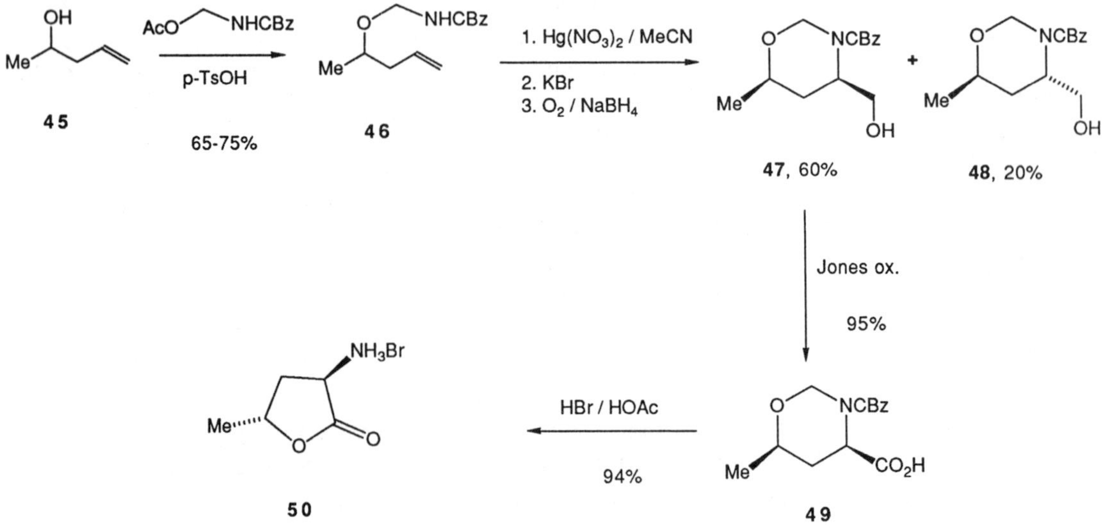

45 → **46**

65-75%

1. Hg(NO$_3$)$_2$ / MeCN
2. KBr
3. O$_2$ / NaBH$_4$

47, 60% + **48**, 20%

Jones ox.

95%

HBr / HOAc

94%

50

49

SCHEME 13

46

1. Hg(NO$_3$)$_2$ / MeCN / HBr
2. KBr
3. O$_2$ / NaBH$_4$

47, 4.5% + **48**, 52%

Jones ox.

97%

HBr / HOAc

88%

52

51

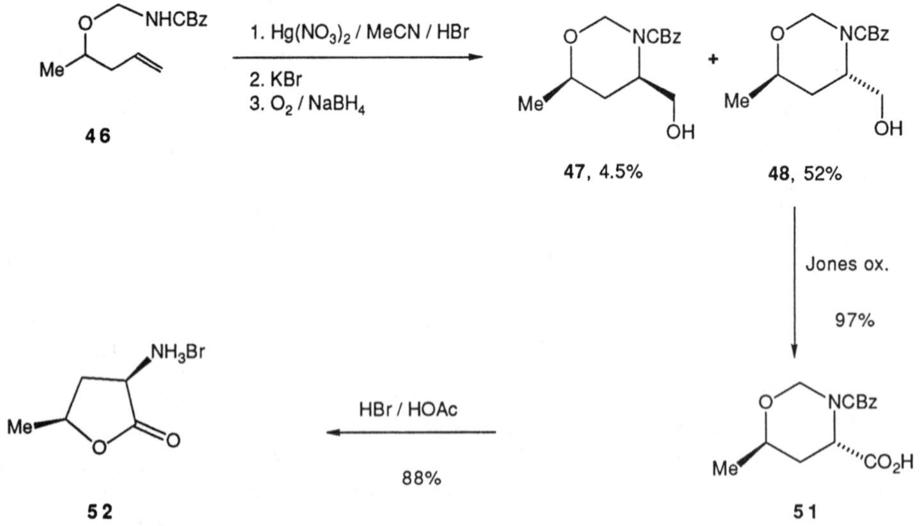

Harding, K.; Marman, T.H.; Nam, D., *Tetrahedron Lett.* (1988) **29**, 1627.

SCHEME 14

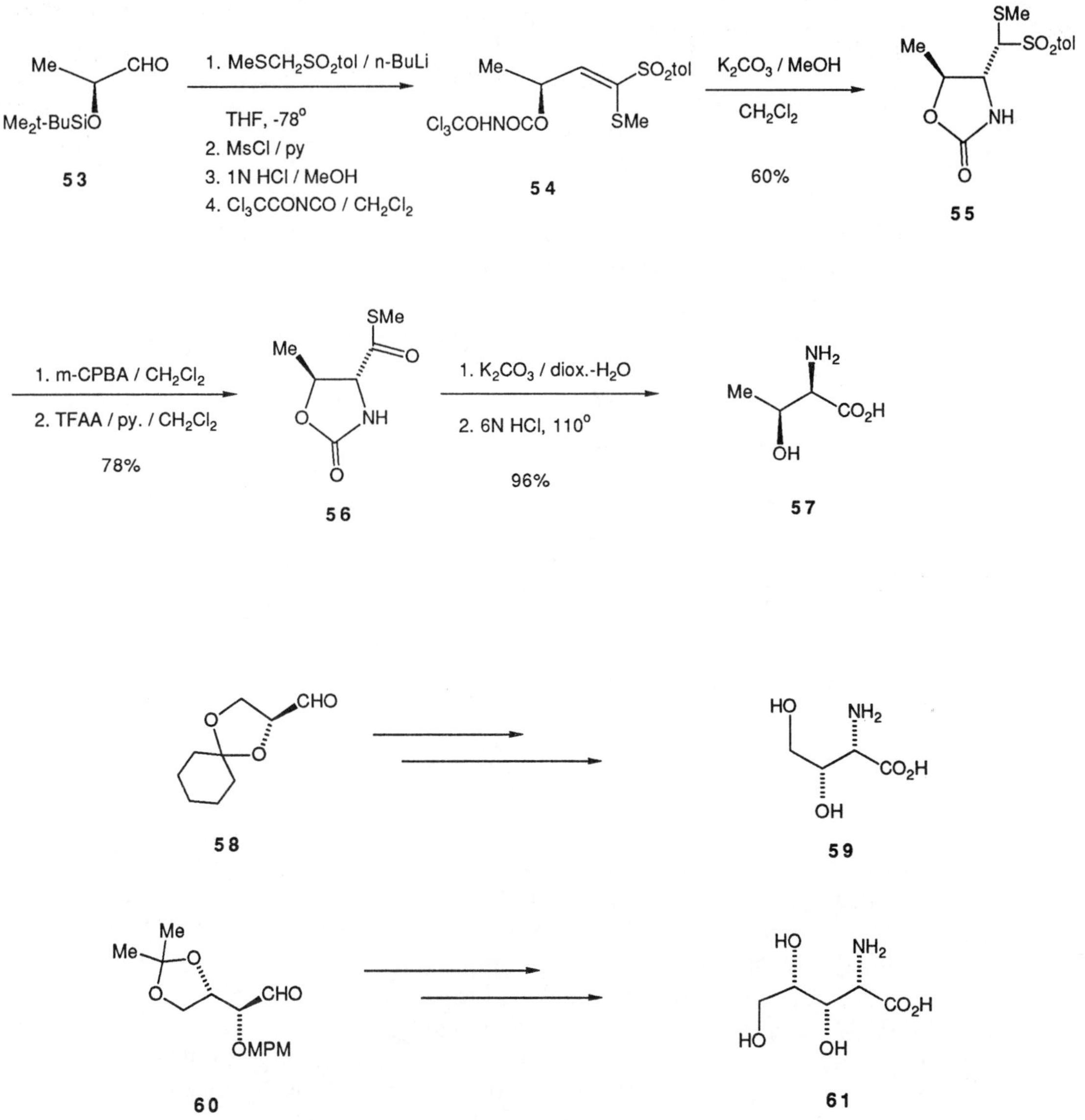
Me
CHO
Me2t-BuSiO
53
1. MeSCH2SO2tol / n-BuLi
THF, -78°
2. MsCl / py
3. 1N HCl / MeOH
4. Cl3CCONCO / CH2Cl2
Me
SO2tol
Cl3COHNOCO
SMe
54
K2CO3 / MeOH
CH2Cl2
60%
SMe
Me
SO2tol
O
NH
O
55
1. m-CPBA / CH2Cl2
2. TFAA / py. / CH2Cl2
78%
SMe
Me
O
O
NH
O
56
1. K2CO3 / diox.-H2O
2. 6N HCl, 110°
96%
NH2
Me
CO2H
OH
57
CHO
O
O
58
HO
NH2
CO2H
OH
59
Me
Me
O
O
CHO
OMPM
60
HO
NH2
CO2H
HO
OH
61

Hirama, M.; Hioki, H.; Ito, S., *Tetrahedron Lett.* (1988) **29**, 3125
Hirama, M.; Hioki, H.; Ito, S.; Kabuto, C., *Tetrahedron Lett.* (1988) **29**, 3121

Miscellaneous Methods

SCHEME 15

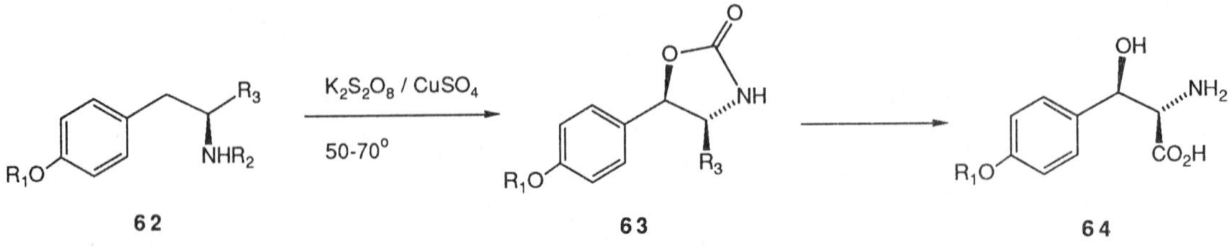

$$62 \quad\xrightarrow[\;50\text{-}70^\circ\;]{K_2S_2O_8 \,/\, CuSO_4}\quad 63 \quad\longrightarrow\quad 64$$

Shimamoto, K.; Ohfune, Y., *Tetrahedron Lett.* (1988) **29**, 5177

R$_1$	R$_2$	R$_3$	YIELD (% 63)	ERYTHRO : THREO
Me	t-BOC	COOMe	55	1 : 49
Me	CBz	COOMe	4	
Me	t-BOC	CH$_2$OAc	0	
Me	t-BOC	CH$_2$OAc	76	1 : 37
Me	t-BOC	CH$_2$OSiMe$_2$t-Bu	0	
Me	t-BOC	H	75	
Ac	t-BOC	CH$_2$OAc	4	
MEM	t-BOC	CH$_2$OAc	0	

yields. Acidic removal of the Schiff base moiety and base removal of the methyl esters provides the β-allenic-α-amino acids **75**. A general experimental procedure is provided along with salient spectroscopic data for the derivatives prepared.

In 1978, Casara and Metcalf [15] prepared a variety of α-alkylated ethynylglycine derivatives as shown in Schemes 18 and 19. It is interesting that the parent amino acid , ethynylglycine (FR900130) was subsequently isolated from *Streptomyces catenulae* by a Fujisawa group in 1980 [16]. The parent amino acid displayed antimicrobial properties but proved to be extremely unstable to purification and handling. The absolute configuration of the natural product is not known. Coupling of bis-(trimethylsilyl)acetylene with the chloro-glycinate **76** furnished the fully protected substance **77** in 65% yield. Dianion formation and alkylation proceeded in a regioselective manner to afford the α-alkylated protected

SCHEME 16

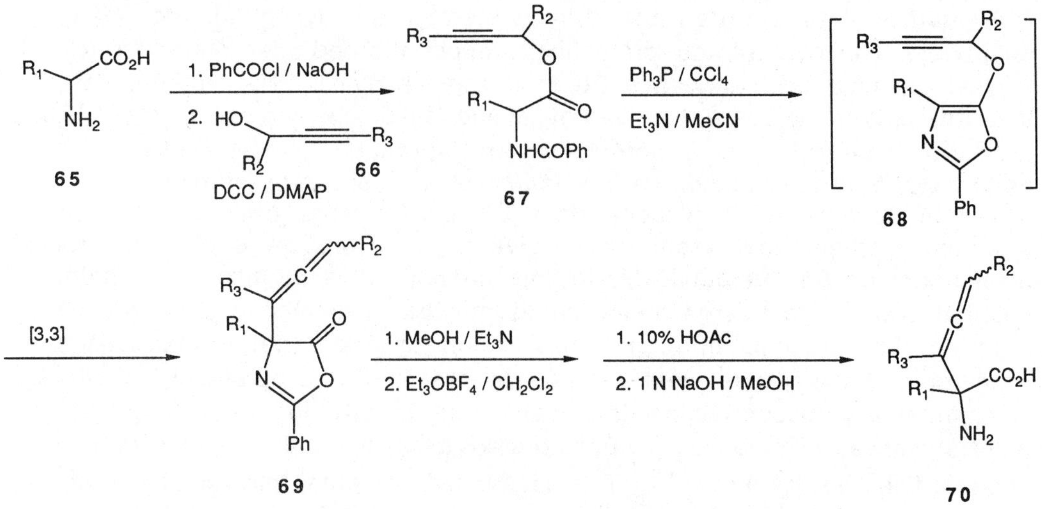

Castelhano, A.L.; Pliura, D.H.; Taylor, G.J.; Hsieh, K.C.; Krantz, A., *J.Am.Chem.Soc.* (1984) **106**, 2734

SCHEME 17

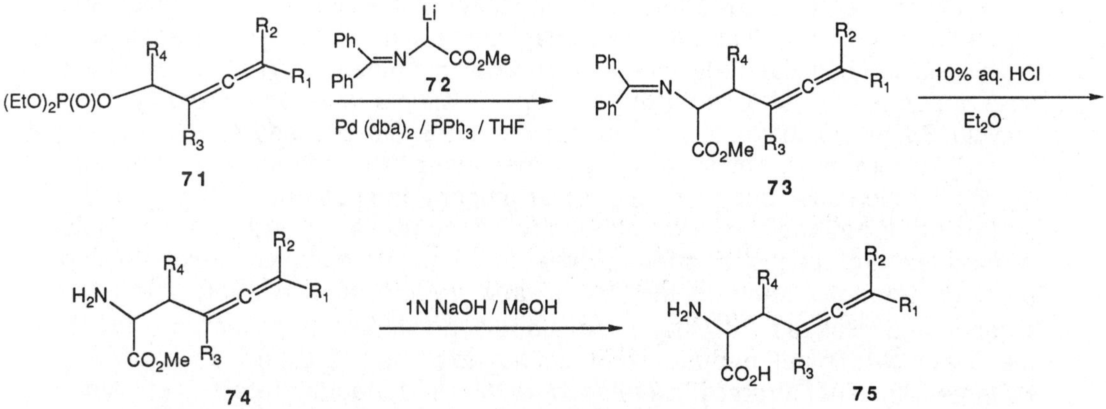

Cazes, B.; Djahanbini, D.; Gore, J.; Genet, J-P.; Gaudin, J-M., *Synthesis* (1988) 983

R_1	R_2	R_3	R_4	YIELD (% 73)	YIELD (%74)	YIELD (%75)
H	H	H	H	66	72	88
Me	H	H	H	64	76	-
H	H	Me	H	81	62	-
Me	Me	H	H	51	83	82
H	H	H	Me	64	74	-

ethynylglycine derivatives **78** in good yields. In one instance, the authors were able to deprotect the alkylation adduct to the free amino acid. Thus, as shown in Scheme 19, conjugate addition of the dianion derived from **77** with methacrylate afforded the fully protected ethynyl glutamate **79** in 65% yield. Removal of the protecting groups afforded the interesting α-ethynyl glutamic acid derivative **80** in unspecified yield. The authors note that attempts to prepare the parent amino acid ethynylglycine from **77** were unsuccessful.   In a complementary approach, these workers [17] performed a regioselective carboxylation of the propargyl amine **81** as shown in Scheme 20. The labile acids **82** and **83** were directly esterified with diazomethane and separated to give a 4 : 1 mixture of **84** and the allenic isomer **85**, respectively. In this context, it is significant to point out that there have been virtually no asymmetric α-amino acid syntheses reported that proceed through carboxylation of appropriately derivatized α-amino carbanion equivalents. This appears to be a potentially useful, complementary disconnection that has been largely ignored. Numerous other syntheses of racemic β,γ-unsaturated α-amino acid derivatives have appeared in the literature [18]. This is an area of great medicinal interest that needs improved and versatile asymmetric synthetic technologies.

Kuzuhara and associates [19] have examined a bio-mimetic transamination of a chiral pyridoxamine derivative (**86**, Scheme 21) with several α-keto acids (**87**). The experimental protocol involves admixture of **86** and **87** in methanol in the presence of zinc perchlorate for 1 day at ambient temperature to afford the optically active amino acid (**89**) and the aldehyde component **88**. Presumably, the aldehyde is recycled to the amine via a chemical reductive amination with ammonia. The %ees' range from moderate to 96% ee (for leucine). A simple method to catalytically recycle **88** would make this a much more attractive procedure.

The N-methyl amino acids are becoming an increasingly important moiety in bioactive peptides and other important substances. Despite the apparent functional group simplicity, N-methylation of amino acid derivatives and peptides often proves to be a difficult, and low yielding process through standard alkylation and reductive alkylation methods. Grieco and Bahsas [20] have developed an interesting method to N-methylate N-terminal peptides and amino acid methyl esters as shown in Scheme 22. The strategy developed involves aqueous hetero-Diels-Alder addition of cyclopentadiene to formaldehyde imines generated *in situ* from the corresponding primary amines (**90**) affording a diastereomeric mixture of the adducts **91**. Triethylsilane reduction of the incipient imine generated by acid-catalyzed retro aza Diels-Alder reaction furnishes the desired mono-N-methylated amino acids and peptides (**92**) in good yields as shown in the Table. The authors note that the procedure is compatible

SCHEME 18

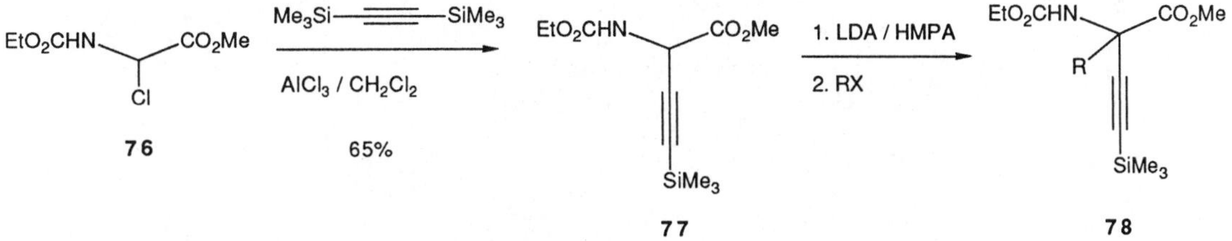

Casara, P.; Metcalf, B.W., *Tetrahedron Lett.* (1978) 1581

RX	YIELD (% 78)
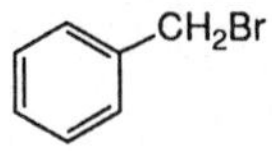 ⌬CH₂Br	75
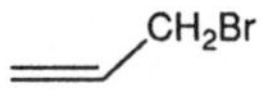 ⟋⟍CH₂Br	70
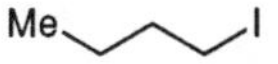 Me⟋⟍⟋I	60

SCHEME 19

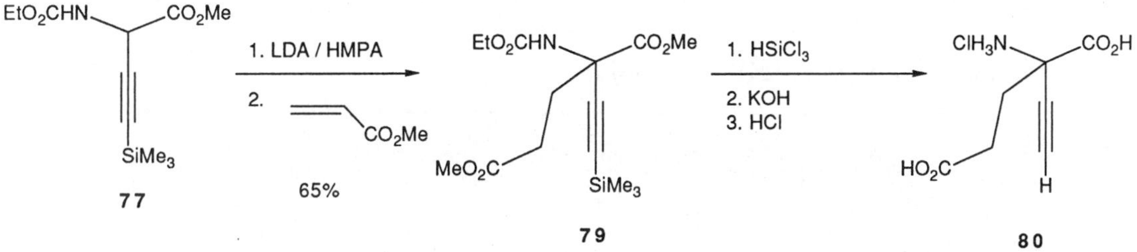

SCHEME 20

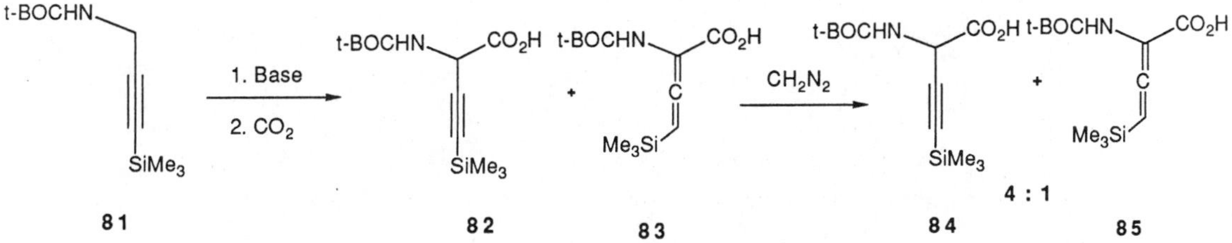

Metcalf, B.W.; Casara, P., *J.Chem.Soc. Chem.Comm.* (1979) 119

Miscellaneous Methods

SCHEME 21

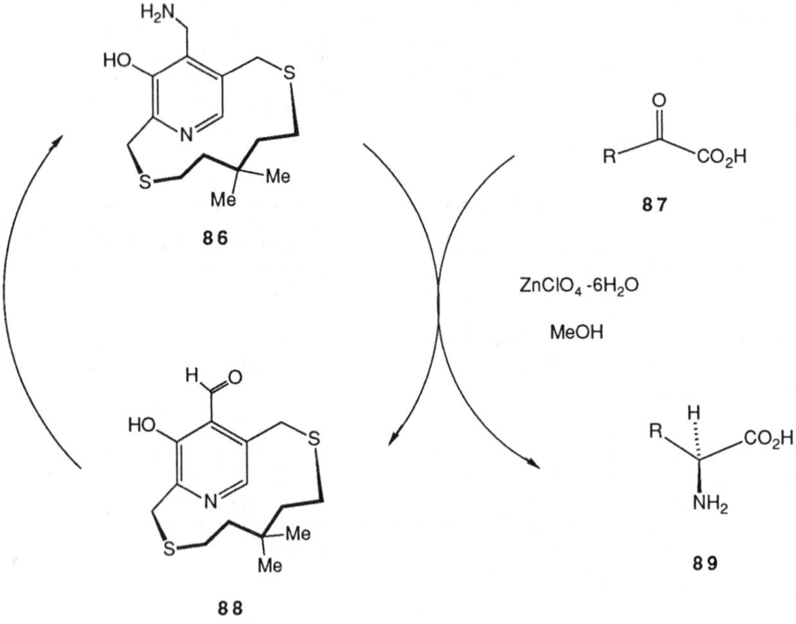

Tachinaba, Y.; Ando, M.; Kuzuhara, H., *Chemistry Lett.* (1982) 1765

R	ABSOLUTE CONFIG. (86)	YIELD (%89)	%ee (89)
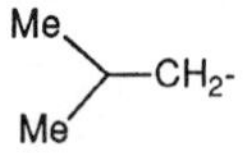 Me₂CH—CH₂- (isobutyl)	R	68	96 (S)
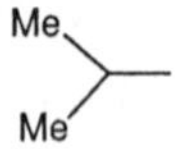 Me₂CH- (isopropyl)	R	57	79 (S)
Me	R	72	69 (S)
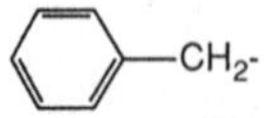 Ph—CH₂- (benzyl)	R	60	61 (S)
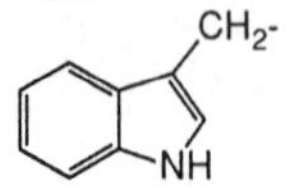 indol-3-yl-CH₂-	S	62	60 (R)

SCHEME 22

SUBSTRATE (90)	ADDUCT (91)	YIELD (%91)	RATIO	YIELD (%92)
H-Leu-OMe-HCl		94	1.7 : 1	91
H-Phe-OMe-HCl		84	1.8 : 1	92
H-Val-OMe-HCl		82	4 : 1	78
H-Tyr-OMe-HCl		87	2 : 1	75
H-Phenylglycine-methylester-HCl		94	1.7 :1	88
H-Ser-OMe-HCl		98	3.3 : 1	83
H-Leu-Phe-OMe-HCl		87	1 : 1	78
H-(Ala)$_2$Ala-OMe-HCl		81	1 : 1	67

Grieco, P.A.; Bahsas, A., *J.Org.Chem.* (1987) **52**, 5746

generated by acid-catalyzed retro aza Diels-Alder reaction furnishes the desired mono-N-methylated amino acids and peptides (**92**) in good yields as shown in the Table. The authors note that the procedure is compatible with unprotected phenols and alcohols and proceeds in an acidic medium which obviates most racemization problems associated with base-labile residues.

Hegedus, et.al.[21], have devised an interesting and highly unusual approach to the synthesis of amino acids based on diastereoselective reactions of chromium carbene complexes. As shown in Scheme 23, the alkoxy carbene **93** is acylated and treated with the optically active *erythro*-diphenyl ethanolamine **94**. Acetate displacement results in the amino carbenes **95**. Photolysis of these substances results in CO migration to afford the incipient chromium-complexed ketenes **96**; subsequent stereospecific lactonization/protonation furnishes the *syn*-oxazinones **97**. The intrinsic stereoelectronic parameters that result in the stereospecific breakdown of **96** are not clear, but will provide a fascinating mechanistic problem to elucidate. These substances can be reductively cleaved to the amino acids (**98**) or, in the case of the arylglycine cases, ring-opened and oxidatively cleaved to the amino acids as described in Chapter 1 (Williams, Scheme 117). In a related approach, the phenylglycinol-derived amino carbenes **101** undergoe a diastereoselective intermolecular addition of *tert*-butyl alcohol to the incipient chromium-complexed ketene formed from photolytic CO migration from **101**. The protected *t*-butyl ester **102** is formed in high yield and >93% de. Alternatively, the methyl residue of **101** can be alkylated and manipulated as above to afford the homologated adducts **103**. The authors' well established history of publishing full papers indicate that complete experimental descriptions of these preliminary studies will be forthcoming.

On the following pages, a series of isotopically labelled amino acids and some chiral 1-aminocycloprpane carboxylic acids are listed along with the references to their preparation. Space did not permit a detailed description of the synthetic schemes deployed; it is hoped that a quick visual reference to these structures, useful for mechanistic and biochemical studies will be of value.

SCHEME 23

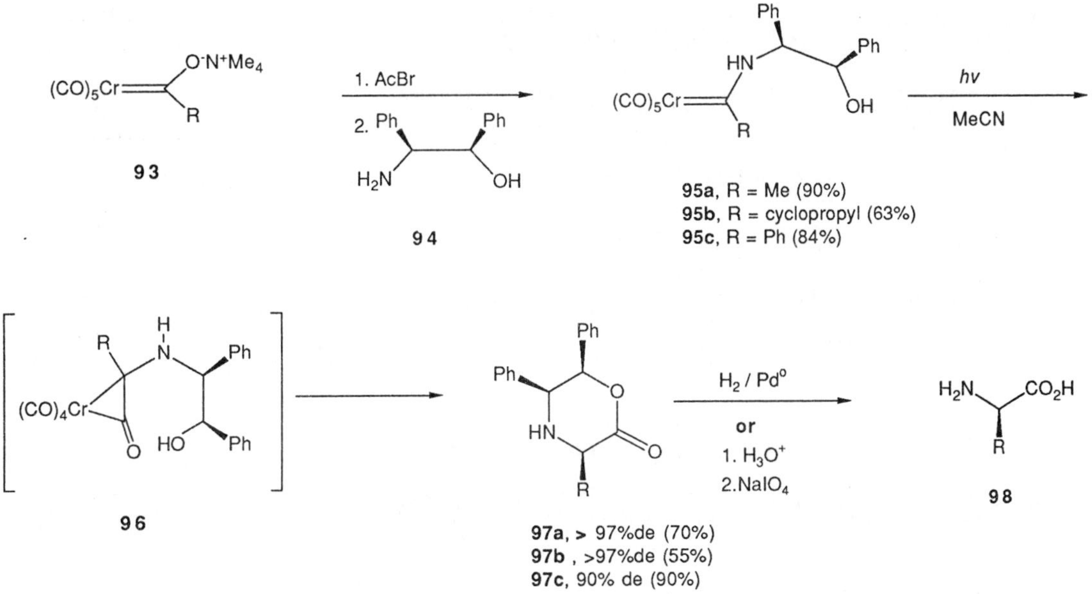
$(CO)_5Cr$
$O^-N^+Me_4$
R
93
1. AcBr
2. Ph Ph
H_2N OH
94
$(CO)_5Cr$
HN
Ph
Ph
OH
R
95a, R = Me (90%)
95b, R = cyclopropyl (63%)
95c, R = Ph (84%)
hv
MeCN
R
H
N
Ph
$(CO)_4Cr$
O
HO
Ph
96
Ph
Ph
O
HN
O
R
97a, > 97%de (70%)
97b , >97%de (55%)
97c, 90% de (90%)
H_2 / Pd°
or
1. H_3O^+
2.$NaIO_4$
H_2N
CO_2H
R
98

Hegedus, L.S.; de Weck, G.; D'Andrea, S., *J.Am.Chem.Soc.* (1988) **110**, 2122

SCHEME 24

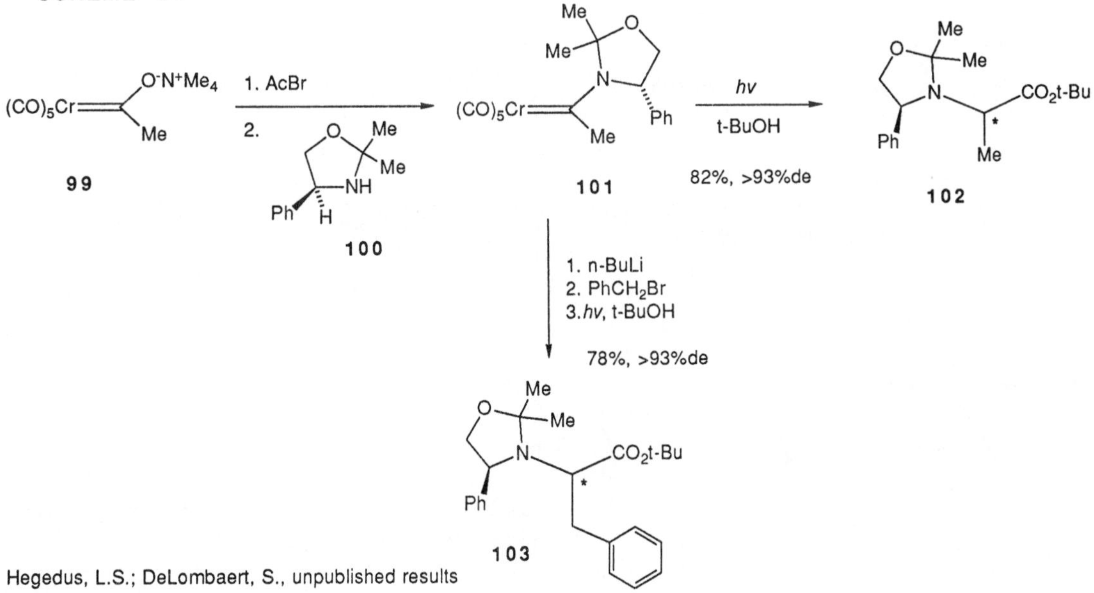
$(CO)_5Cr$
$O^-N^+Me_4$
Me
99
1. AcBr
2.
O
Me
Me
NH
Ph
H
100
Me
O
Me
N
$(CO)_5Cr$
Ph
Me
101
hv
t-BuOH
82%, >93%de
Me
O
Me
N
CO_2t-Bu
Ph
Me
102
1. n-BuLi
2. $PhCH_2Br$
3.*hv*, t-BuOH
78%, >93%de
Me
O
Me
N
CO_2t-Bu
Ph
103

Hegedus, L.S.; DeLombaert, S., unpublished results

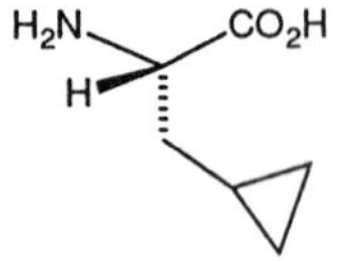

CYCLOPROPYLALANINE

Ohta, T.; Nakajima, S.; Sato, Z.; Aoki, T.; Hatanaka, S.; Nozoe, S., *Chemistry Lett.* (1986) 511.

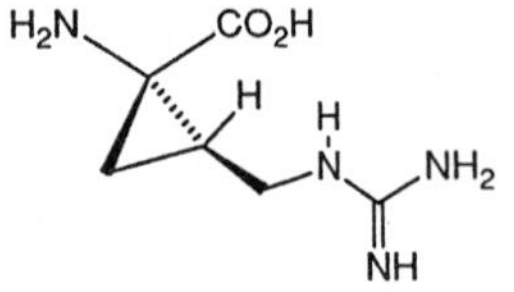

CARNOSADINE

Wakima, T.; Oda, Y.; Fujita, H.; Shiba, T., *Tetrahedron Lett.* (1986) **27**, 2143.

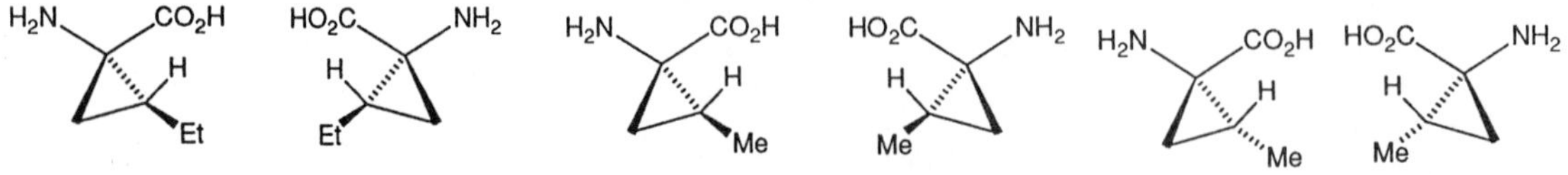

Baldwin, J.E.; Adlington, R.M.; Rawlings, B.J.; Jones, R.H., *Tetrahedron Lett.* (1985) **26**, 485
Baldwin, J.E.; Adlington, R.M.; Rawlings, B.J., *Tetrahedron Lett.* (1985) **26**, 481

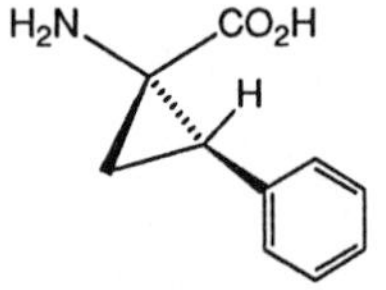

Kimura, H.; Summer, C.H., *J.Org.Chem.* (1983) **48**, 2440.

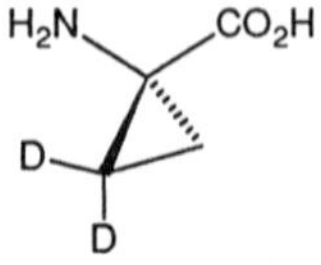

Hill, R.K.; Prakash, S.R.; Weisendanger, R.; Angst, W.; Martinoni, B.; Arigoni, D.;
Liu, H-W.; Walsh, C.T., *J.Am.Chem.Soc.* (1984) **106**, 795

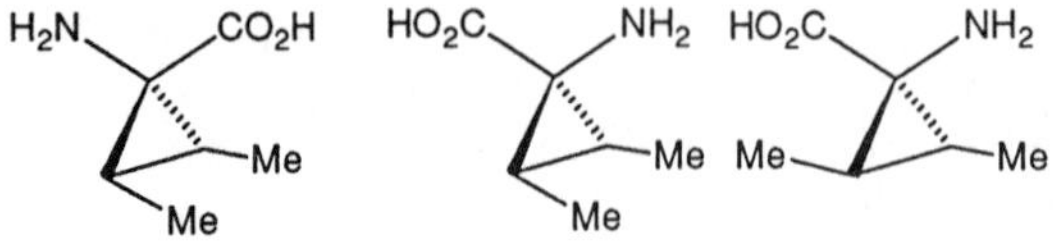

Baldwin, J.E.; Adlington, R.M.; Lajoie, G.A.; Lowe, C.;
Baird, P.D.; Prout, K., J.Chem.Soc.Chem.Comm. (1988) 775

Subramanian, P.K.; Kalvin, D.M.; Ramalingam, K.; Woodard, R.W., *J.Org.Chem.* (1989) **54**, 270.

Son, J-K.; Woodard, R.W., *J.Am.Chem.Soc.* (1989) **111**, 1363

Schwab, J.M.; Ray, T.; Ho, C-K., *J.Am.Chem.Soc.* (1989) **111**, 1057

Chang, M.N.T.; Walsh, C.T., *J.Am.Chem.Soc.* (1980) **102**, 7370

R = CD$_3$, H

Sawada, S.; Nakayama, T.; Esaki, N.; Tanaka, H.; Soda, K.; Hill, R.K., *J.Org.Chem.* (1986) **51**, 3384

References   Chapter   8

1.   Matteson, D.S.; Beedle, E.C., *Tetrahedron Lett.* (1987) **28**, 4499.

2.   Georg, G.; Guan, X.; Kant, J., *Tetrahedron Lett.* (1988) **29**, 403.

3.   Tomioka, K.; Ando, K.; Takemasa, Y.; Koga, K., *J.Am.Chem.Soc.* (1984) **106**, 2718.

4.   O'Donnell, M.J.; Bennett, W.D.; Wu, S., *J.Am.Chem.Soc.* (1989) **111**, 2353.

5.   Dolling, U-H.; Davis, P.; Grabowski, E.J., *J.Am.Chem.Soc.* (1984) **106**, 446.

6.   Ramalingam, K.; Nanjappan, P.; Kalvin, D.M.; Woodard, R.W., *Tetrahedron* (1988) **44**, 5597.

7.   a) Williams, R.M.; Zhai, D.; Sinclair, P.J., *J.Org.Chem.* (1986) **51**, 5021; b) Ramer, S.E.; Cheng, H.; Palcic, M.M.; Vederas, J.C., *J.Am.Chem.Soc.* (1988) **110**, 8526.

8.   Barton, D.H.R.; Crich, D.; Herve, Y.; Potier, P.; Thierry, J., *Tetrahedron* (1985) **41**, 4347.

9.   Afzali-Ardakani, A.; Rapoport, H., *J.Org.Chem.* (1980) **45**, 4817.

10.   Hanessian, S.; Sahoo, S.P., Tetrahedron Lett. (1984) 25, 1425.

11.   Harding, K.E.; Marman, T.H.; Nam, D., *Tetrahedron Lett.* (1988) **29**, 1627.

12.   a) Hirama, M.; Hioki, H.; Ito, S., *Tetrahedron Lett.* (1988) **29**, 3125; b) Hirama, M.; Hioki, H.; Ito, S.; Kabuto, C., *Tetrahedron Lett.* (1988) **29**, 3121.

13.   a) Castelhano, A.L.; Pliura, D.H.; Taylor, G.J.; Hsieh, K.C.; Krantz, A., *J.Am.Chem.Soc.* (1984) **106**, 2734; see also : b) Casara, P.; Jund, K.; Bey, P., *Tetrahedron Lett.* (1984) **25**, 1891.

14.   Cazes, B.; Djahanbini, D.; Gore, J.; Genet, J-P.; Gaudin, J-M., *Synthesis* (1988) 983.

15.   Casara, P.; Metcalf, B.W., *Tetrahedron Lett.* (1978) 1581.

16.   a) Kuroda, Y.; Okuhara, M.; Goto, T.; Iguchi, E.; Kohsaka, M.; Aoki, H.; Imanaka, H., *J.Antibiotics* (1980) **33**, 125; b) Kuroda, Y.; Okuhara, M.; Goto, T.; Kohsaka, M.; Aoki, H.; Imanaka, H., *J.Antibiotics* (1980) **33**, 132.

17.   Metcalf, B.W.; Casara, P., *J.Chem.Soc.Chem.Comm.* (1979) 119.

18.   For other miscellaneous syntheses of racemic β,γ-unsaturated α-amino acids, see : a) Baldwin, J.E.;   Haber, S.B.; Kruse, L.I., *J.Org.Chem.* (1977) **42**, 1239; b) Greenlee, W.J.,*J.Org.Chem.* (1984) **49**, 2632; c) Hudrlik, P.F.; Kulkarni, A.K., *J.Am.Chem.Soc.* (1981) 103, 6251; d) Castelhano, A.L.; Horne, S.; Billedeau, R.; Krantz, A., *Tetrahedron Lett.* (1986) **27**, 2435; e) Lipschutz, B.H.; Huff, B.; Vaccaro, W., *Tetrahedron Lett.* (1986) **27**, 4241; f) Agouridas, K.; Girodeau, J.M.; Pineau, R., *Tetrahedron Lett.* (1985) **26**, 3115; g) Metcalf, B.W.; Bonilavri, E., *J.Chem.Soc. Chem.Comm.* (1978) 914; h) Greenlee, W.J.; Taub, D.; Patchett, A.A., *Tetrahedron Lett.* (1978) 3999; i) Paik, H.; Dowd, P., *J.Org.Chem.* (1986) **51**, 2910; j) Steglich, W.; Wegmann, H., *Synthesis* (1980) 481; k) Thornberry, N.A.; Bull, H.G.; Taub, D.; Greenlee, W.J.; Patchett, A.A.; Cordes, E.H., *J.Am.Chem.Soc.* (1987) **109**, 7543; l) Fitzner, J.N.; Pratt, D.V.; Hopkins, P.B., *Tetrahedron Lett.* (1985) **26**, 1959.

19. Tachinaba, Y.; Ando, M.; Kuzuhara, H., *Chemistry Lett.* (1982) 1765.

20. Grieco, P.A.; Bahsas, A., *J.Org.Chem.* (1987) **52**, 5746.

21. a) Hegedus, L.S.; deWeck, G.; D'Andrea, S., J.Am.Chem.Soc. (1988) 110, 2122; b) Hegedus, L.S., DeLombaert, S., unpublished results.

CHAPTER 9

TOTAL SYNTHESIS OF COMPLEX AMINO ACIDS

Organic chemists have traditionally embraced the total synthesis of natural products as an ultimate and demanding test to define the limitations and strengths of emerging synthetic technology. The adjective *complex* in the Chapter title should not neccessarily be equated with sheer molecular size, number of stereogenic centers, or variety of functionality. Many other factors can be responsible for rendering a target structure "complex". In choosing specific natural products for this chapter, emphasis has been placed on the level of difficulty or "quality" of the synthetic target as it relates to synthetic complexity. Functional group incompatibility, lability to racemization, rearrangements, conformational subtleties , physical properties such as solubility and stereochemical control can all contribute to the level of "complexity".

These challenges can be found in large molecular weight substances, such as the fearsome vancomycin structure as well as small molecular weight substances, such as the notoriously unstable antibiotic ethynyl glycine (FR-900130). For example, the Wild Fire toxin Tabtoxin contains in the side chain, a novel and labile α-hydroxylated β-lactam that readily undergoes rearrangement to the thermodynamically more stable δ-lactam isotabtoxin. Natures' capacity to oxidatively render amino acid side chains into seemingly hopelessly complex functionality is breathtaking. A large class of oxidized cyclic dipeptides, the piperazinediones, further illustrate the challenging complexities an amino acid "side chain" can pose. The epidithiopiperazinediones melinacidin and aranotin being only two highlights of this class of biologically active substances. Paraherquamide is an example of a mixed biogenesis where the oxidative incorporation of isoprene units in the molecule form unique α,α-disubstituted amino acid moieties. Examples abound of depsipeptides comprised of highly unusual amino acid moieties such as the β-chloro dehydroalanine residue present in HV-Toxin M. In this context, it might be noted that none of these examples defining the extremes of "complexity" have yielded to total synthesis as of this writing. Of course, most β-lactam antibiotics are sculpted from simpler amino acid building blocks;

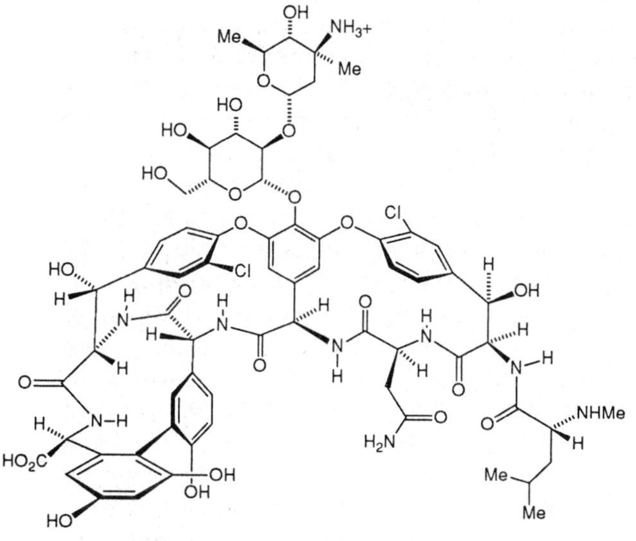

VANCOMYCIN

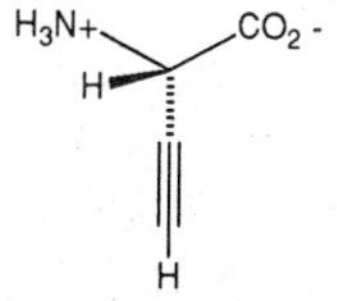

**ETHYNYLGLYCINE
(FR-900130)**

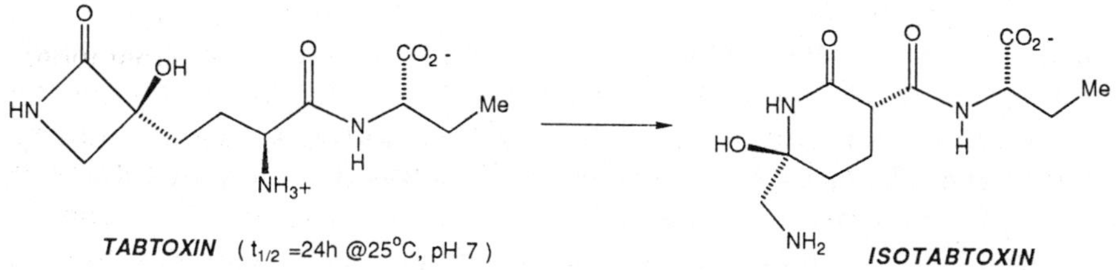

TABTOXIN ($t_{1/2}$ =24h @25°C, pH 7) → **ISOTABTOXIN**

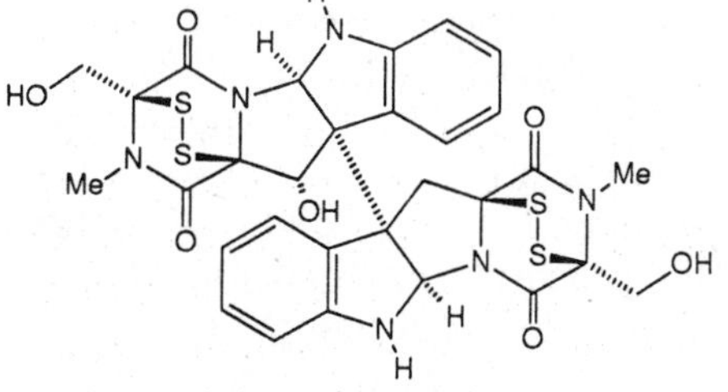

MELINACIDIN III

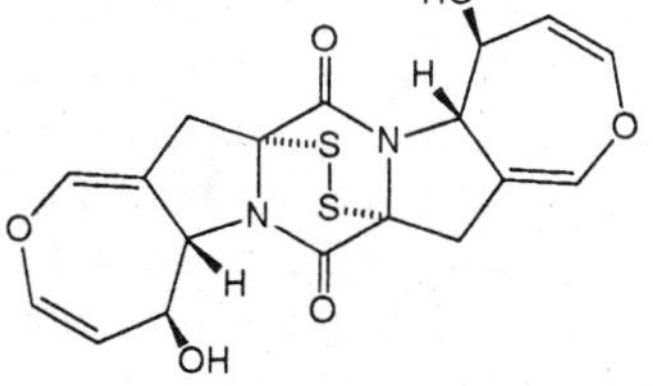

ARANOTIN

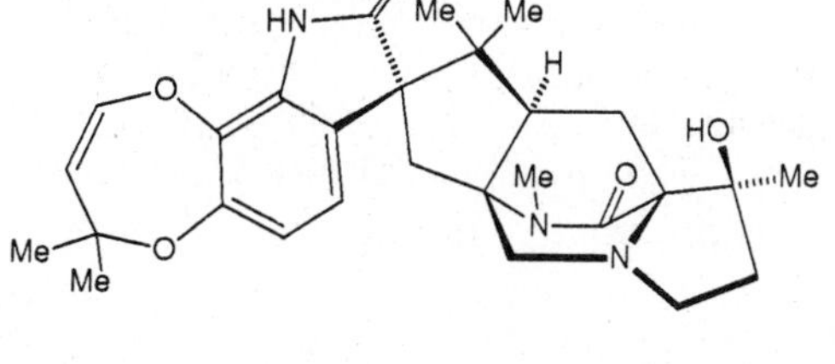

PARAHERQUAMIDE

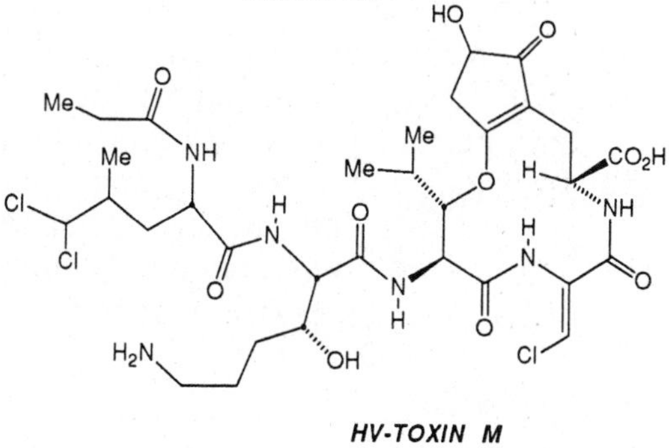

HV-TOXIN M

these will not be covered in this monograph as many excellent works dealing specifically with this important class of compounds are available.

This chapter is by no means, a comprehensive treatise on the total synthesis of naturally-occurring amino acid-containing compounds. The specific examples chosen illustrate some of the more elegant and impressive syntheses that have captured the imagination and devotion of the synthetic community. The schemes that follow also comprises a "gallery" of structures that the reader is invited to brouse through; a deeper appreciation of the creative palette of natures' biosynthetic craftwork on the simple amino acid backbone should inspire additional contributions in this field. At the end of this chapter, a sampling of complex amino acid-containing structures is listed that have not yet yielded to total synthesis. These structures have been compiled to emphasize that this field is, by no means mature and that nature continues to provide the synthetic chemist with an interesting and challenging future. Total syntheses that were detailed in previous chapters of this text will not be repeated in this section.

A. THE KAINIC ACID FAMILY

Kainic acid and closely related congeners domoic acid and acromelic acid, have attracted considerable attention due to their potent neuroexcitatory, insecticidal and anthelmintic properties[1].

Ohfune and Tomita [2] have employed a Diels-Alder strategy to control the relative stereochemistry of the appendages at the 3- and 4-positions as illustrated in their total synthesis of (-)-domoic acid (**10**, Scheme 1). Pyrroglutamic acid was protected and dehydrated to the dienophile **2**; subsequent [4+2] cycloaddition proceeded with the expected facial selectivity to afford the adduct **4**. Oxidative cleavage of the olefin and manipulation of the appendages provided domoic acid (**10**). By employing a Wittig reagent of known absolute stereochemistry, these workers were able to reassign the structure as that depicted. The original structure proposed a Z,Z-geometry for the diene. The present synthesis established the absolute stereochemistry at C-5' as well as redefining the olefin geometry as E,Z.

Oppolzer and Thirring [3] developed an interesting and concise synthesis of (-)-α-kainic acid (**17**) by the use of a stereocontrolled intramolecular ene reaction. Starting with L-glutamic acid, the key diene **14** was constructed in good overall yield and subjected to thermolysis in hot toluene. The ene cyclization reaction proceeded in 75% yield to furnish the desired *syn*-substituted substance **15**. The relative stereochemistry of **15** was rigorously ascertained by conversion of this compound to (-)-α-

SCHEME 1

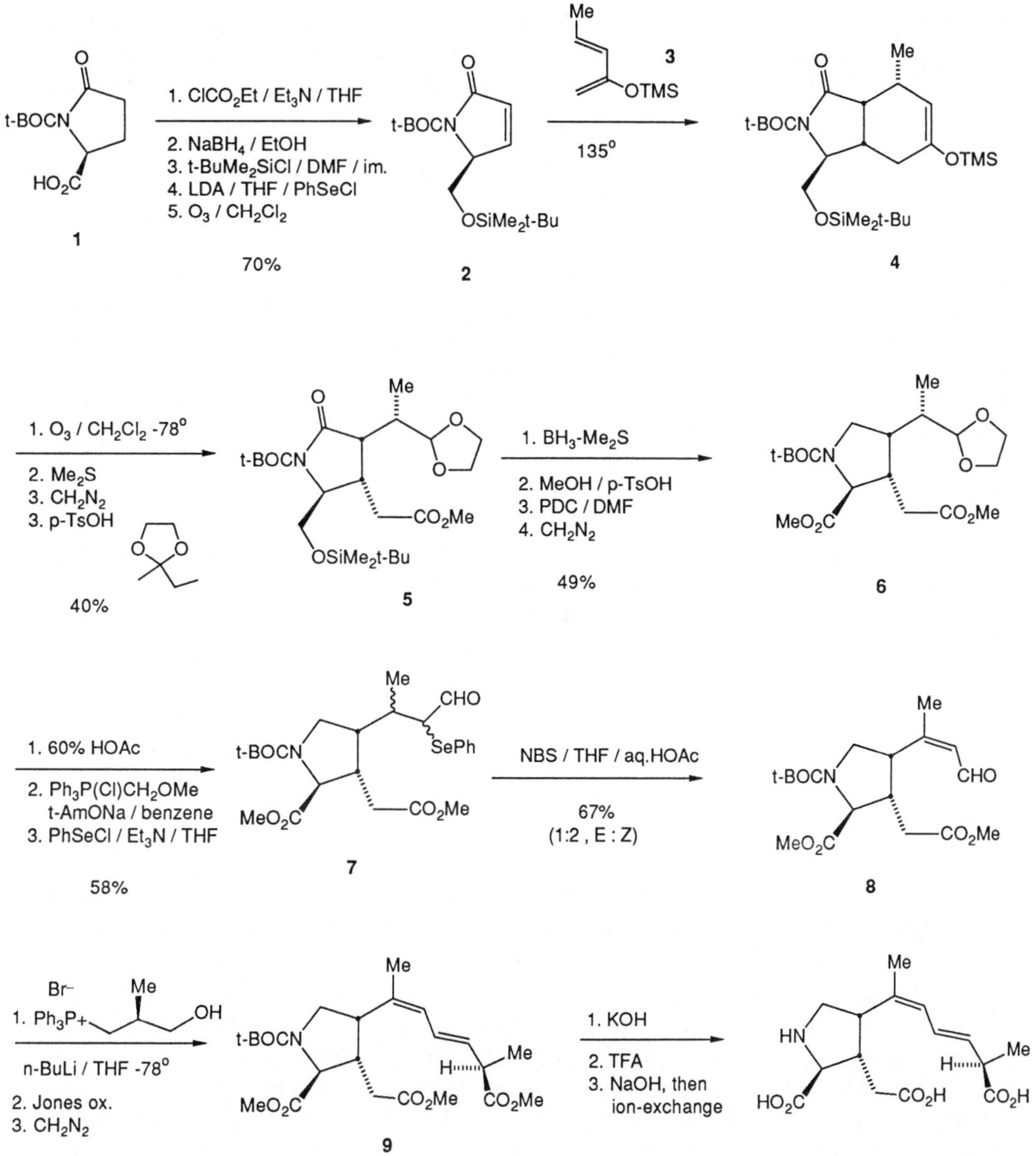
1
1. ClCO2Et / Et3N / THF
2. NaBH4 / EtOH
3. t-BuMe2SiCl / DMF / im.
4. LDA / THF / PhSeCl
5. O3 / CH2Cl2
70%
2
3
OTMS
135°
4
1. O3 / CH2Cl2 -78°
2. Me2S
3. CH2N2
3. p-TsOH
40%
5
1. BH3-Me2S
2. MeOH / p-TsOH
3. PDC / DMF
4. CH2N2
49%
6
1. 60% HOAc
2. Ph3P(Cl)CH2OMe t-AmONa / benzene
3. PhSeCl / Et3N / THF
58%
7
NBS / THF / aq.HOAc
67%
(1:2 , E : Z)
8
1. Ph3P+ (Br−)
n-BuLi / THF -78°
2. Jones ox.
3. CH2N2
35%
9
1. KOH
2. TFA
3. NaOH, then ion-exchange

10, *(-)- DOMOIC ACID*

Ohfune, Y.; Tomita, M., *J.Am.Chem.Soc.* (1982) **104**, 3511.

SCHEME 2

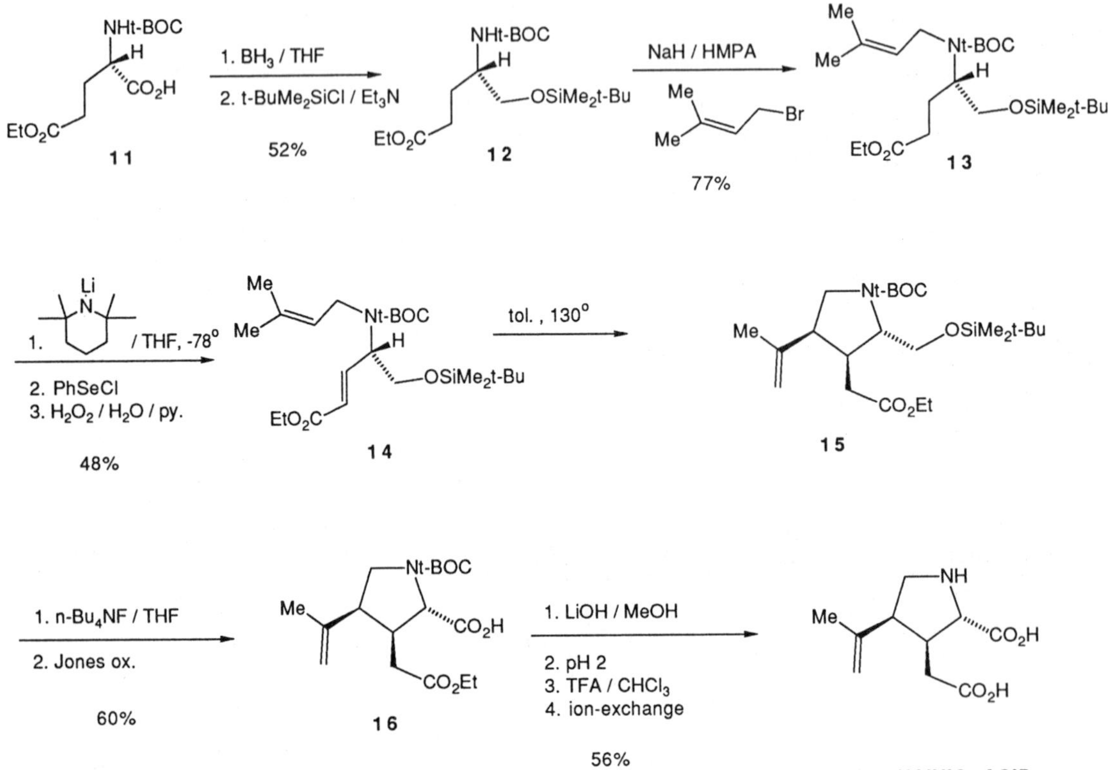

Oppolzer, W.; Thirring, K., *J.Am.Chem.Soc.* (1982) **104**, 4978.

kainic acid (**17**) in six straightforward steps. This work also provided the first rigorous assignment of the absolute stereochemistry to the natural product as that depicted.

    Recently, Knight and associates [4] have developed an enantioselective synthesis of (-)-α-kainic acid starting with L-aspartic acid. The key feature of their strategy illustrated in Scheme 3, is the use of a stereocontrolled enolate Claisen rearrangement to control the relative stereochemistry at the 3,4-positions. The key 9-membered azalactone **23** is constructed from **18** in modest overall yield. Enolate formation in the presence of the silyl trapping reagent proceeded smoothly to furnish the pyrrolidine **25** in 55% yield. The authors note that the base and silyl chloride had to be premixed at low temperature prior to addition of substrate **23** for the Claisen reaction to proceed. Homologation of the acid and oxidation furnished **17** in seven additional steps and good yield.

SCHEME 3

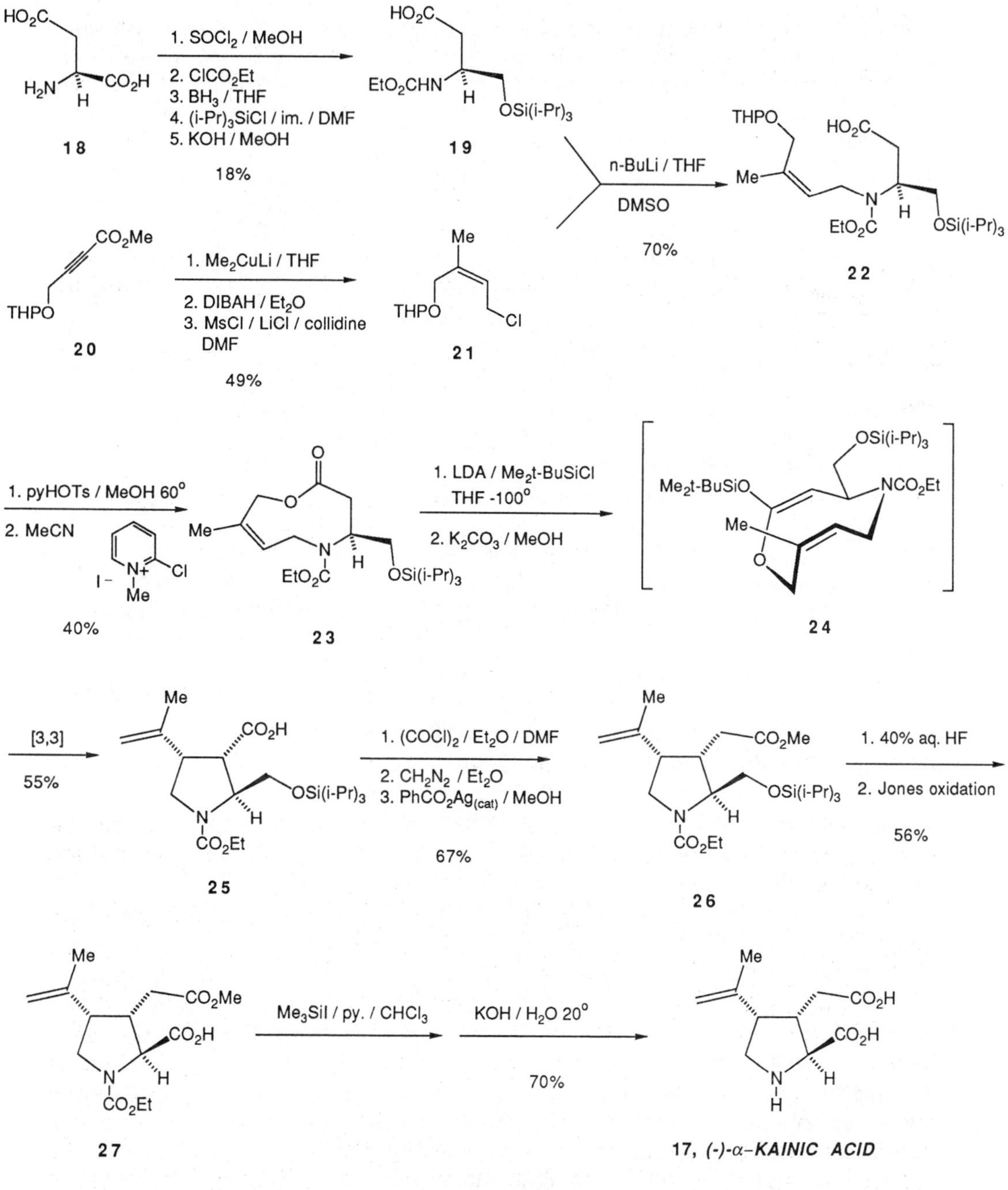

17, *(-)-α-KAINIC ACID*

Cooper, J.; Knight, D.W.; Gallagher, P.T., *J.Chem.Soc. Chem.Comm.* (1987) 1220.

Takano and co-workers[5] have employed an interesting intramolecular hetero-Diels-Alder cyclization to construct (-)-α-kainic acid (Scheme 4). Epoxy alcohol **28** , obtained from diethyl L-tartrate, was converted into the azide **29** with inversion of configuration by the use of diphenylphosphoryl azide and diisopropyl carbodiimide. Conversion to the aldehyde **32** was followed by condensation with Meldrums' acid which generated the key [4+2] cyclization intermediate **33**. This species spontaneously cyclized at 0°C to ambient temperature to the tricyclic adduct **34** with the desired facial selectivity. Hydrolysis of the ketal, hydrogenation and re-protection of the amine furnished **35** in 55% overall yield from **31**. Manipulation of the 3,4-appendages gave (-)-α-kainic acid (**17**); full experimental details accompany this report.

The same group has developed several additional methodologies to construct pyrrolidine-based systems as shown in Schemes 5-7. A simple synthesis of *trans*-4-hydroxyproline[6] (**49**, Scheme 5) features a stereoselective iodo-lactonization of the allylic benzamide **42**. Initial iodo-cyclization via **43** is invoked to produce **44**; subsequent hydration to hemi-ketal **45** frees the amine for closure on the iodomethyl moiety (to **46**). Collapse of the hemi-ketal **46** furnishes the pyrrolidine product **47** isolated from this procedure in 78% yield. Five-step manipulation of the functionality provides 4-hydroxyproline (**49**) in good overall yield.

The utility of azomethine ylides in [1,3] dipolar cycloadditions has recently been applied to elegant total syntheses of Acromelic acid A and Kainic acid by Takano and associates[7]. As shown in Scheme 6, (S)-O-benzylglycidol (**39**), is opened with lithio- 2-methyl-5-ethynylpyridine to furnish **50** (76%). Lindlar reduction of the alkyne (85%) followed by acylation with 2,3-dibromopropionyl chloride and immediate aziridine formation provides the azomethine ylide precursor **52** (65%). Heating **52** in *ortho*-dichloro benzene at 200°C in a sealed tube, generates the azomethine ylide **53** which cyclizes to the desired lactone **54** in 73% yield as a single diastereomer. Manipulation of the lactone appendages affords the diester **55** which is cleanly epimerized in high yield through the agency of sodium hydride and DBU. Oxidation of the pyridine ring and esterification affords the pyridone **57** which has previously been hydrolyzed to Acromelic acid A.

Using essentially the same approach, (-)-α-kainic acid was constructed by these workers [8] as illustrated in Scheme 7. Aziridine **61** required heating to 310°C in xylenes in a sealed tube producing the all *syn*-adduct **63** in 70% yield. The authors postulate that the azomethine ylide adopts conformation **62** which places the substituents in an equatorial disposition resulting in the observed, all *syn*- stereochemistry. As in the case of the acromelic acid synthesis above, inversion of stereochemistry at C-2 is required and was effected with sodium hydride and DBU (**66** to **67**).

SCHEME 4

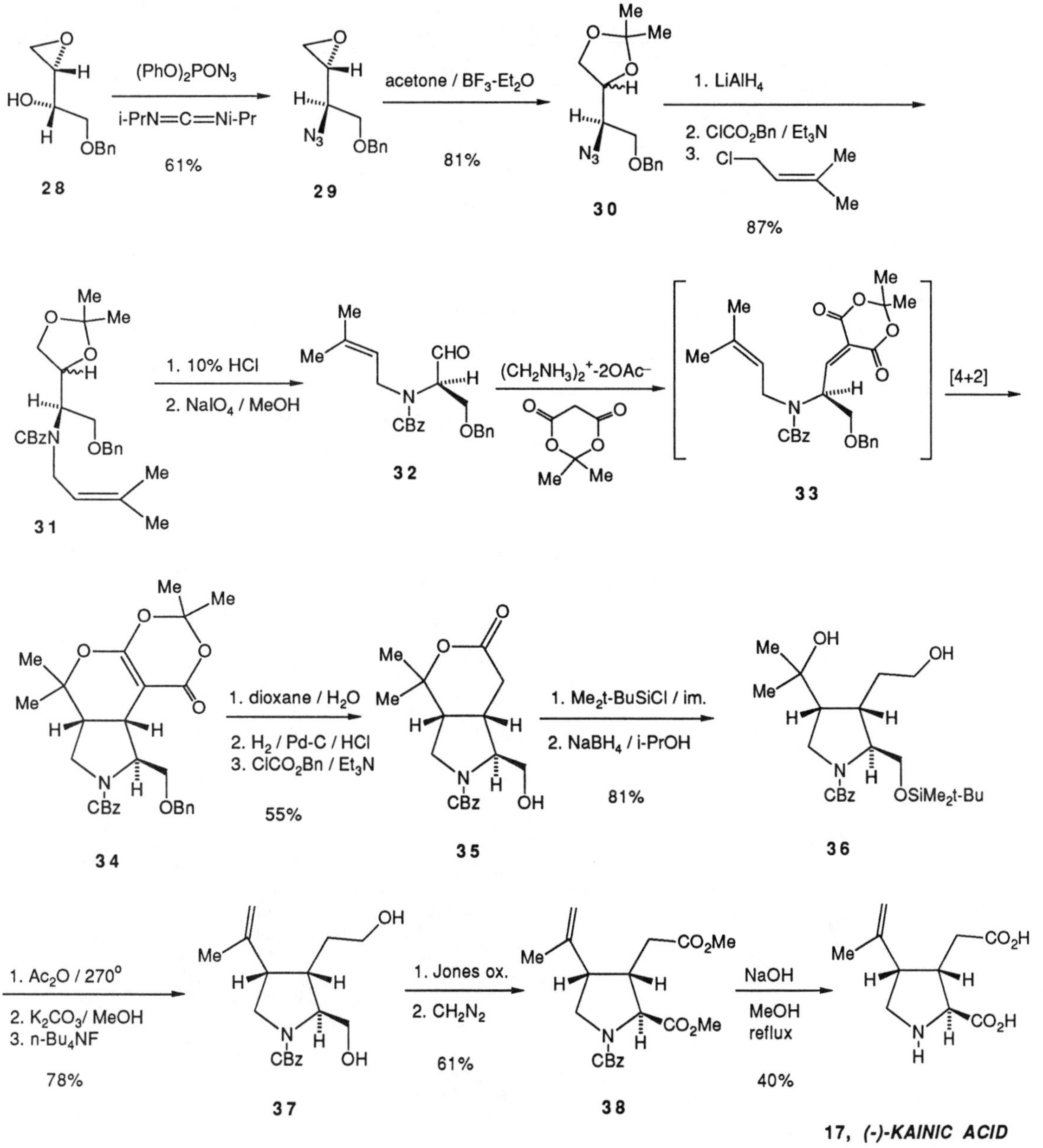

Takano, S.; Sugihara, T.; Satoh, S.; Ogasawara, K., *J.Am.Chem.Soc.* (1988) **110**, 6467.

SCHEME 5

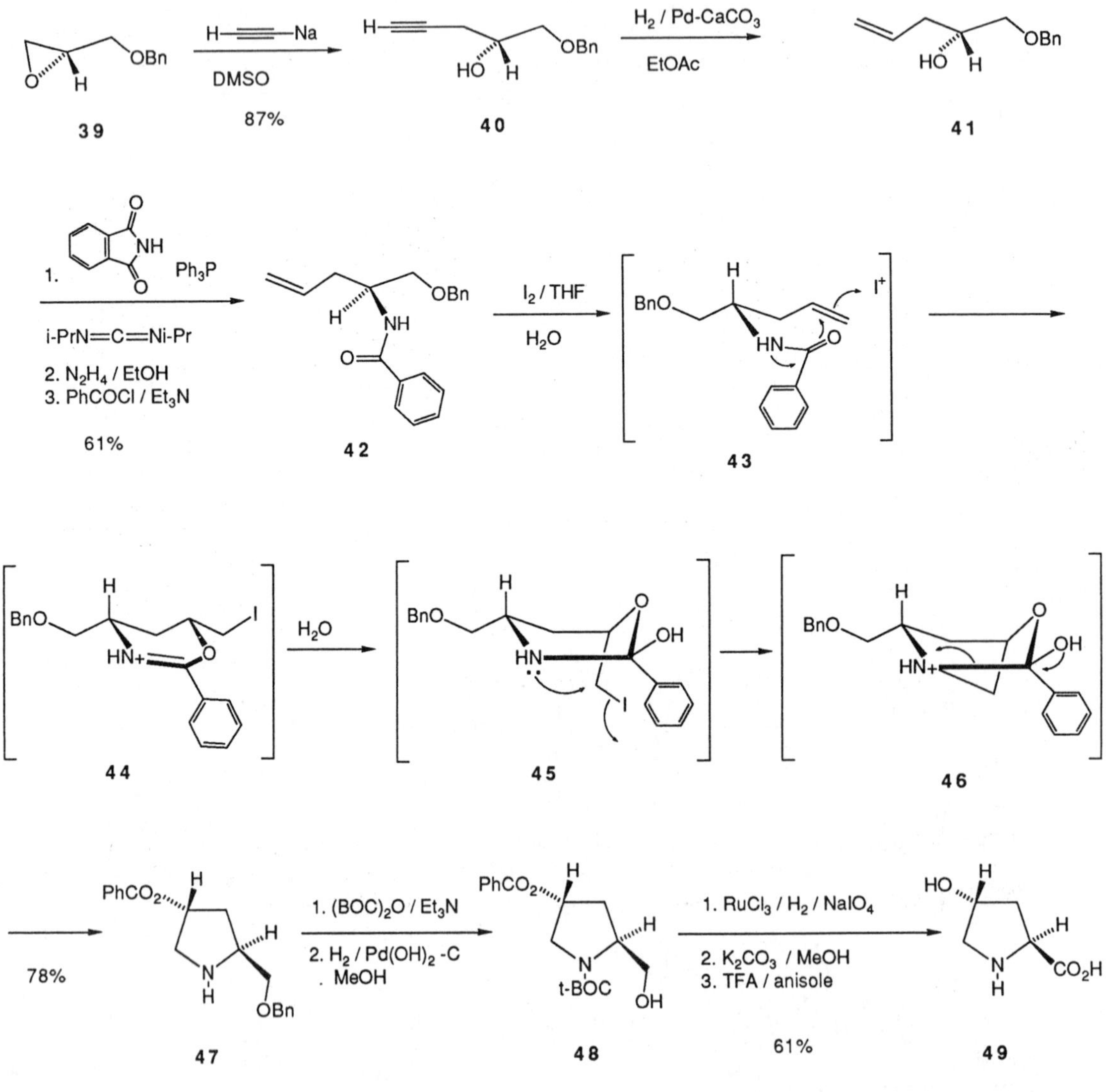

Takano, S.; Iwabuchi, Y.; Ogasawara, K., *J.Chem.Soc. Chem.Comm.* (1988) 1527.

SCHEME 6

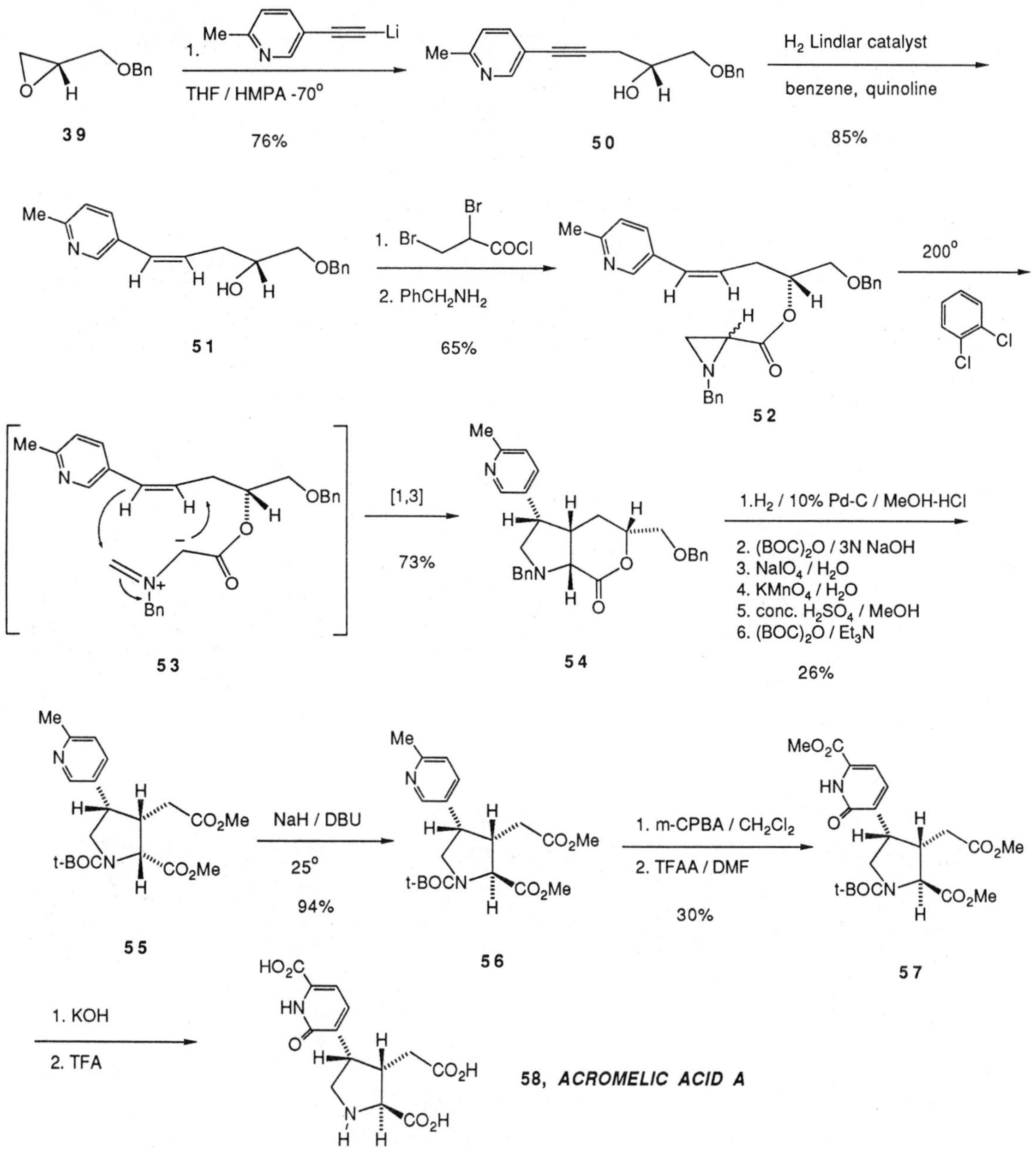

58, *ACROMELIC ACID A*

Takano, S.; Iwabuchi, Y.; Ogasawara, K. *J.Am.Chem.Soc.* (1987) **109**, 5523.

SCHEME 7

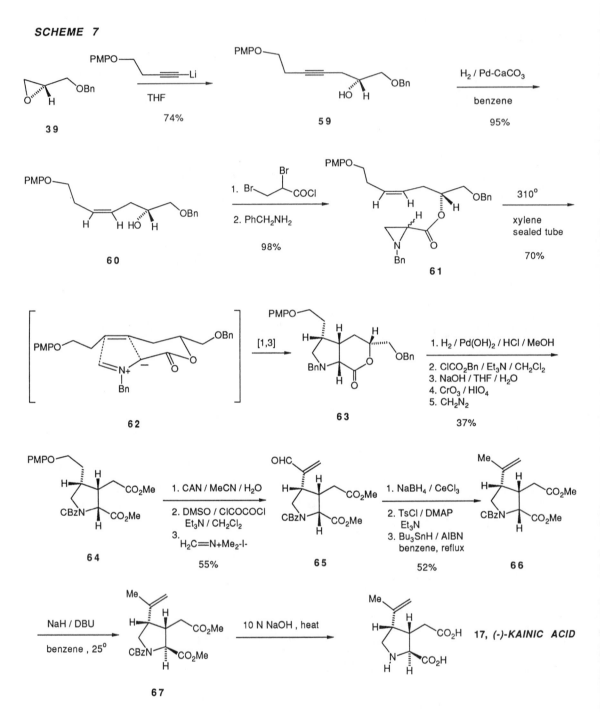

Takano, S.; Iwabuchi, Y.; Ogasawara, K., *J.Chem.Soc. Chem.Comm.* (1988) 1204.

A semi-synthetic approach to the acromelic acids starting with natural kainic acid was carried out by Shirahama and associates [9] as illustrated in Schemes 8 and 9. Kainic acid is now commercially available for ca $7.00 per gram. These workers employed a new pyridine synthesis that employs a Pummerer reaction as a key step. The kaianate isopropenyl moiety is epoxidized and rearranged with strong base to the allylic alcohol **69**. Further manipulation to sulfide **71** is followed by Pummerer rearrangement and aminolysis provides the pyridine **73**. Selenium dioxide oxidation of the methyl group and esterification affords methyl ester **73** which is successively oxidized and hydrolyzed to Acromelic acid A (**58**).

Acromelic acid B (**82**) was constructed in similar fashion from aldehyde **70** by a nine-step construction of the regioisomeric enone sulfide **78**; Pummerer rearrangement and amminolysis provided the isomeric pyridine **79**. Conversion of this substance into acromelic acid B was achieved in nine steps along the same lines as that applied for **58**. The authors note that acromelic acids A and B, isolated from the poisonous mushroom *Clitcybe acromelaga* Ichimura are obtainable in only sub-milligram quantities. Although the syntheses outlined from kainic acid are rather lengthy, the experimental section reports construction of >40 mg of **58** and > 10mg of **82**. While improvements in this Scheme are possible, the present route should allow for the production of sufficient quantities of these potent amino acids for more detailed biological investigations.

Baldwin and Li [10] have recently deployed a concise cobalt-mediated free radical cyclization reaction to construct the pyrrolidine ring systems of kainic acid and acromelic acid. As shown in Scheme 10, the optically active epoxy alcohol **83**, obtained by the Sharpless asymmetric epoxidation, was condensed with dimethallyl isocyanate to afford after protection, **84**. Conversion of the alcohol into the corresponding iodide **85** with complete inversion was effected with triflic anhydride and sodium iodide. Cyclization of **85** with cobaloxime (I) afforded a 5 : 3 mixture of the *syn-* and *anti-* products **86** and **87**, respectively. Compound **86** was transformed into (-)-α-kainic acid (**17**) by a six-step sequence in 27% overall yield from **86**. (+)-α-Allokainic acid was similarly obtained from **87** in 14% overall yield.

These workers [11] applied the same cobaloxime-mediated cyclization reaction to the enantiospecific total synthesis of (+)-acromelic acid A (**58**) as detailed in Scheme 11. The latent pyridone moiety was introduced via the agency of the Wittig reagent **90** providing **91** in good overall yield from the epoxy alcohol **83**. Transformation to the iodide **92** followed by cyclization with cobaloxime provided **93** in 64% yield as a 1:1 mixture of olefin isomers plus 11% of the corresponding C-4 epimer. Removal of the ketal and amminolysis provided the pyridone **95** in 54% yield. Oxidative manipulation of the remaining appendages afforded (+)-acromelic acid A.

SCHEME 8

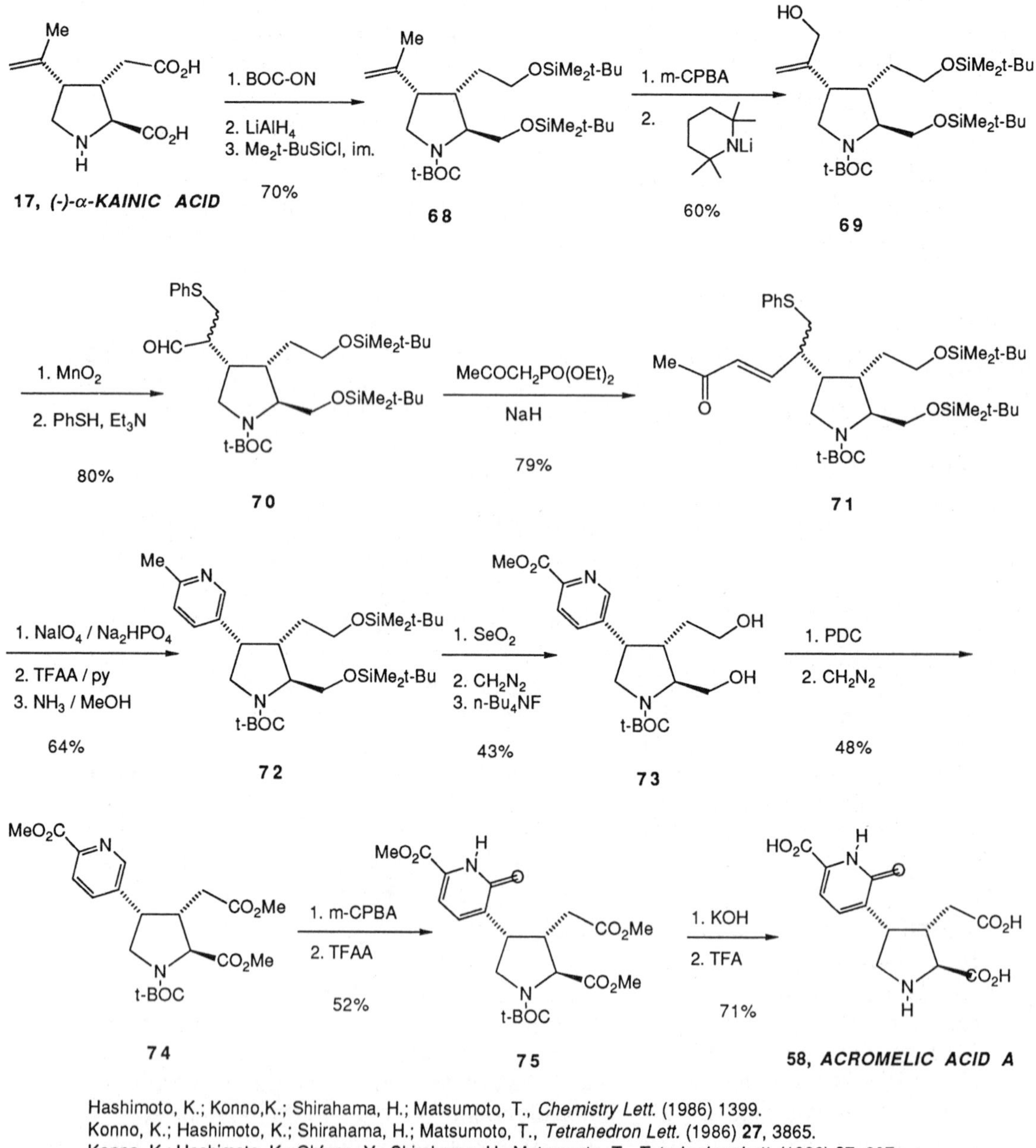

17, (-)-α-KAINIC ACID

1. BOC-ON
2. LiAlH$_4$
3. Me$_2$t-BuSiCl, im.

70%

68

1. m-CPBA
2. [piperidine NLi reagent]

60%

69

1. MnO$_2$
2. PhSH, Et$_3$N

80%

70

MeCOCH$_2$PO(OEt)$_2$

NaH

79%

71

1. NaIO$_4$ / Na$_2$HPO$_4$
2. TFAA / py
3. NH$_3$ / MeOH

64%

72

1. SeO$_2$
2. CH$_2$N$_2$
3. n-Bu$_4$NF

43%

73

1. PDC
2. CH$_2$N$_2$

48%

74

1. m-CPBA
2. TFAA

52%

75

1. KOH
2. TFA

71%

58, ACROMELIC ACID A

Hashimoto, K.; Konno, K.; Shirahama, H.; Matsumoto, T., *Chemistry Lett.* (1986) 1399.
Konno, K.; Hashimoto, K.; Shirahama, H.; Matsumoto, T., *Tetrahedron Lett.* (1986) **27**, 3865.
Konno, K.; Hashimoto, K.; Ohfune, Y.; Shirahama, H.; Matsumoto, T., *Tetrahedron Lett.* (1986) **27**, 607.
* Konno, K.; Hashimoto, K.; Ohfune, Y.; Shirahama, H.; Matsumoto, T., *J.Am.Chem.Soc.* (1988) **110**, 4807.

SCHEME 9

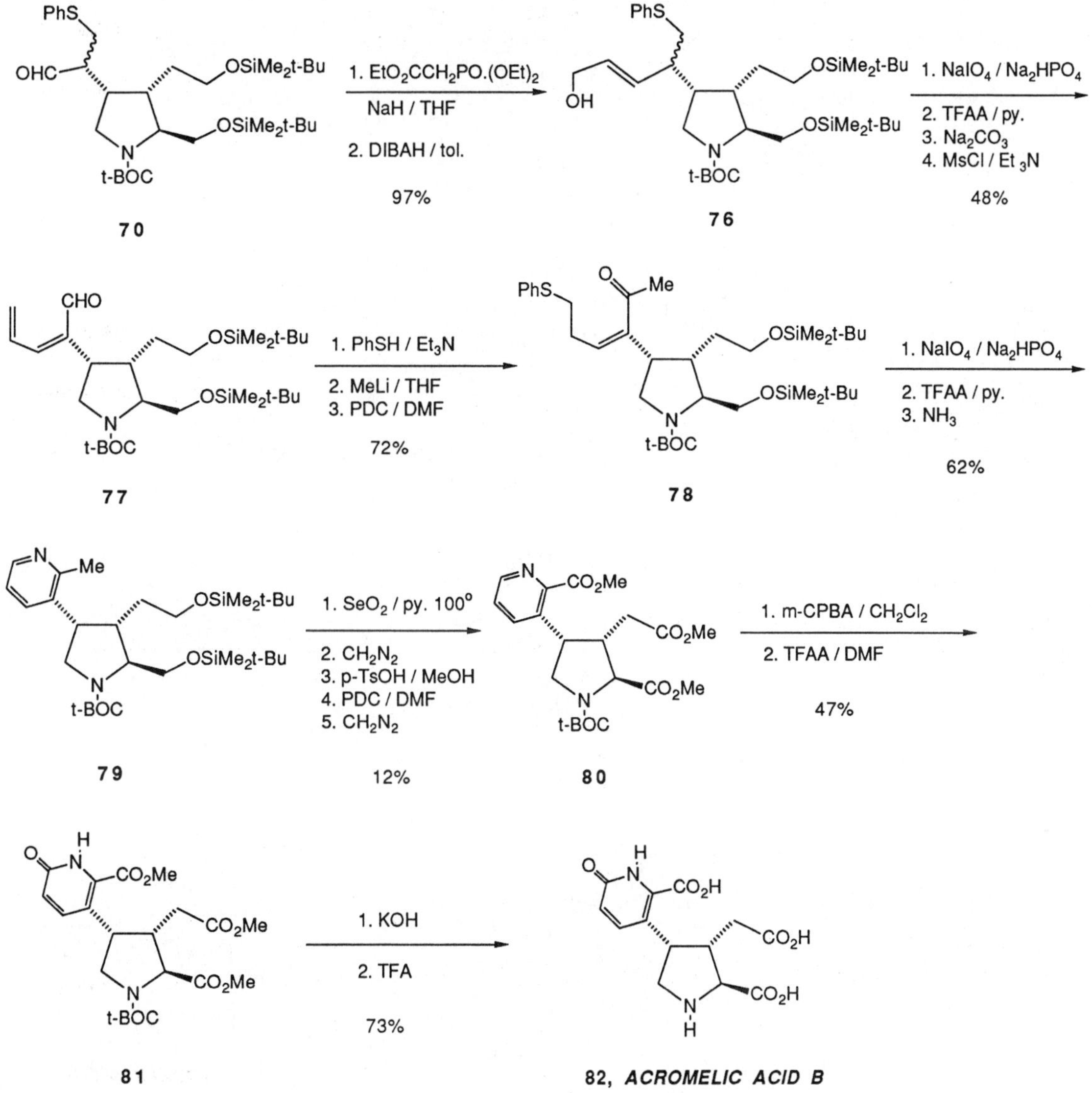

70

1. EtO$_2$CCH$_2$PO.(OEt)$_2$
NaH / THF
2. DIBAH / tol.

97%

76

1. NaIO$_4$ / Na$_2$HPO$_4$
2. TFAA / py.
3. Na$_2$CO$_3$
4. MsCl / Et$_3$N

48%

77

1. PhSH / Et$_3$N
2. MeLi / THF
3. PDC / DMF

72%

78

1. NaIO$_4$ / Na$_2$HPO$_4$
2. TFAA / py.
3. NH$_3$

62%

79

1. SeO$_2$ / py. 100°
2. CH$_2$N$_2$
3. p-TsOH / MeOH
4. PDC / DMF
5. CH$_2$N$_2$

12%

80

1. m-CPBA / CH$_2$Cl$_2$
2. TFAA / DMF

47%

81

1. KOH
2. TFA

73%

82, ACROMELIC ACID B

Hashimoto, K.; Konno,K.; Shirahama, H.; Matsumoto, T., *Chemistry Lett.* (1986) 1399.
Konno, K.; Hashimoto, K.; Shirahama, H.; Matsumoto, T., *Tetrahedron Lett.* (1986) **27**, 3865.
Konno, K.; Hashimoto, K.; Ohfune, Y.; Shirahama, H.; Matsumoto, T., *Tetrahedron Lett.* (1986) **27**, 607.
* Konno, K.; Hashimoto, K.; Ohfune, Y.; Shirahama, H.; Matsumoto, T., *J.Am.Chem.Soc.* (1988) **110**, 4807.

SCHEME 10

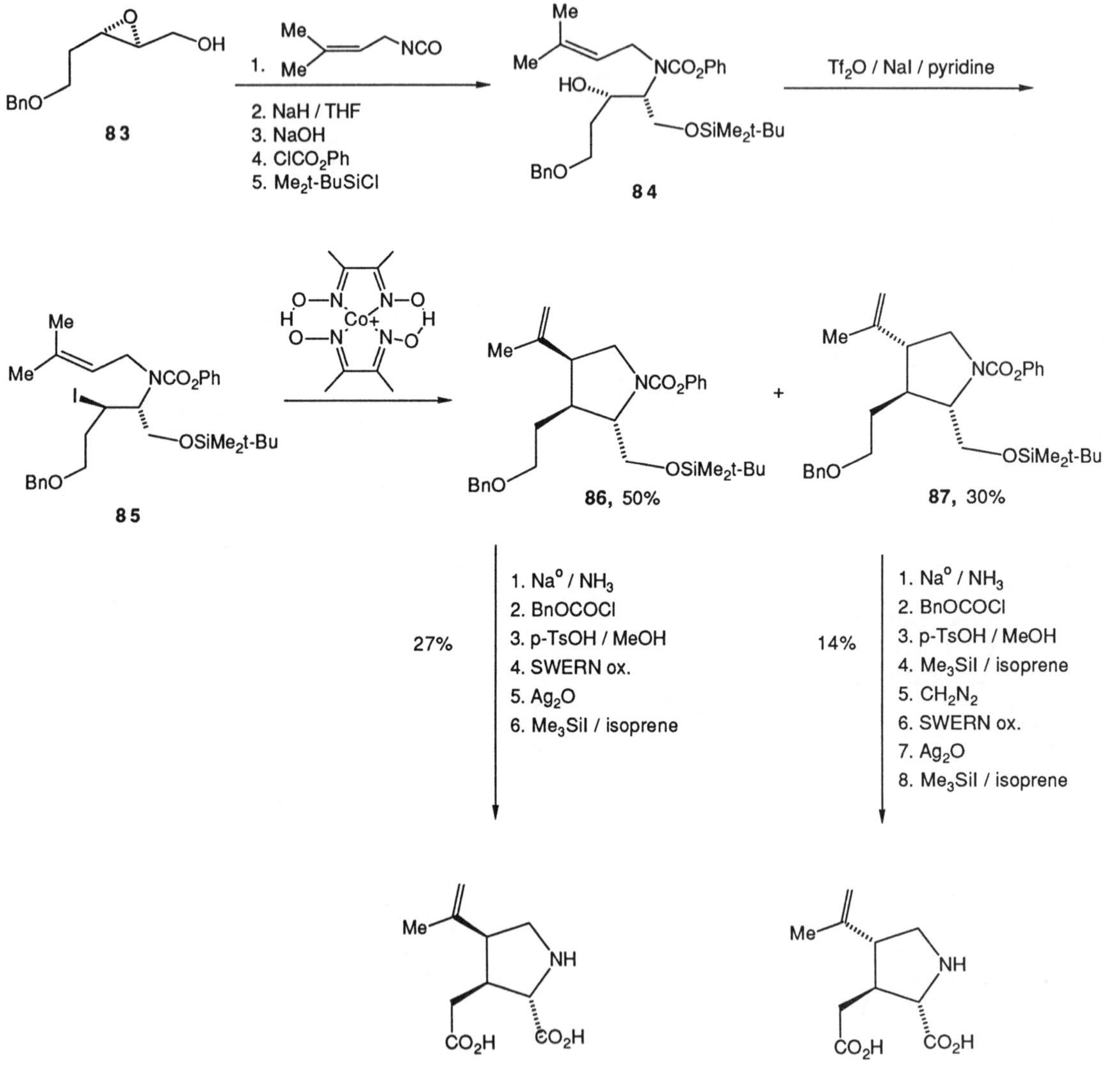

17, *(-)-α–KAINIC ACID*

88, *(+)-α–ALLOKAINIC ACID*

Baldwin, J.E.; Li, C-S., *J.Chem.Soc.Chem.Comm.* (1987) 166.

SCHEME 11

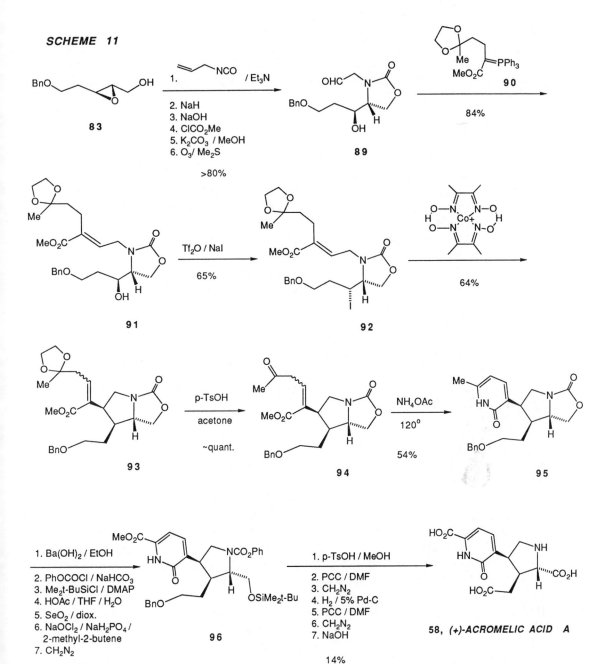

Baldwin, J.E.; Li, C.; *J.Chem.Soc.Chem.Comm.* (1988) 261.

Other syntheses [12] of the kainoids have appeared in the literature and more are sure to follow. The scarcity and functional group complexities of domoic acid and the acromelates relative to the more abundant and simpler kainic acids would seem to demand further improvements in the syntheses of these potent neuroexcitatory amino acids.

B. ACIVICIN (AT-125)

Acivicin (**108**) is an antimetabolite antitumor antibiotic isolated from *Streptomyces sviceus* in 1973 by Martin, et.al. [13]. Acivicin has been shown to inhibit several glutamine-dependent amidotransferases which are important in purine and pyrimidine biosynthesis. This densely functionalized natural product has entered phase II clinical trials as a candidate for synergistic combination chemotherapy. Several total syntheses of acivicin have appeared as well as analogs to probe the SAR of this novel amino acid; these are delineated below.

The first total synthesis of acivicin was reported by Kelly and associates[14] of the Upjohn Co. in 1979. Cyclopentadiene mono-epoxide (**97**) was subjected to amminolysis to give the trans-aminoalcohol **98**; this substance was resolved through the agency of the derived tartrate salts providing both optical isomers of **98**. Protection of the amine and replacement of the hydroxyl group with N-hydroxy phthalimide afforded the syn-adduct **100**. Removal of the phthalimido group , acylation reduction and protection afforded the key substance **102**. Sodium periodate-ruthenium cleavage of the olefin afforded a 1 : 4 mixture of acids **103** : **104**, respectively. The major product (**104**) was hydrogenated, esterified and chlorinated to afford the phthalimido-protected acivicin derivative **107**. Final removal of the phthalimide protection afforded acivicin (**108**).

Baldwin and associates[15] have extensively studied several pathways to access the acivicin structure. The successful total synthesis was reported in a communication in 1981 followed by a full account in 1985. The dehydroglutamate **109** was condensed with O-silyl *para*-methoxybenzylhydroxyl amine (**110**) affording the hydroxamate **111**. Exposure of this substance to dilute base effected ring closure to a ~1:1 diastereomeric mixture of racemic tricholomic acid derivatives **112**. Removal of the *para*-methoxybenzyl group with concomitant dehydration and chlorination via a Von Braun-type procedure afforded the protected acivicin **113** which was chromatographically separated from the undesired diastereomer. Reductive removal of the *ortho*-nitrobenzyl ester and cleavage of the N-CBz group with oxalylchloride afforded racemic acivicin **108**. This substance was subsequently acylated and resolved with hog kidney acylase to provide optically pure acivicin.

SCHEME 12

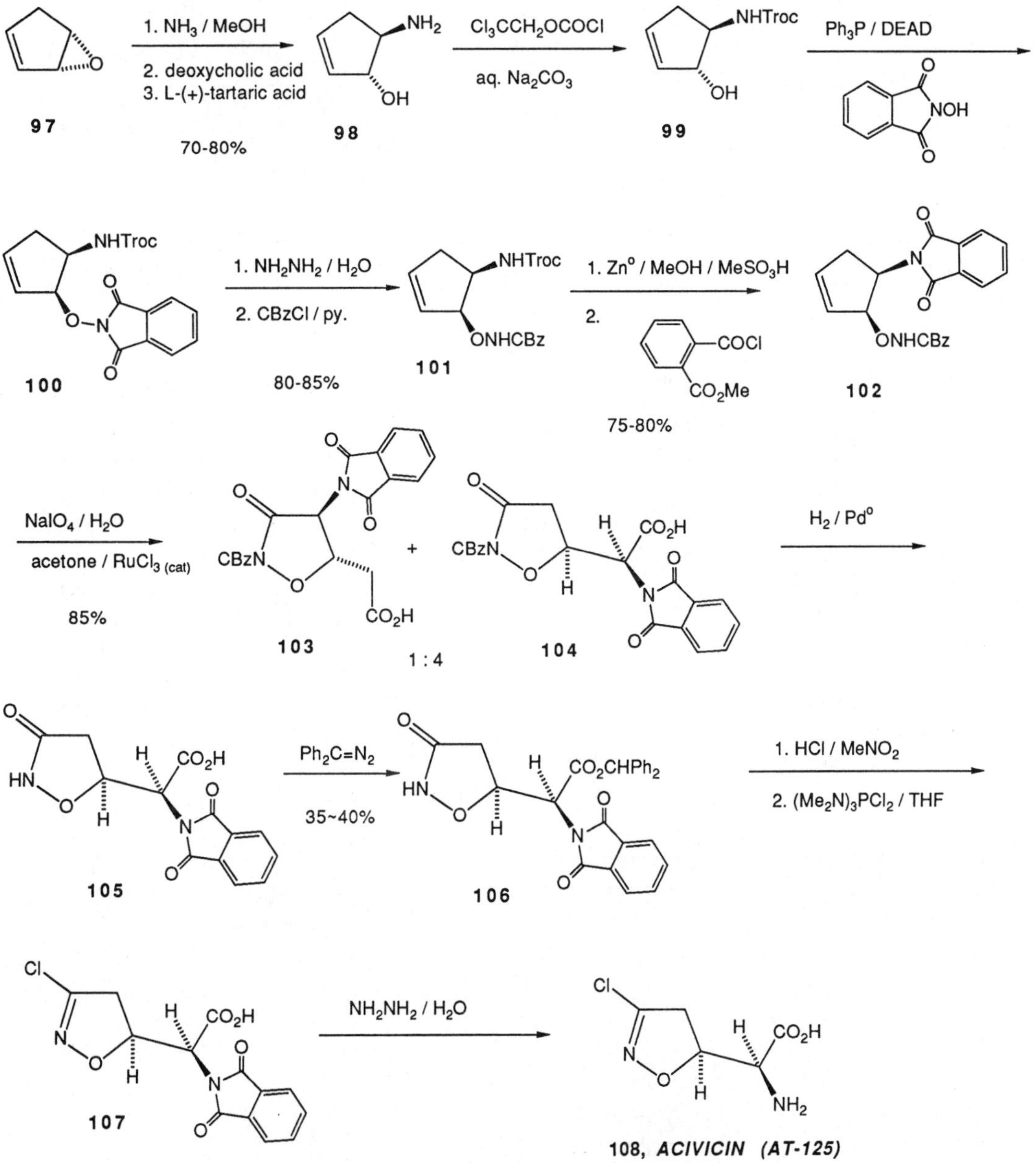

1. NH₃ / MeOH
2. deoxycholic acid
3. L-(+)-tartaric acid

70-80%

97

98

Cl₃CCH₂OCOCl

aq. Na₂CO₃

99

Ph₃P / DEAD

100

1. NH₂NH₂ / H₂O
2. CBzCl / py.

80-85%

101

1. Zn° / MeOH / MeSO₃H
2. —COCl / —CO₂Me

75-80%

102

NaIO₄ / H₂O

acetone / RuCl₃ (cat)

85%

103 + **104**

1 : 4

H₂ / Pd°

105

Ph₂C=N₂

35~40%

106

1. HCl / MeNO₂
2. (Me₂N)₃PCl₂ / THF

107

NH₂NH₂ / H₂O

108, ACIVICIN (AT-125)

Kelly, R.C.; Schletter, I.; Stein, S.J.; Wierenga, W., *J.Am.Chem.Soc.* (1979) **101**, 1054

SCHEME 13

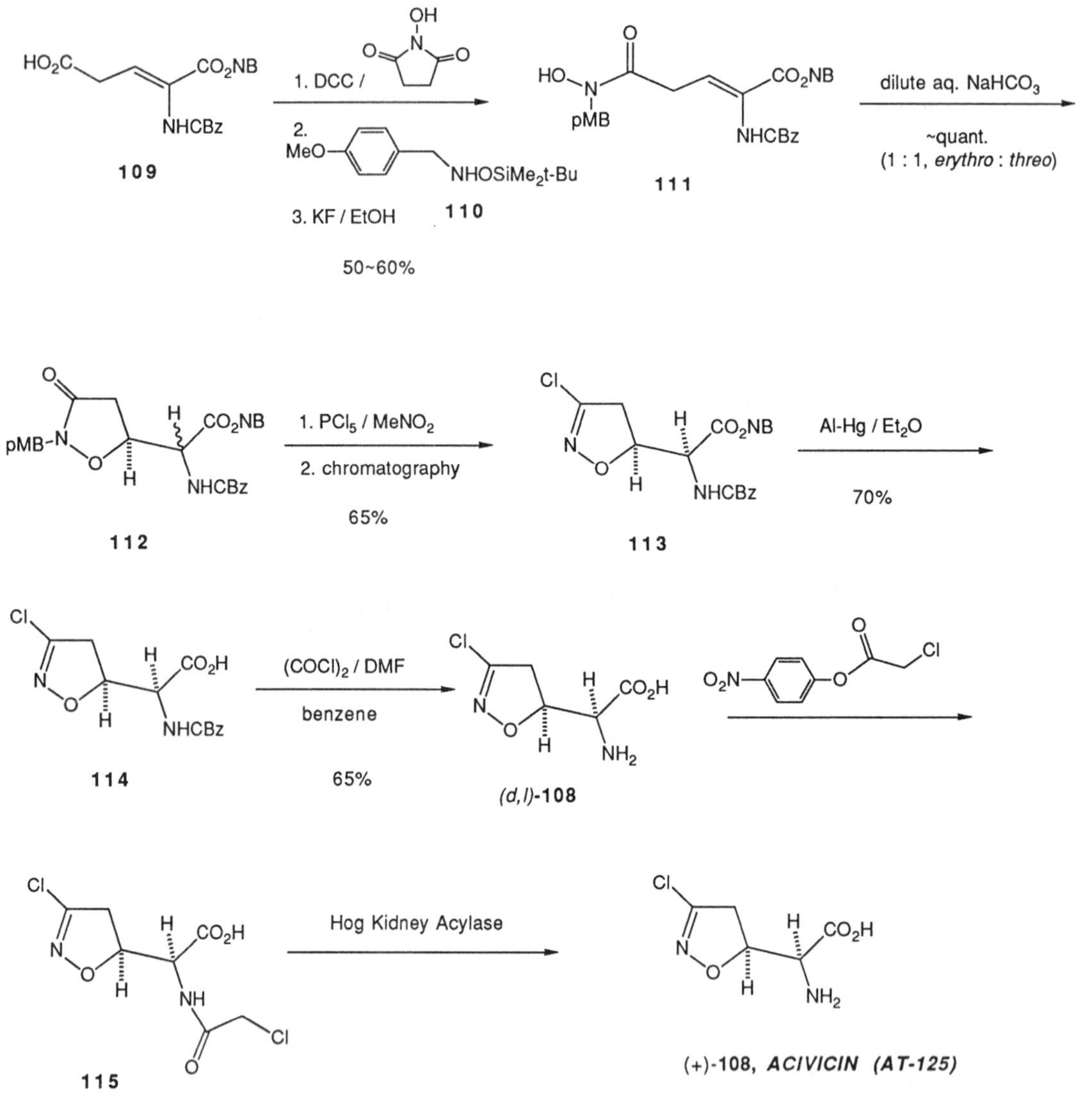

Baldwin, J.E.; Kruse, L.I.; Cha, J-K., *J.Am.Chem.Soc.* (1981) **103**, 942
Baldwin, J.E.; Cha, J-K.; Kruse, L.I., *Tetrahedron* (1985) **41**, 5241
Baldwin, J.E.; Hoskins, C.; Kruse, L., *J.Chem.Soc. Chem.Comm.* (1976) 795.

Silverman and Holladay [16] have also employed a glutamic acid-based entry to construct the tricholomic acid system as shown in Scheme 14. Photolytic chlorination of L-Glu gives a 1:1 diastereomeric mixture of β-chloro glutamates **116** in 33% yield. Acylation and cyclic anhydride formation provides **117** which is ring-opened with N-alkoxy phthalimide to provide the active ester **118**. Treatment of this material with aqueous hydroxylamine at pH 6 afforded the key cyclization precursor **119**. Cyclization of each diastereomer was effected with aqueous triethylamine (pH 11) to afford the N-acylated tricholomic acid **120**. Esterification, dehydration / chlorination and removal of the protecting groups afforded (+)-acivicin (**108**) in good overall yield.

The most concise approach to acivicin was first recognized by Baldwin and associates [15c] in their attempts to add chloronitrile oxide to vinyl glycine. While it was found that chloronitrile oxide underwent (in low yield) [1,3] dipolar cycloaddtion to styrene, these workers were unable to coax vinyl glycine or various acyl and ester derivatives to participate in this reaction. This was attributed to the slow rate of addition relative to furoxan (dimer) formation. These workers found that a high yield (99%) of cycloadduct was formed between chloronitrile oxide and the potassium salt of 2-nitro-3-methyl butenoate; this adduct was subsequently reduced and converted into a diastereomeric mixture of 5-methyl acivicin derivatives. This pioneering contribution paved the way for the eventual realization of this simple approach to the total synthesis of acivicin and several analogs by several other groups.

As shown in Scheme 15, Hagedorn and co-workers [17] found that the considerably more reactive bromonitrile oxide added to d,l-vinyl glycine (**123**) to give very good overall yield of an inseparable mixture (2:5) of the bromo analog of acivicin, **126**.

Wade and associates [18] made an important observation in this area and found that simply adding silver nitrate to dichloroformaldoxime in the presence of various olefins resulted in good yields of [1,3] dipolar cycloaddition adducts. Thus, the reaction of chloronitrile oxide with styrene, which only proceeded in 6% yield as reported by Baldwin [15c] was increased to 73% in the presence of four equivalents of silver nitrate. Application of this procedure to the protected vinyl glycine **128** gave a 37 : 15 ratio (52% combined yield) of the undesired **131** plus the desired **107**, respectively (Scheme 16) . Compound **107** was previously shown by the Upjohn group to be convertible into acivicin by treatment with hydrazine hydrate; a yield for this last deprotection has apparently not been recorded. Full experimental details accompany this work.

In a related approach, Vyas and associates [19] have found a method to convert the bromonitrile oxide adduct into the corresponding chloro derivative as shown in Scheme 17. Z-2-Butene-1,4-diol was condensed with trichloroacetonitrile to give the trichloro acetimidate **133**.

SCHEME 14

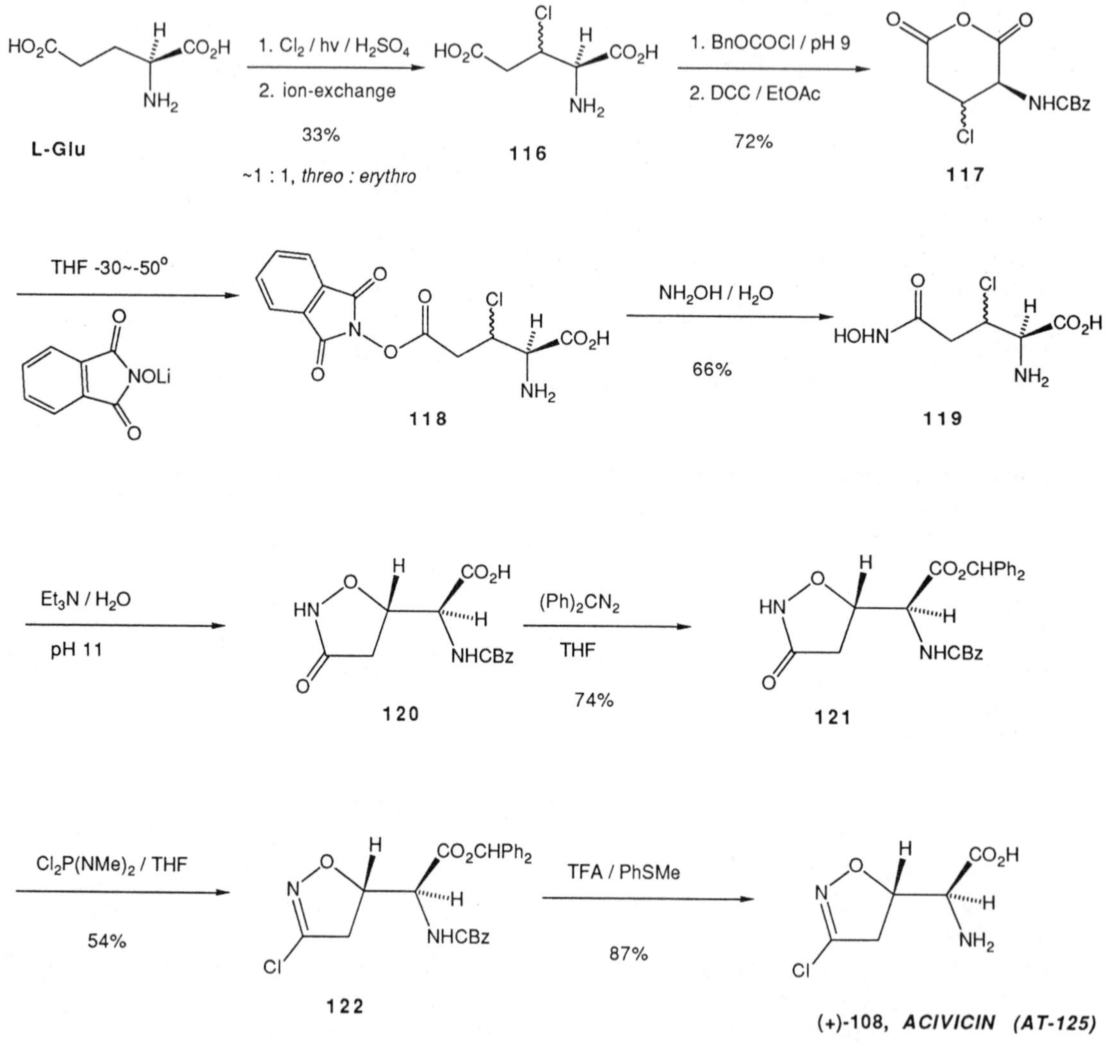

Reaction of **L-Glu** with 1. Cl_2 / hv / H_2SO_4; 2. ion-exchange (33%, ~1 : 1, *threo : erythro*) gives **116**; then 1. BnOCOCl / pH 9; 2. DCC / EtOAc (72%) gives **117**.

THF -30~-50°, with potassium/lithium phthalimide N-OLi gives **118**; then NH_2OH / H_2O (66%) gives **119**.

Et_3N / H_2O, pH 11 gives **120**; then $(Ph)_2CN_2$, THF (74%) gives **121**.

$Cl_2P(NMe)_2$ / THF (54%) gives **122**; then TFA / PhSMe (87%) gives **(+)-108, ACIVICIN (AT-125)**.

Silverman, R.B.; Holladay, M.W., *J.Am.Chem.Soc.* (1981) **103**, 7357.

SCHEME 15

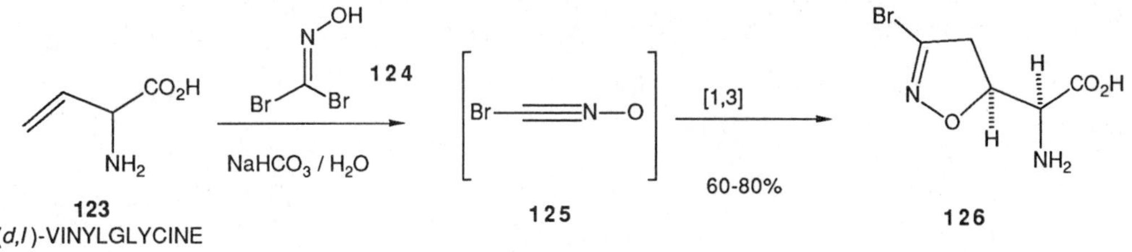

Hagedorn, A.A.; Miller, B.J.; Nagy, J.O., *Tetrahedron Lett.* (1980) **21**, 229.

SCHEME 16

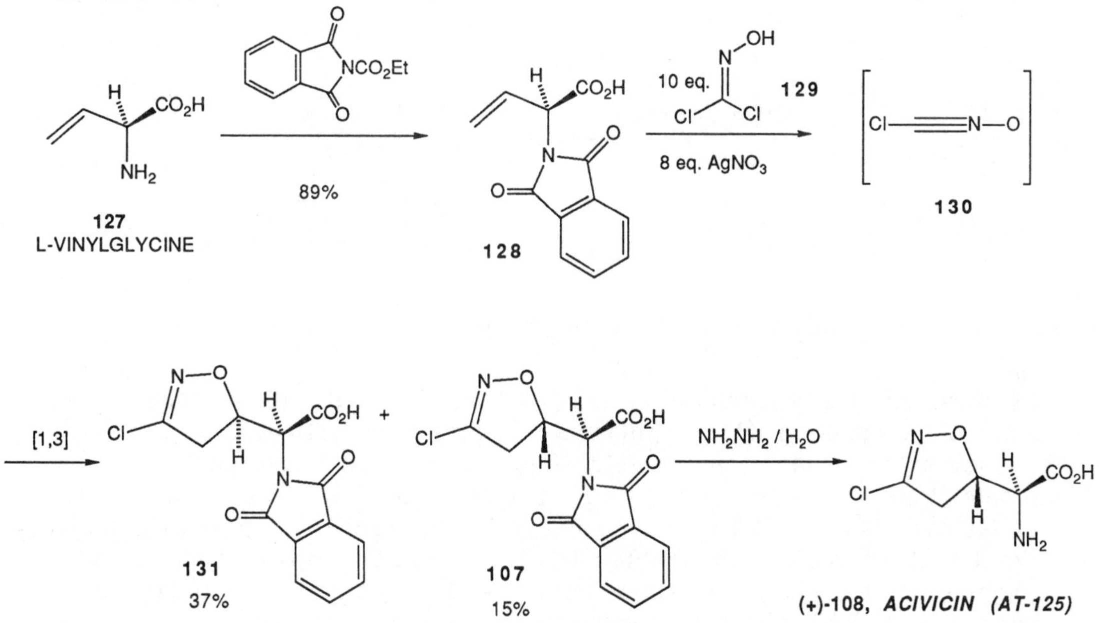

Wade, P.A.; Pillay, M.K.; Singh, S.M., *Tetrahedron Lett.* (1982) **23**, 4563
Wade, P.A.; Singh, S.M.; Pillay, M.K., *Tetrahedron* (1984) **40**, 601.

SCHEME 17

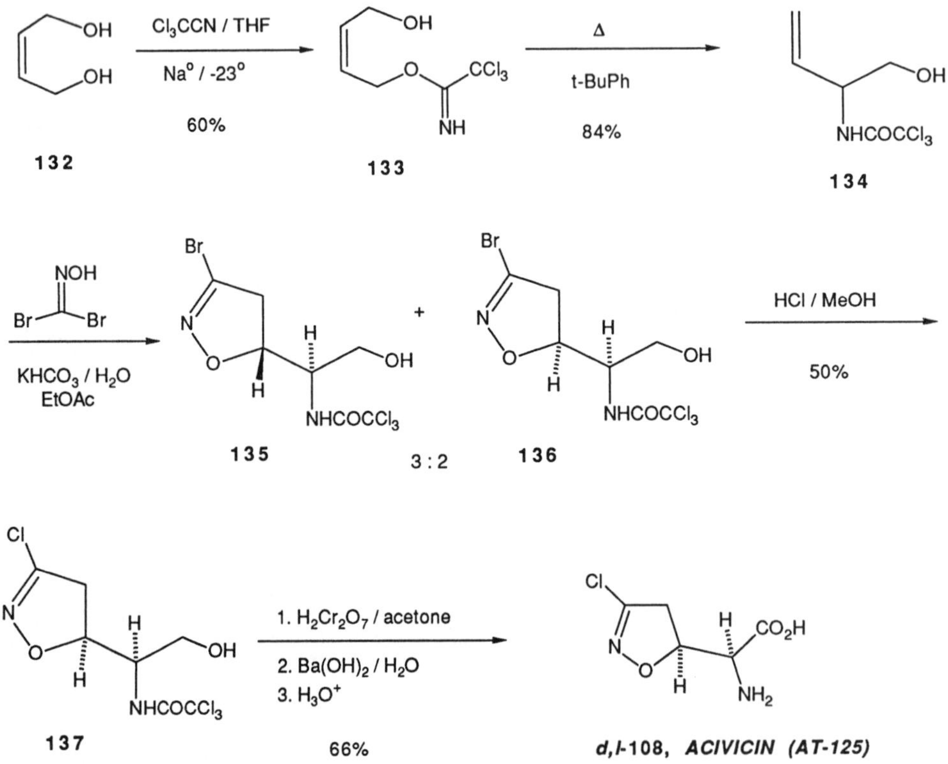

Vyas, D.M.; Chiang, Y.; Doyle, T.W., *Tetrahedron Lett.* (1984) **25**, 487.

Pyrolysis of this substance in tert-butyl benzene at reflux temperature smoothly effected a [3,3] sigmatropic rearrangement to afford the vinyl aminoalcohol **134** in 84% yield. Cycloaddition of this substance with bromonitrile oxide afforded a 3:2 mixture **135:136**, respectively. The undesired isomer **135** was removed by fractional crystallization. Treatment of **136** with methanolic HCl effected replacement of the bromine with a chlorine substituent in 50% yield.   Oxidation and hydrolysis of **137** afforded d,l-acivicin in good overall yield. The authors note that the simplicity and inexpensive reagents employed in this approach render the present synthesis a reasonable candidate for industrial scale synthesis of acivicin and the corresponding bromo analog which has been reported to have nearly identical biological properties to the natural product.

A recent, elegant permutation on the dipolar cycloaddition strategy was reported by Whitney and associates [20] as depicted in Scheme 18. 2,3-O-isopropylidene-5-O-trityl-D-ribose oxime (**138**) was condensed with formaldehyde generating the nitrone **140**. Reaction of this material with the protected vinyl glycine derivative **139** afforded in excellent yield, the adduct **141**. The facial selectivity of this reaction was excellent, giving >19:1 favoring the desired isomer. The employment of double asymmetric induction in this context, provides a significant advantage over the above-described halonitrile oxide additions which generally gave poor facial selectivities. Oxidation of **141** with N-chloro succinimide furnished the isoxazoline **142** which was further chlorinated and deprotected to give (+)-acivicin in good overall yield.

Very recently, an antibiotic structurally related to cycloserine and acivicin, was isolated [21] from *Empedobacter lactamgenus* YK-258 and *Lysobacter albus* YK-422. This substance, called lactivicin, has attracted considerable attention due to the striking similarity in biological activity toward both Gram-positive and Gram-negative bacteria with the β-lactam antibiotics. Additionally, this material was shown to have affinity for penicillin-binding proteins and susceptibility toward β-lactamases. Natsugari and associates [22] have recently reported a total synthesis of lactivicin from S-cycloserine as depicted in Scheme 19. 2-Oxoglutaric acid was esterified and cyclized to the chloro-lactone **146** with thionyl chloride in high yield. Condensation of this substance with N-Cbz-(S)-cycloserine (**147**) furnished the adduct **148** in 60% yield. Hydrogenation of the protecting groups and acetylation provided lactivicin (**150**) in modest yield. The stereoisomeric mixture at the carboxylic acid-bearing stereogenic center is a natural condition for lactivicin which presumably equilibrates through a ring-opened tautomer. These workers also prepared the enantiomorph starting with (R)-cycloserine; as expected, the *enantio-* lactivicin exhibited reduced antimicrobial activity. Preparation of β-lactam analogs **151** and **154** by acylation of **149** with phenylacetyl chloride and 2-(2-chloroacetamidothiazol-4-yl)-(Z)-2-methoxyimino acetyl chloride(**153**), respectively , provided the desired acylated β-lactam analogs. As in the case of lactivicin, the (R)-isomer **152** showed reduced antimicrobial activity. On the other hand, both **151** and **154** exhibited enhanced antimicrobial activities relative to the natural product and is consistent with the hypothesis [23] that these substances are attacking the various penicillin-binding proteins. It is expected that the lactivicin nucleus will be the subject of intensive synthetic and biological study in the coming years.

SCHEME 18

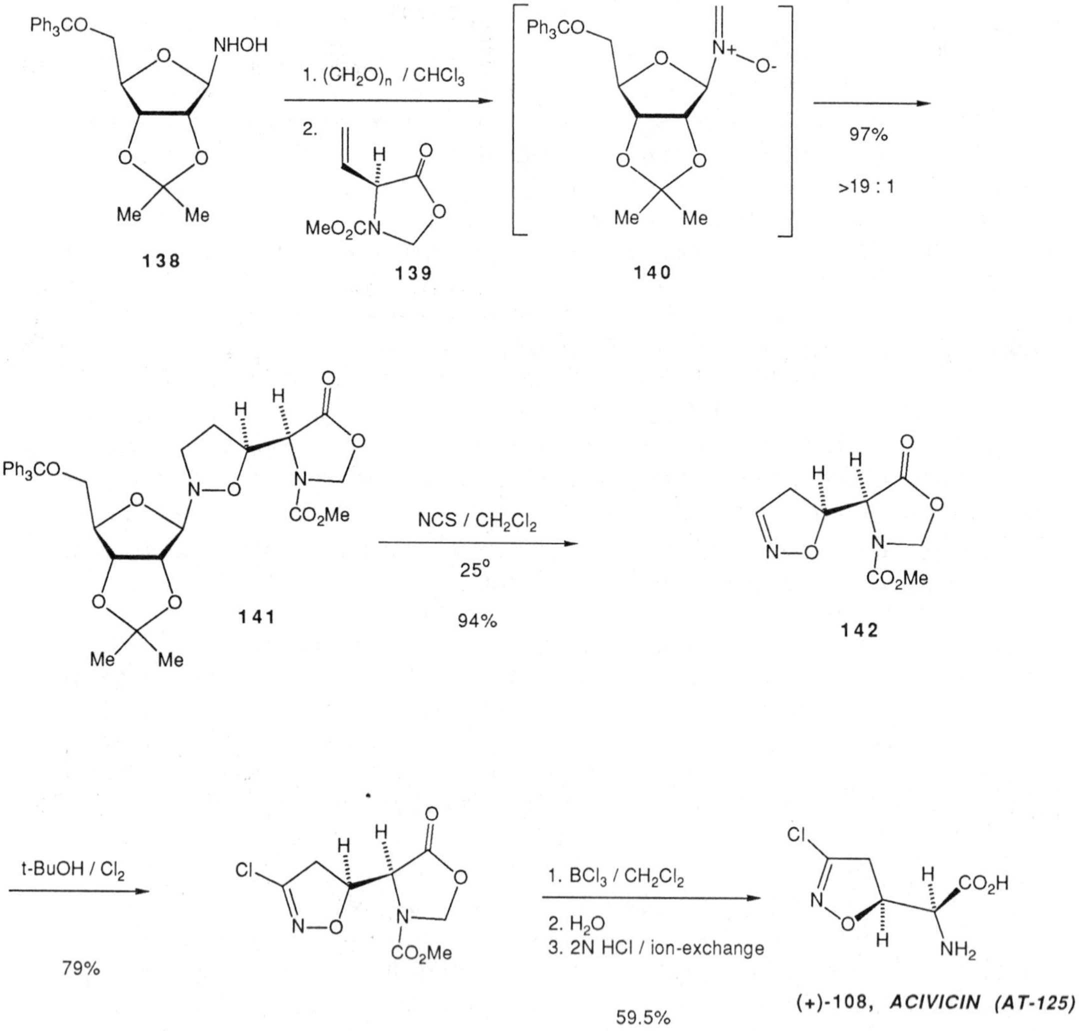

Ph3CO
NHOH
138
1. (CH2O)n / CHCl3
2.
MeO2CN
139
140
97%
>19 : 1
141
NCS / CH2Cl2
25°
94%
142
t-BuOH / Cl2
79%
143
1. BCl3 / CH2Cl2
2. H2O
3. 2N HCl / ion-exchange
59.5%
CO2H
NH2
(+)-108, ACIVICIN (AT-125)

Mzengeza, S.; Yang, C.M.; Whitney, R.A., *J.Am.Chem.Soc.* (1987) **109**, 276
Mzengeza, S.; Whitney, R.A., *J.Org.Chem.* (1988) **53**, 4074.

SCHEME 19

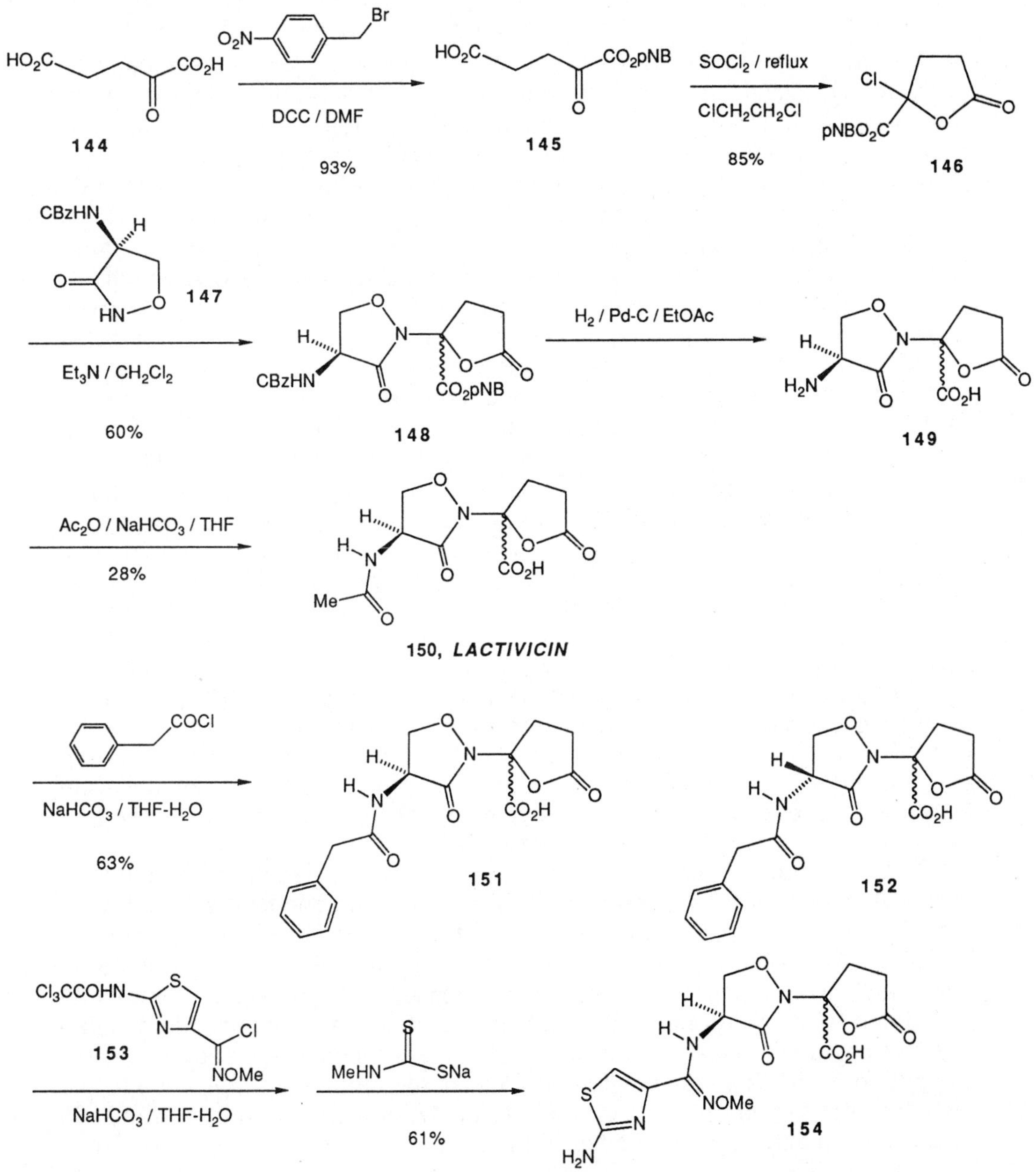

150, *LACTIVICIN*

Natsugari, H.; Kawano, Y.; Morimoto, A.; Yoshioka, K.; Ochiai, M., *J.Chem.Soc.Chem.Comm.* (1987) 62.

C. BULGECIN

A new class of small glycopeptide antibiotics were recently isolated from culture broths of *Pseudomonas acidophila* and *Pseudomonas mesoacidophila* by Shinagawa and associates [24]. This substance was found to synergistically enhance antimicrobial activity by inducing bulge formation in bacteria in cooperation with β-lactam antibiotics, but showed no antibacterial activity by itself. This substance, called bulgecin (**164**) contains the new amino acid (2S,4S,5R)-4-hydroxy-5-hydroxymethylproline (bulgecinine, **163**). As shown in Scheme 20, Shiba and associates [25] have synthesized bulgecinine from D-glucose. After employing well-established methodology to deoxegenate glucose at the 3-position, the 2-hydroxyl group was converted into the azide **157** with inversion and oxidized to the lactone **159**. Ring-opening to the protected amino acid **160** and conversion of the alcohol into the chloride (**161**) was achieved with triphenyl phosphine and carbon tetrachloride. Cyclization of this material sequentially to the lactone **162** and then to the pyrrolidine bulgecinine (**163**) was achieved in good overall yield.

Ohfune and co-workers [26] have extensively examined the halo-lactonization of various unsaturated amino acid derivatives to access a range of important amino acids. These workers have deployed this approach in a concise and elegant construction of bulgecinine. As shown in Scheme 21, protected allyl glycine **165** is homologated to the Z-unsaturated ester **166** in good overall yield. Reduction of the ester and re-oxidation of the amino alcohol moiety to the key amino acid **167** was accomplished in 68% overall yield. Bromolactonization of **167** proceeded in excellent yield and good diastereoselectivity (8.8:1) by treatment with NBS in THF giving the bromolactone **168**. Removal of the t-BOC group effected pyrrolidine cyclization and base treatment afforded bulgecin in high yield. The C-5 epimer of bulgecinine (**171**) was similarly prepared from the E-allylic alcohol **169** (Scheme 22).

In a related study, Ohfune examined the facial selectivity of intramolecular S_N2' cyclization of the allylic chlorides **172** and **173** which were prepared from **167** and **169**, respectively as shown in Scheme 23. Treatment of each isomer separately with silver(I)triflate in the presence of 2,6-lutidine gave, in both cases a 6:1 mixture of *syn*-**175** and *anti*-**176** in good yield. The authors postulate a common intermediate (**174**) from both systems that results in the preponderance of the *syn*-compound **175**. The authors speculate that the urethane moiety may help stabilize the *syn*-orientation of the allylic cation formed in the transition state, presumably through complexation.

Fleet and associates [27] have deployed D-glucuronolactone as a template for the construction of bulgecinine and related pyrrolidines as shown in Scheme 24. The isopropylidene derivative **177** was converted

SCHEME 20

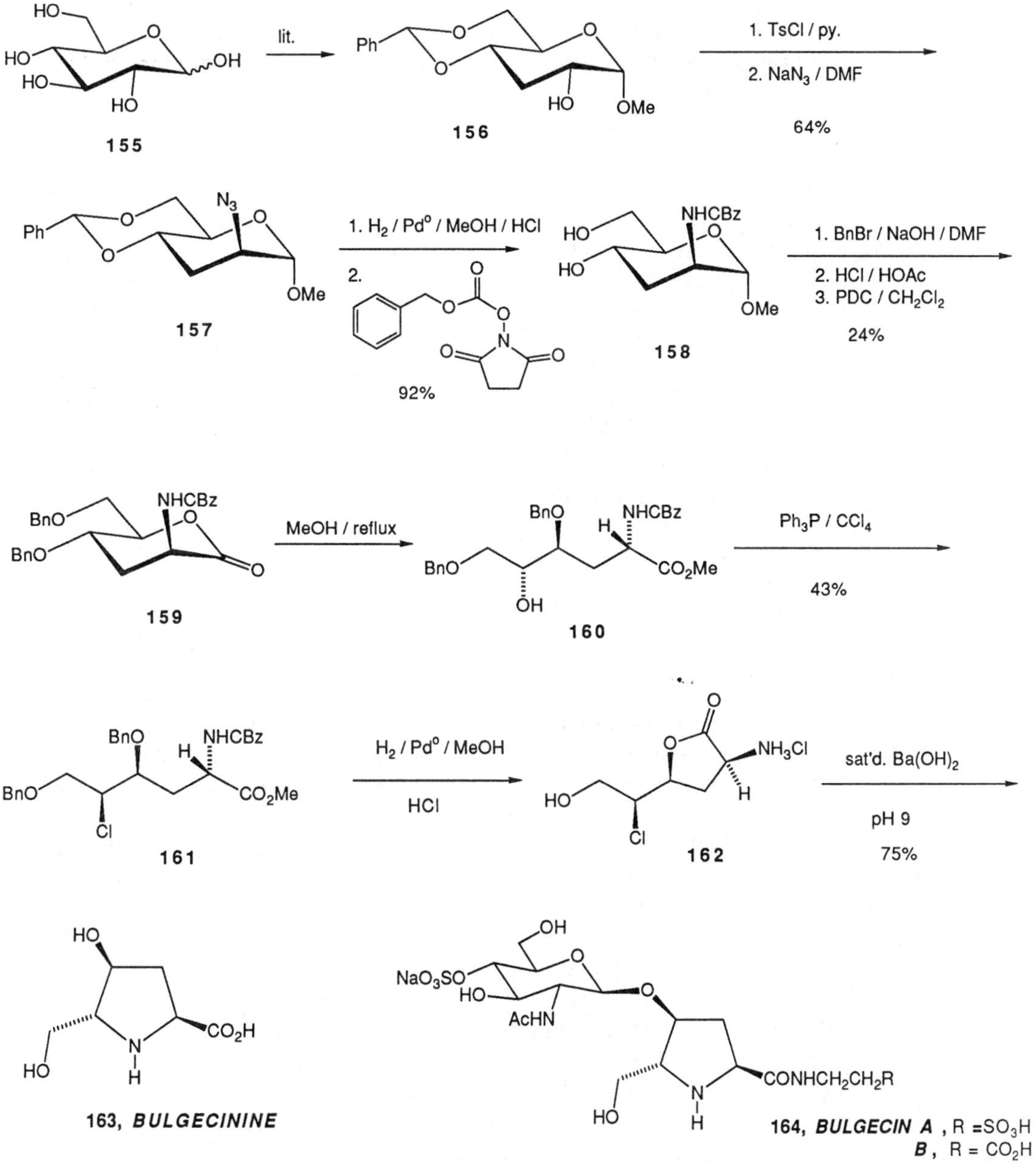

155 → (lit.) → 156 → (1. TsCl / py. 2. NaN$_3$ / DMF, 64%)

157 → (1. H$_2$ / Pd^0 / MeOH / HCl 2. [benzyl N-hydroxysuccinimide carbonate], 92%) → 158 → (1. BnBr / NaOH / DMF 2. HCl / HOAc 3. PDC / CH$_2$Cl$_2$, 24%)

159 → (MeOH / reflux) → 160 → (Ph$_3$P / CCl$_4$, 43%)

161 → (H$_2$ / Pd^0 / MeOH, HCl) → 162 → (sat'd. Ba(OH)$_2$, pH 9, 75%)

163, BULGECININE

164, BULGECIN A, R = SO$_3$H
B, R = CO$_2$H

Wakamiya, T.; Yamanoi, K.; Nishikawa, M.; Shiba, T., *Tetrahedron Lett.* (1985) **26**, 4759

SCHEME 21

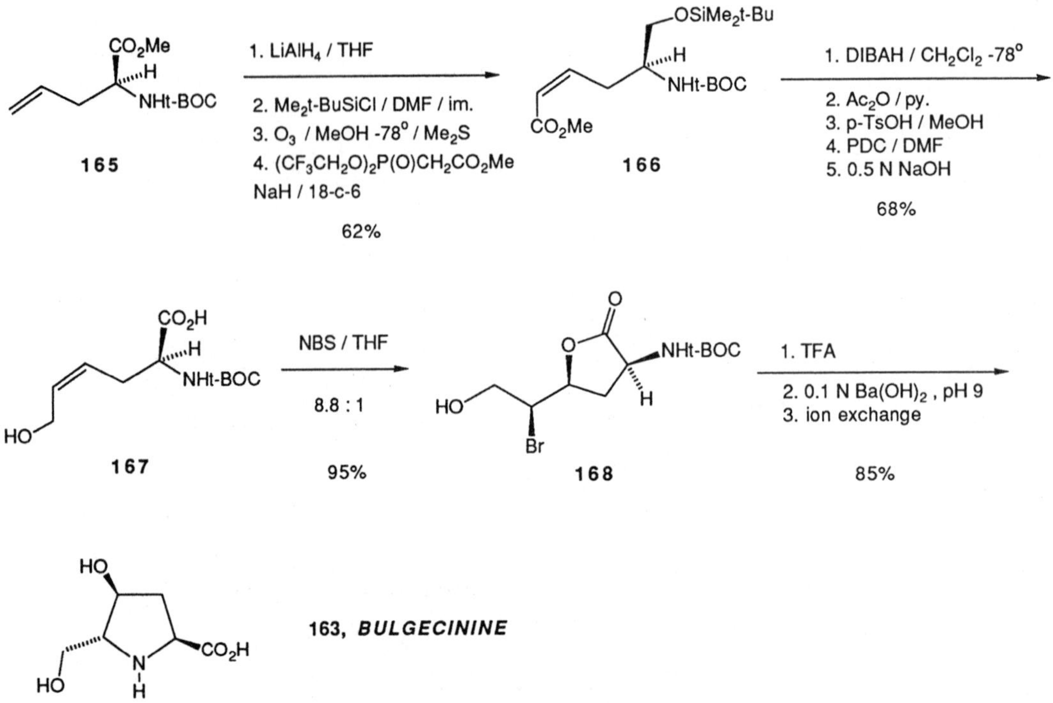
CO2Me
H
NHt-BOC
165
1. LiAlH4 / THF
2. Me2t-BuSiCl / DMF / im.
3. O3 / MeOH -78° / Me2S
4. (CF3CH2O)2P(O)CH2CO2Me
NaH / 18-c-6
62%
OSiMe2t-Bu
H
NHt-BOC
CO2Me
166
1. DIBAH / CH2Cl2 -78°
2. Ac2O / py.
3. p-TsOH / MeOH
4. PDC / DMF
5. 0.5 N NaOH
68%
CO2H
H
NHt-BOC
HO
167
NBS / THF
8.8 : 1
95%
O
O
NHt-BOC
H
HO
Br
168
1. TFA
2. 0.1 N Ba(OH)2 , pH 9
3. ion exchange
85%
HO
CO2H
N
H
HO
163, BULGECININE

SCHEME 22

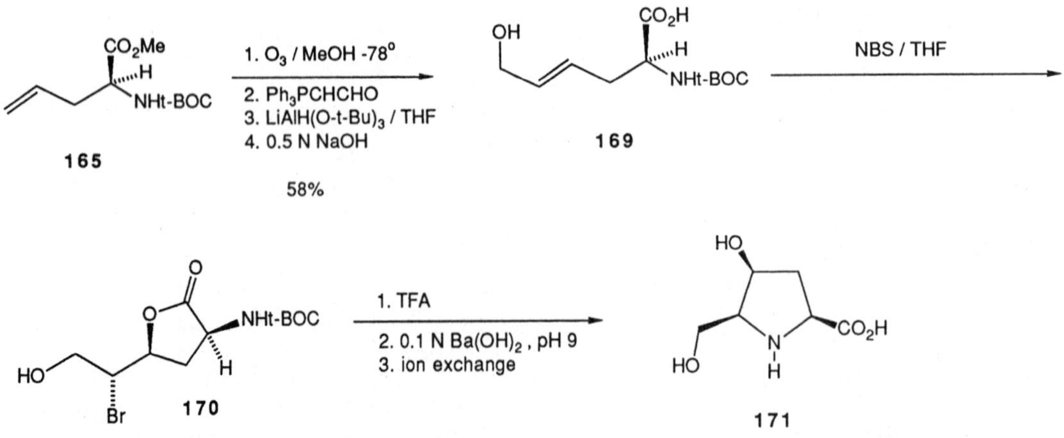
CO2Me
H
NHt-BOC
165
1. O3 / MeOH -78°
2. Ph3PCHCHO
3. LiAlH(O-t-Bu)3 / THF
4. 0.5 N NaOH
58%
OH
CO2H
H
NHt-BOC
169
NBS / THF
O
O
NHt-BOC
H
HO
Br
170
1. TFA
2. 0.1 N Ba(OH)2 , pH 9
3. ion exchange
HO
CO2H
N
H
HO
171

Ohfune, Y.; Hori, K.; Sakaitani, M., *Tetrahedron Lett.* (1986) **27**, 6079
Ohfune, Y.; Kurokawa, N., *Tetrahedron Lett.* (1985) **26**, 5307.

SCHEME 23

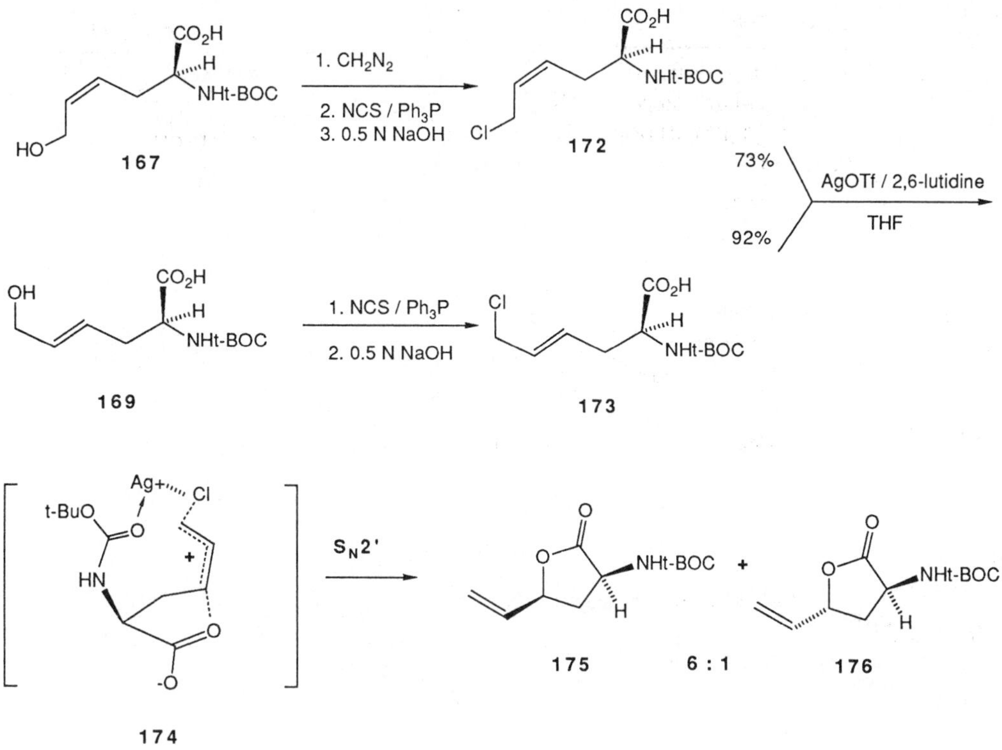

Ohfune, Y.; Hori, K.; Sakaitani, M., *Tetrahedron Lett.* (1986) **27**, 6079

into the corresponding azide **178** with inversion via the triflate. Reduction, acylation, elimination and reduction afforded the dehydro lactone **180** in high overall yield. Catalytic hydrogenation occurs stereospecifically from the least hindered face providing **181**. Activation of the secondary alcohol as the mesylate followed by a similar deprotection/cyclization sequence to those used above afforded bulgecinine (**163**) in 46% overall yield from **180** (26% overall yield from **177**). Compound **180** was also transformed into (2S,4S,5R)-dihydroxypipecolic acid (**184**) by selective mesylation of the primary alcohol and ring closure. The diol **184** is a natural constituent of the leaves of *Derris eliptica*.

SCHEME 24

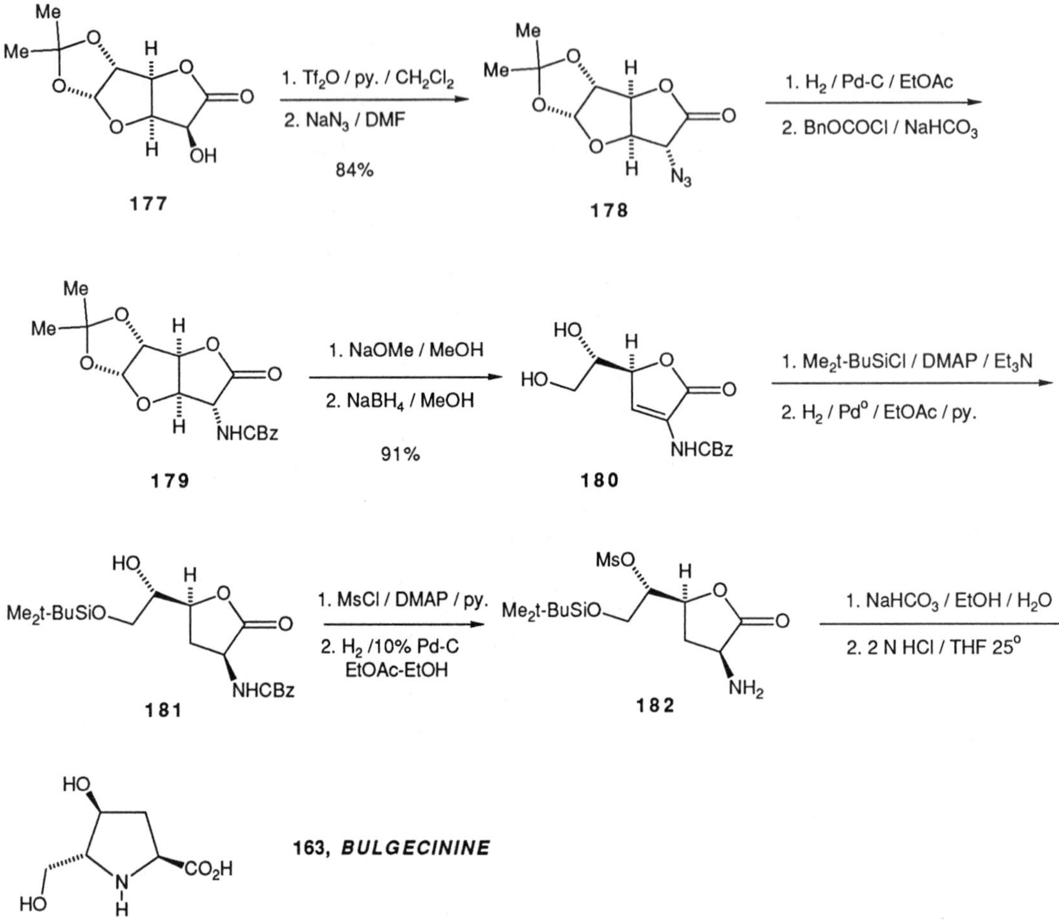

177 → 1. Tf₂O / py. / CH₂Cl₂ → 178

1. Tf$_2$O / py. / CH$_2$Cl$_2$
2. NaN$_3$ / DMF

84%

1. H$_2$ / Pd-C / EtOAc
2. BnOCOCl / NaHCO$_3$

179 →
1. NaOMe / MeOH
2. NaBH$_4$ / MeOH
91%
→ 180

1. Me$_2$t-BuSiCl / DMAP / Et$_3$N
2. H$_2$ / Pd^0 / EtOAc / py.

181 →
1. MsCl / DMAP / py.
2. H$_2$ /10% Pd-C
EtOAc-EtOH
→ 182

1. NaHCO$_3$ / EtOH / H$_2$O
2. 2 N HCl / THF 25^0

163, BULGECININE

SCHEME 25

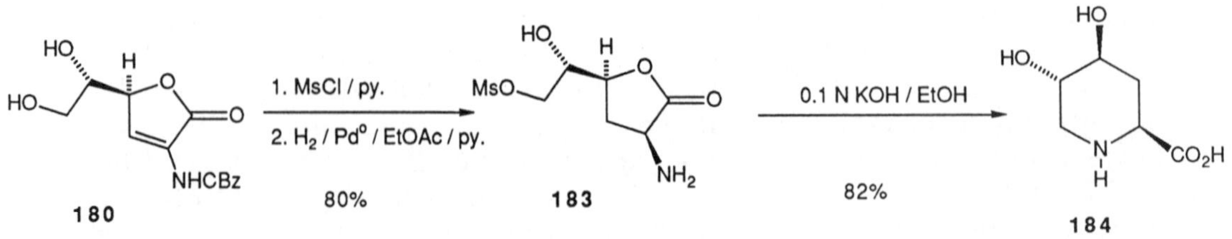

180 →
1. MsCl / py.
2. H$_2$ / Pd^0 / EtOAc / py.
80%
→ 183 →
0.1 N KOH / EtOH
82%
→ 184

Bashyal, B.P.; Chow, H-F.; Fleet, G.W.J., *Tetrahedron Lett.* (1986) **27**, 3205.

SCHEME 26

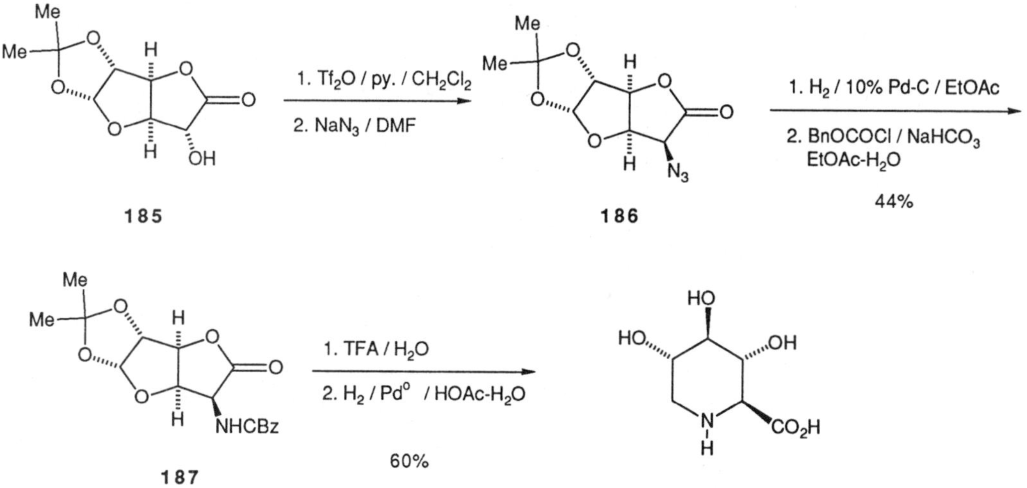

1. Tf$_2$O / py. / CH$_2$Cl$_2$
2. NaN$_3$ / DMF

185 → **186**

1. H$_2$ / 10% Pd-C / EtOAc
2. BnOCOCl / NaHCO$_3$
   EtOAc-H$_2$O

44%

1. TFA / H$_2$O
2. H$_2$ / Pd^0 / HOAc-H$_2$O

60%

187 → **188**

Bashyal, B.P.; Chow, H-F.; Fleet, G.W.J., *Tetrahedron Lett.* (1986) **27**, 3205.

These workers also prepared (2S,3R,4R,5S)-trihydroxypipecolic acid (**188**), a naturally occurring glucuronidase inhibitor obtained from the seeds of *Baphia racemosa*. Starting with the D-ribonolactone derivative **185**, azide formation, reduction, acylation and finally, reductive amination furnished **188**. The same substrate (**185**) was also used to prepare (2R,3S,4R)-dihydroxyproline by a related series of transformations.

D. ECHINOCANDIN

The echinocandins are a novel set of oligopeptide antibiotics isolated from *Aspergillus ruglosus* and *Aspergillus nidulans*. These complex substances have attracted interest due to their structural novelty and their activity against fungi, yeast and candidosis.

Ohfune and Kurokawa [29] have completed a linear total synthesis of echinocandin D (**212**) as shown in Schemes 27-31. The three novel amino acid constituents (**192**, **197** and **204**) were all prepared by manipulation of unsaturated amino acid derivatives. As shown in Scheme 27, homoallylic alcohol **189**, derived from vinyl glycine is stereoselectively

SCHEME 27

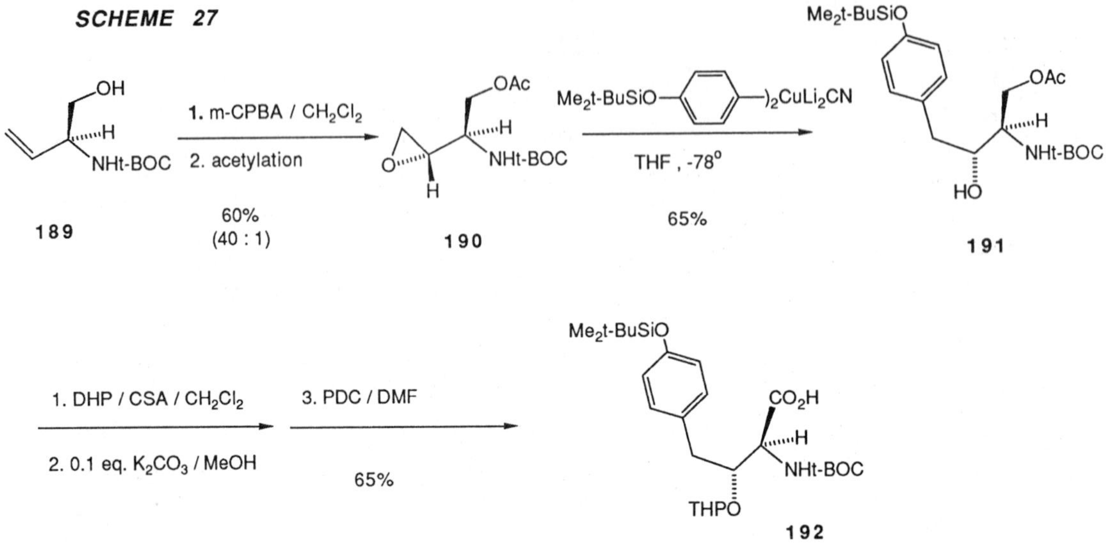
OH
H
NHt-BOC
189
1. m-CPBA / CH2Cl2
2. acetylation
60%
(40 : 1)
OAc
H
NHt-BOC
H
190
Me2t-BuSiO—⟨⟩—)2CuLi2CN
THF , -78°
65%
Me2t-BuSiO
OAc
H
NHt-BOC
HO
191
1. DHP / CSA / CH2Cl2
2. 0.1 eq. K2CO3 / MeOH
3. PDC / DMF
65%
Me2t-BuSiO
CO2H
H
NHt-BOC
THPO
192

SCHEME 28

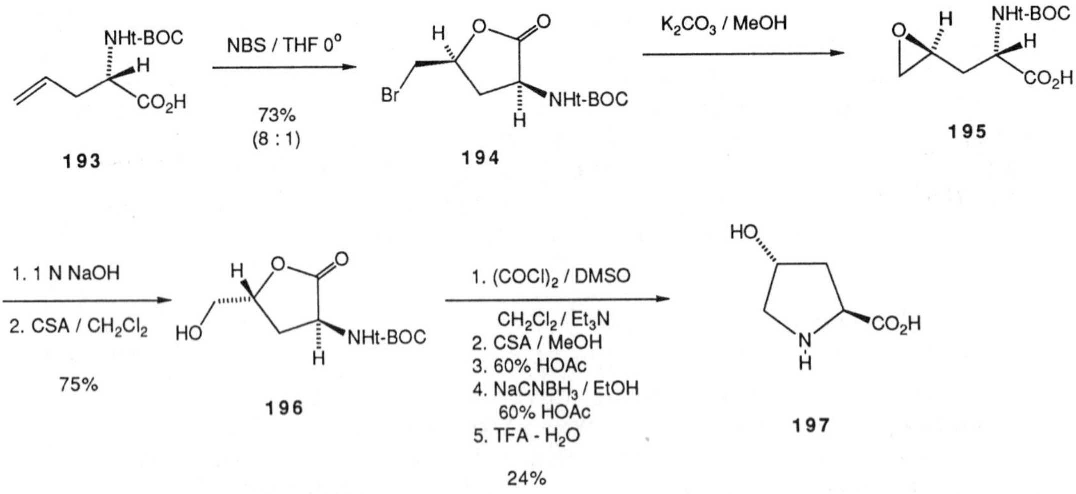
NHt-BOC
H
CO2H
193
NBS / THF 0°
73%
(8 : 1)
H
O
O
Br
NHt-BOC
H
194
K2CO3 / MeOH
H
NHt-BOC
H
CO2H
195
1. 1 N NaOH
2. CSA / CH2Cl2
75%
H
O
O
HO
NHt-BOC
H
196
1. (COCl)2 / DMSO
CH2Cl2 / Et3N
2. CSA / MeOH
3. 60% HOAc
4. NaCNBH3 / EtOH
60% HOAc
5. TFA - H2O
24%
HO
N
H
CO2H
197

Kurokawa, N.; Ohfune, Y., *J.Am.Chem.Soc.* (1986) **108**, 6041
Kurokawa, N.; Ohfune, Y., *J.Am.Chem.Soc.* (1986) **108**, 6043.

SCHEME 29

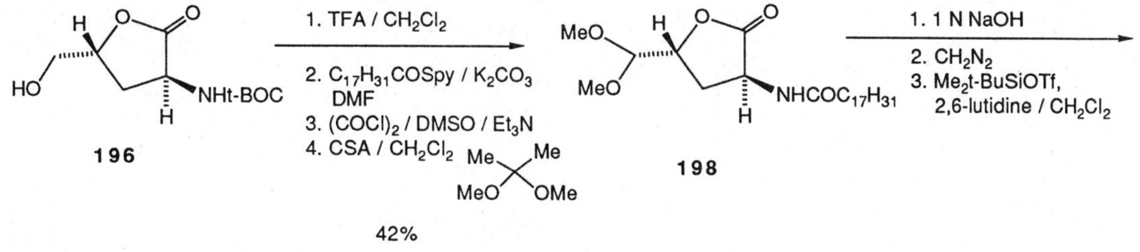

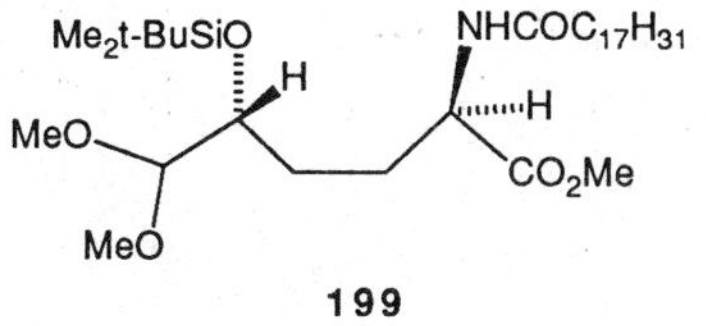

196

1. TFA / CH₂Cl₂
2. C₁₇H₃₁COSpy / K₂CO₃
   DMF
3. (COCl)₂ / DMSO / Et₃N
4. CSA / CH₂Cl₂ Me⤸Me
                 MeO OMe

42%

198

1. 1 N NaOH
2. CH₂N₂
3. Me₂t-BuSiOTf,
   2,6-lutidine / CH₂Cl₂

199

Kurokawa, N.; Ohfune, Y., *J.Am.Chem.Soc.* (1986) **108**, 6041
Kurokawa, N.; Ohfune, Y., *J.Am.Chem.Soc.* (1986) **108**, 6043.

SCHEME 30

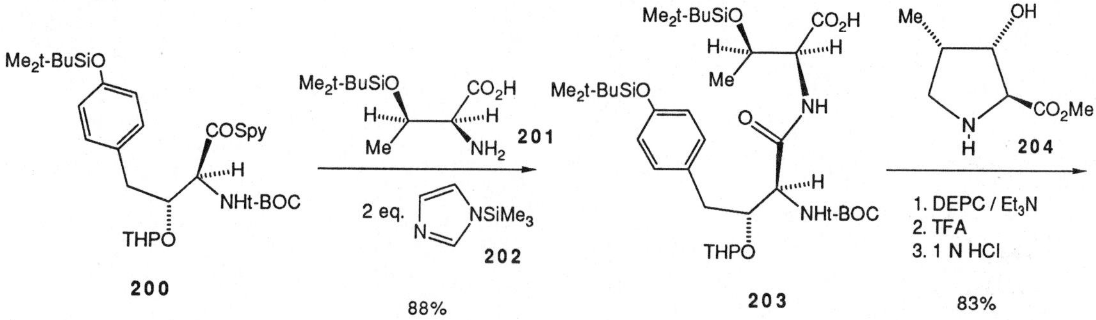

200

2 eq.   201   202

88%

203

204

1. DEPC / Et₃N
2. TFA
3. 1 N HCl

83%

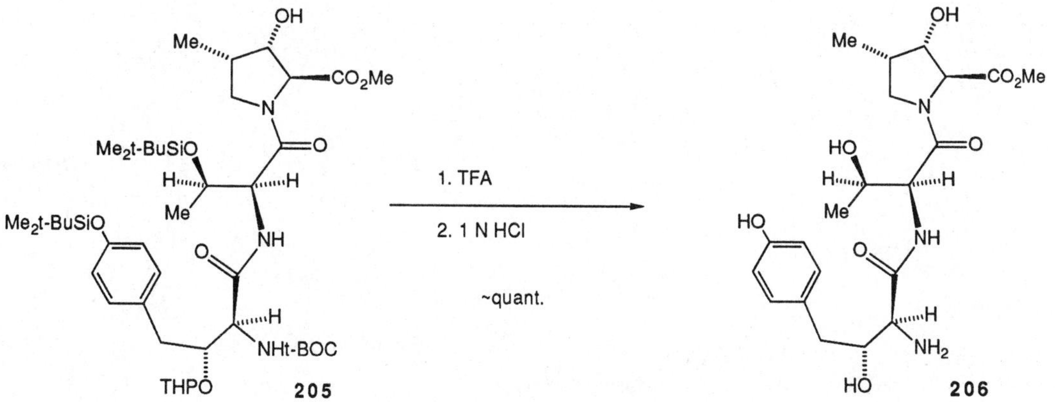

205

1. TFA
2. 1 N HCl

~quant.

206

SCHEME 31

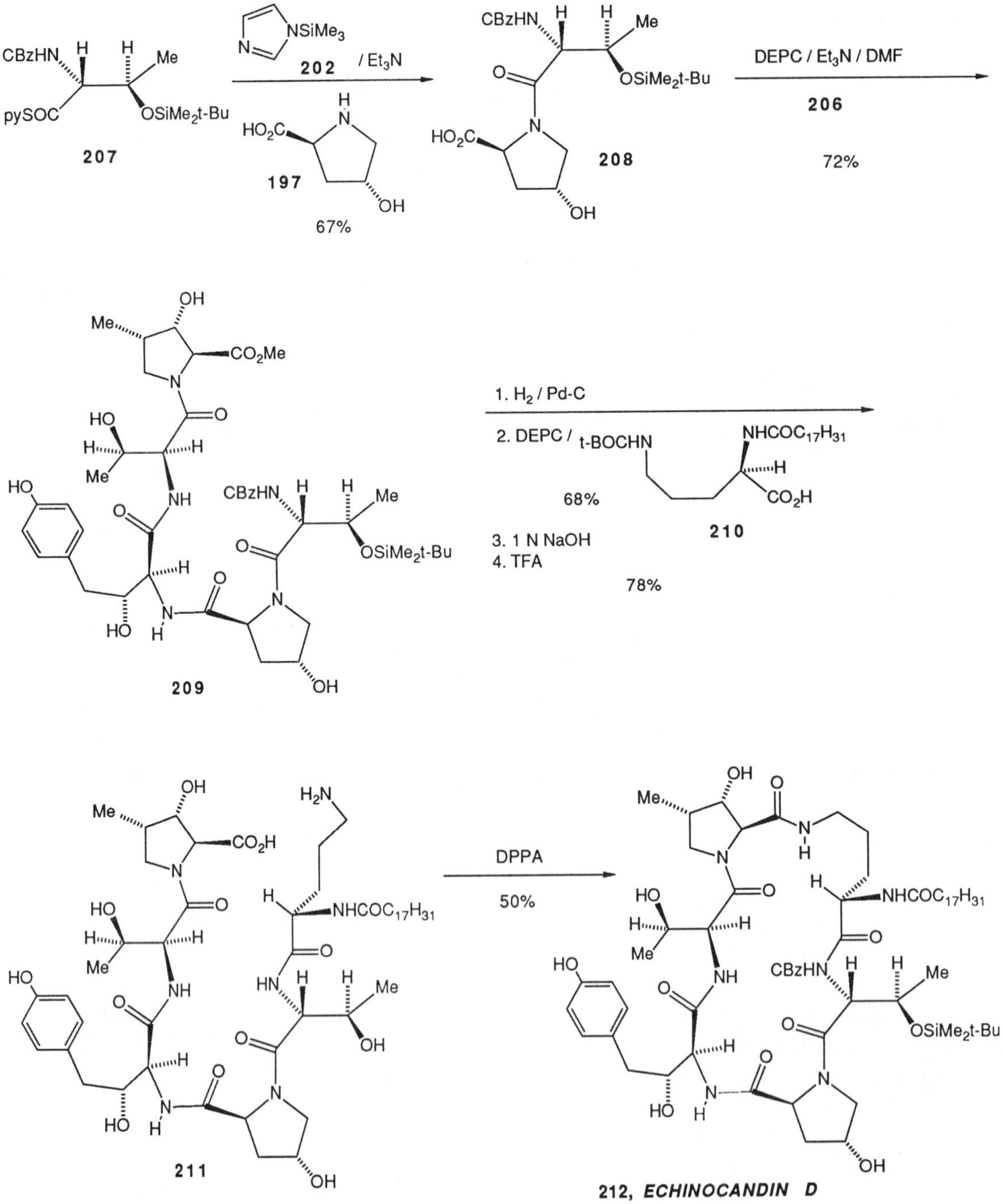

209

211

212, *ECHINOCANDIN D*

Kurokawa, N.; Ohfune, Y., *J.Am.Chem.Soc.* (1986) **108**, 6043 epoxidized (40:1) to afford **190**. Cuprate opening of the epoxide furnishes **191** which is sequentially protected and oxidized to the (2S,3R)-3-hydroxyhomotyrosine derivative **192**.

Using a similar sequence to that employed by these workers on the preparation of bulgecinine (Scheme 21) *trans*-4-hydroxyproline (**197**) was prepared via the bromolactonization strategy as shown in Scheme 28.

The γ-hydroxy amino acid constituent **199** was prepared from lactone **196** as shown in Scheme 29. Removal of the *t*-BOC group from **196** followed by acylation with an activated linoleyl derivative , Swern oxidation and acetalization furnished **198**. Hydrolysis, esterification and silylation furnished the fully protected derivative **199**. While this scheme provides a potential source of the γ,δ-dihydroxyornithine moiety found in echinocandin B, they were unable to effect ring closure to the unusual hemi-amino acetal with this precursor. As will be seen below, ornithine itself was therefore employed to close the 21-membered ring of echinocandin D.

The homotyrosine derivative **192** was converted into the pyridyl thioester **200** and coupled with O-silyl-threonine (**201**) by the agency of trimethylsilyl imidazole to afford the dipeptide **203** in 88% yield (Scheme 30). The synthesis of (2S,3S,4S)-3-hydroxy-4-methylproline (**204**) was previously described in Chapter 4 (Scheme 17) via a nucleophilic amination of an epoxy alcohol. Coupling of **203** and **204** with diethyl phosphorocyanidate (DEPC) gave , after deprotection, tripeptide **206** in high yield.

Coupling of the activated threonine **207** with 4-hydroxyproline (**197**) afforded **208** which was condensed with **206** to give the pentapeptide **209** in 72% yield (Scheme 31). Protecting group manipulation followed by coupling to the acylated ornithine derivative (**210**) gave the pentultimate hexapeptide **211** in good yield. Cyclization of the 21-membered ring was realized with diphenyl phosphoryl azide in 50% yield giving echinocandin D (**212**).

More recently, Evans and Weber [30] recorded a striking and concise total synthesis of echinocandin D and the constituent amino acids via glycine and bromoacetate enolate aldol methodology as described in Scheme 32. Bromoacetate **213** was transformed into the corresponding boron enolate and condensed with methacrolein to afford the aldol adduct **214** in 50% yield. The authors note that examination of the crude aldol mixture revealed that the diastereoselectivity of this process was 97%. Displacement of the azide with clean inversion and removal of the chiral auxilliary furnished **216**. This substance was subjected to a novel intramolecular cycloalkylation via reduction of the olefin with dicyclohexylborane; the incipient aza-borate complex **217** suffered migratory insertion to afford the requisite pyrrolidine **218** in 72% yield. Peptide coupling of **218** with N-t-BOC-threonine **219** furnished **220** after

SCHEME 32

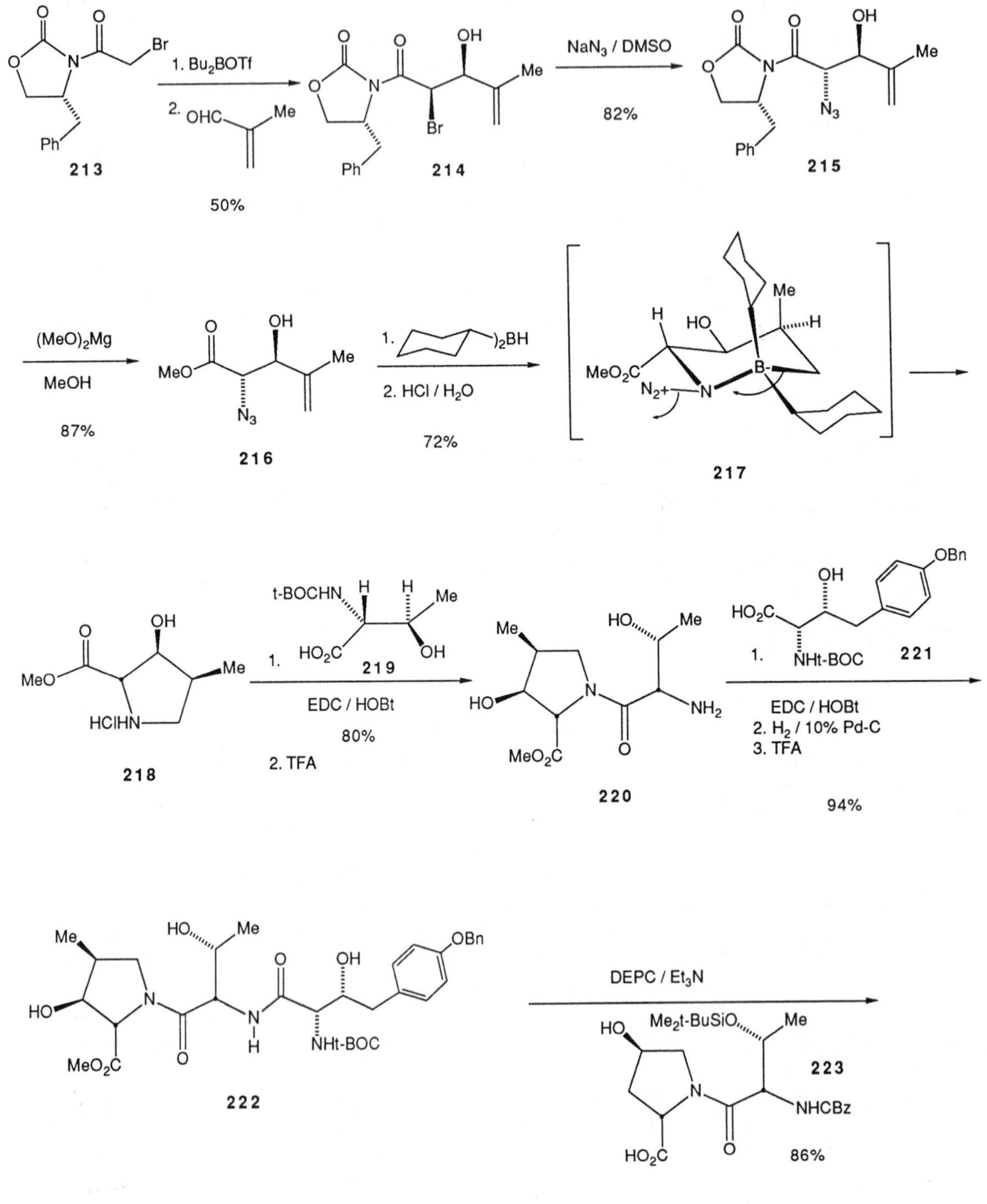

213

1. Bu$_2$BOTf

2. OHC—C(=CH$_2$)Me

50%

214

NaN$_3$ / DMSO

82%

215

(MeO)$_2$Mg

MeOH

87%

216

1. (cyclohexyl)$_2$BH

2. HCl / H$_2$O

72%

217

218

1. 219

t-BOCHN—CH(H)—CH(OH)Me, HO$_2$C

EDC / HOBt

80%

2. TFA

220

1. 221

HO$_2$C—CH(NHt-BOC)—CH(OH)—CH$_2$—C$_6$H$_4$—OBn

EDC / HOBt

2. H$_2$ / 10% Pd-C

3. TFA

94%

222

DEPC / Et$_3$N

223

86%

SCHEME 32 **(con't.)**

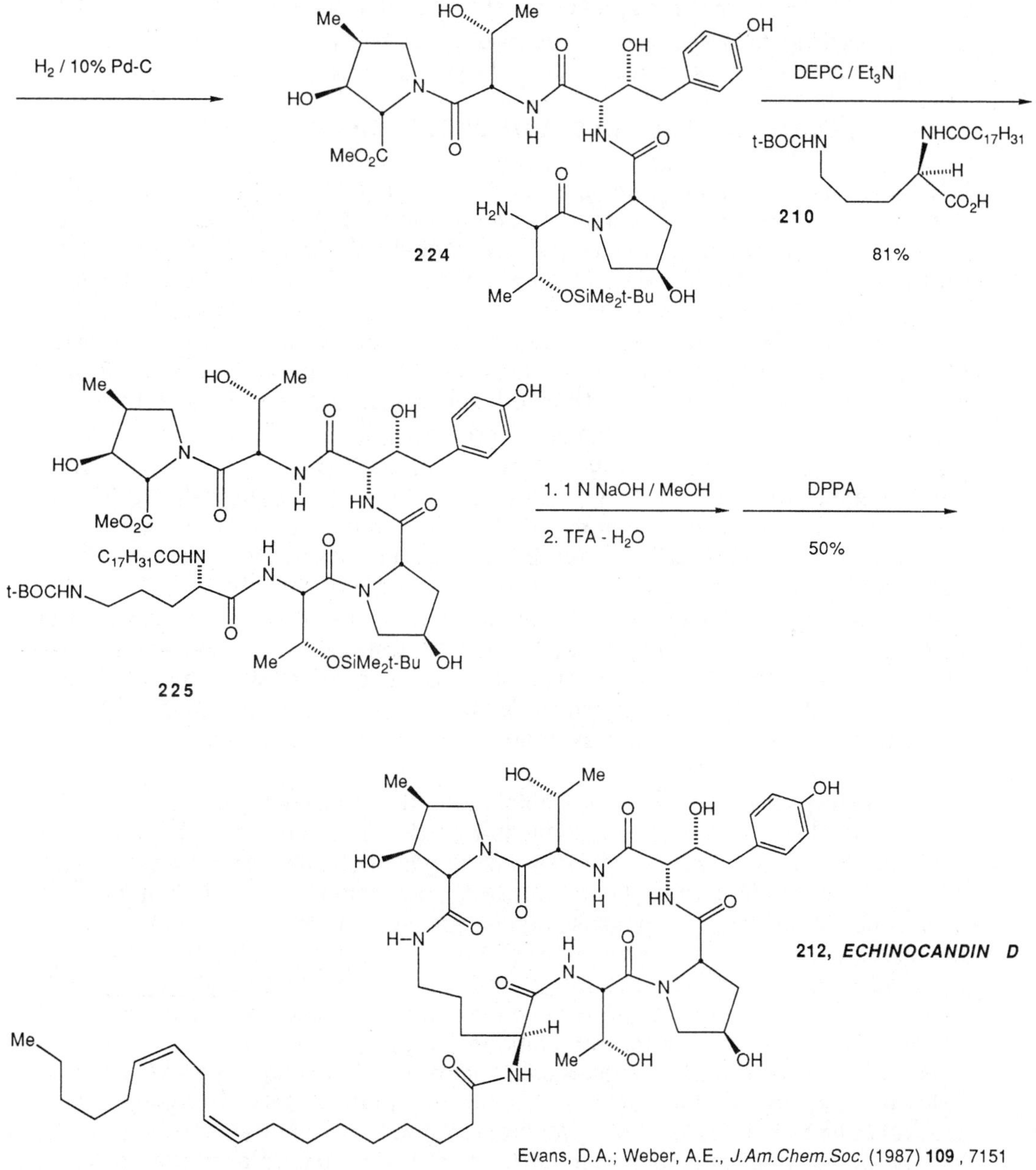
H_2 / 10% Pd-C
224
DEPC / Et_3N
t-BOCHN
$NHCOC_{17}H_{31}$
CO_2H
210
81%
225
1. 1 N NaOH / MeOH
2. TFA - H_2O
DPPA
50%
212, *ECHINOCANDIN D*

Evans, D.A.; Weber, A.E., *J.Am.Chem.Soc.* (1987) **109**, 7151 removal of the BOC group with TFA. The β-hydroxyhomotyrosine derivative (**221**) was synthesized by these workers via an asymmetric glycine enolate aldol strategy and was presented in Chapter 1 (Scheme 45). Coupling of **220** and **221** with EDC furnished tripeptide **222** in excellent yield. Sequential addition of the dipeptide **223** and the acylated ornithine yielded **225** which was cyclized to echinocandin D in 50% yield using DPPA according to the procedure described above by Ohfune [29]. Complete experimental details accompany this work.

E. COMPLEX CYCLIC PEPTIDES AND CYCLODEPSIPEPTIDES

Natures' seemingly limitless supply of highly derivatized peptides and cyclopeptides continues to provide synthetic challenges as well as a multitude of biologically significant substances. A few highlights in this broad area have been selected with an accent on structures containing unusual amino acid constituents.

Grieco and associates[31] have recently completed a total synthesis of the 19-membered ring cyclodepsipeptide (+)-jasplakinolide (**240**) which was isolated from a soft-bodied sponge *Jaspis sp*. This substance contains two unusual amino acids, 2-bromoabrine and β-tyrosine [32] and has been shown to display potent insecticidal, anti-fungal and anthelminthic properties. The synthesis, outlined in Schemes 33-36 is configured around construction of the aliphatic component from the optically active acid **226**; the 2-bromoabrine derivative (**234**) from tryptophan (Scheme 34) via a regioselective bromination ; and homologation of *para*-hydroxyphenylglycine to the β-tyrosine derivative **237** (Scheme 35). Coupling of **234** and **237** with DCC / HOBt affords the dipeptide **238** in 50% yield. Coupling of this unit with the aliphatic component **231** affords **239** which is subsequently cyclized with DCC to give after desilylation, jasplakinolide (**240**) in 31% overall yield from **239**.

These workers [33] have also recently prepared the structurally related cyclodepsipeptide geodiamolide B (**248**) which was isolated from the marine sponge *Geodia sp*. This antifungal antibiotic contains the same 12-carbon hydroxy carboxylic acid (**246**) that constitutes the right half of jasplakinolide as well as possessing the unusual amino acid, 3-bromo-N-methyl-D-tyrosine . As shown in Scheme 37, N-*t*-Boc-D-tyrosine (**241**) is silylated and N-methylated through the agency of *tert*-butyllithium and methyl iodide in THF (71%). Bromination was effected with bromine and mercuric acetate in carbon tetrachloride in good yield to afford the key tyrosine derivative **243**. Peptide coupling of the flanking alanine residues followed by connection of the 12-carbon moiety **246** furnishes the precyclization substrate **247**. Macrolactonization of the corresponding hydroxy acid was effected with DCC in refluxing chloroform but in very

SCHEME 33

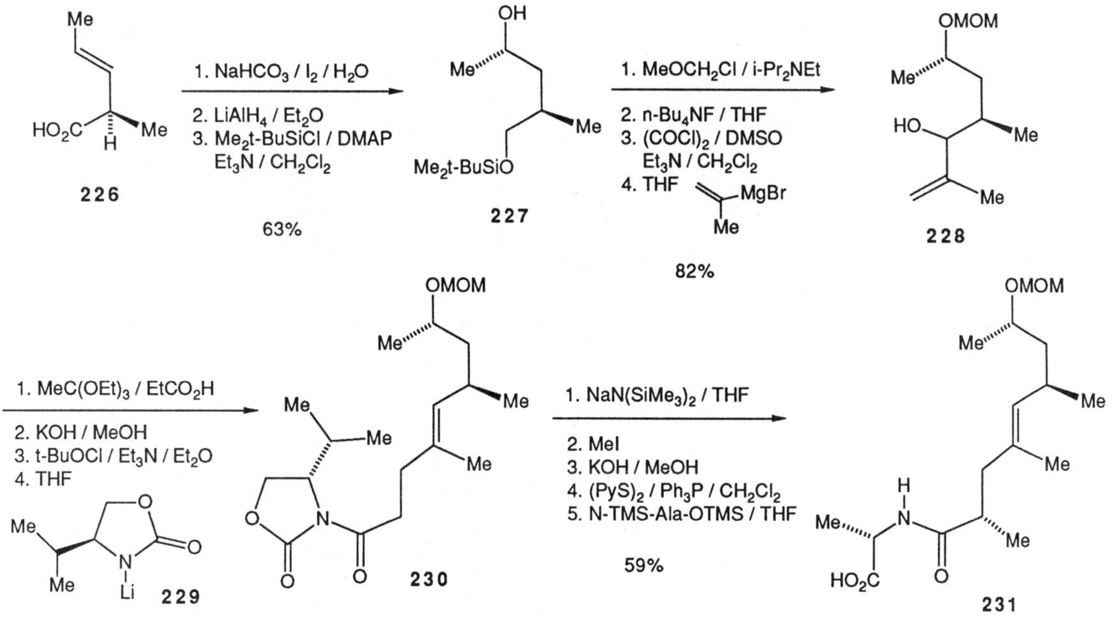
Me
HO2C
H
Me
226
1. NaHCO3 / I2 / H2O
2. LiAlH4 / Et2O
3. Me2t-BuSiCl / DMAP
Et3N / CH2Cl2
63%
OH
Me
Me
Me2t-BuSiO
227
1. MeOCH2Cl / i-Pr2NEt
2. n-Bu4NF / THF
3. (COCl)2 / DMSO
Et3N / CH2Cl2
4. THF
MgBr
Me
82%
OMOM
Me
HO
Me
Me
228
1. MeC(OEt)3 / EtCO2H
2. KOH / MeOH
3. t-BuOCl / Et3N / Et2O
4. THF
Me
O
O
N
Li
Me
229
71%
OMOM
Me
Me
Me
Me
Me
O
N
O
O
230
1. NaN(SiMe3)2 / THF
2. MeI
3. KOH / MeOH
4. (PyS)2 / Ph3P / CH2Cl2
5. N-TMS-Ala-OTMS / THF
59%
OMOM
Me
Me
Me
H
N
Me
Me
HO2C
O
231

SCHEME 34

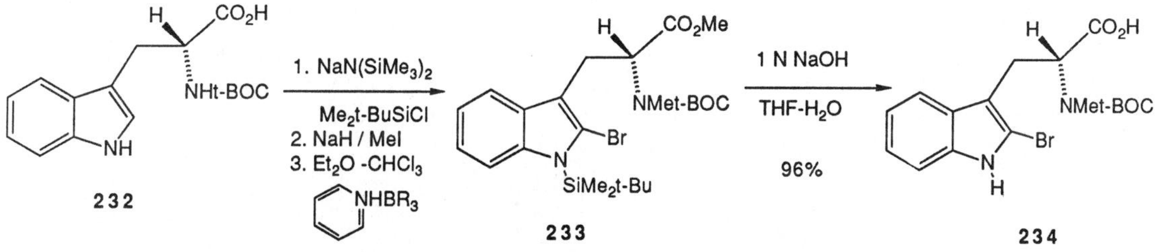
H
CO2H
NHt-BOC
NH
232
1. NaN(SiMe3)2
Me2t-BuSiCl
2. NaH / MeI
3. Et2O -CHCl3
NHBR3
40%
H
CO2Me
NMet-BOC
Br
N
SiMe2t-Bu
233
1 N NaOH
THF-H2O
96%
H
CO2H
NMet-BOC
Br
N
H
234

SCHEME 35

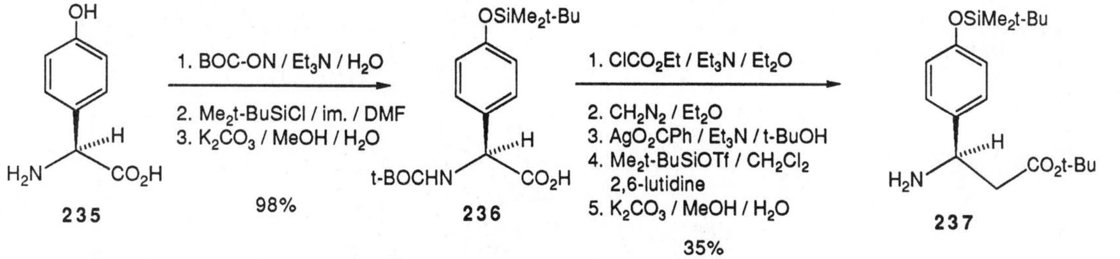
OH
H
H2N
CO2H
235
1. BOC-ON / Et3N / H2O
2. Me2t-BuSiCl / im. / DMF
3. K2CO3 / MeOH / H2O
98%
OSiMe2t-Bu
H
t-BOCHN
CO2H
236
1. ClCO2Et / Et3N / Et2O
2. CH2N2 / Et2O
3. AgO2CPh / Et3N / t-BuOH
4. Me2t-BuSiOTf / CH2Cl2
2,6-lutidine
5. K2CO3 / MeOH / H2O
35%
OSiMe2t-Bu
H
H2N
CO2t-Bu
237

Grieco, P.A.; Hon, Y.S.; Perez-Medrano, A., *J.Am.Chem.Soc.* (1988) **110**, 1630.

SCHEME 36

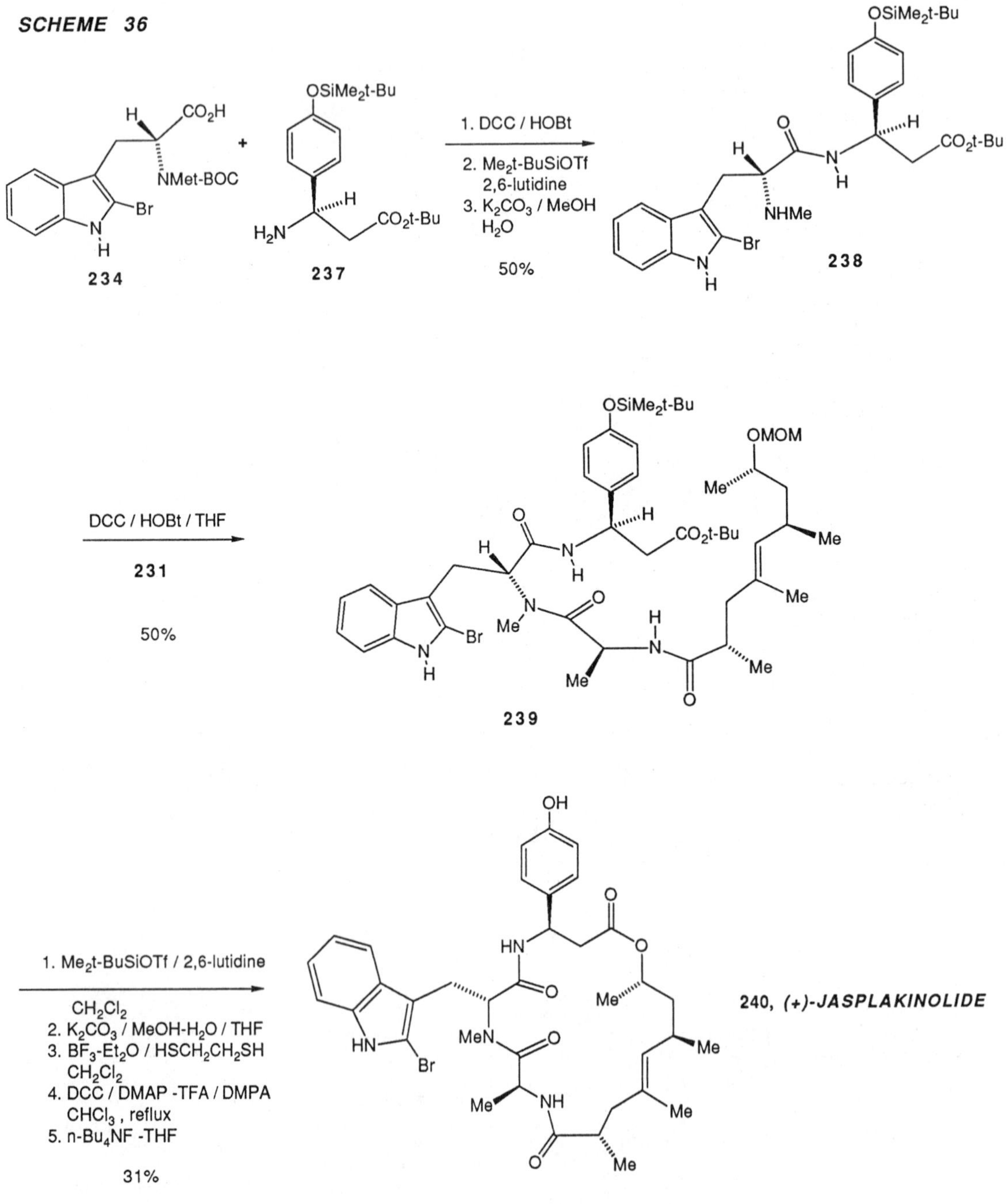

234 + 237 → 238

1. DCC / HOBt
2. Me$_2$t-BuSiOTf
   2,6-lutidine
3. K$_2$CO$_3$ / MeOH
   H$_2$O

50%

DCC / HOBt / THF

231

50%

239

1. Me$_2$t-BuSiOTf / 2,6-lutidine

   CH$_2$Cl$_2$
2. K$_2$CO$_3$ / MeOH-H$_2$O / THF
3. BF$_3$-Et$_2$O / HSCH$_2$CH$_2$SH
   CH$_2$Cl$_2$
4. DCC / DMAP -TFA / DMPA
   CHCl$_3$, reflux
5. n-Bu$_4$NF -THF

   31%

240, (+)-JASPLAKINOLIDE

Grieco, P.A.; Hon, Y.S.; Perez-Medrano, A., *J.Am.Chem.Soc.* (1988) **110**, 1630.

SCHEME 37

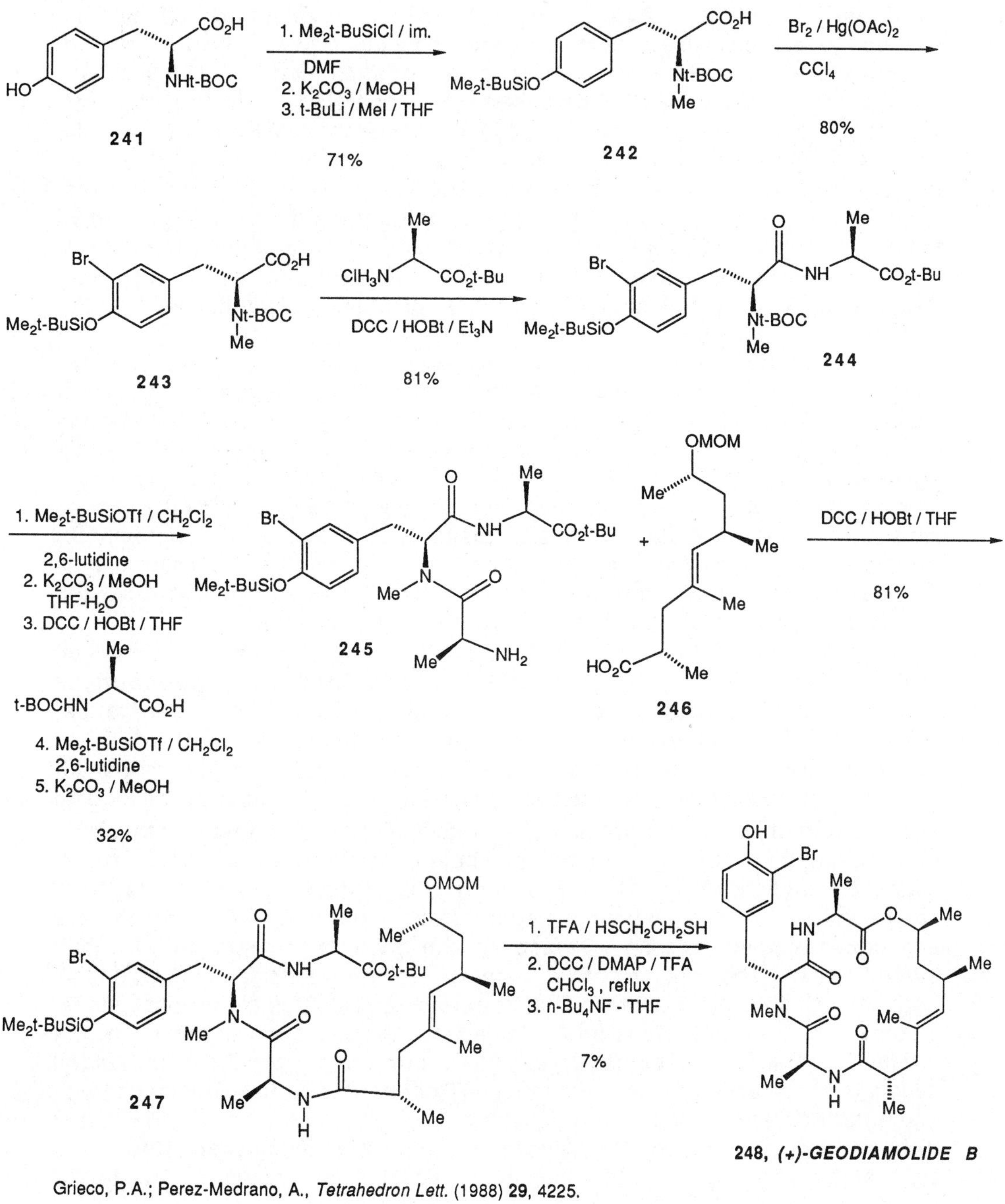

241
1. Me$_2$t-BuSiCl / im.
   DMF
2. K$_2$CO$_3$ / MeOH
3. t-BuLi / MeI / THF

71%

242

Br$_2$ / Hg(OAc)$_2$
CCl$_4$

80%

243

CIH$_3$N   CO$_2$t-Bu

DCC / HOBt / Et$_3$N

81%

244

1. Me$_2$t-BuSiOTf / CH$_2$Cl$_2$
   2,6-lutidine
2. K$_2$CO$_3$ / MeOH
   THF-H$_2$O
3. DCC / HOBt / THF

t-BOCHN   CO$_2$H

4. Me$_2$t-BuSiOTf / CH$_2$Cl$_2$
   2,6-lutidine
5. K$_2$CO$_3$ / MeOH

32%

245

246

DCC / HOBt / THF

81%

247

1. TFA / HSCH$_2$CH$_2$SH
2. DCC / DMAP / TFA
   CHCl$_3$, reflux
3. n-Bu$_4$NF - THF

7%

248, (+)-GEODIAMOLIDE B

Grieco, P.A.; Perez-Medrano, A., *Tetrahedron Lett.* (1988) **29**, 4225.

low yield(15%). Completion of the synthesis of geodiamolide B (**248**) was realized by removal of the phenolic protection.

White and Amedio [34] prepared the congener geodiamolide A (**262**) which only differs from **248** by the substitution of iodine for bromine in the tyrosine residue. The synthesis of the nonenoic acid unit (**255**, Scheme 38) utilizes a Claisen rearrangement of an orthopropionate (**254** to **255**). The Grignard addition to **253** produced an inseparable mixture of isomers which each gave the same 1.5 : 1 ratio of diastereomers **255** which could then be separated by HPLC. N-Methylation of the tyrosine residue was accomplished in excellent yield on the *t*-BOC derivative **257**. The authors comment that the bulk of the iodo-substituent posed an additional strategic problem that was partially accomodated by the employment of a *para*-methoxybenzyl ether for protecting the tyrosine hydroxyl group. Coupling and cyclization as above for **248** provided geodiamolide A (**262**) in 16% yield from **261**.

Synthetic chemists have recently begun examining methodologies that would ultimately be applicable to the complex vancomycin group of antibiotics. The synthesis of both biaryl and biphenyl amino acids bearing a variety of substituents must be embraced to gain entry to this demanding series of structures. Several "simpler" biaryl-containing cyclic peptides and derivatives have been isolated and recently, several of these have succumbed to total synthesis. Elegant solutions to the problem of oxidative coupling of tyrosine derivatives to produce the requisite biaryls are delineated below. The more challenging biphenyl systems and the biaryls derived from phenyl glycine have yet to be conquered.

Piperazinomycin (**271**) is a novel cyclic piperazine antibiotic that was isolated [35] from *Streptoverticillium olivoreticuli* subsp. *neonacticus* and displays activity against fungi and yeasts. This substance, biosynthetically derived by the oxidative coupling of a *cyclo*-Tyr-Tyr precursor, with reduction of the amides is perhaps the simplest model for the more complex biaryls discussed below and the vancomycins alluded to above. Yamamura and associates [36] have disclosed a total synthesis of piperazinomycin that features a thallium trinitrate-mediated biaryl coupling reaction. As shown in Scheme 41, N-formyl-L-tyrosine (**263**) is converted into the dibromo substrates **264** and **265** which are subsequently coupled and cyclized to the piperazinedione **267** in good overall yield. The choice of a piperazinedione for the cyclization as opposed to the piperazine oxidation state was a judicious one since, piperazinediones are well-known to adopt a boat conformation; in the present context, this places the two phenyl rings proximal for biaryl coupling in a pseudoaxial orientation. Treatment of this substance with thallium trinitrate in methanol followed by zinc reduction of the initial dienone adduct provided the bicyclic structure **268** in modest (19%) yield plus small amounts of other oxidized bicyclic derivatives (16% total).

SCHEME 38

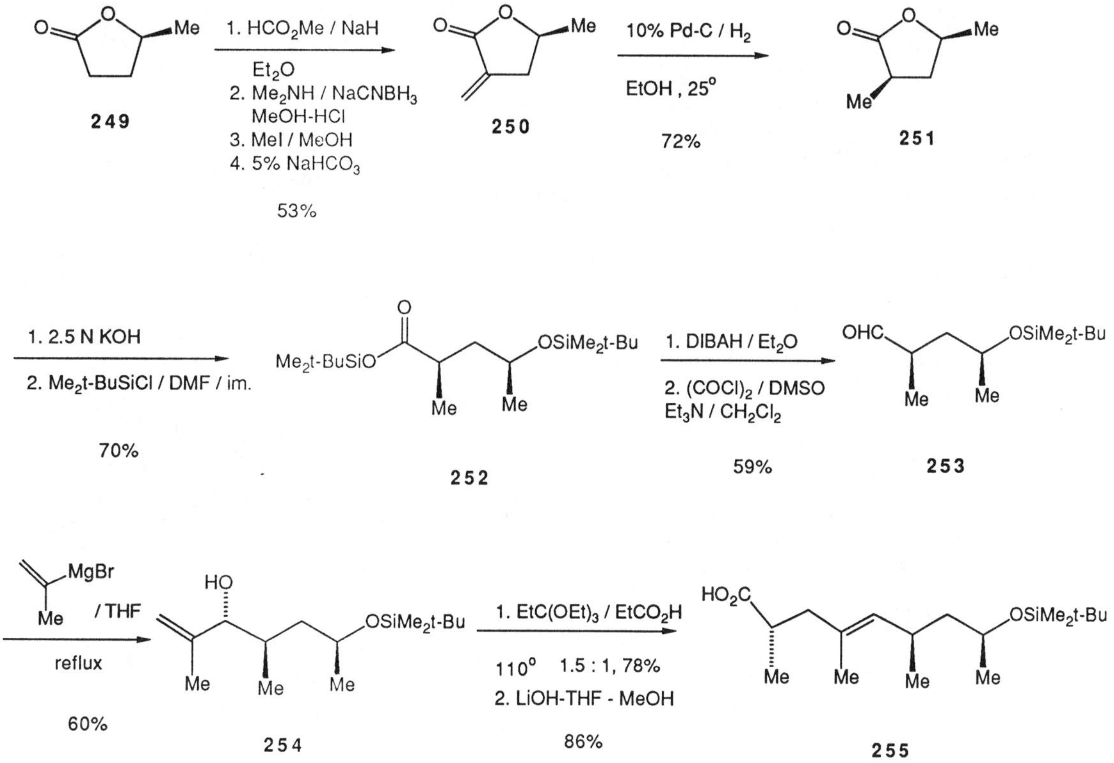

SCHEME 39

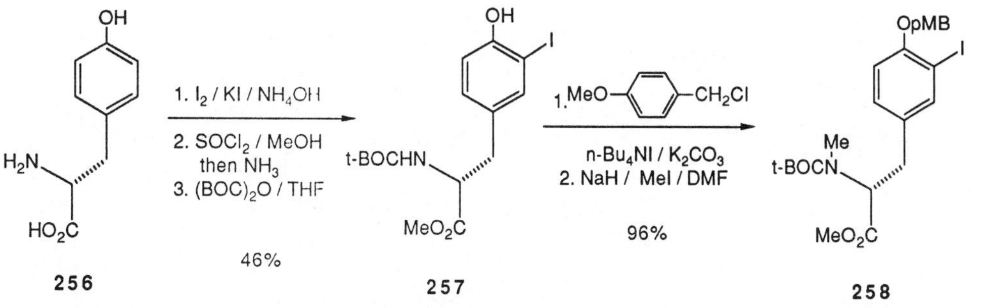

White, J.D.; Amedio, J.C. , *J.Org.Chem.* (1989) **54**, 738.

SCHEME 40

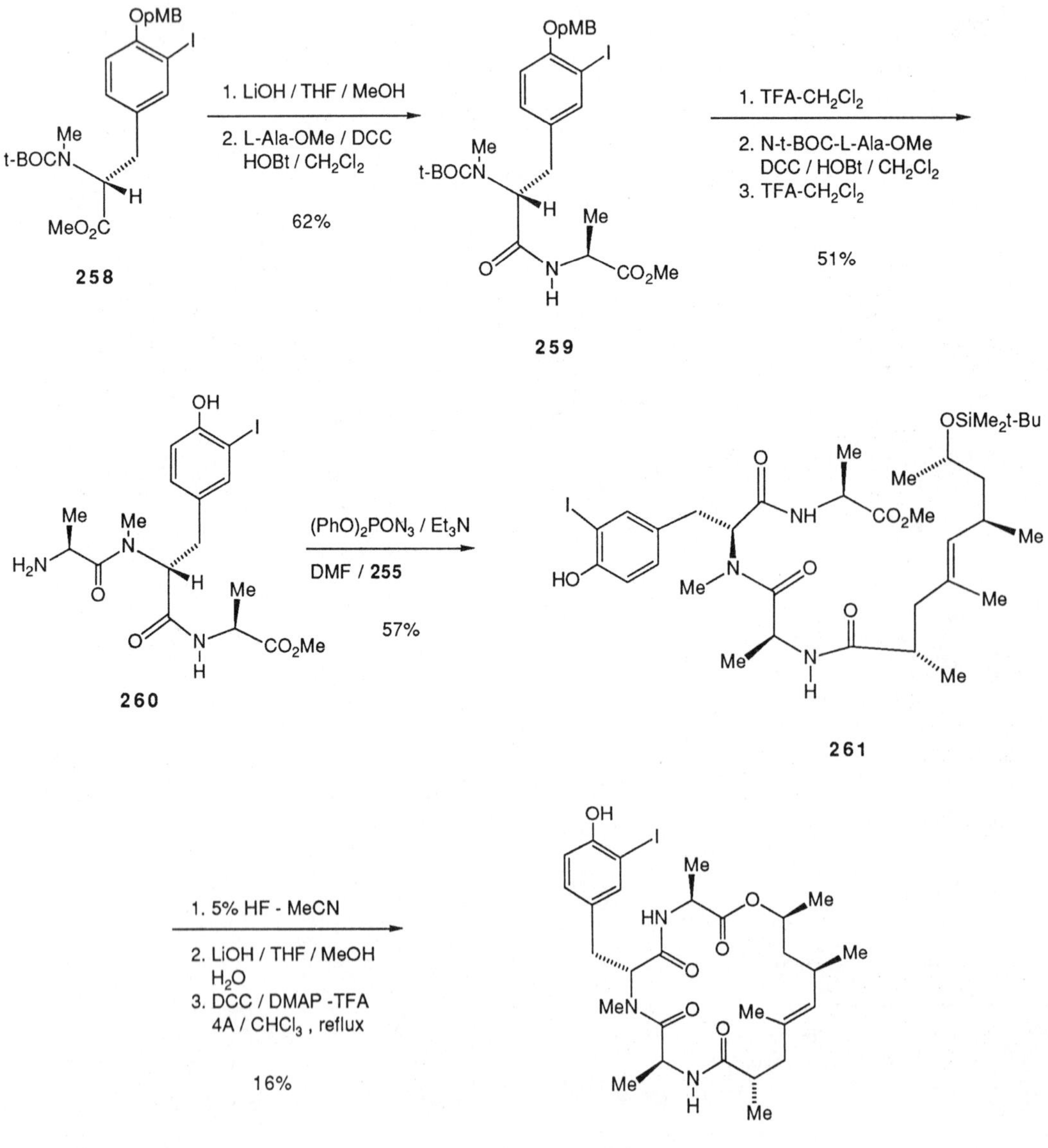

258

1. LiOH / THF / MeOH
2. L-Ala-OMe / DCC
   HOBt / CH₂Cl₂

62%

259

1. TFA-CH₂Cl₂
2. N-t-BOC-L-Ala-OMe
   DCC / HOBt / CH₂Cl₂
3. TFA-CH₂Cl₂

51%

260

(PhO)₂PON₃ / Et₃N

DMF / 255

57%

261

1. 5% HF - MeCN
2. LiOH / THF / MeOH
   H₂O
3. DCC / DMAP -TFA
   4A / CHCl₃ , reflux

16%

262, *GEODIAMOLIDE A*

White, J.D. ; Amedio, J.C., *J.Org.Chem.* (1989) **54**, 738

SCHEME 41

263

1. Br$_2$ / NaOAc
2. 2 N HCl-MeOH
88%

264

1. Br$_2$ / NaOAc
2. 1 N NaOH
3. ion-exchange
88%

265

DCC / Et$_3$N

81%

266

1. 1.5 N HCl-MeOH
2. 0.1 N HOAc / n-BuOH
reflux
85%

267

1. Tl (NO$_3$)$_3$ / MeOH
2. Zn^0 / HOAc / THF
19%

268

1. H$_2$ / 10% Pd-C / NaOAc
MeOH
2. Ac$_2$O / py.
56%

269

1. PhCH$_2$Br / NaH / DMF
2. K$_2$CO$_3$ / MeOH
3. 1. PhCH$_2$Br / NaH / DMF
78%

270

1. NaBH$_4$ -BF$_3$-Et$_2$O
2. 2N HCl - H$_2$O
3. NaHCO$_3$ / H$_2$O
4. H$_2$ / 10% Pd-C / HCl
36%

271, *(+)-PIPERAZINOMYCIN*

Nishiyama, S.; Nakamura, K.; Suzuki, Y.; Yamamura, S., *Tetrahedron Lett.* (1986) **27**, 4481.

Reductive removal of the bromine atoms followed by N-benzylation, sodium borohydride reduction of the amides and final reductive debenzylation afforded (+)-piperazinomycin (**271**).

These workers [37] have applied the same thallium-mediated cyclization to the total synthesis of OF4949-III (**279**), a novel aminopeptidase inhibitor, as shown in Scheme 42. Alternative syntheses of this natural product have been described by Evans and Schmidt and are detailed in Chapters 3 and 6, respectively. Tyrosine is chlorinated, esterified and coupled to N-*t*-BOC-L-asparagine in excellent yield to afford **274**. Coupling of this dipeptide to the dibromotyrosine **275** furnishes **276** which is subjected to TTN oxidative coupling yielding the cyclic material **277** (25% yield). Reduction of this substance with zinc in acetic acid furnishes the interesting biaryl ether **278** which is reduced and deprotected to give OF4949-III (**279**).

Inoue and co-workers[38] have applied the TTN oxidative coupling to the total synthesis of the complex bicyclic hexapeptides deoxybouvardin (**289**) and RA-VII (**287**). These structures have attracted attention [39] due to their potent anti-tumor activity. Structurally, these compounds contain a challenging, strained fourteen-membered ring constituted of the biaryl ether moiety; this array mandates an unusual *cis*-peptide grouping. As shown in Scheme 43, the halogenated tyrosines **280** and **281** are coupled and cyclized with TTN to give the desired cyclic materials **283** and **284** in very low yield (*ca* 20% combined). After zinc reduction, esterification and hydrogenation, the amine **285** was coupled and cyclized to the tetrapeptide **286** in relatively good overall yield providing **287**. Manipulation of the phenolic oxygens gave deoxybouvardin (**289**) and RA-II (**288**).

Schmidt, et.al., have extensively studied the synthesis of various cyclic peptides and depsipeptides. As shown in Scheme 44, the cytostatic cyclic tetrapeptide chlamydocin (**296**) contains the unusual amino acid (2S,9S)-2-amino-8-oxo-9,10-epoxydecanoic acid (Aoe). This subunit was prepared by reaction of the aldehyde **292** with the glycine phosphonate reagent to provide the dehydroamino acid derivative **293**. This material was hydrogenated , re-acylated and coupled with **294** to give a diastereomeric mixture of tetrapeptides **295** which were later chromatographically separated. These workers have developed a very useful method to cyclize these small pepetides efficiently through the agency of the corresponding active pentafluorophenyl esters. Thus, **295** was saponified and converted into the pentafluorophenyl ester; removal of the N-CBz group effected cyclization in 65% overall yield from **295**. Final deketalization, elimination and epoxidation gave chlamydocin (**296**) in good yield.

Schmidt and associates [41] have deployed the excellent pentafluorophenyl ester cyclization method to the total synthesis of several ansa-peptide alkaloids. These strained and densely functionalized

SCHEME 42

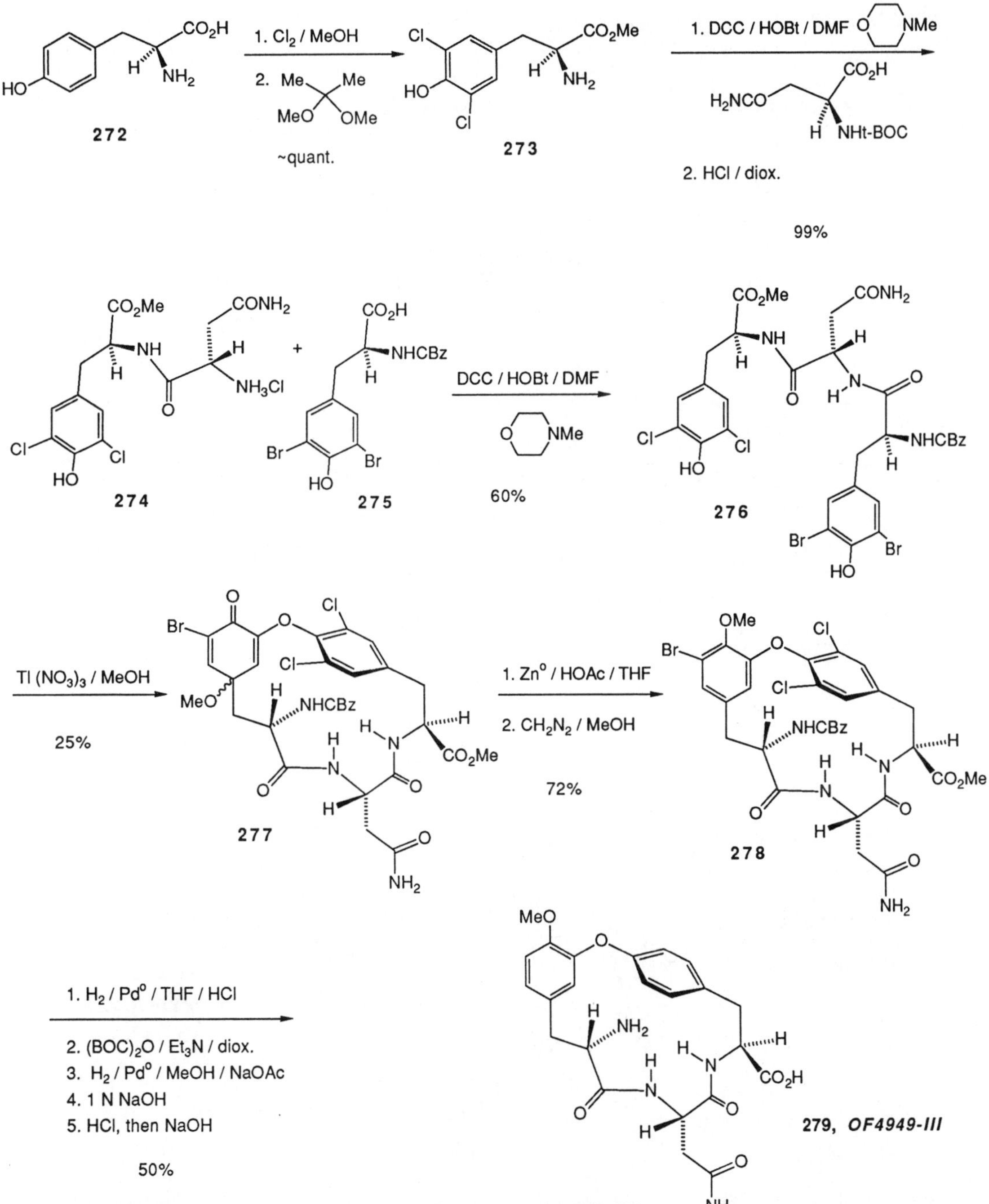

Nishiyama, S.; Suzuki, Y.; Yamamura, S., *Tetrahedron Lett.* (1988) **29**, 559.

SCHEME 43

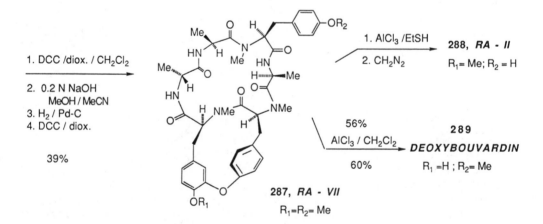

Inaba, T.; Umezawa, I.; Yuasa,M.; Inoue, T.; Mihashi, S.; Itokawa, H.; Ogura, K., *J.Org.Chem.* (1987)**52**, 2957

substances have been isolated from various plants and have displayed properties including complexing alkali earth metals, inhibiting energy transfer processes in chloroplasts and exhibit antibacterial and antifungal activities. As illustrated in Scheme 45, zizyphine A (**307**) was constructed starting from 3-bromodehydroproline methyl ester (**297**). Condensation of this substance with the phenolate **298** gave a *syn-* / *anti-* mixture of the corresponding (aryloxy)prolines which were separated by crystallization. Coupling of the protected fragment **299** with L-proline benzyl ester (**300**) gave the dipeptide **301** in high yield. Manipulation of this substance to the β-keto ester **302** , oximation, reduction, decarboxylation and reduction of the ketone afforded a mixture of amino alcohols **303**. Conversion to the active pentafluorophenyl ester with DCC and pentafluorophenol followed by cyclization in hot dioxane with comcomitant reductive cleavage of the CBz group gave the strained *meta-*ansa13-membered ring product **304** in 60-63% yield. Installation of the Z-eneamide moiety was effected via selenoxide-mediated elimination to **305**. Final deblocking and coupling of the activated Ile ester **306** gave zizyphine A (**307**) in 38% overall yield from **305**. Full experimental details accompany this work.

The 15-membered *meta*-ansa peptide alkaloid mucronin B (**316**) was prepared [42] in a related fashion as shown in Scheme 46. This synthesis features an interesting asymmetric homogeneous hydrogenation to construct the novel disubstituted phenylalanine derivative **311**. Aldehyde **308** is condensed with the glycine phosphonate reagent **309** followed by N-methylation to afford the dehydroamino acid **310** in good yield. Hydrogenation of this substance with rhodium-DIPAMP gave the desired L-amino acid **311** in 97% yield and >98%ee. Manipulation of the carboxyl group to the amino alcohol (**312**) followed by peptide coupling with **313** furnished the key substrate **314**. Cyclization of this material via the pentafluorophenyl ester gave the macrocycle **315** in 79% yield. Selenoxide-mediated elimination furnished the natural product mucronin B (**316**) in 36% yield. Complete experimental details accompany this account.

Joullie and collaborators[43] have prepared the *meta*-ansa peptide alkaloid dihydromauritine A (**322**, Scheme 47) by similar technology. These workers chose to effect macrocyclization via a *para*-nitrophenyl ester; the yield of cyclic material however was only 10%. Full experimental details are included with this work.

Very recently, Yamamura and associates [44] have synthesized the cyclic tripeptide K-13 (**329**) which is produced by *Micromonospora halophytica* subsp. *exilisia* K-13 and is an inhibitor of angiotensin I converting enzyme. Following a similar strategy to that adopted by the same group[37] in the synthesis of OF4949-III (see Scheme 42, above), the tripeptide **327** was oxidatively cyclized with thallium trinitrate to

SCHEME 44

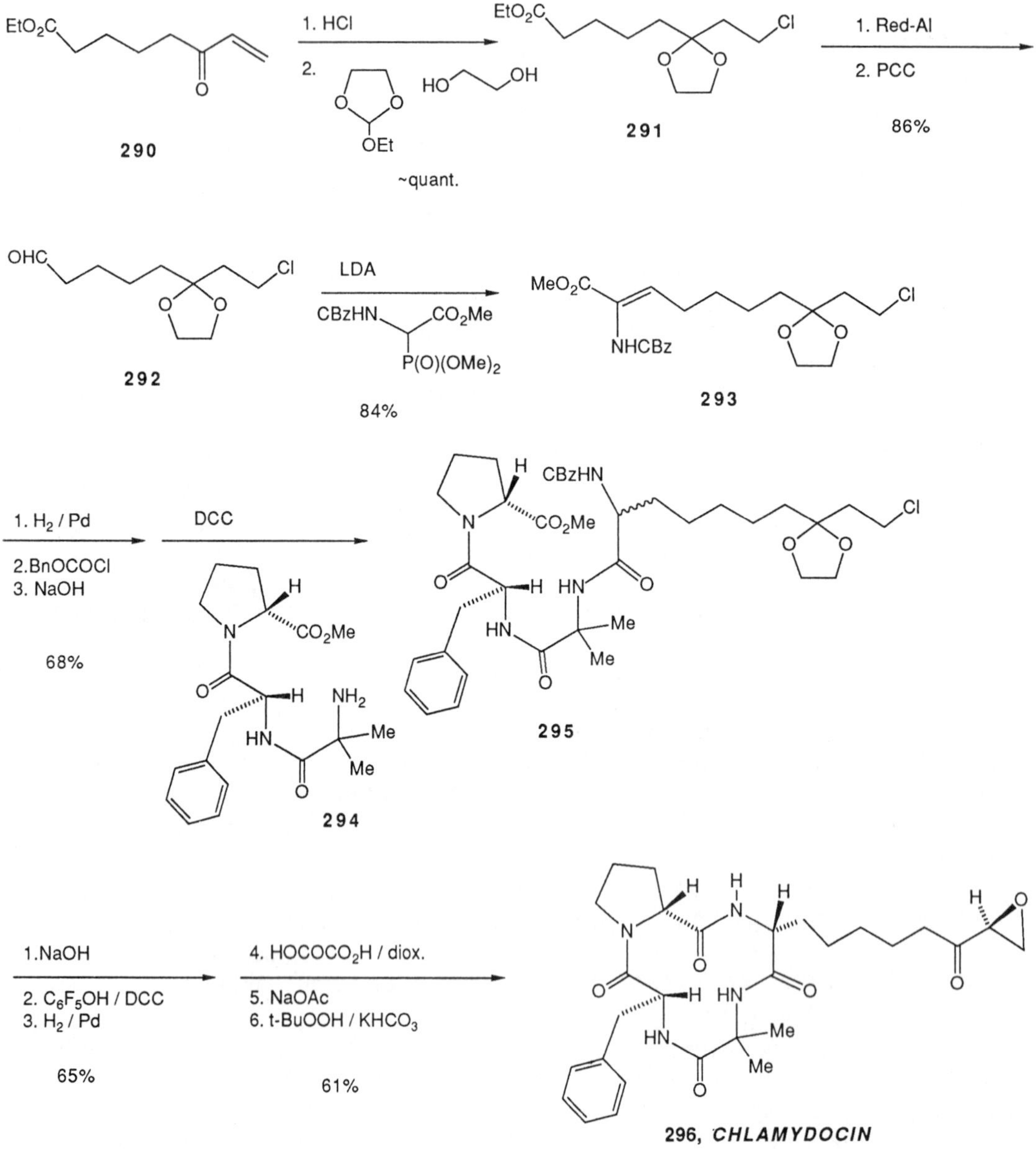

290 → (1. HCl; 2. [dioxolane-OEt], HOCH₂CH₂OH, ~quant.) → 291 → (1. Red-Al; 2. PCC, 86%) →

292 → (LDA; CBzHN–CH(CO₂Me)–P(O)(OMe)₂, 84%) → 293

293 → (1. H₂/Pd; 2. BnOCOCl; 3. NaOH, 68%) DCC → 294, 295

294, 295 → (1. NaOH; 2. C₆F₅OH/DCC; 3. H₂/Pd, 65%) → (4. HOCOCO₂H/diox.; 5. NaOAc; 6. t-BuOOH/KHCO₃, 61%) → 296, *CHLAMYDOCIN*

Schmidt, U.; Beuttler, T.; Lieberknecht, A.; Griesser, H., *Tetrahedron Lett.* (1983) **24**, 3573

SCHEME 45

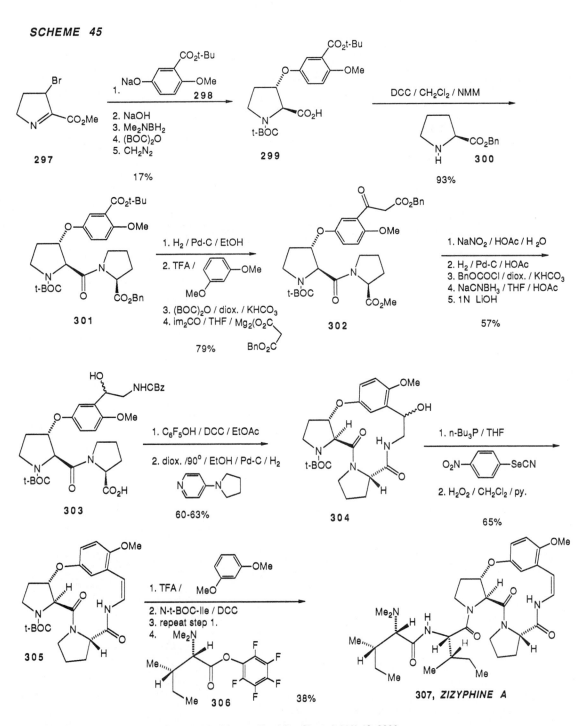

Schmidt, U.; Lieberknecht, A.; Bokens, H.; Griesser, H., *J.Org.Chem.* (1983) **48**, 2680.

SCHEME 46

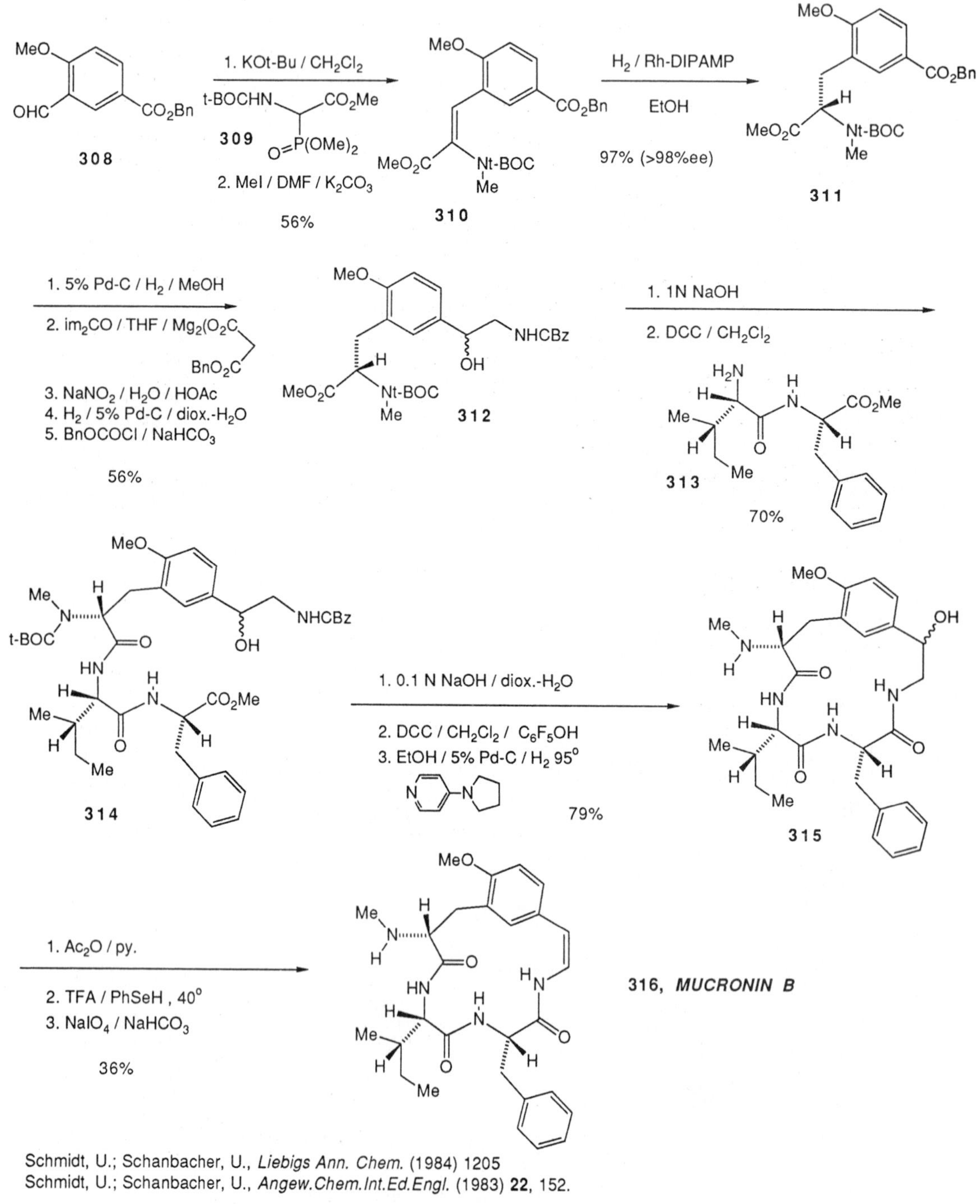

316, *MUCRONIN B*

Schmidt, U.; Schanbacher, U., *Liebigs Ann. Chem.* (1984) 1205
Schmidt, U.; Schanbacher, U., *Angew.Chem.Int.Ed.Engl.* (1983) **22**, 152.

SCHEME 47

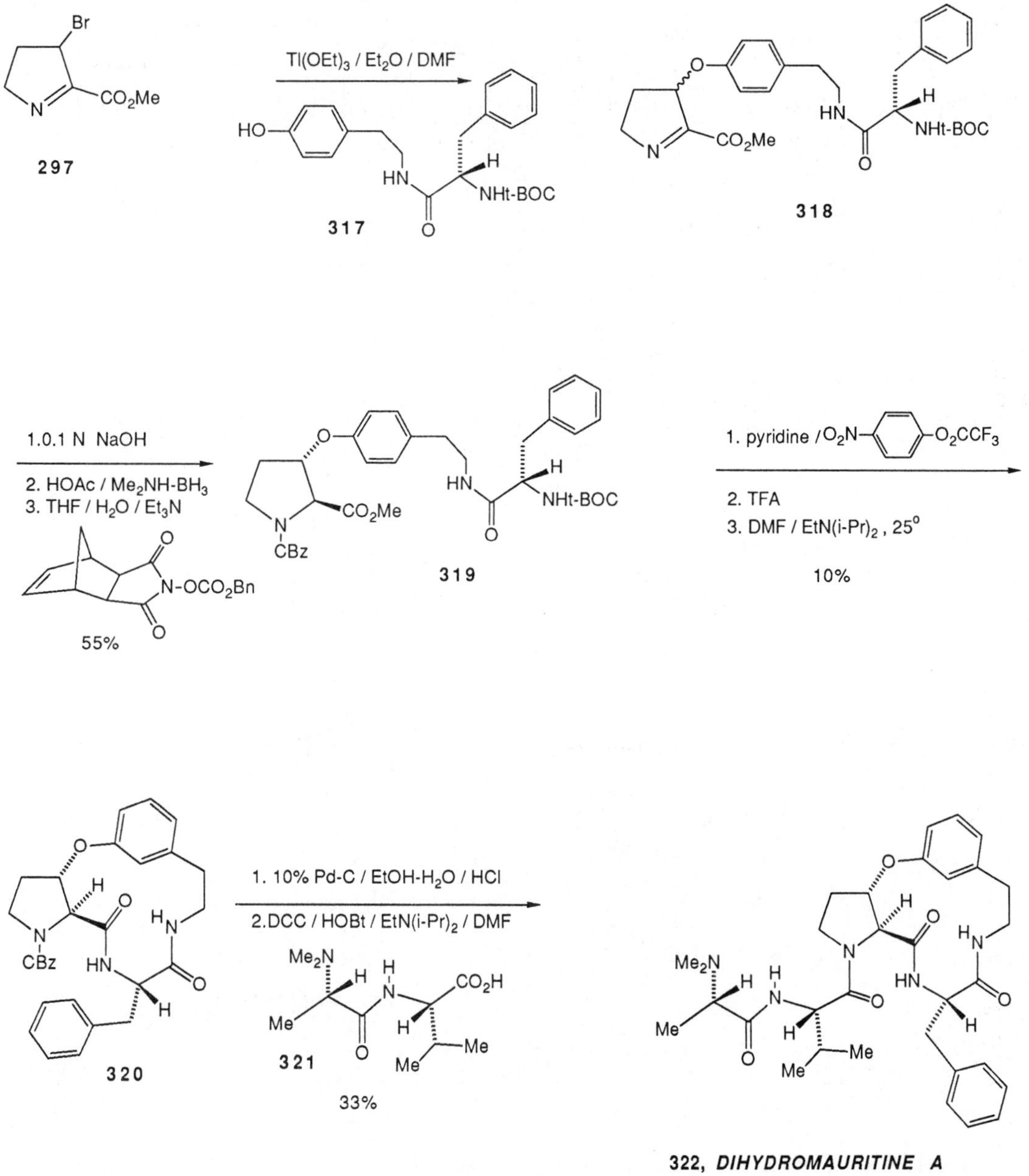

Nutt, R.K.; Chen, K-M.; Joullie, M.M., *J.Org.Chem.*(1984) **49**, 1013

SCHEME 48

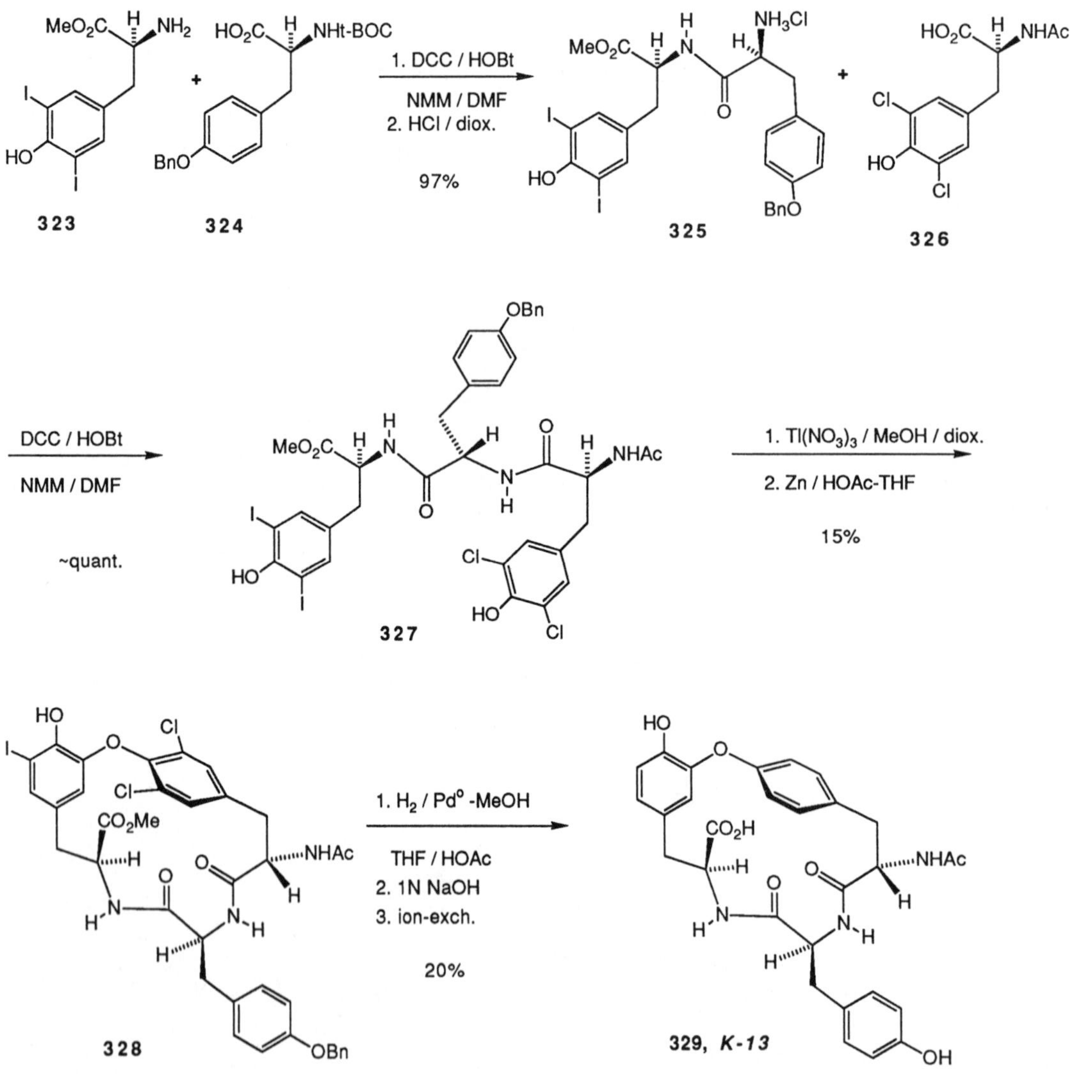

Nishiyama, S.; Suzuki, Y.; Yamamura, S.; *Tetrahedron Lett.* (1989) **30**, 379.

afford, after zinc reduction, the cyclic substance **328** in low (15%) yield. This compound was further hydrogenated and hydrolyzed to K-13 (**329**). The authors propose that this approach is biomimetic; it is indeed very reasonable to assume that a naturally derived tripeptide L-Tyr-L-Tyr-L-Tyr is oxidatively cyclized in the biogenesis of K-13.

F. CYCLOSPORINE

One of the most exciting recent developments in peptide-based immunosuppressives was the discovery of the cyclic undecapeptide cyclosporine. This complex substance is a secondary metabolite isolated from *Tolypocladium inflatum* Gams and has been used since 1978 for the prevention of graft rejection following human organ and bone marrow transplants [45]. This substance contains several known N-methylated amino acids plus the unusual amino acid (4R)-4-((E)-2-butenyl)-4-N-dimethyl-L-threonine (MeBmt, **340**). The first synthesis of MeBmt was reported by Wenger [45] in 1983 and was soon followed by the incorporation of this amino acid in a total synthesis of cyclosporine. As shown in Scheme 49, tartaric acid was converted into the C_2-symmetric epoxide **333** and ring-opened with methyl cuprate to afford **334** in good overall yield. Homologation to **336** proceeds smoothly in seven steps which is followed by conversion to the key aldehyde **337** used for the subsequent Strecker synthesis. Reaction of **337** with potassium cyanide and methylammonium chloride affords the corresponding cyanohydrin as a mixture of diastereomers. Reaction of this material with carbonyl diimidazole furnishes the oxazolidinone **338** as a 6 : 1 mixture of *cis* : *trans* (relative to the ring) isomers. The undesired sense of the relative stereochemistry of this substance is adjusted by conversion to the imidate which epimerizes to the thermodynamically more stable *trans* isomer **339**. Final hydrolysis of this compound to MeBmt (**340**) is accomplished in excellent yield. Complete experimental details accompany this practical synthesis.

Many subsequent syntheses of MeBmt have since appeared in the literature. In addition to the syntheses detailed by other workers in Chapter 1, two additional syntheses of this interesting amino acid have recently appeared. As shown in Scheme 50, Rich and Tung [46] employed an alternate Strecker strategy to access MeBmt. Sharpless epoxidation of the dienol **341** afforded the epoxy ether **342** in good yield. Cuprate opening of the epoxide proceeds with good C-3 selectivity furnishing **343** in 80% yield. Reaction of **343** with methyl isocyanate and acidic removal of the trityl group provided **344**. Swern oxidation and acetylation of the incipient carbinolamine provided the Strecker substrate **345**. Treatment of this material with trimethylsilyl cyanide gave **346** as a 1.6 : 1 ratio of

SCHEME 49

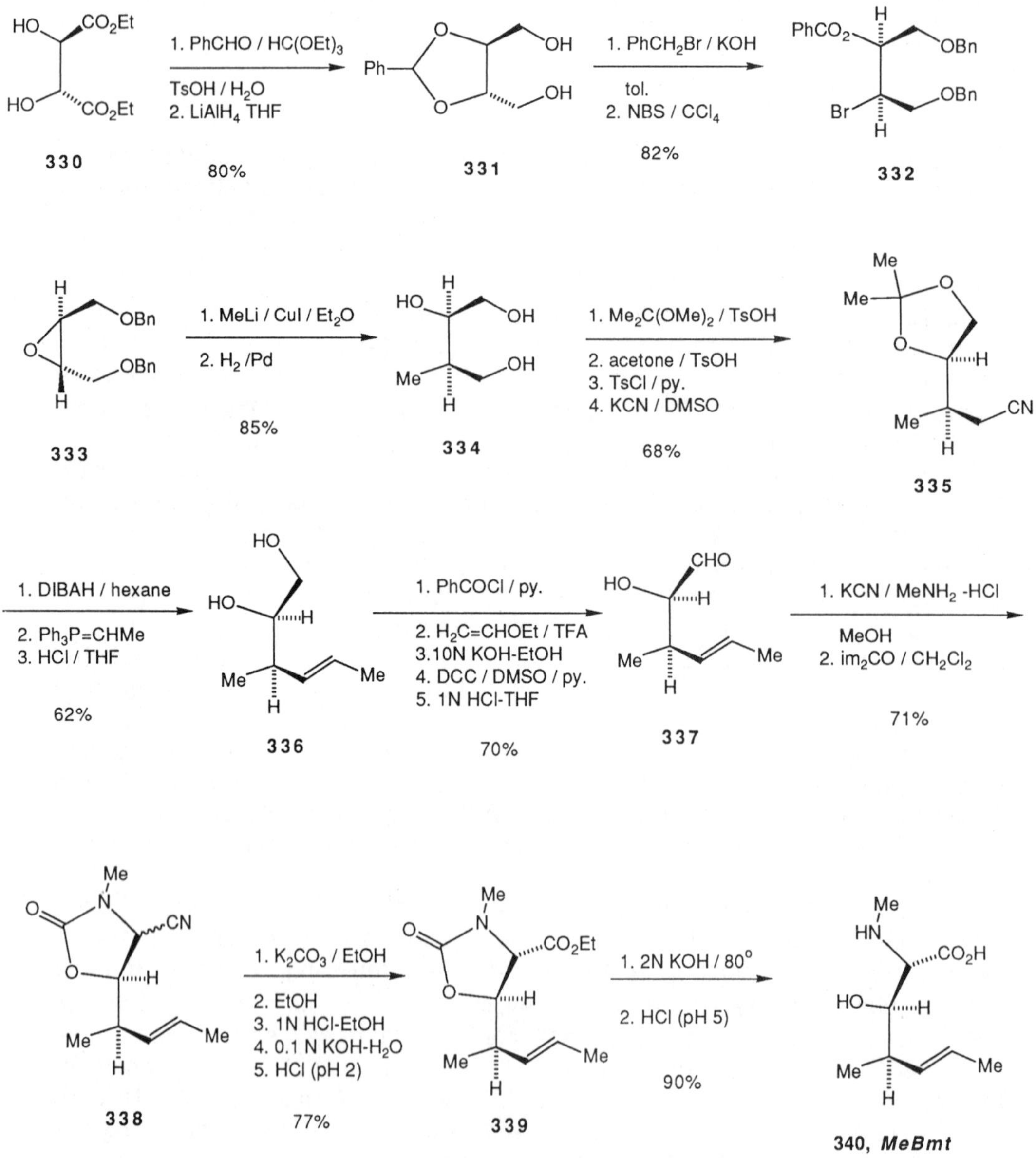

Compound labels and reagents in the scheme:

330 → 1. PhCHO / HC(OEt)$_3$, TsOH / H$_2$O; 2. LiAlH$_4$ THF (80%) → **331** → 1. PhCH$_2$Br / KOH, tol.; 2. NBS / CCl$_4$ (82%) → **332**

333 → 1. MeLi / CuI / Et$_2$O; 2. H$_2$ /Pd (85%) → **334** → 1. Me$_2$C(OMe)$_2$ / TsOH; 2. acetone / TsOH; 3. TsCl / py.; 4. KCN / DMSO (68%) → **335**

1. DIBAH / hexane; 2. Ph$_3$P=CHMe; 3. HCl / THF (62%) → **336** → 1. PhCOCl / py.; 2. H$_2$C=CHOEt / TFA; 3. 10N KOH-EtOH; 4. DCC / DMSO / py.; 5. 1N HCl-THF (70%) → **337** → 1. KCN / MeNH$_2$ -HCl, MeOH; 2. im$_2$CO / CH$_2$Cl$_2$ (71%)

338 → 1. K$_2$CO$_3$ / EtOH; 2. EtOH; 3. 1N HCl-EtOH; 4. 0.1 N KOH-H$_2$O; 5. HCl (pH 2) (77%) → **339** → 1. 2N KOH / 80°; 2. HCl (pH 5) (90%) → **340, *MeBmt***

Wenger, R.M., *Helv.Chim.Acta* (1983) **66**, 2672
Wenger, R.M., *Helv.Chim.Acta* (1985) **24**, 77.

SCHEME 50

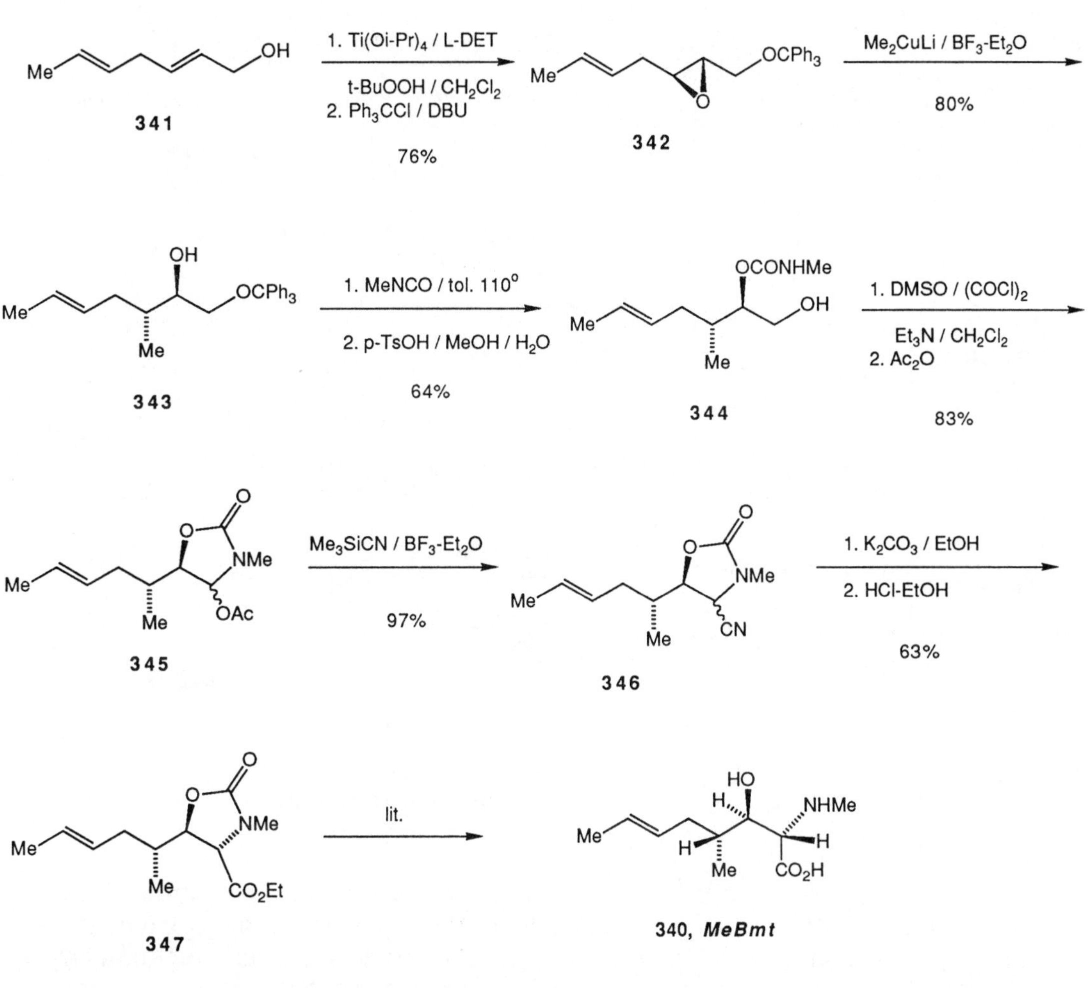

341

1. Ti(Oi-Pr)₄ / L-DET

t-BuOOH / CH₂Cl₂
2. Ph₃CCl / DBU

76%

342

Me₂CuLi / BF₃-Et₂O

80%

343

1. MeNCO / tol. 110°

2. p-TsOH / MeOH / H₂O

64%

344

1. DMSO / (COCl)₂

Et₃N / CH₂Cl₂
2. Ac₂O

83%

345

Me₃SiCN / BF₃-Et₂O

97%

346

1. K₂CO₃ / EtOH

2. HCl-EtOH

63%

347

lit.

340, *MeBmt*

Tung, R.D.; Rich, D.H., *Tetrahedron Lett.* (1987) **28**, 1139.

SCHEME 51

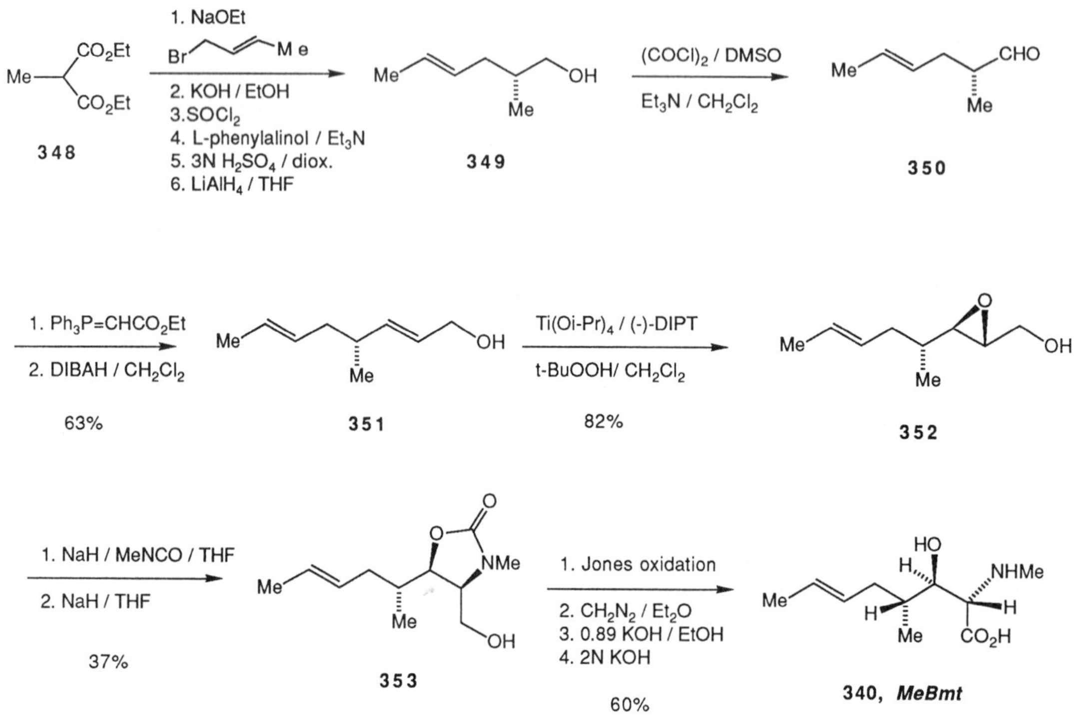

Rao, A.V.R.; Dhar, T.G.M.; Chakraborty, T.K.; Gurjar, M.K., *Tetrahedron Lett.* (1988) **29**, 2069.

cis : trans isomers. This compound was converted into the carboethoxy oxazolidinone **347** according to the method described above by Wenger; the authors note that this material is obtained in 94% ee which presumably reflects the maximum level of asymmetric induction achieved in the Sharpless epoxidation of **341**. Enantiomerically pure material may be obtained by crystallization with (+)-ephedrine.

Rao and collaborators[47] have also employed the Sharpless asymmetric epoxidation to access MeBmt as shown in Scheme 51. Ethyl 2-carboethoxypropionate (**348**) was alkylated and resolved via the corresponding phenylalaninol amide. Following reduction, Swern oxidation of **349** afforded the aldehyde **350** which was homologated to the allylic alcohol **351**. Sharpless asymmetric epoxidation of **351** gave **352** which was intramolecularly ring-opened with methyl isocyanate to give the

SCHEME 52

Synthesis of Cyclosporine   Step sequence for peptide bond formation

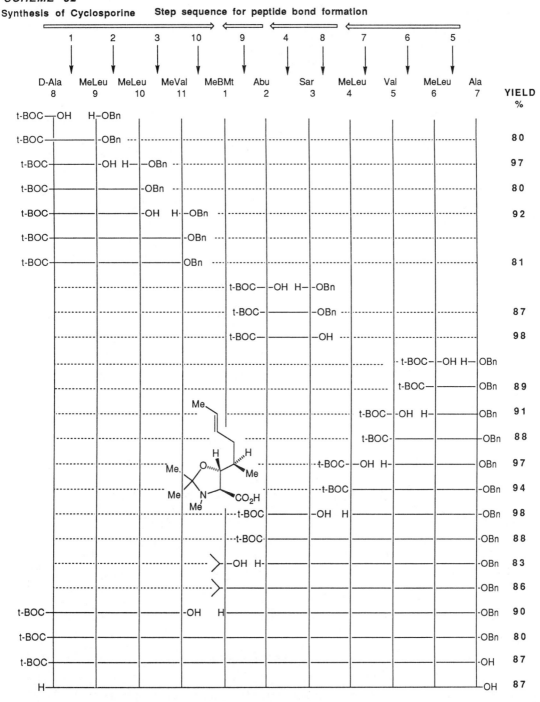

Wenger, R. M., *Helv.Chim.Acta* (1984) **67**, 502; Wenger, R.M., *Angew.Chem.Int.Ed.Engl.* (1985) **24**, 77

SCHEME 52 (con't)

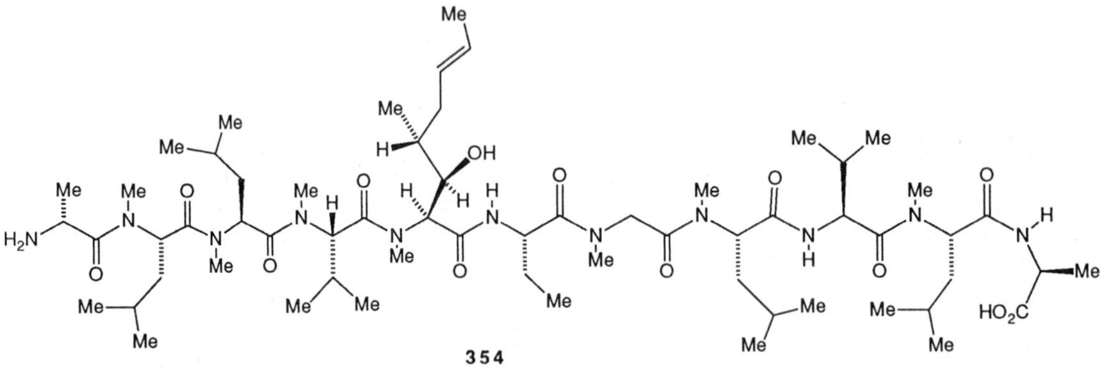

354

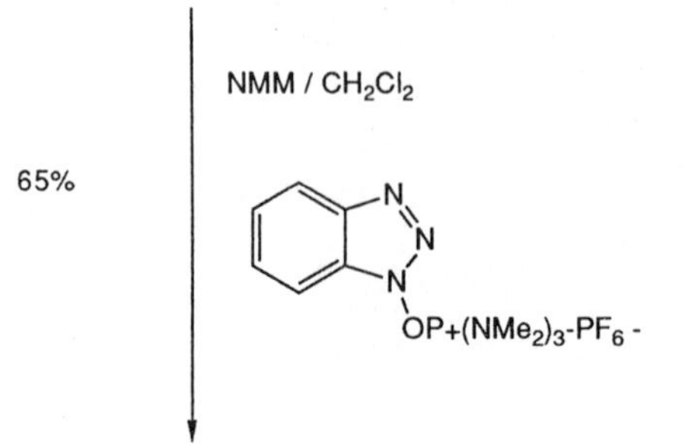

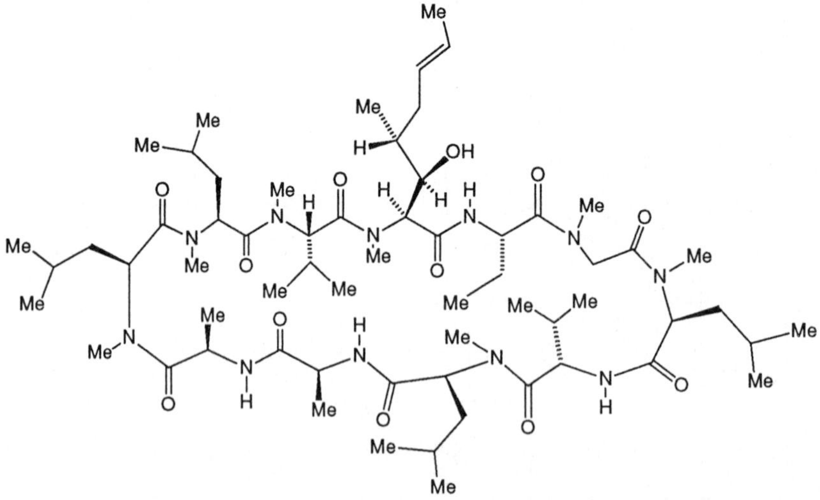

NMM / CH₂Cl₂

65%

355, CYCLOSPORINE

Wenger, R.M., *Helv.Chim.Acta* (1984) **67**, 502; Wenger, R.M., *Angew.Chem.Int.Ed.Engl.* (1985) **24**, 77.

oxazolidinone **353** in modest yield. Jones oxidation, esterification and hydrolysis afforded MeBmt (**340**) in good yield.

The total synthesis of cyclosporine was reported in a beatiful full paper [45] by Wenger in 1984 and is detailed in Scheme 52. The yields for the segment condensations of the peptide fragments are all very good to excellent. The strategy adopted chose to connect as the pentultimate cyclization step, the only two non-N-methylated amino acids, L-Ala and D-Ala as the carboxy and amino terminal residues, respectively. This strategy relied on the assumption that intramolecular hydrogen bonds of the linear peptide should resemble those of the natural product and would stabilize the folded conformation to facilitate macrocyclization. Additionally, alternative peptide bond formation between any of the bulkier N-methylated amino acids would almost certainly be more difficult. Formation of the peptide fragments was achieved with the mixed pivalic anhydride method in the presence of a tertiary amine base, such as N-methylmorpholine. Incorporation of the amino acid MeBmt was performed by protecting the hydroxy and amino functions as a dimethyloxazolidine derivative. This cyclic protecting group obviates problems of racemization associated with the carboxyl activation step since the *anti-* configuration of the ring is maintained as the thermodynamically most stable isomer. The linear undecapeptide **354** was treated with (1H-1,2,3-benzotriazol--1-yloxy)tris(dimethylamino)-phosphonium hexafluorophosphate and N-methylmorpholine in methylene chloride at room temperature at 0.0002M concentration. Cyclosporine (**355**) was obtained in crystalline form in 62% yield by this procedure. Wenger also prepared several analogs of cyclosporine to begin developing a structure/activity profile of this substance. It was found that MeBmt was essential for potency; saturation of the olefin and modification of the hydroxyl lowered activity. The author notes that MeBmt itself however, has no immunosuppressive activity.

G. MISCELLANEOUS

Many interesting and structurally diverse amino acids and unusual amino acid-containing natural products have recently attracted the attention of the synthetic community. Below, are detailed just a few special highlights; many more challenging and important substances await the hand of the experimentalist.

Hazimycin factors 5 (**359**) and 6 (**360**) are two recently identified isonitrile-containing antifungal antibiotics obtained from *Pseudomonas* sp. SCC 1411. The structures have been determined by X-ray analysis and total synthesis [48] as detailed in Schemes 53 and 54. The X-ray analysis of hazimycin factor 5 revealed that this substance existed in the solid state as an equal mixture of (R,R) and (S,S) stereoisomers; hazimycin factors 5

SCHEME 53

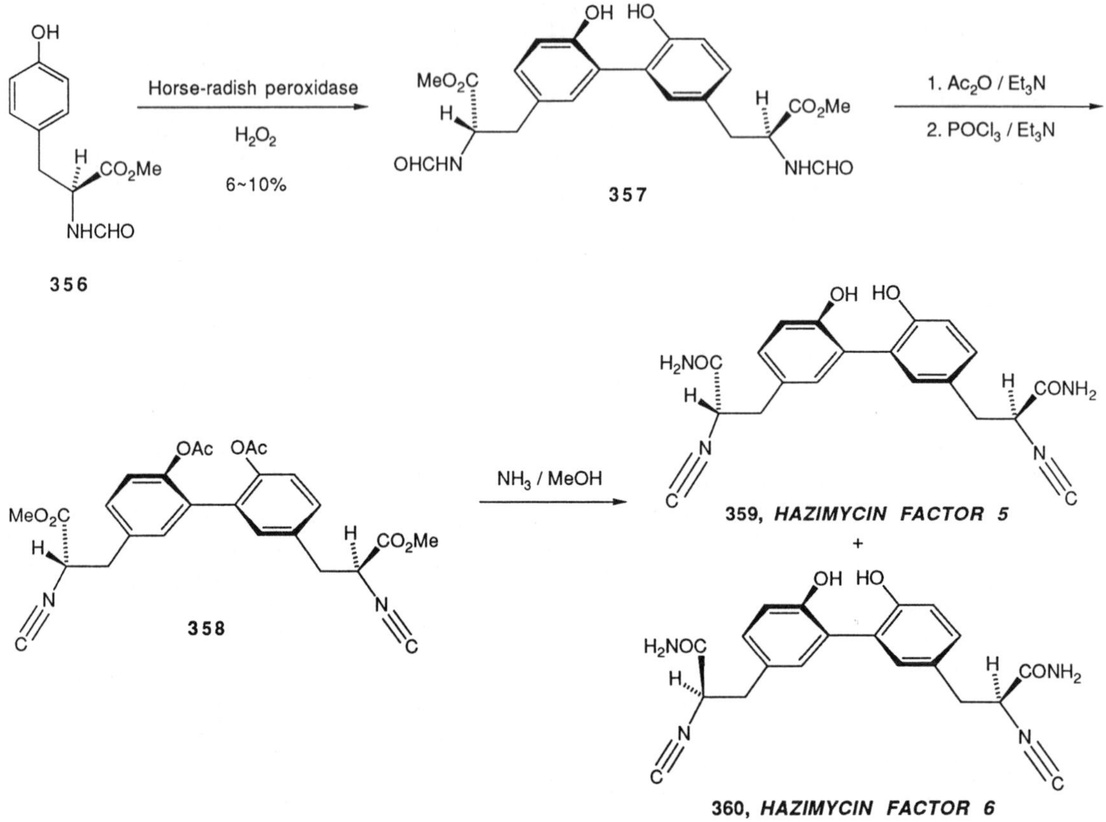

356 → (Horse-radish peroxidase, H₂O₂, 6~10%) → 357 → (1. Ac₂O / Et₃N, 2. POCl₃ / Et₃N)

358 → (NH₃ / MeOH) → 359, HAZIMYCIN FACTOR 5 + 360, HAZIMYCIN FACTOR 6

SCHEME 54

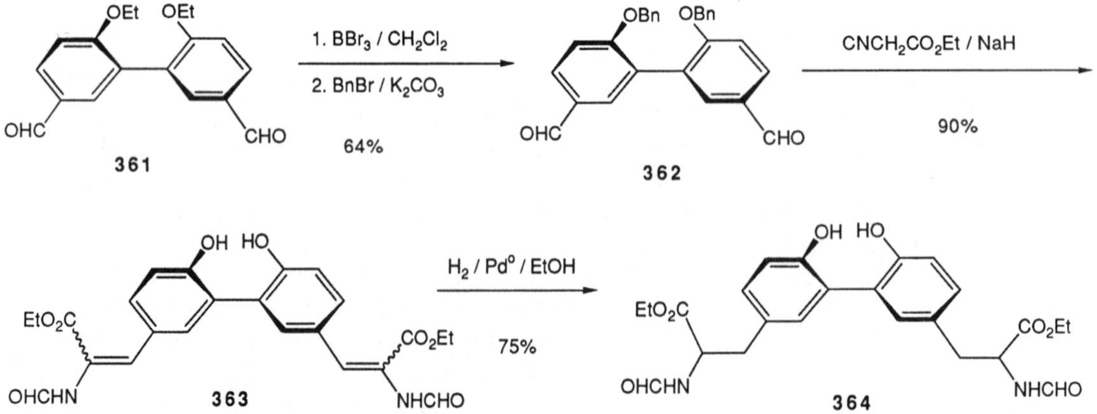

361 → (1. BBr₃ / CH₂Cl₂, 2. BnBr / K₂CO₃, 64%) → 362 → (CNCH₂CO₂Et / NaH, 90%)

363 → (H₂ / Pd° / EtOH, 75%) → 364

Wright, J.J.K.; Cooper, A.B.; McPhail, A.T.; Merrill, Y.; Nagabhushan, T.L.; Puar, M.S., *J.Chem.Soc.Chem.Comm.* (1982) 1188 and 6 are interconvertible in the presence of water with equilibrium being reached in several days. Hazimycin factor 6 possesses the R,S (meso) stereochemistry. Two approaches were employed for the total synthesis of these biaryls. The most direct approach involved enzymatic oxidative coupling of N-formyl tyrosine methyl ester (**356**) with horse-radish peroxidase in the presence of hydrogen peroxide. The dimeric substance **357** was obtained in low yields (6 to 10%); the overall convenience of this direct approach however, offsets the low yield of coupling. Isonitrile formation under standard conditions followed by amminolysis gave (in unspecified yield) equal parts of **359** and **360**. In an alternate route, the known dialdehyde **361** was prepared and transformed into **362**. Condensation of this material with the anion of ethyl isocyanoacetate gave a diastereomeric mixture of the adducts **363** in high yield. Hydrogenation of **363** gave the diethyl analog of **357** in good yield.

Rhizobitoxine (**370**) is a β,γ-unsaturated amino acid with an unusual γ-enol ether moiety that was isolated from *Rhizobium japonicum*[49]. This compound inactivates β-cystathionase in plants and bacteria ; it inhibits the conversion of methionine into ethylene in plants; and was shown to cause the symptoms of rhizobial-induced chlororis in *Glycine max.* L.Merrill (soybean)[50]. A Hoffmann-La Roche group elucidated the absolute stereochemistry of this substance employing a chiroptical method based on measuring the CD spectra of the derived 3,5-diphenyl-5-hydroxy-2-pyrrolin-4-ones[51a]. The same group also reported the total synthesis[51b] of this amino acid and the simpler congener[52] L-2-amino-4-methoxy-trans-but-3-enoic acid (**376**) as shown in Schemes 55 and 56, respectively. The latter substance is a natural metabolite produced by *Pseudomonas aeruginosa* and is an inhibitor of ethylene biosynthesis. L-Homoserine was protected and oxidized to the aspartic acid β-semialdehyde derivative **366**. This substance was converted into the corresponding dimethoxy acetal with trimethyl orthoformate in methanol and then into the mixed acetic / methoxy acetal through the agency of acetic anhydride. Heating this material under reduced pressure at 180°C effected elimination to the β,γ-unsaturated enol ether **367** obtained as a 3 : 2, E : Z mixture. Dichlorobis(benzonitrile)palladium (II) effected the enol ether exchange with the amino alcohol **368** providing **369** in 26% yield. Finally, dissolving metal reduction of all benzyl protecting groups afforded rhizobitoxine (**370**) in high yield. The authors make a special note that the dissolving metal reduction step must be performed by inverse addition and buffered with acetamide to preclude racemization of the base-sensitive β,γ-unsaturated amino acid. The simpler amino acid **376** was synthesized along similar lines as shown in Scheme 56. The racemic E-olefin **374** was resolved with Hog kidney acylase to provide after deacetylation, natural **376** in ~80% ee. Complete experimental details accompany this work.

SCHEME 55

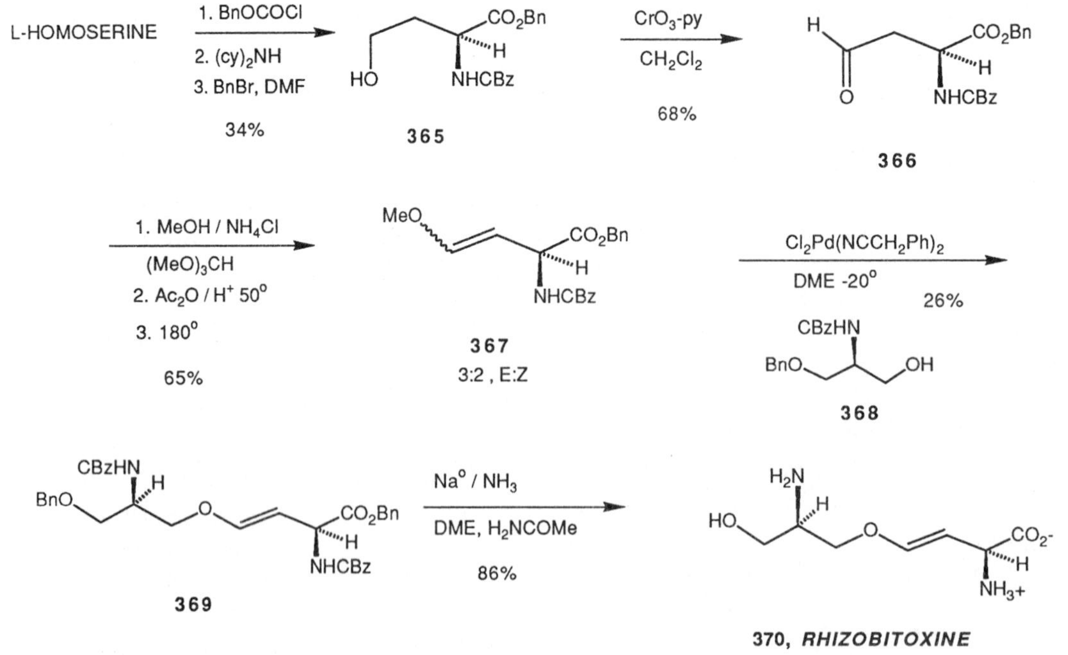

Keith, D.D.; Tortora, J.A.; Ineichen, K.; Leingruber, W. *Tetrahedron* (1975) **31**, 2633

SCHEME 56

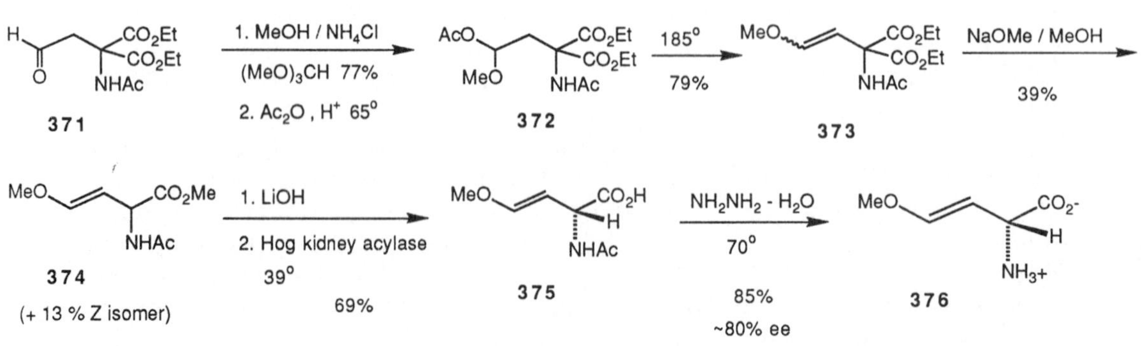

Keith, D.D.; Tortora, J.A.; Yang, R., *J.Org. Chem.* (1978) **43**, 3711.

The legume *Atelia herbert smithii*, found in Costa Rica, produces defensive substances that cause over one hundred predators to avoid these seeds. It has been proposed [53,54] that the novel amino acids 2,4-methanoproline (**380**), the related alcohol **391**, 2,4-methanoglutamic acid (**392**) and the labile amide, 2,4-methanopyroglutamic acid **389** may be responsible for the avoidance activity. It was proposed that amide **389** may be a biogenetic precursor (not isolated) of the other three amino acids and could be responsible for the biological activity since this strained amide should be a highly reactive acylating agent (IR 1720 cm^{-1}). Synthesis of 2,4-methanoproline was reported concomitantly by Pirrung [53] and Clardy [54] via an intramolecular photo [2+2] cycloaddition reaction. As shown in Scheme 57, methyl-2-benzamido-3-chloropropionate (**377**) was dehydrohalogenated and allylated to form **378**. Photolysis of this substance gave the photo adduct **379** in 87% yield. Acidic hydrolysis gave 2,4-methanoproline (**380**) in high yield. The remaining three amino acids were prepared by Clardy and Hughes[54] as shown in Scheme 59. Initial attempts to provide **389** via photocycloaddition of **383** (Scheme 58) proceeded in low yields and proved impractical. Oxidation of **385** prepared from 2,4-methanoproline with ruthenium teraoxide gave the key acid **386** in 66% yield. Deprotection of **386** directly furnished 2,4-methanoglutamic acid (**392**) and reduction afforded **391**. The strained amide **389** was prepared by treatment of **386** with thionyl chloride and cyclization to the amide **388**. Basic hydrolysis to **389** proceeded cleanly, but this compound proved difficult to separate from **392** which formed rapidly from **389**. Complete experimental details accompany this work [54].

Kurokawa and Ohfune have recently reported [55] the synthesis of α-(methylenecyclopropyl) glycine (**398**) as shown in Scheme 60. This amino acid is a natural constituent of the seeds of the *Sapindaceae* family and causes hypoglycemic symptoms in animals. β-Acetoxy allyl glycine (**393**) was obtained from allyl glycine via a selenium dioxide-mediated oxidation. Rearrangement of this substance with a palladium(II) catalyst provided the β,γ-unsaturated acetate **394** in 60% yield. Cyclopropanation of this material gave a 1 : 1 mixture of **395** and **396** in good yield. Removal of the acetate and selenium-mediated oxidative/elimination followed by deprotection provided α-(methylenecyclopropyl) glycine (**398**); yields for the transformation of **397** to **398** were not provided. The conformationally restricted analog of glutamic acid (**400**) was prepared by conversion of **396** to the alcohol **399**, Jones oxidation and hydrolysis. Yields for this sequence were not provided.

Ohfune and Yamanoi [56] have reported a more extensive synthetic study of the α-(carboxycyclopropyl) glycines as shown in Scheme 61. Interest in these compounds stems from recent hypotheses concerning the active conformations of glutamic acid in binding to various neural receptors. These workers have synthesized all possible L-stereoisomers

SCHEME 57

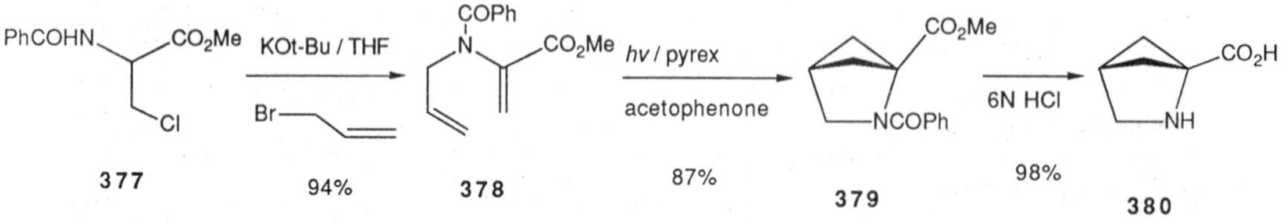

377   94%   **378**   87%   **379**   98%   **380**

SCHEME 58

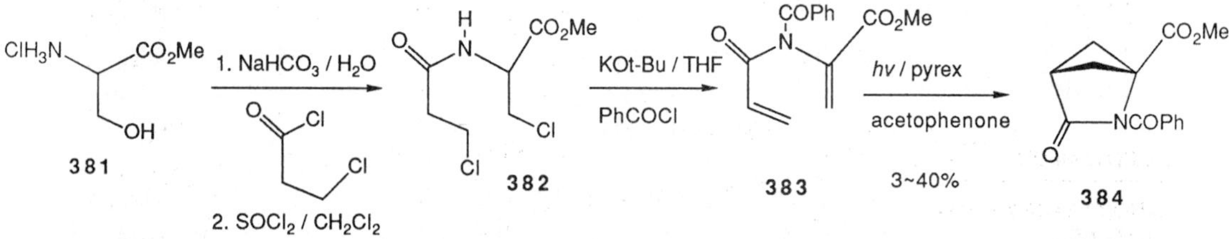

381   **382**   **383**   3~40%   **384**

SCHEME 59

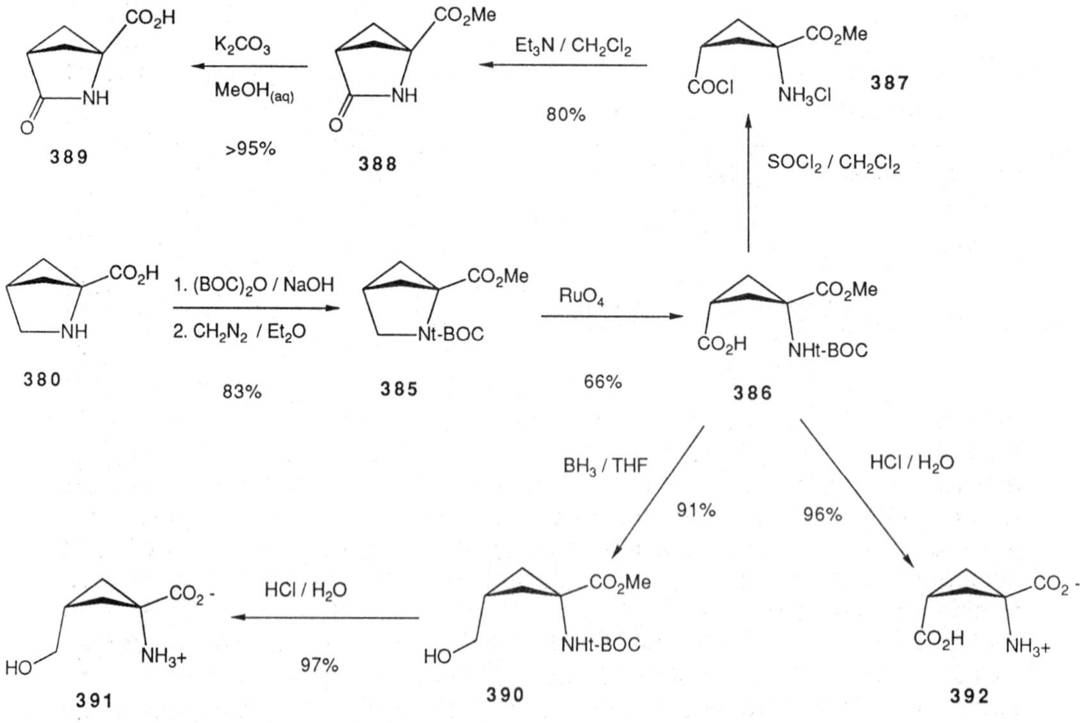

389   >95%   **388**   80%   **387**

380   83%   **385**   66%   **386**

91%   96%

391   97%   **390**   **392**

Hughes, P.; Clardy, J., *J.Org. Chem.* (1988) **53**, 4793; Hughes, P.; Martin, M.; Clardy, J., *Tetrahedron Lett.* (1980) **21**, 4579; Pirrung, M.C., *Tetrahedron Lett.* (1980) **21**, 4577.

SCHEME 60

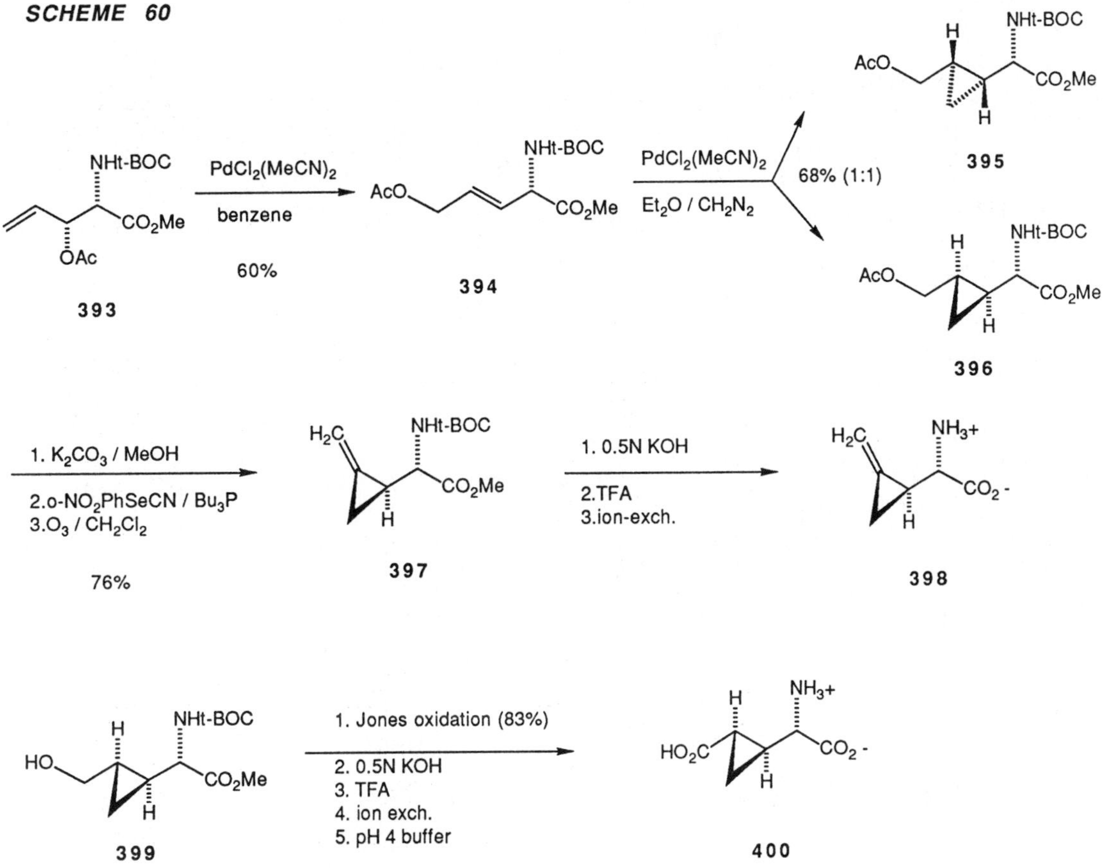

Kurokawa, N.; Ohfune, Y., *Tetrahedron Lett.* (1985) **26**, 83.

of these conformationally restricted glutamate analogs in hopes of providing a working model for the relationship between conformation and activity of glutamic acid. Amino alcohol **401** was acylated and protected to form **402**. Diazotization and intramolecular cyclopropanation gave a 6:1 mixture of **403** (plus a diastereomer) in 43% overall yield from **402**. This substance was in turn, deprotected, acylated, oxidized and hydrolyzed to furnish the cis isomer **405**. To access the other isomers, intermolecular cyclopropanation of **406** gave a 1.2 : 3.5 : 1 : 1 mixture of **407** : **408** : **409** : **410**, respectively, in 41% combined yield. After removal of the silyl ether, **408** and **409** were separated from the other components by medium-pressure chromatography. Isomers **407** and **410** were separated

SCHEME 61

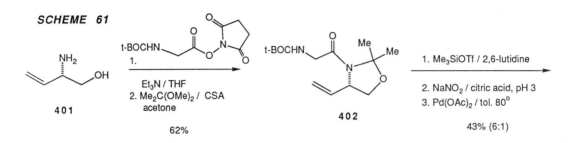

Yamanoi, K.; Ohfune, Y., *Tetrahedron Lett.* (1988) **29**, 1181.

SCHEME 62

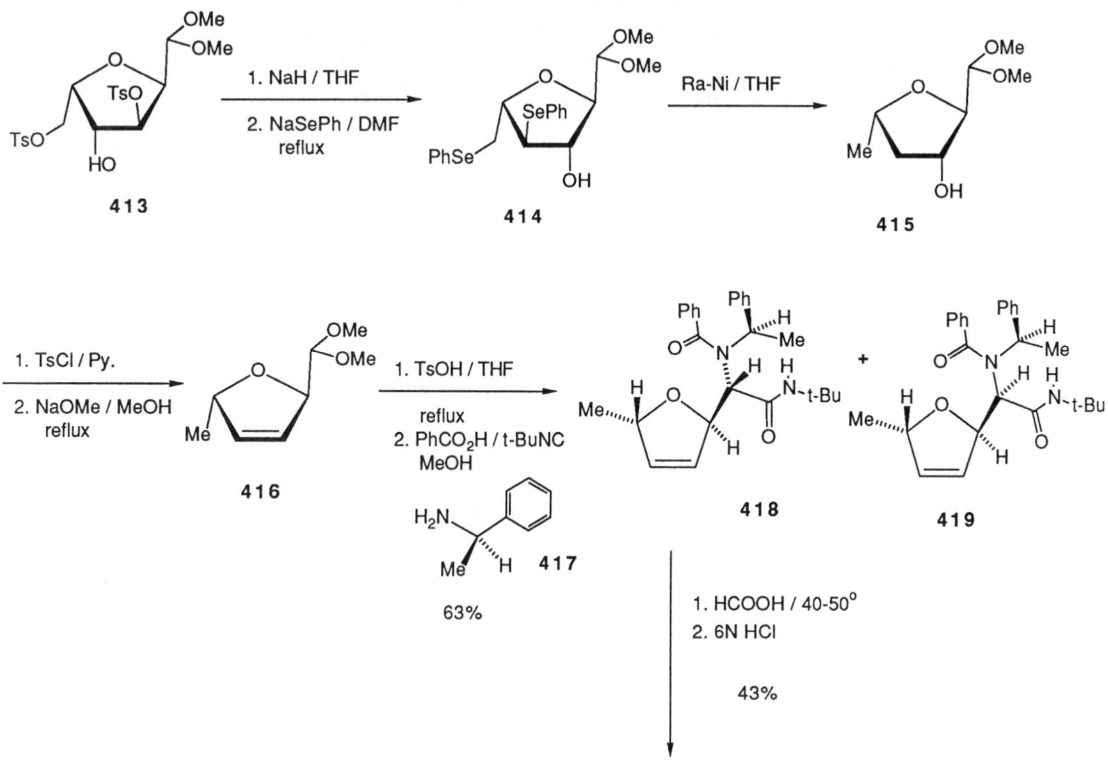

Joullie, M.M.; Wang, P.C.; Semple, J.E.,
J.Am.Chem.Soc (1980) **102**, 887
Semple, J.E.; Wang, P.C.; Lysenko, Z.; Joullie, M.M.,
J.Am.Chem.Soc. (1980) **102**, 7505

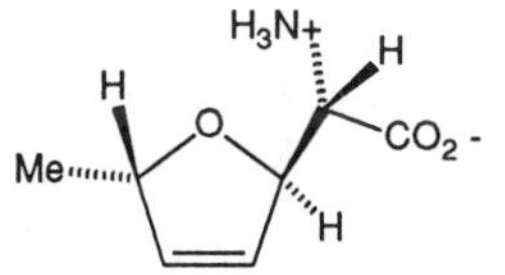

420, *FURANOMYCIN*

by conversion of **410** (R=Et) to the corresponding δ-lactone (CSA/CH₂Cl₂); hydrolysis and esterification provided **410** (R=Me) which was converted to **411**. This work also reports some preliminary bioassays of these substances toward periodically oscillating neuron (PON) which is sensitive to β-hydroxy-L-glutamic acid.

Furanomycin (**420**) is a naturally occurring antibiotic produced by *Streptomyces threomyceticus*. Joullie and collaborators [57] have prepared this compound from the furanose **413** which is derived from glucose; this work also corrected the previously assigned stereochemistry to furanomycin. Conversion of the di-tosylate **413** to the bis-selenide **414** proceeds via the 2,3-epoxide which is opened regioselectively. Reduction and elimination affords substrate **416** which was subjected to the Ugi four component condensation reaction. The two diastereomers **418** and **419** are produced in equal amounts and were separated by chromatography. Deprotection of the naturally configured isomer (**418**) afforded

SCHEME 63

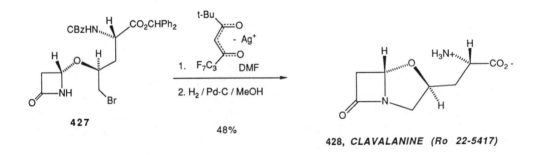

DeBernardo, S.; Tengi, J.P.; Sasso, G.J.; Weigele, M., *J.Org.Chem.* (1985) **50**, 3457

furanomycin in 43% yield. Complete experimental details accompany this work.

A Hoffmann-La Roche group [58] has recently isolated a very unique β-lactam antibiotic produced by *Streptomyces clavuligerus*; the structure and absolute stereochemistry of this substance called clavalanine (**428**, Ro 22-5417) was ascertained by chiroptical methods and total synthesis. This β-lactam antibiotic is structurally related to clavulanic acid, but has the unusual (S)-stereochemistry at the ring juncture. As a consequence of this odd stereochemical difference with all the β-lactam antibiotics, this substance does not interfere with peptidoglycan synthesis and is neither a substrate nor inhibitor of β-lactamases. It was shown that clavalanine is an antimetabolite of O-succinylhomoserine and thus, intervenes in the biosynthesis of methionine. The total synthesis [59] of this compound is detailed in Scheme 63. The D-xylose derivative **421** was reduced to **422** and converted into the *para*-chlorophenyl sulfonate **423**. Oxidation to the lactone and S_N2 displacement of the sulfonate followed by reduction and acylation provided **424**. It might be noted that lactone **424** was also recently synthesized by Williams, et.al. [60] by a more direct route and was detailed in Chapter 1. Ring-opening of **424** to the hydroxy acid, followed by esterification with diphenyldiazomethane and sulfonylation afforded the key protected amino acid **425**. Coupling of **425** with acetoxy azetidinone **426** in the presence of palladium acetate followed by lithium bromide displacement of the sulfonate provided **427** as a 70 : 30, (S) : (R) mixture. Cyclization to the bicyclo [3.2.0] β-lactam was achieved with silver 2,2-dimethyl-6,6,7,7,8,8,8-heptafluoro-3,5-octanedioate in DMF; final debenzylation afforded clavalanine (**428**). These workers also synthesized a stereoisomer possessing the (R)-stereochemistry at the ring juncture; not surprisingly, this isomer displayed β-lactamase inhibitory properties. It is also mentioned that clavalanine is a weak antibiotic but shows reasonable anti-fungal activity. Complete experimental details accompany these reports [58,59].

A Fujisawa group [61] recently discovered the unusual peptide immunomodulator FR900490 (**436**) as a metabolite of *Discosia* sp. F-11809. This substance contains L-His and L-Asn plus the unusual amino acid 2,3-diaminobutanoic acid. The diamino acid was prepared from the optically active N-tosyl aziridine **429** obtained from L-threonine. Nucleophilic ring-opening of the aziridine with L-His afforded the two regioisomeric ring-cleavage products **430** and **431**. These were not separated, but allowed to react with sodium in liquid ammonia to afford the amino acids **432** (27%) and **433**(44%), respectively which could be separated by chromatography. Coupling of the major isomer **433** with the

SCHEME   64

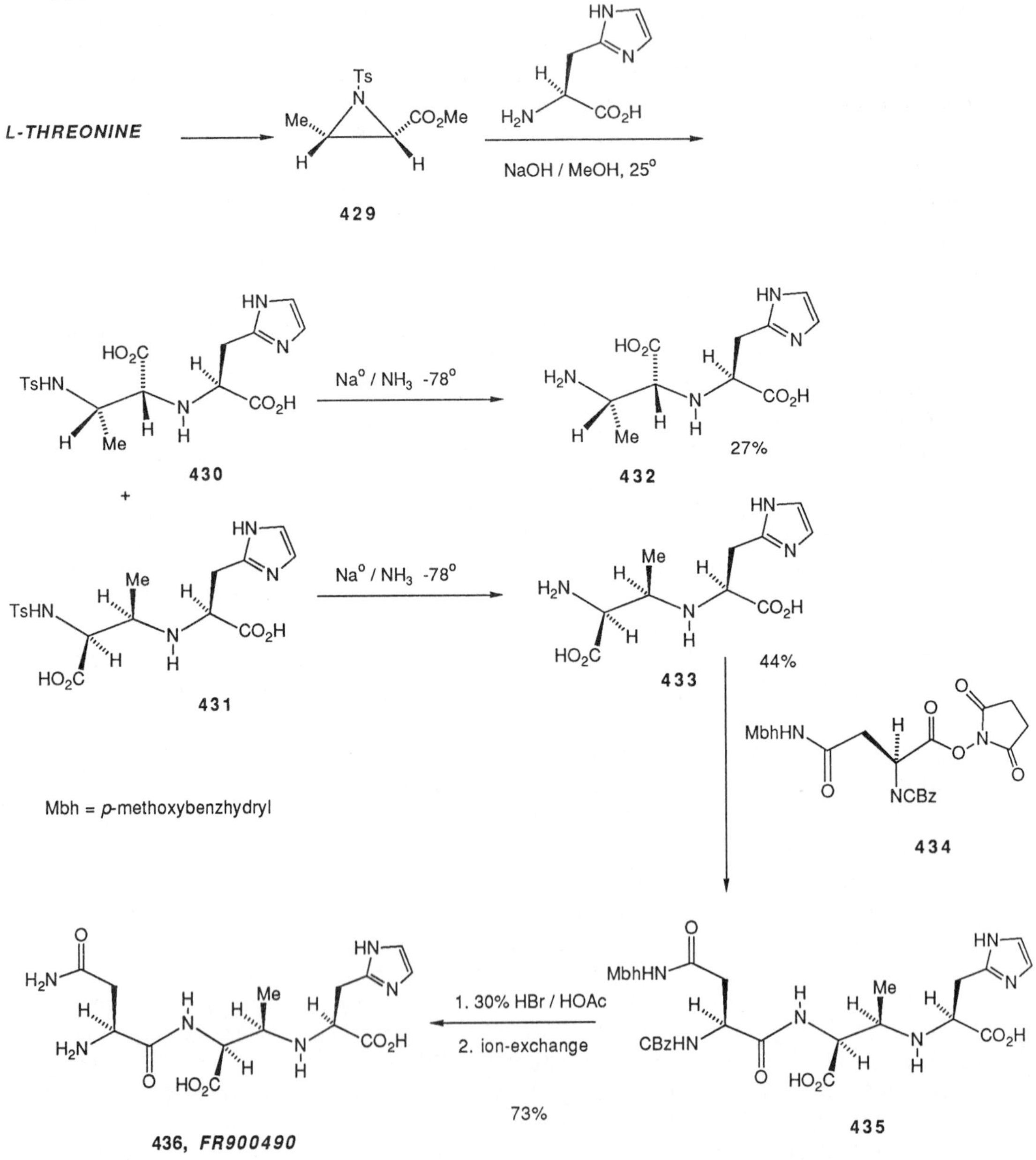

Mbh = *p*-methoxybenzhydryl

436, FR900490

Shigematsu, N.; Setoi, H.; Uchida, I.; Shibata, T.; Terano, H.; Hashimoto, M., *Tetrahedron Lett.* (1988) **29**, 5147 activated asparagine derivative **434** afforded the peptide **435**. Cleavage of the protection and ion-exchange purification afforded the natural substance **436** in 73% yield. The authors note that FR900490 has the capacity to restore the depression of bone marrow cell maturation by immunosuppressive factors in tumor-bearing mouse serum.

It is hoped that the preceding selection of total synthesis highlights and the structures selected in the following pages which have not yet yielded to total synthesis, will convince the reader that many elegant and beautiful accomplishments have been realized in recent years and that many more await the emergence of new technologies and strategies. Any experimentalist who has travelled in the area of amino acid synthesis can attest to the reality that these substances are a treacherous and demanding class of compounds that require great experimental skill, patience and careful strategic planning. It is certain that the synthesis of complex amino acids and physiologically important compounds containing unusual amino acids , whether they be of natural origin or designed by the scientist to perform a specific function, will play an increasingly important role in all disciplines of biology and chemistry.

References   Chapter   9

1. For an overview of the biological properties of these substances, see: Johnson, R.; Koerner, J.F., *J.Med.Chem.* (1988) **31**, 2057.

2. Ohfune, Y.; Tomita, M., *J.Am.Chem.Soc.* (1982) **104**, 3511.

3. a)Oppolzer, W.; Thirring, K., *J.Am.Chem.Soc.* (1982) **104**, 4978; for a racemic synthesis of kainic acid and (+)-*allo*-kainic acid, see: b) Oppolzer, W.; Andres, H., *Helv.Chim.Acta.* (1979) **62**, 2282; c) Oppolzer, W.; Robbiani, C.; Battig, K., *Helv.Chim.Acta.* (1980) **63**, 2015.

4. Cooper, J.; Knight, D.W.; Gallagher, P.T., *J.Chem.Soc.Chem.Comm.* (1987) 1220.

5. Takano, S.; Sugihara, T.; Satoh, S.; Ogasawara, K., *J.Am.Chem.Soc.* (1988) **110**, 6467.

6. Takano, S.; Iwabuchi, Y.; Ogasawara, K., *J.Chem.Soc.Chem.Comm.* (1988) 1527.

7. Takano, S.; Iwabuchi, Y.; Ogasawara, K., *J.Am.Chem.Soc.* (1987) **109**, 5523.

8. Takano, S.; Iwabuchi, Y.; Ogasawara, K., *J.Chem.Soc.Chem.Comm.* (1988) 1204.

9. a) Konno, K.; Hashimoto, K.; Ohfune, Y.; Shirahama, H.; Matsumoto, T., *J.Am.Chem.Soc.* (1988) **110**, 4807; b) Konno, K.; Hashimoto, K.; Ohfune, Y.; Shirahama, H.; Matsumoto, T., *Tetrahedron Lett.* (1986) **27**, 607; c) Hashimoto, K.; Konno, K.; Shirahama, H.; Matsumoto, T., *Chemistry Lett.* (1986) 1399; d) Konno, K.; Hashimoto, K.; Shirahama, H.; Matsumoto, T., *Tetrahedron Lett.* (1986) **27**, 3865.

10. Baldwin, J.E.; Li, C-S., *J.Chem.Soc.Chem.Comm.* (1987) 166.

11. Baldwin, J.E.; Li, C-S., *J.Chem.Soc.Chem.Comm.* (1988) 261.

12. Other kainoid syntheses, see: a) DeShong, P.; Kell, D.A., *Tetrahedron Lett.* (1986) **27**, 3979; b) Kraus, G.A.; Nagy, J.O., *Tetrahedron* (1985) **41**, 3537; c) Yoo, S-E.; Lee, S-H.; Kim, N-J., *Tetrahedron Lett.* (1988) **29**, 2195.

13. Martin, D.G ; Duchamp, D.J.; Chidester, C.G., *Tetrahedron Lett.* (1973) 2549.

14. Kelly, R.C.; Schletter, I.; Stein, S.J.; Wierenga, W., *J.Am.Chem.Soc.* (1979) **101**, 1054.

15. a) Baldwin, J.E.; Kruse, L.I.; Cha, J-K., *J.Am.Chem.Soc.* (1981) **103**, 942; b) Baldwin, J.E.; Cha, J-K.; Kruse, L.I., *Tetrahedron* (1985) **41**, 5241; c) Baldwin, J.E.; Hoskins, C.; Kruse, L.I., *J.Chem.Soc.Chem.Comm.* (1976) 795.

16. Silverman, R.B.; Holladay, M.W., *J.Am.Chem.Soc.* (1981) **103**, 7357.

17. Hagedorn, A.A.; Miller, B.J.; Nagy, J.O., *Tetrahedron Lett.* (1980) **21**, 229.

18. a) Wade, P.A.; Pillay, M.K.; Singh, S.M.; *Tetrahedron Lett.* (1982) **23**, 4563; b) Wade, P.A.; Singh, S.M.; Pillay, M.K., *Tetrahedron* (1984) **40**, 601.

19. Vyas, D.M.; Chiang, Y.; Doyle, T.W., *Tetrahedron Lett.* (1984) **25**, 487.

20. a) Mzengeza, S.; Yang, C.M.; Whitney, R.A., *J.Am.Chem.Soc.* (1987) **109**, 276; b) Mzengeza, S.; Whitney, R.A., *J.Org.Chem.* (1988) **53**, 4074.

21. Harada, S.; Tsubotani, S.; Hida, T.; Ono, H.; Akazaki, H., *Tetrahedron Lett.* (1986) **27**, 6229.

22. Natsugari, H.; Kawano, Y.; Morimoto, A.; Yoshioka, K.; Ochiai, M., *J.Chem.Soc.Chem.Comm.* (1987) 62.

23. For a related study on a monobactam mimic of this system, see: Baldwin, J.E.; Ng, S.C.; Pratt, A.J., *Tetrahedron Lett.* (1987) **28**, 4319.

24. Shinagawa, S.; Maki, M.; Kintaka, K.; Imada, A.; Asai, M., *J.Antibiotics* (1985) **38**, 17.

25. Wakamiya, T.; Yamanoi, K.; Nishikawa, M.; Shiba, T., *Tetrahedron Lett.* (1985) **26**, 4759.

26. a) Ohfune, Y.; Hori, K.; Sakaitani, M., *Tetrahedron Lett.* (1986) **27**, 6079; b) Ohfune, Y.; Kurokawa, N., *Tetrahedron Lett.* (1985) **26**, 5307.

27. a) Bashyal, B.P.; Chow, H-F.; Fleet, G.W.J., *Tetrahedron Lett.* (1986) **27**, 3205; b) Dho, J.C.; Fleet, G.W.J.; Peach, J.M.; Prout, K.; Smith, P.W., *Tetrahedron Lett.* (1986) **27**, 3203.

28. Benz,v.F.; Knusel, F.; Nuesch, J.; Treichler, H.; Voser, W.; Nyfeler, R.; Keller-Schierlein, W., *Helv.Chim.Acta.* (1974) **57**, 2459.

29. a) Kurokawa, N.; Ohfune, Y., *J.Am.Chem.Soc.* (1986) **108**, 6041; b)Kurokawa, N.; Ohfune, Y., *J.Am.Chem.Soc.* (1986) **108**, 6043.

30. Evans, D.A.; Weber, A.E.; *J.Am.Chem.Soc.* (1987) **109**, 7151.

31. Grieco, P.A.; Hon, Y.S.; Perez-Medrano, A., *J.Am.Chem.Soc.* (1988) **110**, 1630.

32. For an alternate synthesis of the aliphatic component of jasplakinolide, see: Schmidt, U.; Siegel, W.; Mundinger, K., *Tetrahedron Lett.* (1988) **29**, 1269; for an alternate synthesis of the tripeptide moiety, see: Kato, S.; Hamada, Y.; Shioiri, T., *Tetrahedron Lett.* (1988) **29**, 6465.

33. Grieco, P.A.; Perez-Medrano, A., *Tetrahedron Lett.* (1988) **29**, 4225.

34. White, J.D.; Amedio, J.C., *J.Org.Chem.* (1989) **54**, 738.

35. a) Tamai, S.; Kaneda, M.; Nakamura, S., *J.Antibiotics* (1982) **35**, 1130; b) Kaneda, M.; Tamai, S.; Nakamura, S.; Hirata, T.; Kushi, Y.; Suga, T., *J.Antibiotics* (1982) **35**, 1137.

36. Nishiyama, S.; Nakamura, K.; Suzuki, Y.; Yamamura, S., *Tetrahedron Lett.* (1986) **27**, 4481.

37. Nishiyama, S.; Suzuki, Y.; Yamamura, S., *Tetrahedron Lett.* (1988) **29**, 559.

38. Inaba, T.; Umezawa, I.; Yuasa, M.; Inoue, T.; Mihashi, S.; Itokawa, H.; Ogura, K., *J.Org.Chem.* (1987) **52**, 2957.

39. For an approach to deoxybouvardin and some interesting cyclic biaryl analogs, see: Boger, D.L.; Yohannes, D., *J.Org.Chem.* (1988) **53**, 487.

40. Schmidt, U.; Beuttler, T.; Lieberknecht, A.; Griesser, H., *Tetrahedron Lett.* (1983) **24**, 3573.

41. a) Schmidt, U.; Lieberknecht, A.; Bokens, H.; Griesser, H., *J.Org.Chem.* (1983) **48**, 2680; b) Schmidt, U.; Griesser, H.; Lieberknecht, A.; Talbiersky, J., *Angew.Chem.Int.Ed.Engl.* (1981) **20**, 280; c) Schmidt, U.; Lieberknecht, A.; Bokens, H.; Griesser, H., *Angew.Chem.Int.Ed.Engl.* (1981) **20**, 1026; d) Schmidt, U.; Lieberknecht, A.; Griesser, H.; Hausler, J.; *Angew.Chem.Int.Ed.Engl.* (1981) **20**, 281.

42.   a) Schmidt, U.; Schanbacher, U., *Liebigs Ann.Chem.* (1984) 1205; b) Schmidt, U.;
       Schanbacher, U., *Angew.Chem.Int.Ed.Engl.* (1983) **22**, 152; c) Schmidt, U.; Lieberknecht, A.;
       Griesser, H.; Talbiersky, J., *J.Org.Chem.* (1982) **47**, 3261.

43.   Nutt, R.K.; Chen, K-M.; Joullie, M.M., *J.Org.Chem.* (1984) **49**, 1013.

44.   Nishiyama, S.; Suzuki, Y.; Yamamura, S., *Tetrahedron Lett.* (1989) **30**, 379.

45.   a) Wenger, R.M., *Helv.Chim.Acta* (1984) **67**, 502; b) Wenger, R.M., *Helv.Chim.Acta* (1983)
       66, 2308; c) Wenger, R.M., *Angew.Chem.Int.Ed.Engl.* (1985) **24**, 77, and references cited
       therein.

46.   a)Tung, R.D.; Rich, D.H., *Tetrahedron Lett.* (1987) **28**, 1139; b) for related preparations of
       MeBmt analogs, see: Sun, C-Q.; Rich, D.H., *Tetrahedron Lett.* (1988) **29**, 5205.

47.   Rao, A.V.R.; Dhar, T.G.M.; Chakraborty, T.K.; Gurjar, M.K., *Tetrahedron Lett.* (1988) **29**,
       2069.

48.   Wright, J.J.K.; Cooper, A.B.; McPhail, A.T.; Merrill, Y.; Nagabhushan, T.L.; Puar, M.S.,
       J.Chem.Soc. Chem.Comm. (1982) 1188.

49.   Owens, L.D.; Wright, D.A., *Plant Physiol.* (1965) **40**, 927.

50.   a) Pruess, D.; Scanell, J., *Adv.Appl.Microbiol.* (1974) **17**, 29; b)Owens, L.D.; Guggenheim, S.;
       Hilton, J., *Biochim.Biophys.Acta.* (1968) **158**, 219; c) Giovanelli, J.; Owens, L.D.; Mudd,
       S.H., *Biochim.Biophys. Acta.* (1971) **227**, 671; d) Owens, L.D.; Lieberman, M.; Kunishi, A.,
       Plant Physiol. (1971) **48**, 1.

51.   a) Keith, D.D.; DeBernardo, S.; Weigele, M., *Tetrahedron* (1975) **31**, 2629; b) Keith, D.D.;
       Tortora, J.A.; Ineichen, K.; Leimgruber, W., *Tetrahedron* (1975) **31**, 2633.

52.   Keith, D.D.; Tortora, J.A.; Yang, R., *J.Org.Chem.* (1978) **43**, 3711.

53.   Pirrung, M.C., *Tetrahedron Lett.* (1980) **21**, 4577.

54.   a) Hughes, P.; Martin, M.; Clardy, J., *Tetrahedron Lett.* (1980)**21**, 4579; b) Hughes, P.;
       Clardy, J., *J.Org.Chem.* (1988) **53**, 4793.

55.   Kurokawa, N.; Ohfune, Y., *Tetrahedron Lett.* (1985) **26**, 83.

56.   Yamanoi, K.; Ohfune, Y., *Tetrahedron Lett.* (1988) **29**, 1181.

57.   a) Joullie, M.M.; Wang, P.C.; Semple, J.E., *J.Am.Chem.Soc.* (1980) **102**, 887; b) Semple,
       J.E.; Wang, P.C.; Lysenko, Z.; Joullie, M.M., *J.Am.Chem.Soc.* (1980) **102**, 7505.

58.   a)Pruess, D.L.; Kellett, M., *J.Antibiotics* (1983) **36**, 208; b) Evans, R.H.; Ax, H.; Jacoby, A.;
       Williams, T.H.; Jenkins, E.; Scannell, J.P., *J.Antibiotics* (1983) **36**, 213; c) Muller, J-C.;
       Toome, V.; Pruess, D.L.; Blount, J.F.; Weigele, M., *J.Antibiotics* (1983) **36**, 217.

59.   DeBernardo, S.; Tengi, J.P.; Sasso, G.J.; Weigele, M., *J.Org.Chem.* (1985)**50**, 3457.

60.   Williams, R.M.; Sinclair, P.J.; Zhai, D.; Chen, D., *J.Am.Chem.Soc.* (1988) **110**, 1547.

61.   Shigematsu, N.; Setoi, H.; Uchida, I.; Shibata, T.; Terano, H.; Hashimoto, M., *Tetrahedron Lett.*
       (1988) **29**, 5147.

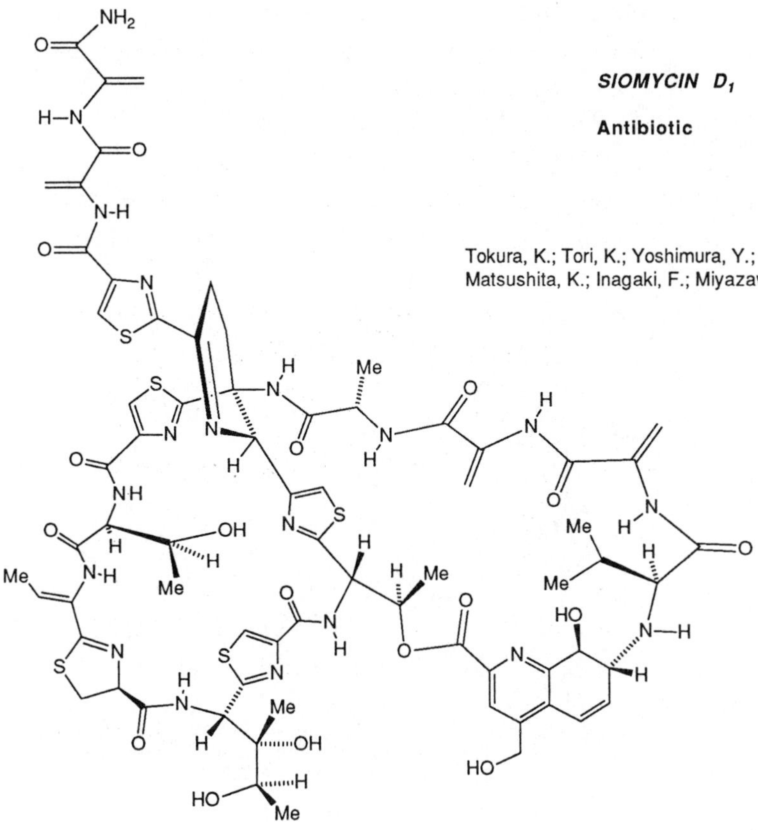

SIOMYCIN D₁

Antibiotic

Tokura, K.; Tori, K.; Yoshimura, Y.; Okabe, K.; Otsuka, H.; Matsushita, K.; Inagaki, F.; Miyazawa, T., J.Antibiotics (1980) 33, 1563.

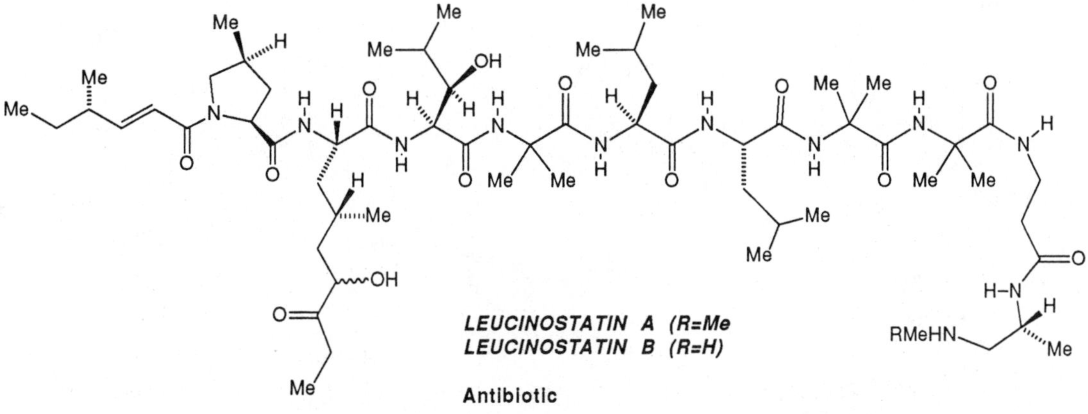

LEUCINOSTATIN A (R=Me
LEUCINOSTATIN B (R=H)

Antibiotic

Fukushima, K.; Arai, T.; Mori, Y.; Tsuboi, M.; Suzuki, M., *J.Antibiotics* (1983) **36**, 1606

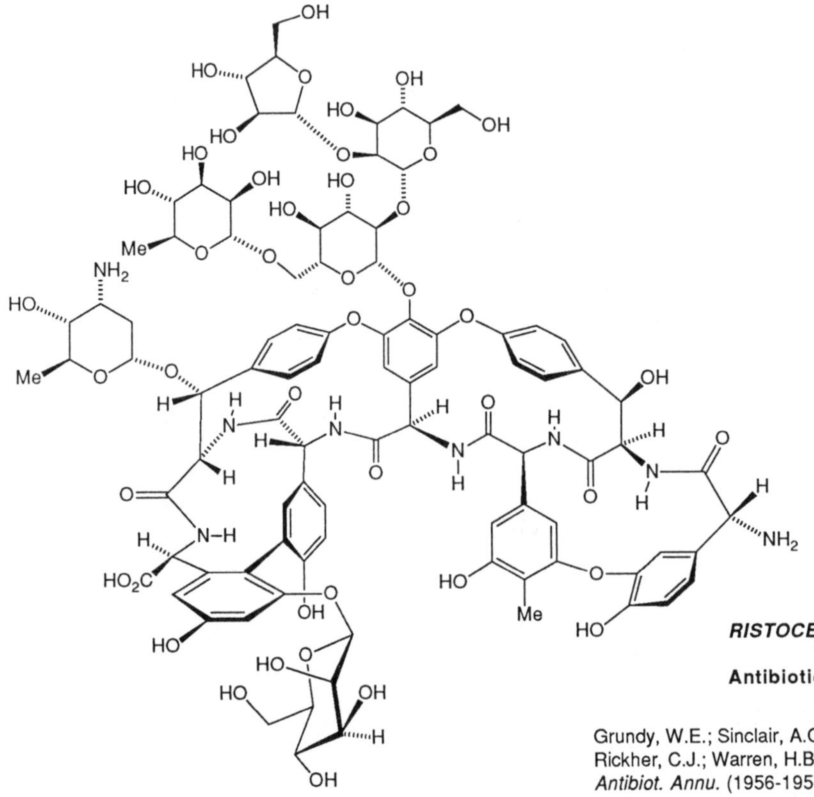

Total Synthesis of Complex Amino Acids

RISTOCETIN A

Antibiotic

Grundy, W.E.; Sinclair, A.C.; Theriault, R.J.; Goldstein, A.W.; Rickher, C.J.; Warren, H.B.; Oliver, T.J.; Sylvester, J.C., *Antibiot. Annu.* (1956-1957) 699.

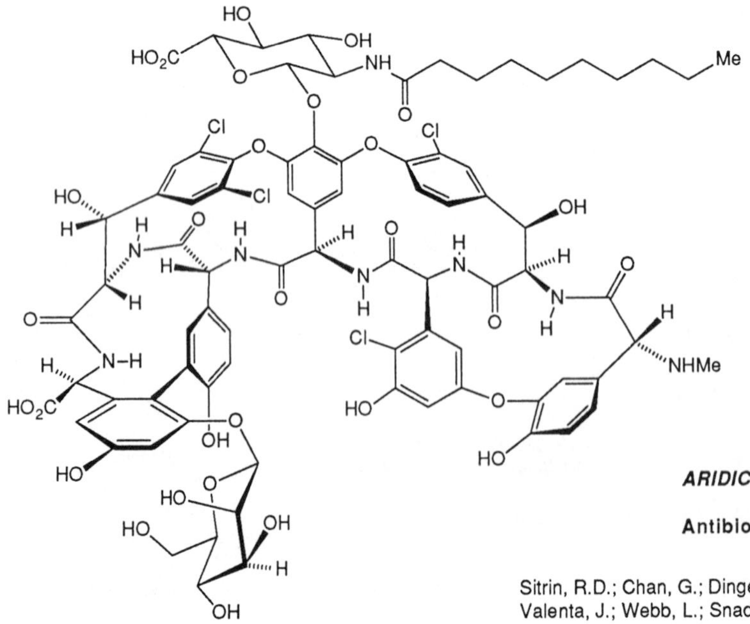

ARIDICIN A

Antibiotic

Sitrin, R.D.; Chan, G.; Dingerdissen, J.; Holl, W.; Hoover, J.R.E.; Valenta, J.; Webb, L.; Snader, K.M., *J.Antibiotics* (1985) **38**, 561

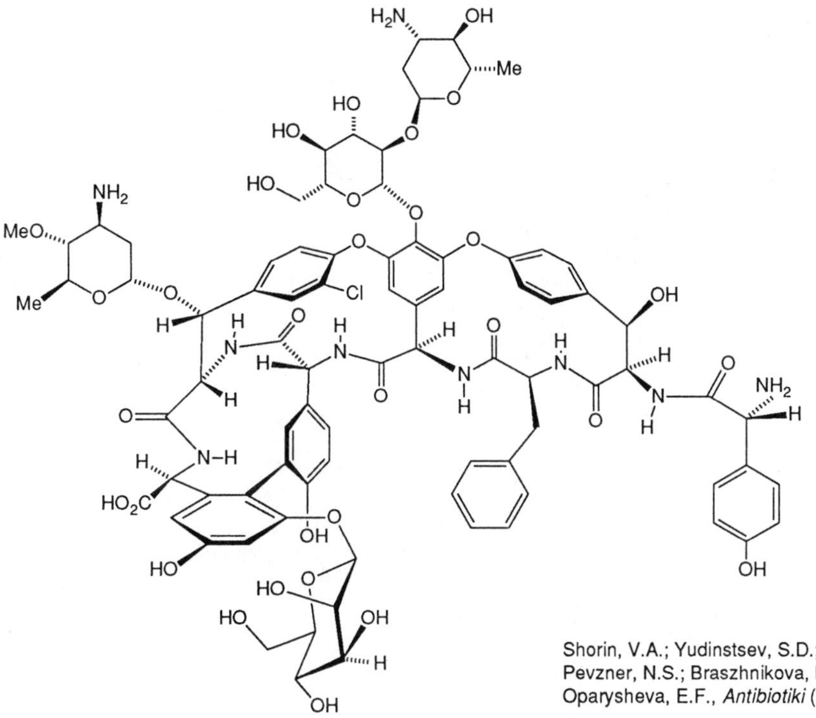

ACTINOIDIN A

Antibiotic

Shorin, V.A.; Yudinstsev, S.D.; Kunrat, I.A.; Goldberg, L.E.; Pevzner, N.S.; Braszhnikova, M.G.; Lomakina, N.N.; Oparysheva, E.F., *Antibiotiki* (1957) **2**, 44.

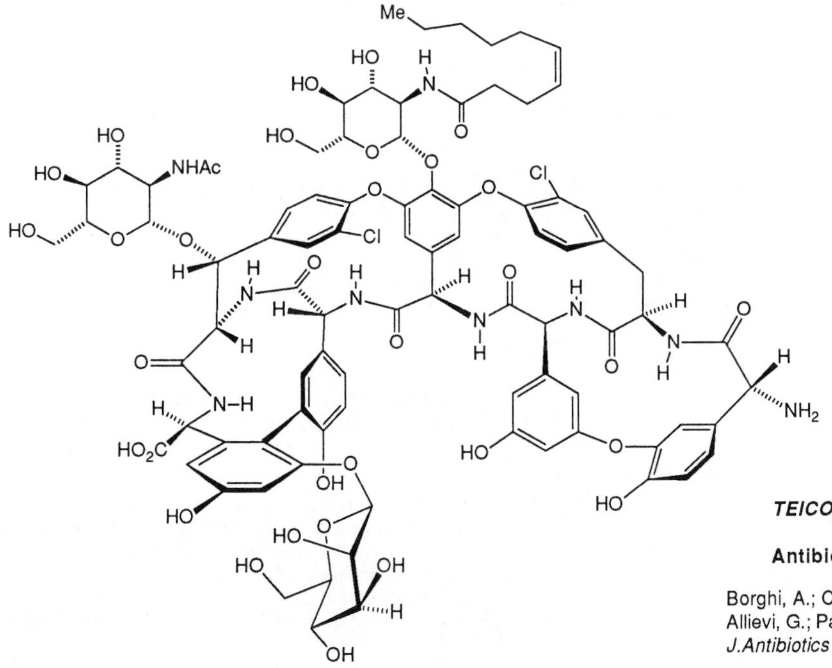

TEICOPLANIN A2-1

Antibiotic

Borghi, A.; Coronelli, C.; Faniuolo, L.; Allievi, G.; Pallanza, R.; Gallo, G.G., *J.Antibiotics* (1984) **37**, 615.

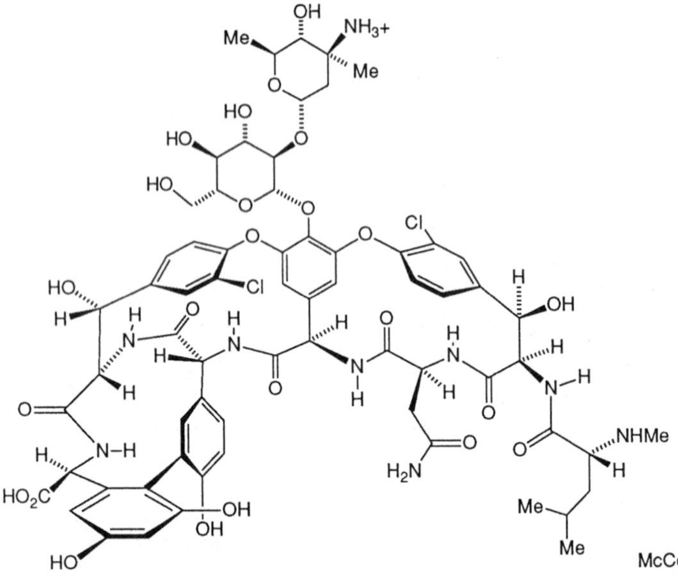

VANCOMYCIN

Antibiotic

McCormick, M.H.; Stark, W.M.; Pittenger, G.E.; Pittenger, R.C.; McGuire, J.M., *Antibiot. Annu.* (1955-56) 606.

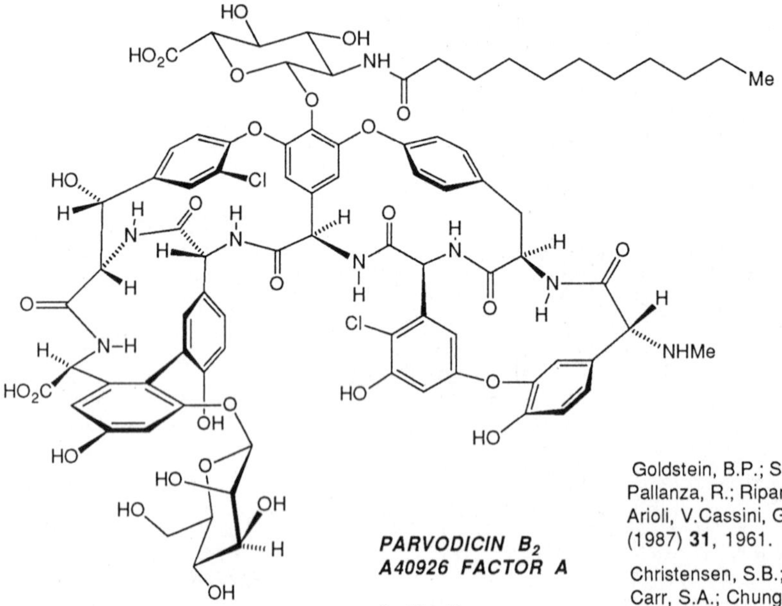

PARVODICIN B₂
A40926 FACTOR A

Antibiotic

Goldstein, B.P.; Selva, E.; Gastaldo, L.; Berti, M.; Pallanza, R.; Ripamonti, F.; Ferrari, P.; Denaro, M.; Arioli, V.Cassini, G., *Antimicrob. Agents Chemother* (1987) **31**, 1961.

Christensen, S.B.; Allaudeen, H.S.; Burke, M.R.; Carr, S.A.; Chung, S.K.; DePhillips, P.; Dingerdissen, J.J.; DiPaolo, M.; Giovenella, A.J.; Heald, S.L.; Killmer, L.B.; Mico, B.A.; Mueller, L.; Pan, C.H.; Poehland, B.J.; Rake, J.B.; Roberts, G.D.; Shearer, M.C.; Sitrin, R.D.; Nisbet, L.J.; Jeffs, P.W., *J.Antibiotics* (1987) **40**, 970.

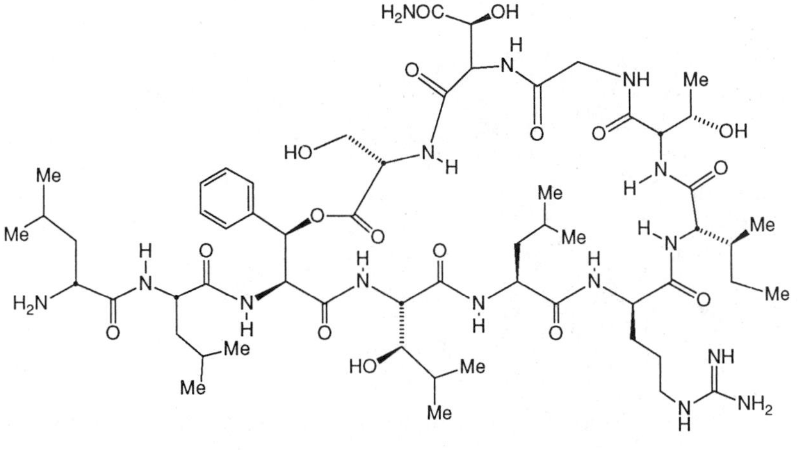

LYSOBACTIN

Antibacterial

O'Sullivan, J.; McCullough, J.E.; Tymiak, A.A.; Kirsch, D.R.; Trejo, W.H.; Principie, P.A., *J.Antibiotics* (1988) **41**, 1740

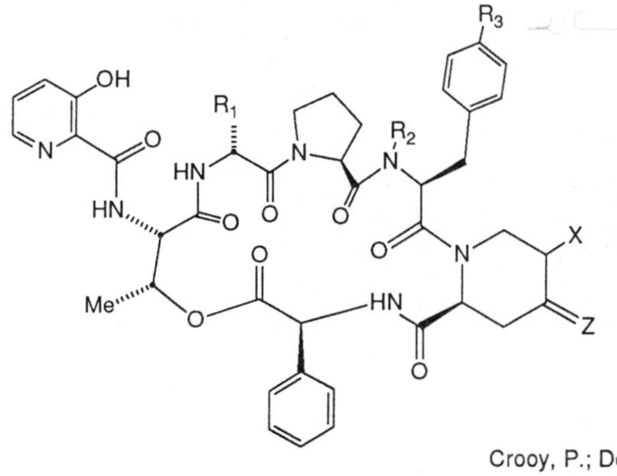

NAME	R₁	R₂	R₃	Z	X
PATRICIN B	Et	Me	H	H,H	H
VIRGINIAMYCIN S₁	Et	Me	H	O	H
VIRGINIAMYCIN S₂	Me	Me	H	O	H
VIRGINIAMYCIN S₃	Et	H	H	O	H
VIRGINIAMYCIN S₄	Et	Me	H	O	OH
STREPTOGRAMIN B	Et	Me	NMe₂	O	H
OSTREOGRYCIN B₁	Me	Me	NMe₂	O	H
OSTREOGRYCIN B₂	Et	Me	NHMe	O	H
OSTREOGRYCIN B₃	Et	Me	NHMe	O	OH
VERNAMYCIN	Me	Me	NMe₂	O	H

Antibiotic

Crooy, P.; DeNeys, *J.Antibiotics* (1972) **25**, 371

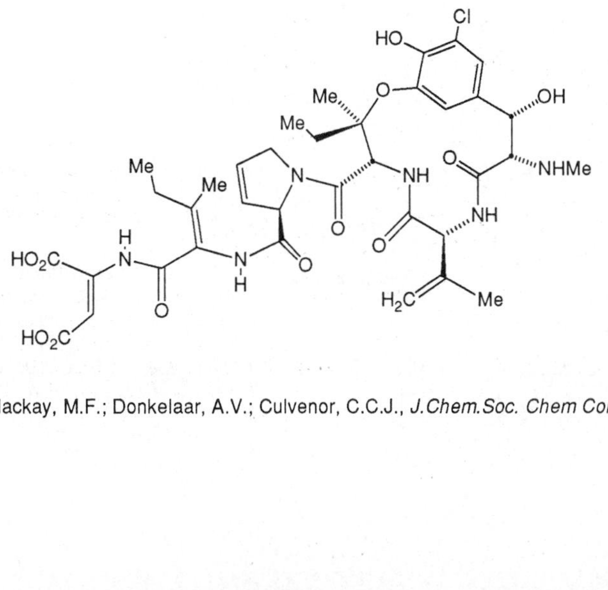

PHOMPOSIN A

Mycotoxin

Mackay, M.F.; Donkelaar, A.V.; Culvenor, C.C.J., *J.Chem.Soc. Chem Comm.* (1986) 1219

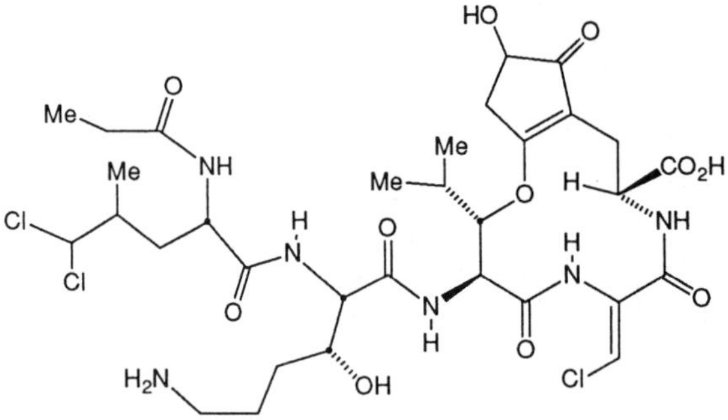

HV-TOXIN M

Cytotoxin

Kono, Y.; Kinoshita, T.; Takeuchi, S.; Daly, J.M., *Agric.Biol.Chem.* (1986) **50**, 2689

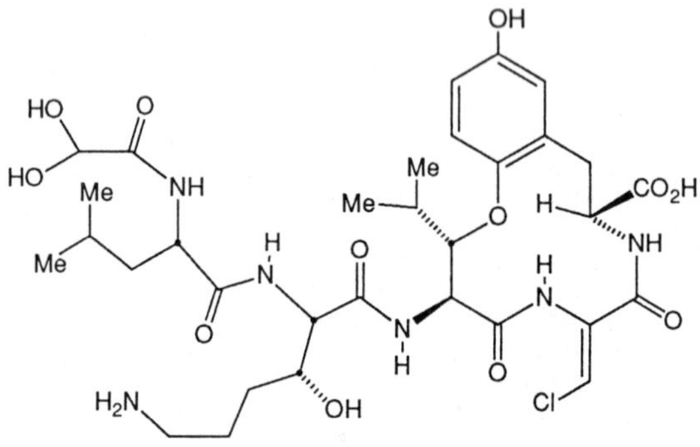

VICTORICIN C

Cytotoxin

Wolpert, T.J.; Macko, V.; Acklin, W.; Jaun, B.; Arigoni, D., *Experientia* (1986) **42**, 1296

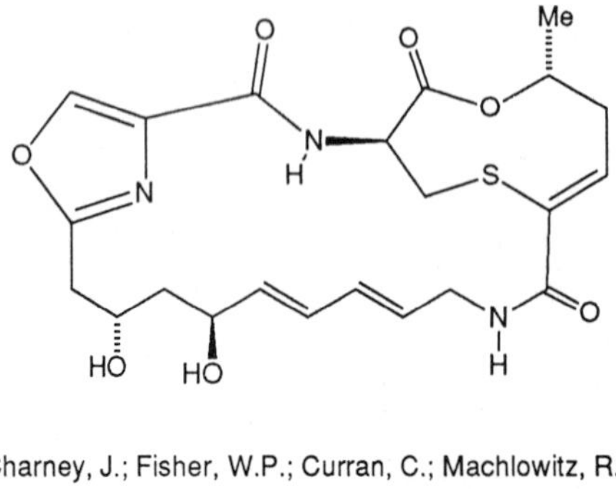

GRISEOVIRIDIN

Antibiotic

Charney, J.; Fisher, W.P.; Curran, C.; Machlowitz, R.A.; Tyteli, A.A., *Antibiot.Chemother.* (1953) **3**, 1283

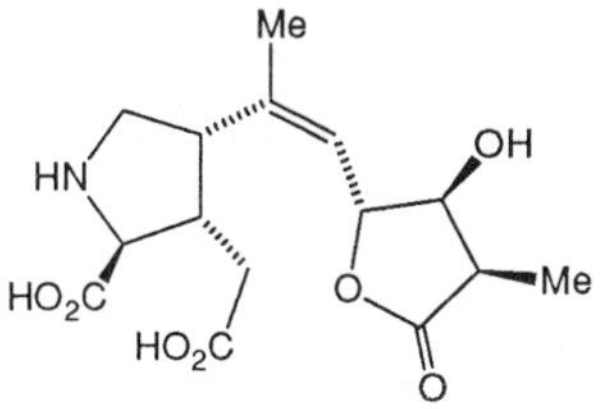

DOMOILACTONE A

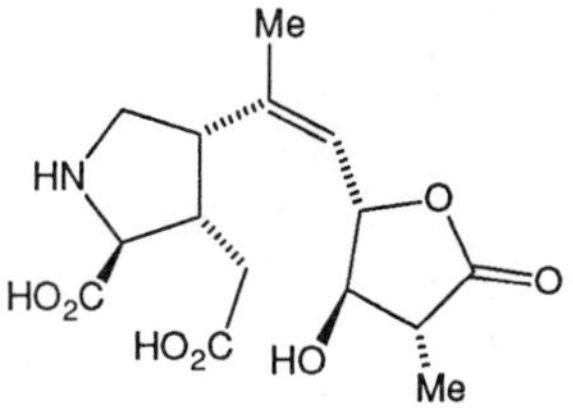

DOMOILACTONE B

Maeda, M.; Kodama, T.; Tanaka, T.; Yoshizumi, H.;
Takemoto, T.; Nomoto, K.; Fujita, T., Tetrahedron Lett. (1987) 28, 633

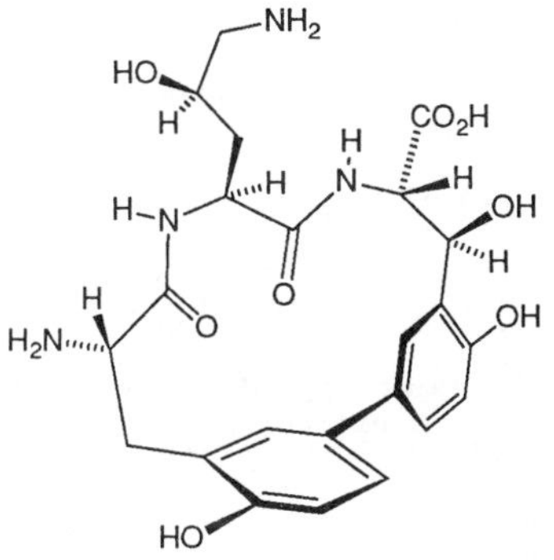

WS-43708A

Antibiotic

Uchida, I.; Ezaki, M.; Shigematsu, N.; Hashimoto, M., *J.Org.Chem.* (1985) **50**, 1342
Kannan, R.; Williams, D.H., *J.Org.Chem.* (1987) **52**, 5435.

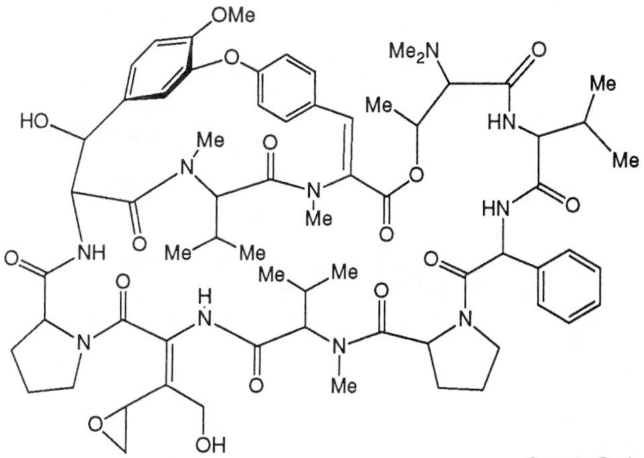

DITYROMYCIN

Antibiotic

Omura, S.; Iwai, Y.; Hirano, A.; Awaya, J.; Suzuki, Y.; Matsumoto, K.
Agric.Biol.Chem. (1977) **41**, 1827
Teshima, T.; Nishikawa, M.; Kubota, I.; Shiba, T.; Iwai, Y.; Omura, S.,
Tetrahedron Lett. (1988) **29**, 1963.

SOAA—N*

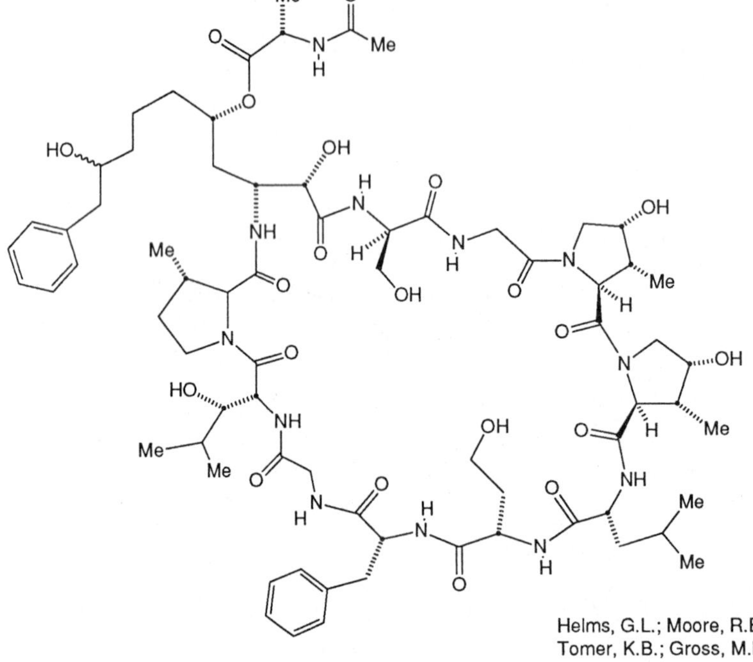

Total Synthesis of Complex Amino Acids

SCYTONEMIN A

Calcium Antagonist

Helms, G.L.; Moore, R.E.; Niemczura, W.P.; Patterson, G.M.L.
Tomer, K.B.; Gross, M.L., *J.Org.Chem.* (1988) **53**, 1298

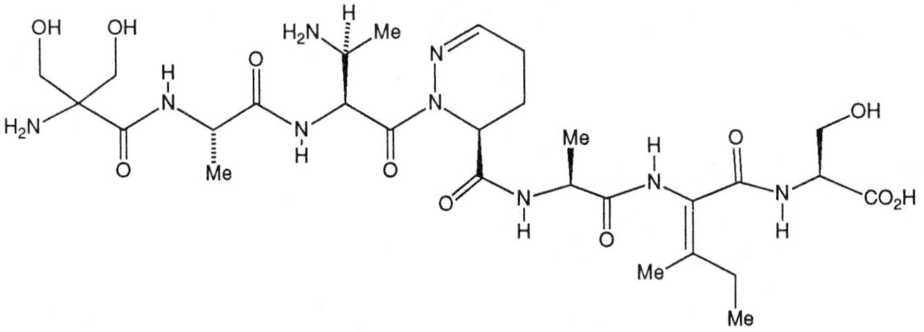

ANTRIMYCIN

Antibiotic

Morimoto, K.; Shimada, N.; Naganawa, H.; Takita, T.; Umezawa, H.,
J.Antibiotics (1981) **34**, 1615.

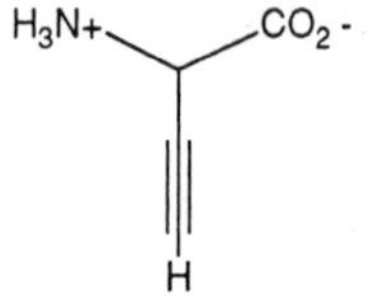

FR-900130

Antibiotic

Kuroda, Y.; Okuhara, M.; Goto, T.; Iguchi, E.; Kohsaka, M.; Aoki, H.; Imanaka, H.,
J.Antibiotics (1980) **33**, 125.

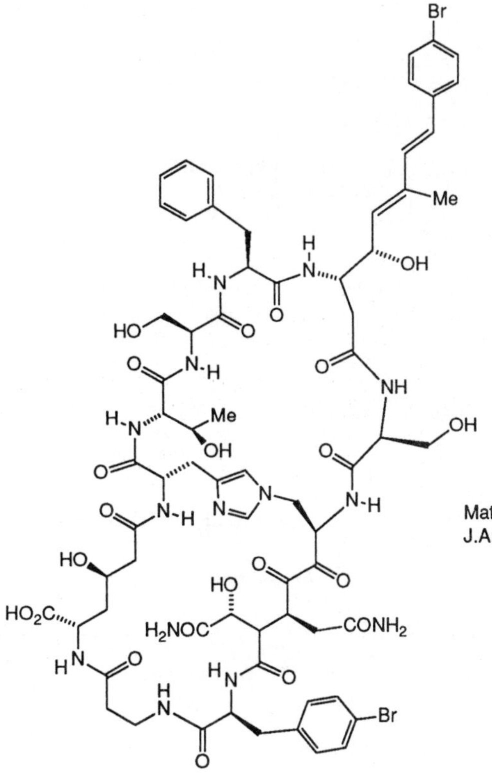

THEONELLAMIDE F

Antifungal

Matsunaga, S.; Fusetani, N.; Hashimoto, K.; Walchli, M.,
J.Am.Chem.Soc.. (1989) 111, 2582

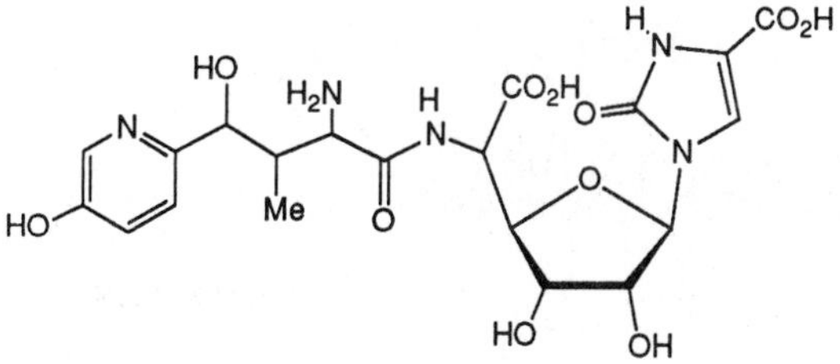

NEOPOLYOXIN B

Antifungal

Kobinata, K.; Uramoto, M.; Nishii, M.; Kusakabe, H.; Nakamura, G.;
Isono, K., *Agric.Biol.Chem.* (1980)44, 1709

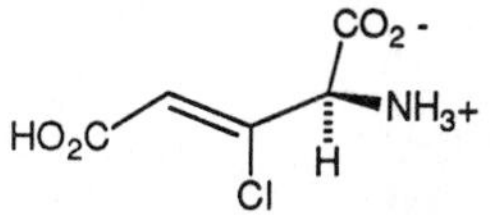

[R-(Z)]-4 AMINO-3-CHLORO-2-PENTENEDIOIC ACID (ACPA)
FR900148
Antibiotic

Chaiet, L.; Arison, B.H.; Monaghan, R.L. ; Springer, J.P.; Smith, J.L.; Zimmerman, S.B.,
J.Antibiotics (1984) **37**, 207
Kuroda, Y.; Okuhara, M.; Goto, T.; Yameshita, E.; Iguchi, E.; Kohsaka, M.; Aoki, H.; Imanaka, H.,
J.Antibiotics (1980) **33**, 259

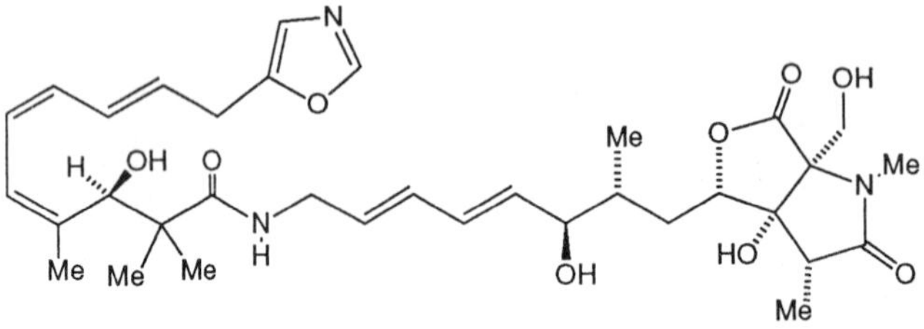

NEOOXAZOLOMYCIN

Anti-tumor Antibiotic

Takahashi, K.; Kawabata, M.; Uemura, D.; Iwadare, S.;Mitomo, R.; Nakano, F.; Matsuzaki, A., *Tetrahedron Lett* (1985) **26**, 1077

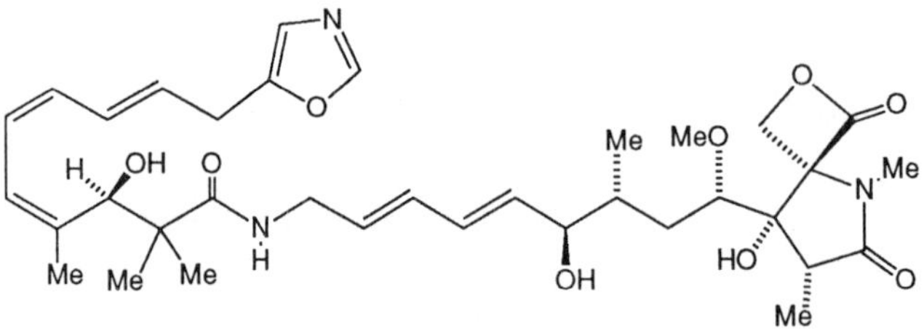

OXAZOLOMYCIN

Anti-tumor Antibiotic

Mori, T.; Takahashi, K.; Kashiwabara, M.; Uemura, D.; Katayama, C.; Iwadare, S.; Shizuri, Y.; Mitomo, R.; Nakano, F.; Matsuzaki, C., *Tetrahedron Lett.* (1985) **26**, 1073

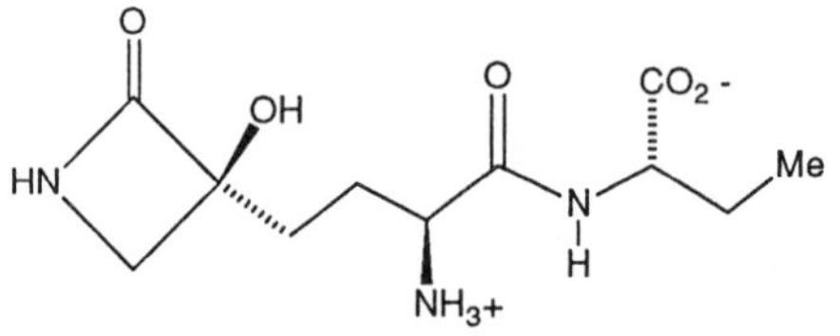

TABTOXIN

Exotoxin

Woolley, D.W.; Pringle, R.B.; Braun, A.C., *J.Biol. Chem.* (1952) **197**, 409. Muller, B.; Hadener, A.; Tamm, *Helv.Chim.Acta.* (1987) **70**, 412

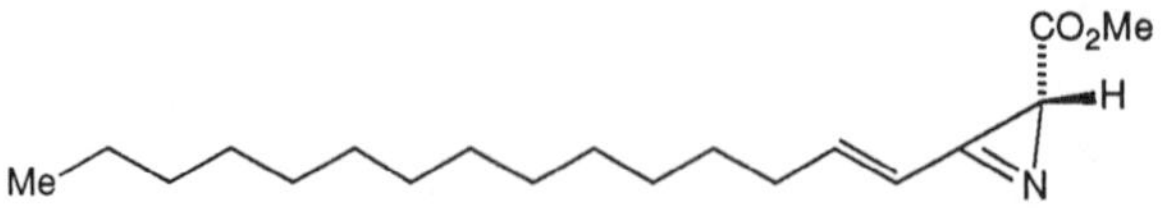

DYSIDAZIRINE

Anti-tumor Antibiotic

Molinski, T.F.; Ireland, C.M., *J.Org.Chem.* (1988) **53**, 2103

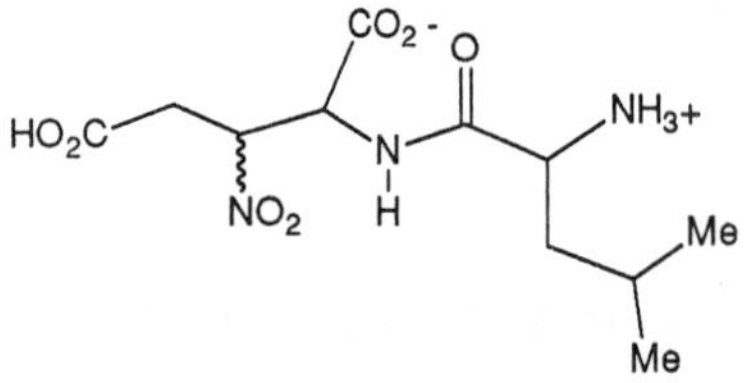

NITROPEPTIN

Glutamine Antimetabolite

Ohba, K.; Nakayama, H.; Furihata, K.; Shimazu, A.; Endo, T.; Seto, H.; Otake, N., *J.Antibiotics* (1987) **40**, 709.

ALAHOPCIN (NOURSEIMYCIN)

Antibiotic

Higashide, E.; Kanamaru, T.; Fukase, H.; Horii, S., *J.Antibiotics* (1985) **38**, 296
Horii, S.; Fukase, H.; Higashide, E.; Yoneda, M., *J.Antibiotics* (1985) **38**, 302.

THERMOCYMOCIDIN

Bagli, J.F.; Kluepfell, D.; St.Jacques, M., *J.Org.Chem.* (1973) **38**, 1253.

N-(2,6-DIAMINO-6-HYDROXYMETHYLPIMELYL)-L-ALANINE

Antibiotic

Shoji, J.; Hinoo, H.; Kato, T.; Nakauchi, K.; Matsuura, S.; Mayama, M.;
Yasuda, Y.; Kawamura, Y., *J.Antibiotics* (1981) **34**, 374.

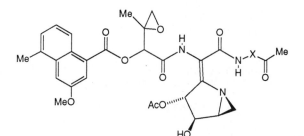

AZINOMYCIN A (X=CH₂)
AZINOMYCIN B (X= C=CHOH)

Anti-tumor Antibiotic

Nagaoka, K.; Matsumoto, M.; Oono, J.; Yokoi, K.; Ishizeki, S.;
Nakashima, T., *J.Antibiotics* (1986) **39**, 1527

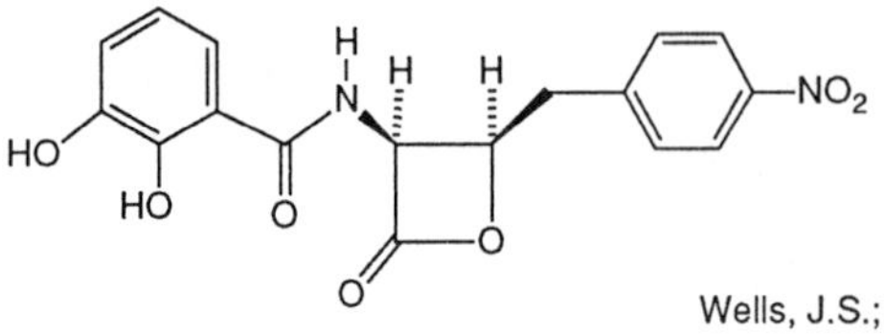

OBAFLUORIN

Antibiotic

Wells, J.S.; Trejo, W.H.; Principe, P.A.; Sykes, R.B., *J.Antibiotics* (1984) **37**, 802.

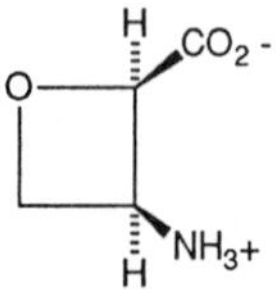

OXETIN

Antibiotic / Herbicidal

Omura, S.; Murata, M.; Imamura, N.; Iwai, Y.; Tanaka, H.; Furusaki, A.; Matsumoto, T., *J.Antibiotics* (1984) **37**, 1324.

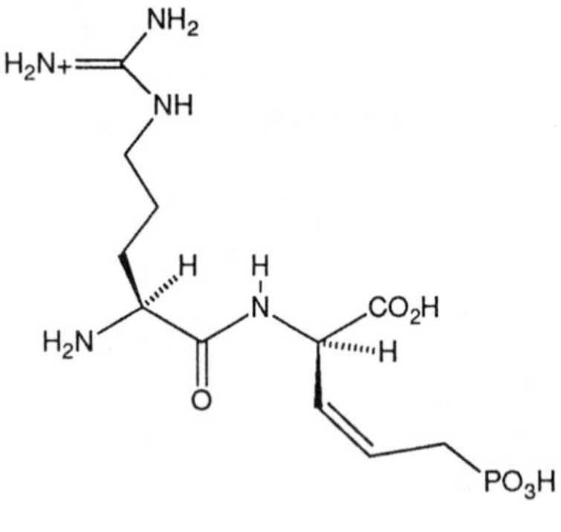

RHIZOCTICIN A

Antifungal

Rapp, C.; Jung, G.; Kugler, M.; Loeffler, W. , *Liebigs Ann.Chem* (1988) 655.

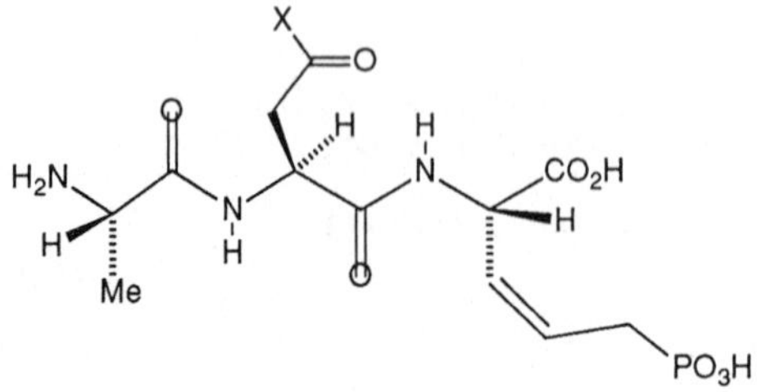

PLUMBEMYCIN A (X=OH)
PLUMBEMYCIN B (X=NH₂)

L-Threonine Antagonists

Park, B.K.; Hirota, A.; Sakai, H., *Agric.Biol.Chem.* (1977) **41**, 573.

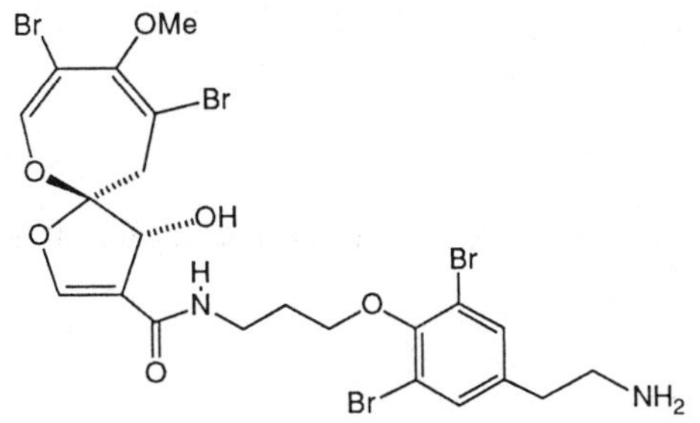

PSAMMAPLYSIN A

Antibiotic

Roll, D.M.; Chang, C.W.J.; Scheuer, P.J.; Gray, G.A.; Shoolery, J.N.; Matsumoto, G.K.; VanDuyne, G.D.; Clardy, J., *J.Am.Chem.Soc.* (1985) **107**, 2916.

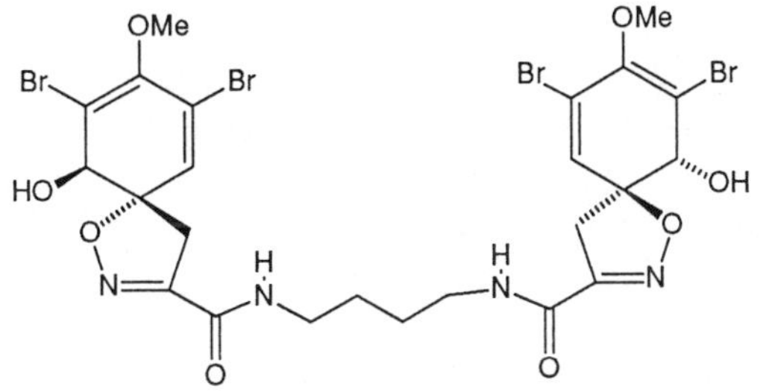

AEROTHIONIN

Antibiotic

McMillan, J.A.; Paul, I.C.; Goo, Y.M.; Rinehart, K.L., *Tetrahedron Lett* (1981) **22**, 39

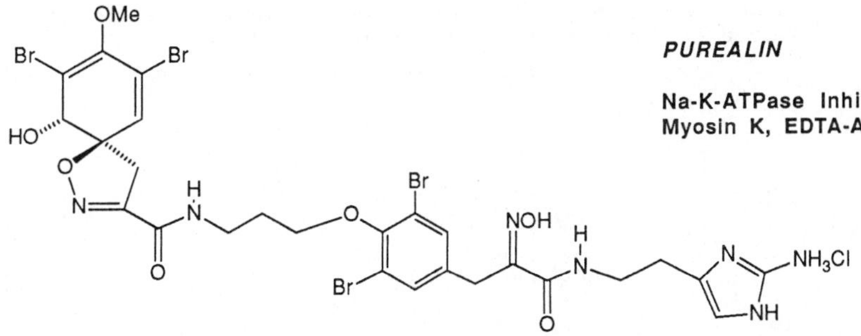

PUREALIN

Na-K-ATPase Inhibition
Myosin K, EDTA-ATPase Activator

Nakamura, H.; Wu, H.; Kobayashi, J.; Nakamura, Y.; Ohizumi, Y.; Hirata, Y., *Tetrahedron Lett.* (1985) **26**, 4517.

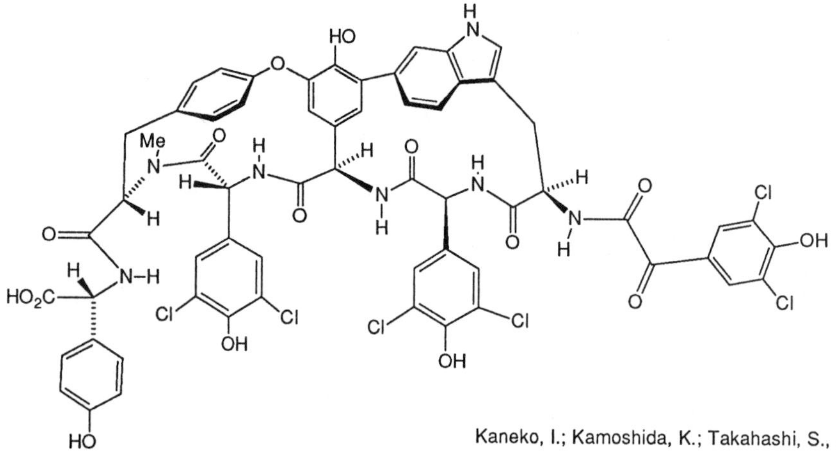

COMPLESTATIN

Anti-Complement

Kaneko, I.; Kamoshida, K.; Takahashi, S., *J.Antibiotics* (1989) **42**, 236

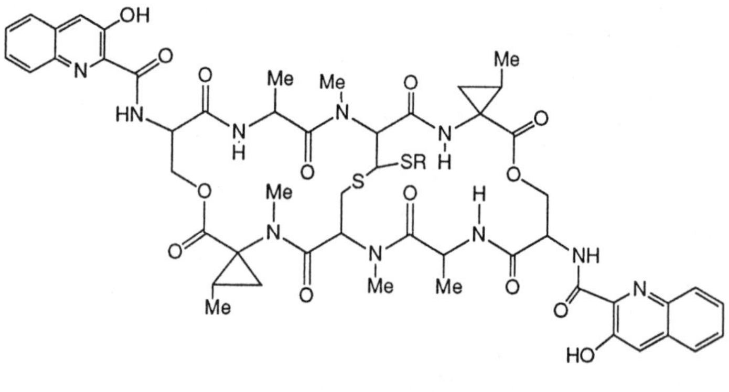

UK-63-052

Antibiotic

Rance, M.J.; Ruddock, J.C.; Pacey, M.S.; Cullen, W.P.; Huang,L.H.; Jefferson, M.T.; Whipple, E.B.; Maeda, H.; Tone, J., *J.Antibiotics* (1989) **42**, 206

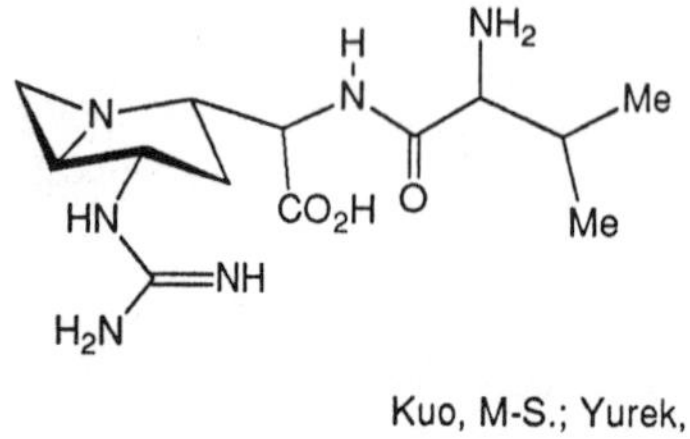

FICELLOMYCIN

Antibiotic

Kuo, M-S.; Yurek, D.A.; Mizsak, S.A., *J.Antibiotics* (1989) **42**, 357

AUTHOR INDEX

SUBJECT INDEX